Lecture Notes in Computer Science 3646

Commenced Publication in 1973
Founding and Former Series Editors:
Gerhard Goos, Juris Hartmanis, and Jan van Leeuwen

A. Fazel Famili   Joost N. Kok
José M. Peña   Arno Siebes
Ad Feelders (Eds.)

Advances in Intelligent Data Analysis VI

6th International Symposium on
Intelligent Data Analysis, IDA 2005
Madrid, Spain, September 8-10, 2005
Proceedings

 Springer

Volume Editors

A. Fazel Famili
IIT/ITI - National Research Council Canada, Ottawa University
School of Information Technology and Engineering
1200 Montreal Rd, M-50, Ottawa, ON K1A 0R6, Canada
E-mail: fazel.famili@nrc-cnrc.gc.ca

Joost N. Kok
Leiden University, Leiden Institute of Advanced Computer Science
Niels Bohrweg 1, 2333 CA Leiden, The Netherlands
E-mail: joost@liacs.nl

José M. Peña
Universidad Politécnica de Madrid, DATSI - Facultad de Informática
Campus de Montegancedo S/N, Boadilla del Monte, 28660 Madrid, Spain
E-mail: jmpena@fi.upm.es

Arno Siebes
Ad Feelders
Utrecht University, Department of Information and Computing Sciences
PO Box 80.089, 3508 TB Utrecht, The Netherlands
E-mail: {arno, ad}@cs.uu.nl

Library of Congress Control Number: 2005931595

CR Subject Classification (1998): H.3, I.2, G.3, I.5.1, I.4.5, J.2, J.1, J.3

ISSN	0302-9743
ISBN-10	3-540-28795-7 Springer Berlin Heidelberg New York
ISBN-13	978-3-540-28795-7 Springer Berlin Heidelberg New York

Springer is a part of Springer Science+Business Media

springeronline.com

© Springer-Verlag Berlin Heidelberg 2005

Typesetting: Camera-ready by author, data conversion by Scientific Publishing Services, Chennai, India
Printed on acid-free paper   SPIN: 11552253   06/3142   5 4 3 2 1 0

Preface

One of the superb characteristics of intelligent data analysis (IDA) is that it is an interdisciplinary field in which researchers and practitioners from a number of areas are involved in a typical project. This also creates a challenge in which the success of a team depends on the participation of users and domain experts who need to interact with researchers and developers of any IDA system. All this is usually reflected in successful projects and of course in the papers that were evaluated by this year's Program Committee from which the final program has been developed.

In our call for papers, we solicited papers on (i) applications and tools, (ii) theory and general principles, and (iii) algorithms and techniques. We received a total of 184 papers, reviewing these was a major challenge. Each paper was assigned to three reviewers. In the end 46 papers were accepted, all of which were included in the proceedings and presented at the conference.

This year's papers reflect the results of applied and theoretical research from a number of disciplines all of which are related to the field of intelligent data analysis. To have the best combination of theoretical and applied research and also provide the best focus, we divided this year's IDA program into tutorials, invited talks, panel discussions and technical sessions.

We managed to organize two excellent tutorials on the first day by Luc De Raedt and Kristian Kersting, entitled *Probabilistic Inductive Logic Programming*, and by Bruno Apolloni, Dario Malchiodi and Sabrina Gaito, entitled *Statistical Bases of Machine Learning*. Our invited speakers were Prof. Ivan Bratko from the Jozef Stefan Institute in Slovenia, and Prof. Alex Freitas from the University of Kent.

We wish to express our sincere thanks to many people who worked hard for the IDA conference to happen in Madrid. Special thanks to tutorial, publicity, local organization, and panel chairs who were in charge of a large portion of our responsibilities. We would also like to thank Xiaohui Liu and Michael Berthold who worked as advisors to this conference, and the members of the Local Organizing Committee for their hard work. Finally, we are grateful to the members of our Program Committee; without their help it would have been impossible to put together such a valuable program.

September 2005

A. Fazel Famili,
José Maria S. Peña,
Joost Kok,
Arno Siebes,
Ad Feelders

Organization

Conference Organization

General Chair

A. Fazel Famili
National Research Council
Ottawa, Canada

Program Chairs

José M. Peña
Universidad Politécnica de Madrid
Madrid, Spain

Arno Siebes
Utrecht University
Utrecht, The Netherlands

Joost Kok
Leiden University
Leiden, The Netherlands

Tutorial Chair

Pedro Larrañaga
EHU-Universidad del País Vasco
San Sebastián, Spain

Publication Chair

Ad Feelders
Utrecht University
Utrecht, The Netherlands

Publicity Chairs

Jorge Muruzábal
Universidad Rey Juan Carlos
Madrid, Spain

Julián Sánchez
Quinao S.L.
Madrid, Spain

Local Organization Chair

Víctor Robles
Universidad Politécnica de Madrid
Madrid, Spain

Panel Chair

Sofian Maabout
LaBRI-Université Bordeaux
Bordeaux, France

Local Committee

**Universidad Politécnica de Madrid
Madrid, Spain**

María S. Pérez
Vanessa Herves
Francisco Rosales
Antonio García
Óscar Cubo
Pilar Herrero
Antonio LaTorre
Alberto Sánchez

**Universidad Rey Juan Carlos
Madrid, Spain**

Susana Vegas
Andrés L. Martinez

Program Committee

Niall Adams, Imperial College London, UK
Riccardo Bellazzi, University of Pavia, Italy
Bettina Berendt, Humboldt University of Berlin, Germany
Michael Berthold, University of Konstanz, Germany
Hans-Georg Beyer, Vorarlberg University of Applied Sciences, Austria
Jean-François Boulicaut, INSA Lyon, France
Christian Borgelt, Otto-von-Guericke-Universität Magdeburg, Germany
Hans-Dieter Burkhard, Humboldt Universität Berlin, Germany
Luis M. de Campos, Universidad de Granada, Spain
Fazel Famili, Institute for Information Technology, NRC, Canada
Giuseppe Di Fatta, University of Konstanz, Germany
Fridtjof Feldbusch, University of Karlsruhe, Germany
Ingrid Fischer, Friedrich-Alexander-Universität Erlangen-Nürnberg, Germany
Douglas Fisher, Vanderbilt University, USA
Peter Flach, University of Bristol, UK
Eibe Frank, University of Waikato, New Zealand
Karl A. Fröschl, ec3 – eCommerce Competence Center, Vienna DC, Austria
Gabriela Guimaraes, CENTRIA UNL, Portugal
Lawrence O. Hall, University of South Florida, USA
Pilar Herrero, Universidad Politécnica de Madrid, Spain
Tom Heskes, Radboud University Nijmegen, The Netherlands
Alexander Hinneburg, University of Halle, Germany
Frank Hoeppner, University of Wolfenbuettel, Germany
Adele Howe, Colorado State University, USA
Klaus-Peter Huber, SAS Institute, Germany
Anthony Hunter, University College London, UK
Alfred Inselberg, Tel Aviv University, Israel
Bert Kappen, Radboud University Nijmegen, The Netherlands
Frank Klawonn, University of Wolfenbuettel, Germany
Joost N. Kok, Leiden University, The Netherlands
Walter Kosters, Leiden University, The Netherlands
Rudolf Kruse, Otto-von-Guericke-Universität Magdeburg, Germany
Pedro Larrañaga, Universidad del País Vasco, Spain
Hans-Joachim Lenz, Freie Universität Berlin, Germany
Xiaohui Liu, Brunel University, UK
Sofian Maabout, LaBRI-Université Bordeaux, France
Rainer Malaka, European Media Laboratory, Heidelberg, Germany
Jorge Muruzábal, Universidad Rey Juan Carlos, Spain
Susana Nascimento, CENTRIA-Universidade Nova de Lisboa, Portugal
Detlef Nauck, BTexact Technologies, UK
Tim Oates, University of Maryland Baltimore County, USA
Simon Parsons, Brooklyn College, City University of New York, USA
José M. Peña, Universidad Politécnica de Madrid, Spain

María S. Pérez, Universidad Politécnica de Madrid, Spain
Bhanu Prasad, Florida A&M University, USA
Víctor Robles, Universidad Politécnica de Madrid, Spain
Lorenza Saitta, Università del Piemonte Orientale, Italy
Paola Sebastiani, Boston University School of Public Health, USA
Arno Siebes, Universiteit Utrecht, The Netherlands
Maarten van Someren, University of Amsterdam, The Netherlands
Myra Spiliopoulou, Otto-von-Guericke-Universität Magdeburg, Germany
Martin Spott, BTexact Technologies, UK
Reinhard Viertl, Vienna University of Technology, Austria
Richard Weber, University of Chile, Chile
Stefan Wrobel, Fraunhofer AIS & University of Bonn, Germany
Mohammed Zaki, Rensselaer Polytechnic Institute, USA

Referees

Silvia Acid
David Auber
Roland Barriot
Concha Bielza
Bouchra Bouqata
Kai Broszat
Andres Cano
Javier G. Castellano
Nicolas Cebron
Víctor Uc Cetina
T.K. Cocx
Nuno Correia
Óscar Cubo
Santiago Eibe
Lukas C. Faulstich
Juan M. Fernández-Luna
Fulvia Ferrazzi
Manuel Gómez
Daniel Goehring
Edgar de Graaf
J.M. de Graaf
Jose A. Gámez
Mark Hall
Alexander Hinneburg
Susanne Hoche
Geoff Holmes
Rainer Holve
Tamás Horváth
Juan F. Huete

Fabien Jourdan
Florian Kaiser
Joerg Kindermann
Christine Koerner
Antonio LaTorre
Marie-Jeanne Lesot
Andres L. Martinez
Michael Mayo
Thorsten Meinl
Ernestina Menasalvas
Dagmar Monett
Serafín Moral
Siegfried Nijssen
Juan A. Fernández del Pozo
Simon Price
Jose M. Puerta
Simon Rawles
Frank Rügheimer
Lucia Sacchi
Alberto Sánchez
Karlton Sequeira
Zujun Shentu
David James Sherman
Hendrik Stange
Micheal Syrjakow
Xiaomeng Wang
Bernd Wiswedel
Marta Elena Zorrilla

Table of Contents

Probabilistic Latent Clustering of Device Usage
  Jean-Marc Andreoli, Guillaume Bouchard 1

Condensed Nearest Neighbor Data Domain Description
  Fabrizio Angiulli 12

Balancing Strategies and Class Overlapping
  Gustavo E.A.P.A. Batista, Ronaldo C. Prati,
  Maria C. Monard 24

Modeling Conditional Distributions of Continuous Variables in
Bayesian Networks
  Barry R. Cobb, Rafael Rumí, Antonio Salmerón 36

Kernel K-Means for Categorical Data
  Julia Couto 46

Using Genetic Algorithms to Improve Accuracy of Economical Indexes
Prediction
  Óscar Cubo, Víctor Robles, Javier Segovia,
  Ernestina Menasalvas 57

A Distance-Based Method for Preference Information Retrieval in
Paired Comparisons
  Esther Dopazo, Jacinto González-Pachón, Juan Robles 66

Knowledge Discovery in the Identification of Differentially Expressed
Genes
  A. Fazel Famili, Ziying Liu, Pedro Carmona-Saez, Alaka Mullick ... 74

Searching for Meaningful Feature Interactions with Backward-Chaining
Rule Induction
  Doug Fisher, Mary Edgerton, Lianhong Tang, Lewis Frey,
  Zhihua Chen 86

Exploring Hierarchical Rule Systems in Parallel Coordinates
  Thomas R. Gabriel, A. Simona Pintilie, Michael R. Berthold 97

Bayesian Networks Learning for Gene Expression Datasets
  Giacomo Gamberoni, Evelina Lamma, Fabrizio Riguzzi,
  Sergio Storari, Stefano Volinia 109

Pulse: Mining Customer Opinions from Free Text
Michael Gamon, Anthony Aue, Simon Corston-Oliver,
Eric Ringger 121

Keystroke Analysis of Different Languages: A Case Study
Daniele Gunetti, Claudia Picardi, Giancarlo Ruffo 133

Combining Bayesian Networks with Higher-Order Data Representations
Elias Gyftodimos, Peter A. Flach 145

Removing Statistical Biases in Unsupervised Sequence Learning
Yoav Horman, Gal A. Kaminka 157

Learning from Ambiguously Labeled Examples
Eyke Hüllermeier, Jürgen Beringer 168

Learning Label Preferences: Ranking Error Versus Position Error
Eyke Hüllermeier, Johannes Fürnkranz 180

FCLib: A Library for Building Data Analysis and Data Discovery Tools
Wendy S. Koegler, W. Philip Kegelmeyer 192

A Knowledge-Based Model for Analyzing GSM Network Performance
Pasi Lehtimäki, Kimmo Raivio 204

Sentiment Classification Using Information Extraction Technique
Jian Liu, Jianxin Yao, Gengfeng Wu 216

Extending the SOM Algorithm to Visualize Word Relationships
Manuel Martín-Merino, Alberto Muñoz 228

Towards Automatic and Optimal Filtering Levels for Feature Selection
in Text Categorization
E. Montañés, E.F. Combarro, I. Díaz, J. Ranilla 239

Block Clustering of Contingency Table and Mixture Model
Mohamed Nadif, Gérard Govaert 249

Adaptive Classifier Combination for Visual Information Processing
Using Data Context-Awareness
Mi Young Nam, Phill Kyu Rhee 260

Self-poised Ensemble Learning
Ricardo Ñanculef, Carlos Valle, Héctor Allende,
Claudio Moraga 272

Discriminative Remote Homology Detection Using Maximal Unique
Sequence Matches
    Hasan Oğul, Ü. Erkan Mumcuoğlu   283

From Local Pattern Mining to Relevant Bi-cluster Characterization
    Ruggero G. Pensa, Jean-François Boulicaut   293

Machine-Learning with Cellular Automata
    Petra Povalej, Peter Kokol, Tatjana Welzer Družovec, Bruno Stiglic.   305

MDS_{polar}: A New Approach for Dimension Reduction to Visualize High
Dimensional Data
    Frank Rehm, Frank Klawonn, Rudolf Kruse   316

Miner Ants Colony: A New Approach to Solve a Mine Planning
Problem
    María-Cristina Riff, Michael Moossen, Xavier Bonnaire   328

Extending the GA-EDA Hybrid Algorithm to Study Diversification
and Intensification in GAs and EDAs
    V. Robles, J.M. Peña, M.S. Pérez, P. Herrero, O. Cubo   339

Spatial Approach to Pose Variations in Face Verification
    Licesio J. Rodríguez-Aragón, Ángel Serrano, Cristina Conde,
    Enrique Cabello   351

Analysis of Feature Rankings for Classification
    Roberto Ruiz, Jesús S. Aguilar-Ruiz, José C. Riquelme,
    Norberto Díaz-Díaz   362

A Mixture Model-Based On-line CEM Algorithm
    Allou Samé, Gérard Govaert, Christophe Ambroise   373

Reliable Hierarchical Clustering with the Self-Organizing Map
    Elena V. Samsonova, Thomas Bäck, Joost N. Kok,
    Ad P. IJzerman   385

Statistical Recognition of Noun Phrases in Unrestricted Text
    José I. Serrano, Lourdes Araujo   397

Successive Restrictions Algorithm in Bayesian Networks
    Linda Smail, Jean Pierre Raoult   409

Modelling the Relationship Between Streamflow and Electrical
Conductivity in Hollin Creek, Southeastern Australia
    Jess Spate   419

Biological Cluster Validity Indices Based on the Gene Ontology
Nora Speer, Christian Spieth, Andreas Zell 429

An Evaluation of Filter and Wrapper Methods for Feature Selection in
Categorical Clustering
Luis Talavera 440

Dealing with Data Corruption in Remote Sensing
Choh Man Teng 452

Regularized Least-Squares for Parse Ranking
Evgeni Tsivtsivadze, Tapio Pahikkala, Sampo Pyysalo,
Jorma Boberg, Aleksandr Mylläri, Tapio Salakoski 464

Bayesian Network Classifiers for Time-Series Microarray Data
Allan Tucker, Veronica Vinciotti, Peter A.C. 't Hoen,
Xiaohui Liu 475

Feature Discovery in Classification Problems
Manuel del Valle, Beatriz Sánchez, Luis F. Lago-Fernández,
Fernando J. Corbacho 486

A New Hybrid NM Method and Particle Swarm Algorithm for
Multimodal Function Optimization
Fang Wang, Yuhui Qiu, Yun Bai 497

Detecting Groups of Anomalously Similar Objects in Large Data Sets
Zhicheng Zhang, David J. Hand 509

Author Index 521

Probabilistic Latent Clustering of Device Usage

Jean-Marc Andreoli and Guillaume Bouchard

Xerox Research Centre Europe, Grenoble, France
FirstName.LastName@xrce.xerox.com

Abstract. We investigate an application of Probabilistic Latent Semantics to the problem of device usage analysis in an infrastructure in which multiple users have access to a shared pool of devices delivering different kinds of service and service levels. Each invocation of a service by a user, called a job, is assumed to be logged simply as a co-occurrence of the identifier of the user and that of the device used. The data is best modelled by assuming that multiple latent variables (instead of a single one as in traditional PLSA) satisfying different types of constraints explain the observed variables of a job. We discuss the application of our model to the printing infrastructure in an office environment.

1 Introduction

It is nowadays common that printing devices in an office or a workplace be accessed through the local network instead of being assigned and directly connected to individual desktops. As a result, a large amount of information can easily be collected about the actual use of the whole printing infrastructure, rather than individual devices. To be useful, this data needs to be analysed and presented in a synthetic way to the administrators of the infrastructure. We are interested here in analysing the correlation between users and devices in the data, ie. how the printing potential of users translates into actual use of the devices. We assume here that users are not strongly constrained in their use, the extreme case being when any user is allowed to print anything on any device in the infrastructure. The expected outcome of such an analysis may be diverse. For example, the administrator could discover communities of device usage, corresponding to different physical or virtual locations of the users at the time of the jobs, and, from these, form hypotheses on the actual behaviour of the users, both in the case of normal functioning of the infrastructure and in case of exceptions (device down or not working properly). This in turn could lead to more refined decisions as to the organisation of the infrastructure and to the instructions given to its users. It could also help work around failures of devices inside the infrastructure, by redirecting a job sent to a failing device toward a working one chosen in accordance with the community to which the job belongs.

A study on inhabitant-device interactions [6] shows that the recorded device usage can be mined to discover significant patterns, which in turn could be used to automate device interactions. To the authors knowledge, generic user-device interaction analysis in the presence of devices delivering possibly multiple services or levels of service has not been studied extensively.

A.F. Famili et al. (Eds.): IDA 2005, LNCS 3646, pp. 1–11, 2005.

Problem statement. Our overall goal is to analyse usage data in an infrastructure consisting of a set of independent devices offering services of different or identical classes, and operated by a set of independent users. An interaction between a user and a device is called a job. The usage data consists of a log of these jobs over a given period of time. More precisely, we make the following assumptions.

– Let N_U, N_D, N_K denote the number of, respectively, users, devices and service classes, assumed invariable over the analysed period. Each user, resp. device, resp. service class, can therefore be identified by a number $u \in \{1, \ldots, N_U\}$, resp. $d \in \{1, \ldots, N_D\}$, resp. $k \in \{1, \ldots, N_K\}$. Each user, device, service class also has a print name, for display and reference purpose.
– Each device offers services of one or more classes. This is captured in a boolean matrix f of dimension $N_K \times N_D$ where f_{kd} is 1 if device d offers the service class k and 0 otherwise. This matrix is assumed static over the analysed period.
– All the jobs are recorded over the analysed period. Let N be the number of recorded jobs. Each job can therefore be identified by an index $i \in \{1, \ldots, N\}$. Each job i contributes exactly one entry in the log, consisting of the pair (u_i, d_i) identifying the user and device involved in that job. Thus the data is entirely defined by the matrix n of dimension $N_U \times N_D$ where n_{ud} is the number of jobs by user u on device d.

A printing infrastructure in an office is a typical example where our method applies. In that case, a service class could be a particular type of printing. For simplification purpose, in the examples, we consider only two service classes: black&white ($k = 1$) and colour ($k = 2$). Note that a colour printer can always also perform black&white jobs, meaning that if $f_{2d} = 1$, then $f_{1d} = 1$.

Outline of the method. The purpose of our analysis is essentially to discover clusters in the usage data. Since the observed data correspond to co-occurrences of discrete variables, we have chosen an aspect model, which is an instance of latent class models [1], nowadays often referred to as Probabilistic Latent Semantics Analysis (PLSA) [5]. This model is particularly relevant here as its basic assumption has a straightforward interpretation in our context. Indeed, the PLSA assumption is that the data can be generated according to a process that first selects a (latent) cluster, then a user and a device, in such a way that, conditionally to the cluster, the choices of user and device are independent. There is a natural interpretation of such clusters as communities of usage which are associated to physical or virtual locations within the infrastructure. The PLSA assumption means that at a given location, users tend to choose devices in the same way, which is quite reasonable. For example, in an office infrastructure comprising multiple floors, each floor can correspond to a community, whose users share the same perception of the infrastructure and tend to choose printers in a similar fashion. PLSA clustering therefore offers a powerful tool to discover such communities of usage.

However, another important determining factor for the choice of device is the nature of the job to be performed. This information may not be directly available

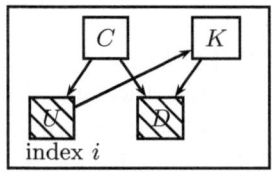

Fig. 1. Graphical representation of the variables dependencies. Observed variables are shaded.

from the logs, still it can be partially inferred from the knowledge of the service class supported by the chosen device. For example, a job sent to a non-colour printer is certainly a black&white job (assuming that users normally do not make mistakes by launching a job on a device not supporting the service class of that job). As a consequence, the basic PLSA model in which a single latent variable (the cluster) explains the observed ones must be extended to account for the presence of additional latent variables with specific constraints attached to them. Some hierarchical extensions to PLSA have already been proposed [4] but where the two factors are assumed to be independent. Here, the service class of the job is an additional latent variable, and, unlike the cluster, its range is known in advance (it is the set of possible service classes supported by the devices of the infrastructure) and its dependency to the chosen device is constrained by the knowledge of the service classes supported by each device. Studying this multi-factor constrained latent structure motivates our investigation.

2   Definition of the Model and Parameter Estimation

The random variables of the model. The recorded jobs are assumed to be independent and identically distributed. We consider 4 random variables that are instantiated for each job: two observed variables U and D defining the user id and the device id, and two latent (or unobserved) discrete variables C and K corresponding to the index of a job cluster and the job service class. We consider that the instantiation of these variables comes from the following generative process: 1. Generate the cluster index C, 2. Generate the user id U, 3. Generate the job service class K depending only on the user, 4. Generate the device choice D depending only on the cluster and the job service class. This process is equivalent to assuming that C is independent to K conditionally to U. A possible factorisation[1] of the joint distribution is $p(U, D, C, K) = p(C)p(U|C)p(K|U)p(D|C, K)$ which is illustrated in the graphical model of Figure 1. Let $\pi^{(C)}$ be the parameters of the multinomial distributions $p(C)$, ie. a vector of proportions of dimension N_C that sums to 1. The other parameters are conditional discrete distributions: $p(U|C)$, $p(K|U)$ and $p(D|C, K)$ are parameterised by the conditional probability tables $\pi^{(U)}$, $\pi^{(K)}$ and $\pi^{(D)}$, respectively. The distribution of the devices is

---

[1] Another equivalent factorisation is $p(U, D, C, K) = p(U)p(C|U)p(K|U)p(D|C, K)$, where the generative process starts with the choice of a user and then a cluster.

constrained by the knowledge of the service classes they support: $f_{kd} = 0$ implies $\pi_{dck}^{(D)} = 0$ for all $c \in \{1, \cdots, N_C\}$. Writing $\theta = (\pi^{(C)}, \pi^{(U)}, \pi^{(K)}, \pi^{(D)})$ the set of parameters involved in the model, the joint distribution is $p(u, d, c, k|\theta) = \pi_c^{(C)} \pi_{uc}^{(U)} \pi_{ku}^{(K)} \pi_{dck}^{(D)}$. The maximum likelihood estimator is not always satisfactory when the number of jobs is small. We use instead a bayesian framework by defining a prior distribution on the parameters. Since they corresponds to conditional probability tables, we assume Dirichlet priors: $\pi_{\cdot \text{pa}(X)}^{(X)} \sim \mathcal{D}(m_j^{(X)}, j = 1, \ldots, N_j)$ where X denotes one of the variables U, K, D and C and pa(X) denotes the parents of variable X. In the application below, the hyper-parameter $m^{(K)}$ is set according to the expected device usage and the others are set to 0.5 (Jeffrey's uninformative prior). In particular, it may happen that during the analysed period, a given user u never performs jobs of a given service class k (eg. never prints in colour), in which case the maximum likelihood estimator will yield $\pi_{ku}^{(K)} = 0$, meaning that user u *never* uses service class k. The prior knowledge on the users' needs in terms of service classes can be used to compensate for insufficient data. In the printer example below, the expected B&W/colour job ratio will be used to define $m^{(K)}$. These can be seen as pseudo-counts of usage of each service class given *a priori* for a "prototypical" user.

Parameter estimation. The MAP estimator $\hat{\theta} = \text{argmax}_\theta \, p(\theta|\mathbf{x})$, where $\mathbf{x}$ denotes the observed data ie., here, the raw data matrix $n_{..}$, is obtained using the EM algorithm [3]. For space reasons, the EM update equations are omitted here. As usual with that algorithm, some care has to be taken in the initialisation. If the number of clusters N_C is known, the MAP estimator can be computed directly. If it is unknown, the MAP estimator must be computed for each possible value of N_C, and the model maximising the BIC score [8] is chosen. This criterion is given by:

$$\text{BIC}(N_C) = \log p(\mathbf{x}|\hat{\theta}; N_C) + \log p(\hat{\theta}; N_C) - \frac{\nu(N_C)}{2} \log N$$

Here, $\log p(\mathbf{x}|\hat{\theta}; N_C)$ is the likelihood of the estimated parameter, $p(\hat{\theta}; N_C)$ is the probability *a priori* of the estimated parameters and $\nu(N_C)$ is the number of free parameters of the model. The selected number of clusters $\widehat{N_C}$ is the one that maximises $BIC(N_C)$. To compute a set of models with different complexities N_C, we first initialise a model with a relatively large complexity, and then decrease it step-by-step until having only one cluster. For each intermediate step, the BIC criterion is computed at the MAP solution obtained by the EM algorithm. Instead of re-initialising the model at each step, we use for level c the $c + 1$ different initialisations that are obtained by removing one cluster from the model learnt at level $c + 1$.

3   Exploitation of the Model

There are various ways in which the probabilistic model, once estimated, can be used. We consider two in particular: outlier detection and smoothing.

Outlier detection. An outlier[2] is a user whose usage profile observed in the log does not match its expected value by the model. Identifying outliers can help an administrator to understand individual needs that are not provided for by the current configuration of the infrastructure. Recall that the raw usage data is given by matrix n_{ud} which gives the number of jobs involving user u and device d. It is the realisation of the random variable $X_{ud} = \sum_{i=1}^{N} \mathbb{I}\{U_i = u, D_i = d\}$. Let n_{ud}^* be its expectation according to the model. We have $n_{ud}^* = \boldsymbol{E}\left[X_{ud}\right] = N\,\boldsymbol{p}(u,d|\hat{\theta})$. The matrix n^* is the smoothed version of $n_{..}$ in which information orthogonal to the model space is considered as noise and eliminated. One possible way to compute outliers is to define a quality-of-fit measure of each user and then find the user above a given threshold. The standard chi-squared statistic is used to test if the actual usage of the devices fits that estimated by the model: $\chi_u^2 = \sum_{d=1}^{N_D} \left(n_{ud}^* - n_{ud}\right)^2/n_{ud}^*$. A user is considered an outlier whenever χ_u^2 is superior to the inverse cumulative distribution of the chi-squared law with $N_D - 1$ degrees of freedom.

Smoothing. Any statistic computed from the raw data matrix $n_{..}$ can now be applied to the smoothed data matrix $n_{..}^*$, yielding more precise information:

- *Correction of the primary devices* A good way to check the benefit of smoothing is to look at the primary device of a user u for a service class k, which is defined by $r_{ku} = \mathrm{argmax}_d\, n_{ud}f_{kd}$. Its smoothed version is given by $r_{ku}^* = \mathrm{argmax}_d\, n_{ud}^* f_{kd}$. The users for which $r_{ku} \neq r_{ku}^*$ have a non-standard behaviour which may be of interest to the administrator.
- *Visualisation of the infrastructure* A useful tool for an administrator is a 2D map of the infrastructure s/he administrates. Even if it does not correspond exactly to the map of the physical setting, such a low dimensional representation provides the administrator with a synthetic view of the overall infrastructure usage. A map of users and devices based on matrix $n_{..}^*$ instead of $n_{..}$ is particularly interesting as $n_{..}$ contains outliers which usually have a strong impact on the dimension reduction algorithms. The use of a smoothed version of the data increases the precision and clarity of the map.
- *Estimating redirections in the infrastructure* Another important tool for administrators is the redirection matrix of the infrastructure for each of the N_K service classes. This matrix gives for each device d and service class k the device choice distribution conditionally to the fact that d is out of order. There are various ways of computing this matrix, taking as input a variant of the data matrix which gives, for a user u, a device d and a service class k, an *a priori* estimate of the number of jobs of service class k involving user u and device d, computed by $n_{udk} = n_{ud}m_k^K f_{kd}/\sum_{k'=1}^{N_K} m_{k'}^K f_{k'd}$. Whatever the algorithm to compute the redirection matrix, more precise results are to be expected if the smoothed version of matrix $n_{...}$ is used instead of its raw version.

---

[2] We consider here only *user* outliers. Other types of outliers, eg. devices, can also be treated in the same way.

4   Experiment on a Print Infrastructure Usage Log

Printing logs from an office infrastructure were used to test our model. About 30 000 jobs were logged over a 5 month period, involving 124 users and 22 printers (5 of them colour). The initial number of clusters was the number of observed primary device configurations and was equal to $N_C^o = 21$. From an initial solution including all the previous configurations, the step-by-step procedure described above learnt 21 models with decreasing complexity. Figure 2(a) shows that there is clearly a minimum of the BIC score within the range of estimated models. The optimal value is $\widehat{N_C} = 13$ clusters. This number of clusters is relatively stable when considering only subsets of the data: from 5 000 to 30 000 jobs, the same number of clusters was selected.

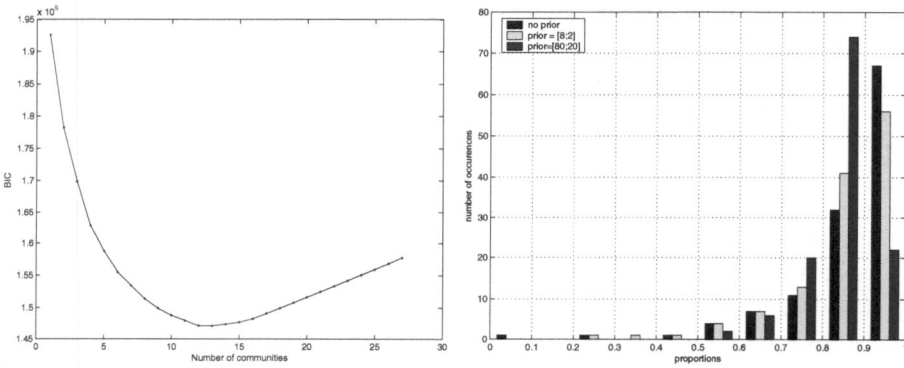

Fig. 2. (a)BIC score of the model for various numbers of clusters. (b) Effect of the prior on the estimations.

For most of the parameters, we used uninformative priors, since the amount of data was sufficient. The only parameter with informative prior was $\pi^{(K)}$. To check the effect of the priors on the estimation, we tried three different values of the hyper-parameter m^K. We compared $m^K = (1,1)$, ie. no prior, equivalent to maximum likelihood estimation, $m^K = (8,2)$, ie. small prior, and $m^K = (80, 20)$, ie. strong prior. The ratio 80/20 means that B&W jobs are *a priori* considered 4 times more frequent than colour jobs for any user. The histogram of the values $\pi_{u1}^{(K)}$ is represented on Figure 2(b) and shows that the priors prevent the parameters from being 0 (only colour jobs) or 1 (no colour job). With the small prior, a user having 25% only of B&W jobs still appears. This corresponds to a user who generally prints B&W jobs to a colour printer. In the sequel, we use the results obtained with the "strong" prior.

Discussion of the results. Most of the cluster parameters are summarised in Table 1. Among the 13 clusters, we can see that the first 4 B&W/Colour pairs represent nearly 50% of the jobs. Some remarks:

Table 1. Summary of the estimated parameters for each job cluster c. The "B&W printer" ($k = 1$) and "colour printer" ($k = 2$) are the printers that are most used, as given by $\text{argmax}_d\, \pi_{dck}^{(D)}$, where the percentage indicates how often this "preferred" printer is chosen. The % column gives the probability of each cluster, as given by $\pi_c^{(C)}$, and the "main users" are the users u corresponding to the 5 biggest values of $\pi_{cu}^{(U)}$.

cluster	B&W printer	colour printer	%	user IDs (% of usage)				
C1	Pre(99%)	Lib(98%)	12.7	ej(13%)	cu(9%)	bw(8%)	cm(8%)	el(8%)
C2	Stu(100%)	Lib(100%)	10	be(16%)	ds(9%)	cp(7%)	au(7%)	dc(6%)
C3	Tim(85%)	Ver(99%)	15.6	db(9%)	ar(9%)	bm(8%)	az(8%)	er(7%)
C4	Vog(99%)	Rep(52%)	13.8	cg(25%)	aw(20%)	ei(18%)	dy(15%)	ep(4%)
C5	Hol(100%)	Lib(100%)	7.7	ch(51%)	ay(31%)	bs(13%)	ec(2%)	bw(0%)
C6	Her(98%)	Tel(98%)	7	ef(26%)	dq(18%)	ce(11%)	dt(10%)	dm(8%)
C7	Geo(97%)	Ver(96%)	5.6	ac(65%)	bv(31%)	dx(2%)	eq(0%)	ec(0%)
C8	Bib(99%)	Rep(100%)	6.8	ag(42%)	bu(38%)	dh(10%)	ec(9%)	et(0%)
C9	Mes(73%)	Ver(84%)	4.5	dx(72%)	em(26%)	ba(0%)	do(0%)	bt(0%)
C10	Lem(97%)	Rep(100%)	3.5	an(92%)	ei(5%)	et(1%)	ch(0%)	bt(0%)
C11	Hod(89%)	Ver(69%)	5.5	eq(20%)	et(14%)	cy(13%)	cc(12%)	ek(9%)
C12	Mid(76%)	Fig(91%)	1.7	da(99%)	ba(0%)	do(0%)	dx(0%)	em(0%)
C13	Sta(99%)	Tel(95%)	5.6	av(12%)	de(10%)	ea(10%)	bh(8%)	cz(8%)

- Each cluster is dominated by the use of a "preferred" printer. One example is cluster $C2$, where 100% of the jobs are sent to printer Stu for B&W printing and Lib for colour printing.
- As an exception, cluster $C4$ associated to the B&W printer Vog has two main colour printer (Lib and Rep) with equal importance. The reason of this behaviour cannot be found in the model, but indicates to the administrator that there is a non-standard use of colour printers among the users of Vog.
- Clusters $C3$ and $C12$ contain colour printers (Lib at 4.7% and Ver at 2%) among the B&W printers. This may indicate the use of colour printer when the nearest B&W device is unavailable.
- There are two clusters composed of only one user: "an" in C10 and "da" in C12. In fact, these users have a specific position in the company, and each of them has her own printer, resp. Lem and Mid. These users are not considered outliers since they print a sufficient number of jobs to create individual clusters.

Many other informations about the print usage can be extracted from a deeper analysis of the parameters, depending on the infrastructure administrator's goal.

Outlier identification. We applied the method described in Section 3. Only 3 users were rejected from the 80% confidence test: "bx", "aw" and "bd". User "bx" is in fact a generic login for a group of people. Users "aw" and "bd" are using specific printers Pho and Leq that are rarely used by other users. They were not put into a specific cluster and are therefore considered as outliers from a usage point of view.

User printing profiles

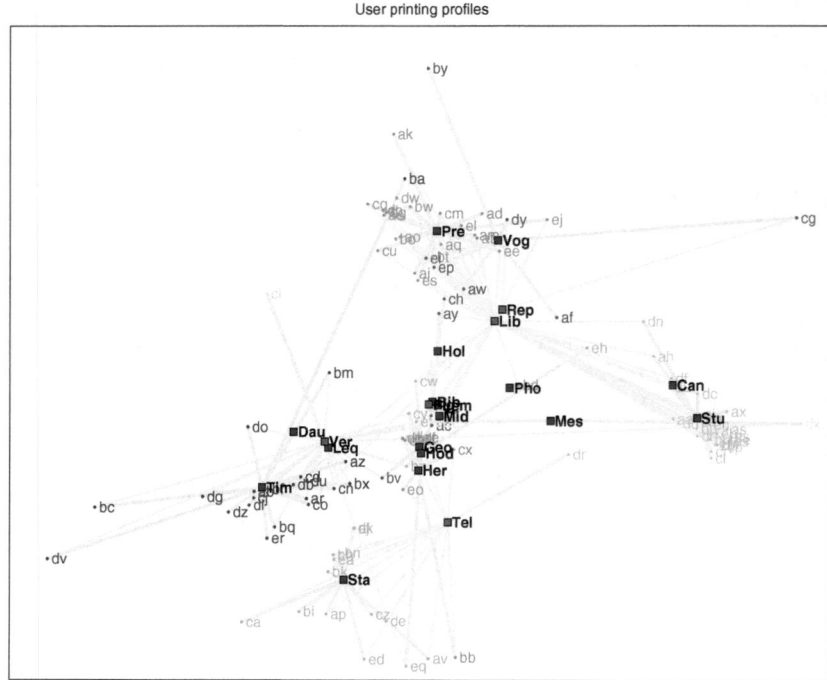

User smoothed printing profiles

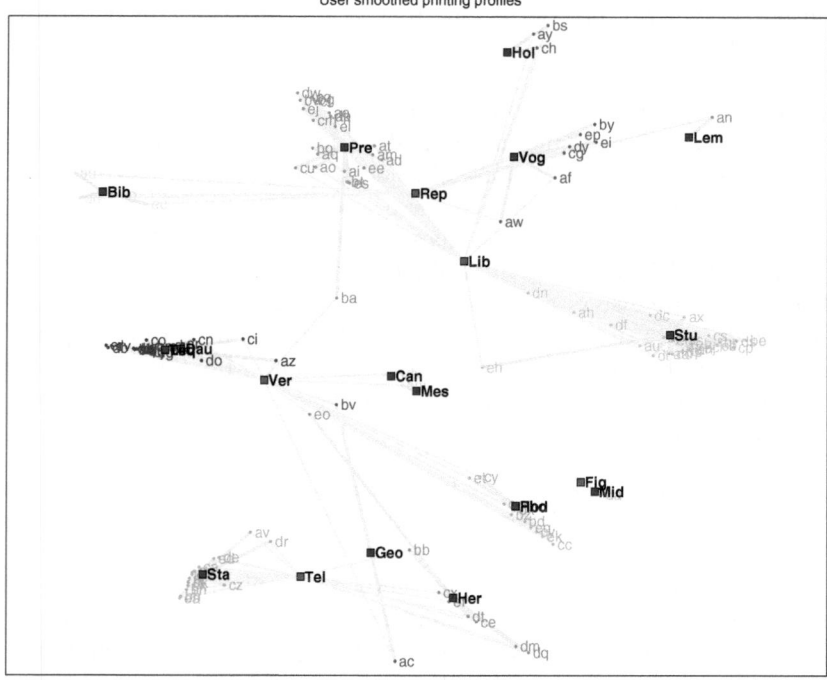

Fig. 3. Low-dimensional representations of the printers and their users

Table 2. Users for which the estimated primary printer is different from the observed one

B&W jobs

az	Ver →Tim
ba	Lib →Pre
bd	Pho→Hod
ci	Ver →Tim
dr	Tel →Sta
eh	Hod→Stu
es	Lib →Pre

colour jobs

al	Lib→Tel	co	Lib→Ver
aw	Lib→Rep	cv	Lib→Ver
ba	Lib→Ver	cw	Lib→Ver
bd	Lib→Ver	db	Lib→Ver
bu	Lib→Rep	dj	Lib→Tel
by	Lib→Rep	dk	Lib→Ver
cc	Lib→Ver	ek	Lib→Ver
cf	Lib→Tel	eo	Tel→Ver

Correction of the primary devices. Following the method of Section 3, Table 2 lists, for each of the two service classes k (B&W and colour), the users whose estimated primary printer differs from their observed one (ie. $r_{ku}^* \neq r_{ku}$). The "raw" primary device r_{ku} is on the left-hand side of the arrow while the "smoothed" one r_{ku}^* is on the right-hand side. In the B&W case, some colour printers[3] such as Lib or Ver are replaced by a more suitable B&W printer. In the colour case, printer Lib is often replaced by another colour printer which is generally closer to the users. The specific role of printer Lib may be due to the fact that it has a high-quality output, contrary to other colour printers. Our model is in fact biased in that case, as it does not distinguish within the service classes the speed or quality of individual printers. This could be improved by introducing more service classes.

Visualisation of the infrastructure. We tried several dimension reduction techniques. PLSA is sometime referred to as a multinomial PCA (mPCA). With our model, the user repartition $\pi^{(U)}$ can also be interpreted as latent coefficients and plotted if we set $N_C = 2$. However, this technique (as well as standard PCA) gave unsatisfactory results, due to the fact that the first two eigenvalues of the covariance matrix contain less that 50% of the data variance. We also tried *Kernel PCA* with a Gaussian Kernel, but the amount of explained information remained below 60%, which cannot yield a reliable map. Instead, we used a simple non-linear dimensionality reduction technique called *Sammon's mapping* [7] applied to the raw data matrix $n_{..}$ (Figure 3, upper map) and to its smoothed version $n_{..}^*$ (Figure 3, lower map), and compared the results. In both maps, service classes (B&W and colour) are represented with different colours. Each user has two links: one to its primary B&W printer and one to its primary colour printer. The colour of the users is given by the primary B&W printer. The global distortion value of the dimension reduction equals 9.9% with the raw data matrix and 4.0% with the smoothed matrix. The printer positions were computed on the reduced space by a weighted means of the community positions. The map based on smoothed data is much more readable than the original one using the

---

[3] Recall that colour printers also support the B&W service class and hence can appear as primary printers for that class

Table 3. The redirection matrices for colour service class based on, respectively, raw and smoothed data

Raw data						Smoothed data				
	Lib	Ver	Rep	Tel	Fig		Lib	Ver	Rep	Tel	Fig
Lib	0	42	55	2	1	Lib	0	23	75	1	0
Ver	74	0	21	0	4	Ver	69	0	26	1	5
Rep	82	18	0	0	0	Rep	90	10	0	0	0
Tel	88	12	0	0	0	Tel	71	29	0	0	0
Fig	25	72	3	0	0	Fig	19	76	5	0	0

raw data. In the latter, users are spread out around their "preferred" printer, but the relation between clusters in confused and hidden by undesired links between users and wrongly estimated "preferred" color printer (e.g. the links to printer Lib). Because of this noise effect, the map does not concentrate the information into clearly distinct clusters. In the smoothed data map, on the other hand, clusters of usages are more visible. Moreover, the different builings and floors of the actual office environment are more separated, mainly due to the corrective effect of the "preferred" user printers.

Estimating printer redirections. The expression $R_{dd'k} \propto \sum_u n_{udk} n_{ud'k} \mathbb{I}\{d \neq d'\}$ is one way to compute the redirection matrix for the service class k. This formula can be justified by assuming that the choice of the redirection printer d' conditionally to the initial printer d follows a multinomial distribution with parameters proportional to $n_{ud} \mathbb{I}\{d \neq d'\}$. Looking at the raw redirection matrix in the colour case on Table 3, printer Lib is redirected at 42% onto printer Ver which is in another building. This quantity is decreased to 23% using the smoothed matrix, while Rep is increased from 55% to 75%, which is more sensible since Rep is much closer to Lib (in the same building). We see that the model uses information about the B&W printers proximity to guess proximity of colour printers. This is of great interest because B&W data is more abundant, leading to an increased precision of the knowledge of the B&W behaviour, which indirectly increases the precision of the estimation of the redirection in the colour case.

5   Conclusion

In this paper, we proposed to analyse usage data in an infrastructure consisting of users operating devices offering services of different classes. We defined precisely the assumptions on the available data, then built a probabilistic latent class model to cluster the jobs (a job is an interaction user-device). From this model, multiple analysis tools were derived that can help administrators monitor the usage. Instead of studying each user profile individually, the model gives a small number of relevant usage patterns which "compress" the probability distribution into a small number of parameters. One important feature of the proposed model is that it takes into account the device functionalities, without assuming that the specific functionality required by each job is observed.

The case study on an office printing infrastructure showed relevant informations about the actual usage of the printers. The model efficiently summarised the whole printing behaviour of the employees, identified non-standard printer usage and proposed changes to the "preferred" user printers that are coherent with the other profiles. The model was used as input to build a map of the printer and user positions, much more readable than those obtained by model-free dimensionality reduction techniques. Finally, the data smoothed by the model gave more sensible results in the estimation of the redirection matrix.

This approach can be generalised to other applications areas. The ability to isolate independent factors can be useful if the observed data is generated by several sources, as with ICA extensions to PLSA [2].

References

1. T.W. Anderson. Some scaling methods and estimation procedures in the latent class model. In U. Grenander, editor, *Probability and Statistics*. John Wiley & Sons, 1959.
2. W. Buntine. Variational extensions to em and multinomial pca. In *Proc. of 13th European Conference on Machine Learning (ECML 2002), Helsinki, Finland, August 19-23*, pages 23–34, 2002.
3. A. Dempster, N. Laird, and D. Rubin. Maximum likelihood from incomplete data via the EM algorithm. *Journal of the Royal Statistical Society*, B 39:1–38, 1977.
4. E. Gaussier and C. Goutte. Probabilistic models for hierarchical clustering and categorisation : Applications in the information society. In *Proceedings of the Intl. Conf. on Advances in Infrastructure for Electronic Business, Education, Science and Medicine on the Internet, L'Aquila, Italy*, 2002.
5. T. Hofmann. Probabilistic latent semantic analysis. In *Proc. of Uncertainty in Artificial Intelligence, UAI'99*, Stockholm, 1999.
6. E.O. Heierman III and D.J. Cook. Improving home automation by discovering regularly occurring device usage patterns. In *Proceedings of the 3rd IEEE International Conference on Data Mining (ICDM 2003), Melbourne, Florida, USA*, pages 537–540, 2003.
7. J. W. Sammon. A nonlinear mapping for data structure analysis. *IEEE Transactions on Computers*, 18(5):401–409, 1969.
8. G. Schwartz. Estimating the dimension of a model. *The Annals of Statistics*, 6(2):461–464, 1978.

Condensed Nearest Neighbor
Data Domain Description

Fabrizio Angiulli

ICAR-CNR
Via Pietro Bucci, 41C
87036 Rende (CS), Italy
angiulli@icar.cnr.it

Abstract. A popular method to discriminate between normal and abnormal data is based on accepting test objects whose nearest neighbors
distances in a reference data set lie within a certain threshold. In this
work we investigate the possibility of using as reference set a subset of
the original data set. We discuss relationship between reference set size
and generalization, and show that finding the minimum cardinality reference consistent subset is intractable. Then, we describe an algorithm
that computes a reference consistent subset with only two reference set
passes. Experimental results confirm the effectiveness of the approach.

1 Introduction

Data domain description, also called one-class classification, is a classification
technique whose goal is to distinguish between objects belonging to a certain
class and all the other objects of the space. The task that it is needed to solve in
one-class classification is the following: given a data set of objects, called training
or reference set, belonging to a certain object space, find a description of the
data, i.e. a rule partitioning the object space in an accepting region, containing
the objects belonging to the class represented by the training set, and a rejecting
region, containing all the other objects. Data domain description is related to
outlier or novelty detection, as the description of the data is then used to detect
the objects deviating significantly from the training data.

Given a data set, also called reference set, of objects from an object space,
and two parameters k and θ, we call Nearest Neighbor Domain Description rule
(NNDD) the classifier that associates to each object p a feature vector $\delta(p) \in \mathbb{R}^k$,
whose elements are the distances of p to its first k nearest neighbors in the reference set, and accepts p iff $\delta(p)$ belongs to the hyper-sphere (according to one
of the L_r Minkowski's metrics, $r \in \{1, 2, \ldots, \infty\}$) centered in the origin of $\mathbb{R}^k$
and having radius θ, i.e. iff $\|\delta(p)\|_r \leq \theta$. The contribution of this work can be
summarized as follows. We define the concept of *reference consistent subset* for
the NNDD rule, that is a subset of the reference set that correctly classifies all
the objects in the reference set, and we discuss relationship between the VC
dimension of the NNDD classifier and size of the reference set, concluding that
replacing the original reference set with a reference consistent subset improves

A.F. Famili et al. (Eds.): IDA 2005, LNCS 3646, pp. 12–23, 2005.

both space requirements, response time, and generalization. We show that finding the minimum cardinality reference consistent subset is a computationally demanding task, and we provide the algorithm CNNDD that computes a reference consistent subset with only *two data set passes*. Experimental results show that the CNNDD algorithm achieves notable training set reduction and sensibly improves accuracy over the the NNDD rule. Finally, we compare the CNNDD algorithm with a related nearest neighbor based approach.

Literature related to this work can be grouped into three main categories: nonparametric binary classification using the nearest neighbor rule, one-class classification, and outlier detection. Next, we briefly describe these approaches.

In the nonparametric binary classification problem we have available a training set $\{(x_1, y_1), \ldots, (x_n, y_n)\}$ of n pairs (x_i, y_i), $1 \le i \le n$, where x_i is an object from an object space and $y_i \in \{-1, 1\}$ is the corresponding class label. The *nearest neighbor rule* (1-NN-rule) [7] assigns to a new object q the label y_j, where x_j is the nearest neighbor of q in $\{x_1, \ldots, x_n\}$ according to a certain metric. This rule is based on the property that the nearest neighbor x_j of q contains at least half of the total discrimination information contained in an infinite-size training set [3, 14, 4]. The generalization of the 1-NN-rule, the k-NN-rule, in which a new pattern q is classified into the class with the most members present among its k nearest neighbors in $\{x_1, \ldots, x_n\}$, has the property that its probability of error asymptotically approaches the Bayes error [5].

There exists several approaches to one-class classification. In the nearest neighbor one-class classification method NN-d [16], a test object p is accepted if the distance to its nearest neighbor q in the training set is less or equal than the distance from q to its nearest neighbor in the training set. This measure is comparable with the Local Outlier Factor [2] used to detect outliers. The k-center method covers the data set with k balls with equal radii [20]. Ball centers are placed on training objects such that the maximum distance of all minimum distances between training objects and the centers is minimized. One-class classification techniques based on Support Vector (SV) Machines extend the SV algorithm to the case of unlabelled data [13, 15].

Research on outlier detection in data mining focuses in providing techniques for identifying the most deviating objects in an input data set. Distance-based outlier detection has been introduced in [11]: a point in a data set is a $DB(c, d)$-outlier with respect to parameters c and d, if at least fraction c of the points in the data set lies greater than distance d from it. This definition generalizes several discordancy tests to detect outlier given in statistics and it is suitable when the data set does not fit any standard distribution. The definition of [12] is closely related to the previous one: given a k and n, a point p is an outlier if no more than $n-1$ other points in the data set have a higher value for D^k than p, where $D^k(p)$ denotes the distance of the kth nearest neighbor of a point p. In order to take into account the sparseness of the neighborhood of a point, [1] considers for each point p the measure $w_k(p)$, denoting the sum of the distances to its k nearest neighbors. [6] provides further algorithms for distance-based anomaly detection. We point out that the measure $\|\delta(p)\|_r$ here used, generalizes all the distance-based measures, since $D^k(p) = \|\delta(p)\|_\infty$, and $w_k(p) = \|\delta(p)\|_1$.

The rest of the paper is organized as follows. In Section 2 we formally define the NNDD rule and the concept of reference consistent subset. Generalization of the rule is discussed in Section 3. In Section 4 we describe the algorithm CNNDD. Finally, Section 5 reports experimental results.

2   The NNDD Rule

In the following we denote with U a set of objects, with d a distance on U, with D a set of objects from U, with k a positive integer number, with θ a positive real number, and with r a Minkowski metric L_r, $r \in \{1, 2, \ldots, \infty\}$.

Given an object p of U, the *kth nearest neighbor* $nn_{D,d,k}(p)$ of p in D according to d is the object q of D such that there exists exactly $k-1$ objects s of D with $d(p, s) \leq d(p, q)$. In particular, if $p \in D$, then $nn_{D,d,1}(p) = p$. The k *nearest neighbors distances vector* $\delta_{D,d,k}(p)$ of p in D is

$$\delta_{D,d,k}(p) = (d(p, nn_{D,d,1}(p)), \ldots, d(p, nn_{D,d,k}(p))).$$

The *Nearest Neighbor Domain Description rule* (NNDD for short) $\text{NNDD}_{D,d,k,\theta,r}$ according to D, d, k, θ, r, is the function from U to $\{-1, 1\}$ such that

$$\text{NNDD}_{D,d,k,\theta,r}(p) = \text{sign}(\theta - \|\delta_{D,d,k}(p)\|_r),$$

where $\text{sign}(x) = -1$ if $x \leq 0$, and $\text{sign}(x) = 1$ otherwise.

Intuitively, the NNDD rule returns 1 when the object belongs to the class represented by D, while it returns -1 when the object does not belong to that class. In the special case $k = 1$ and $\theta = 0$, the rule accepts an object p iff $p \in D$, while for $k = 1$ and $\theta > 0$, the rule accepts an object if it lies in the neighborhood of radius θ of some object in D.

Let f be $\text{NNDD}_{D,d,k,\theta,r}$. The *accepting region* $\mathcal{R}(f)$ of f is the set $\{x \in U \mid f(x) = 1\}$. The *rejecting region* $\overline{\mathcal{R}}(f)$ of f is the set $U \setminus \mathcal{R}(f)$. An object $x \in \overline{\mathcal{R}}(f)$ is said to be an *outlier*. The *empirical risk*, or *training set error*, of the NNDD classifier f is the quantity

$$R^{emp}(f) = \frac{|D \cap \overline{\mathcal{R}}(f)|}{|D|}.$$

The empirical risk is directly proportional to the value of k and inversely proportional to the value of θ. Indeed, $\|\delta_{D,d,k-1}(p)\|_r \leq \|\delta_{D,d,k}(p)\|_r$, for $k > 1$. In particular, $R^{emp}(f)$ is certainly zero for $k = 1$ or for arbitrarily large values of θ.

When the reference set D is large, space requirements to store D and time requirements to find the nearest neighbors of an object in D increase. In the spirit of the reference set thinning problem for the k-NN-rule [9, 17], next we define the concept of NNDD reference consistent subset, and then we show that finding a minimum NNDD reference consistent subset is NP-hard.

A NNDD *reference consistent subset* of D w.r.t. d, k, θ, r, is a subset S of D such that

$$(\forall p \in D)(\text{NNDD}_{D,\text{d},k,\theta,r}(p) = \text{NNDD}_{S,\text{d},k,\theta,r}(p)),$$

i.e. a subset of D that correctly classifies the objects in D.

The complexity of finding a minimum reference consistent subset is related to the complexity of the following decision problem: given an integer number m, $1 \le m \le |D|$, the NNDD *minimum reference consistent subset problem* $\langle D, \text{d}, k, \theta, r, m \rangle$ is: does there exist a NNDD reference consistent subset S of D w.r.t. d, k, θ, r such that $|S| \le m$?

Theorem 1. *Let* $r \in \mathbb{N}^+$ *denote a finite Minkowski metrics* L_r. *Then the* $\langle D, \text{d}, k, \theta, r, m \rangle$ *problem is NP-complete.*

Proof. (Membership) Given a subset S of D, having size $|S| \le m$, we can check in polynomial time that, for each $p \in D$, $\text{NNDD}_{D,\text{d},k,\theta,r}(p) = \text{NNDD}_{S,\text{d},k,\theta,r}(p)$.

(Hardness) The proof is by reduction to the *Dominating Set Problem* [8]. Let $G = (V, E)$ be an undirected graph, and let $m \le |V|$ be a positive integer. The *Dominating Set Problem* is: is there a subset $U \subseteq V$, called *dominating set* of G, with $|U| \le m$, such that for all $v \in (V - U)$ there exists $u \in U$ with $\{u, v\} \in E$?

Let $G = (V, E)$ be an undirected graph. Define the metric d_V on the set V of nodes of G as follows: $\text{d}_V(u, v) = 1$, if $\{u, v\} \in E$, and $\text{d}_V(u, v) = 2$, otherwise. Let $\theta_{k,r}$ be $(1 + 2^r(k - 1))^{1/r}$. Now we prove that G has a dominating set of size m iff $\langle V, \text{d}_V, k, \theta_{k,r}, r, m \rangle$ is a "yes" instance.

First, we note that, for each $v \in V$, $\|\delta_{V,\text{d}_V,k}(v)\|_r \le (0 + 2^r(k - 1))^{1/r} \le \theta_{k,r}$.

($\Rightarrow$) Suppose that G has a dominating set U such that $|U| \le m$. Then U is a reference consistent subset of V w.r.t. $\text{d}_V, k, \theta_{k,r}, r$. Indeed, let v a generic object of V. If $v \in U$, then $\|\delta_{U,\text{d}_V,k}(v)\|_r \le (0 + 2^r(k - 1))^{1/r} < \theta_{k,r}$, otherwise $v \notin U$ and $\|\delta_{U,\text{d}_V,k}(v)\|_r \le (1 + 2^r(k - 1))^{1/r} \le \theta_{k,r}$.

($\Leftarrow$) Suppose that there exists a reference consistent subset U of V such that $|U| \le m$. By contradiction, assume that there exists $v \in (V - U)$ such that, for each $u \in U$, $\{v, u\} \notin E$. Then, $\|\delta_{U,\text{d}_V,k}(v)\|_r \ge 2k^{1/r} > \theta_{k,r}$, and U is not a reference consistent subset of V. It follows immediately that U is a dominating set for G. □

Theorem 1 also holds for the special case $k = 1$ and $r = \infty$. It follows immediately from Theorem 1 that the problem of computing the minimum size reference consistent subset is NP-hard.

3   SRM and NNDD Rule

Given a set $\mathcal{F}$ of functions from U to $\{-1, 1\}$ and a set of examples $\{(x_i, y_i) \in U \times \{-1, 1\} \mid 1 \le i \le n\}$ generated from an unknown probability distribution $P(x, y)$, the goal of nonparametric binary classification is to find a function $f \in \mathcal{F}$ providing the smallest possible value for the risk $R(f) = \int |f(x) - y| \text{d}P(x, y)$. But $R(f)$ is unknown, since $P(x, y)$ is unknown. According to the *Structural*

Risk Minimization (SRM) principle, for any $f \in \mathcal{F}$, with a probability of at least $1 - \nu$, the bound $R(f) \leq R^{emp}(f) + \varepsilon(f, h, \nu)$ holds, where $\nu \in [0,1]$, R^{emp} is the *empirical risk*, or *training set error*, defined as $R^{emp}(f) = \frac{1}{n} \sum_{i=1}^{n} |f(x_i) - y_i|$, $\varepsilon(f, h, \nu)$ is the *confidence term*, and h is the Vapnik-Chervonenkis (VC) dimension of (a subset of) $\mathcal{F}$ [18]. The VC dimension $h = \text{VCdim}(\mathcal{F})$ of a set $\mathcal{F}$ of binary classifiers, is the maximal number h of objects that can be separated into two classes in all possible 2^h ways using functions from $\mathcal{F}$. It can be shown that the confidence term $\varepsilon(f, h, \nu)$ monotonically increases with increasing VC dimension h [19]. In order to bound the confidence term, one induces a structure of nested subsets $\mathcal{F}_1 \subset \mathcal{F}_2 \subset \ldots \subset \mathcal{F}_n \subset \ldots$ of $\mathcal{F}$, such that $\text{VCdim}(\mathcal{F}_1) \leq \text{VCdim}(\mathcal{F}_2) \leq \ldots \leq \text{VCdim}(\mathcal{F}_n) \leq \ldots$ and chooses the function $f \in \mathcal{F}_i$, $i \in \mathbb{N}^+$, such that $R(f)$ is minimal.

Next we define such a structure for sets of NNDD classifiers. Let $F_n^{k_0}$ denote the set composed by the NNDD classifiers $\text{NNDD}_{D,d,k_0,\theta,r}$ on the object space U with metric d, having the same value k_0 for k, and the same value $r \in \{1, 2, \ldots, \infty\}$ for r, and such that $|D| \leq n$. Clearly, $F_n^k \subset F_{n+1}^k$, for each $n \geq 1$. Next, we give bounds for the VC dimension of the sets F_n^k.

Theorem 2. $\text{VCdim}(F_n^k) \geq n/k$.

Proof. For any set $X = \{x_1, \ldots, x_h\}$ of h distinct objects of U, with class labels $y_i \in \{-1, 1\}$, $i = 1, \ldots, h$, we can build the set $D = \bigcup_{y_i=1} \{x_i, \ldots, x_i\}$, where each x_i is repeated k times. Clearly, the classifier $\text{NNDD}_{D,d,k,0,r}$ correctly classifies the objects in the set X, and D is such that $|D| \leq kh$. □

Theorem 3. Let U be a normed linear space, and let $r \in \{1, \infty\}$. Then $\text{VCdim}(F_n^k) \leq 2\binom{n}{k}$.

Proof. First, we prove the following claim.

Claim. Let U be a normed linear space, with norm $\| \cdot \|_U$, and let $d(x, y) = \|x - y\|_U$ be the distance induced by the norm of U. Let D be a subset of U of size n, and let f be $\text{NNDD}_{D,d,k,\theta,r}$, $r \in \{1, \infty\}$. Then $\mathcal{R}(f)$ is the union of at most $m = \binom{n}{k}$ convex sets.

Proof. First, we note that $\mathcal{R}(f) = \bigcup_{i=1}^{m} \mathcal{R}(\text{NNDD}_{E_i,d,k,\theta,r})$, where E_i, $1 \leq i \leq m$, is one of the subset of D having cardinality k. Indeed, $x \in \mathcal{R}(f)$ iff there exists $E_i \subseteq D$, $|E_i| = k$, such that $\|\delta_{E_i,d,k}(x)\|_r \leq \theta$. It remains to show that each $R_i = \mathcal{R}(\text{NNDD}_{E_i,d,k,\theta,r})$ is convex. We have to prove that, for each $x, y \in R_i$, the object $z = (1 - \lambda)x + \lambda y$, $0 < \lambda < 1$, is such that $z \in R_i$, i.e. that $\|\delta_{E_i,d,k}(z)\|_r \leq \theta$. Let $E_i = \{e_{i,1}, \ldots, e_{i,k}\}$. For $r = 1$, we have:
$$\|\delta_{E_i,d,k}(z)\|_1 = \sum_{j=1}^{k} \|e_{i,j} - z\|_U = \sum_{j=1}^{k} \|(1 - \lambda)e_{i,j} + \lambda e_{i,j} - (1 - \lambda)x - \lambda y\|_U$$
$$\leq \sum_{j=1}^{k} [(1 - \lambda)\|e_{i,j} - x\|_U + \lambda \|e_{i,j} - y\|_U] = (1 - \lambda)\sum_{j=1}^{k} \|e_{i,j} - x\|_U +$$
$$\lambda \sum_{j=1}^{k} \|e_{i,j} - y\|_U = (1 - \lambda)\|\delta_{E_i,d,k}(x)\|_1 + \lambda \|\delta_{E_i,d,k}(y)\|_1 \leq (1 - \lambda)\theta + \lambda\theta \leq \theta.$$
The proof for $r = \infty$ is analogous. □

Now we can resume to the main proof. Let x_0 be an object of U, and let m be $\binom{n}{k}$. Consider the set $X = \{x_i = x_{i-1} + x_0 \mid 1 \leq i \leq 2m+1\}$ of $2m+1$ objects of U, with class labels $y_{2i-1} = 1$, $y_{2i} = -1$, $1 \leq i \leq m$, $y_{2m+1} = 1$. Suppose that there exists $f_n \in F_n^k$ such that f_n classifies X correctly. Then, it is the case that there exist $i^*, j^* \in \{1, \ldots, m\}$, such that $R_{j^*} = \mathcal{R}(\text{NNDD}_{E_{j^*},\text{d},k,\theta,r}) \supseteq \{x_{2i^*-1}, x_{2i^*+1}\}$. As R_{j^*} is convex, this implies that $x_{2i^*} \in R_{j^*}$, and hence to $\mathcal{R}(f_n)$, a contradiction.   □

From bounds provided by Theorems 2 and 3 the following relationship between the VC dimensions of sets of NNDD classifiers on a normed linear space hold: $\text{VCdim}(F_n^k) \leq \text{VCdim}(F_{2k\binom{n}{k}}^k)$. Thus, in order to minimize the risk, given a data set D and values of k_0 and θ_0 for k and θ respectively, one can choose the smallest reference consistent subset S of D w.r.t. d, k_0, θ_0, r, and then build the classifier $\text{NNDD}_{S,\text{d},k_0,\theta_0,r}$. Indeed, by definition of reference consistent subset, S is the smallest subset, among all the subsets of D, such that $R^{emp}(\text{NNDD}_{D,\text{d},k_0,\theta_0,r}) = R^{emp}(\text{NNDD}_{S,\text{d},k_0,\theta_0,r})$. Thus, replacing the reference set D with a small reference consistent subset of D has a twofold usefulness: both response time and generalization of the classifier are improved. Analogously to the argumentation of [10] for the nearest neighbor rule, this establishes the link between reference set thinning for NNDD classifiers and the SRM principle.

4 The Condensed NNDD Rule

In this section we describe the algorithm CNNDD that computes a reference consistent subset $RefSet$ of a given dataset with only *two* data set passes. The algorithm, shown in Figure 1, receives in input the dataset $DataSet$ and parameters d, k, θ, and r. Let f denote the classifier $\text{NNDD}_{DataSet,\text{d},k,\theta,r}$. We recall that $RefSet$ must be such that, for each p of $DataSet$, the property

$$f(p) = \text{NNDD}_{RefSet,\text{d},k,\theta,r}(p) \qquad (1)$$

holds. $InRefSet$ and $OutRefSet$ are sets used to partition the objects of the reference consistent subset $RefSet$ as described in the following. Each object p_j in $OutRefSet$ has associated two heaps, δ_j and δ_j^t, storing, respectively, the k nearest neighbors of p_j in $RefSet$, and the k nearest neighbors of p_j in $DataSet$.

1st phase: first dataset pass. During this step $OutRefSet$ stores the objects of $RefSet$ such that $\|\delta_j\|_r = \|\delta_{RefSet,\text{d},k}(p_j)\|_r > \theta$, while $InRefSet$ contains the remaining objects of $RefSet$. The heap δ, associated to the current data set object p_i, stores the k nearest neighbors of p_i in $RefSet$. Hence, $\|\delta\|_r = \|\delta_{RefSet,\text{d},k}(p_i)\|_r$. For each incoming data set object p_i, the distances among p_i and the objects p_j of $OutRefSet$ are computed and the heaps δ_j^t are updated. Next, until the value $\|\delta\|_r$ remains above the threshold θ, the distances among p_i and the objects p_j of $InRefSet$ are computed. After having compared p_i with the objects in $InRefSet$, if $\|\delta\|_r$ remains above θ, then

Algorithm CNNDD($DataSet$,d,k,θ,r)
$InRefSet = \emptyset$; $OutRefSet = \emptyset$; // — First data set pass —
for each (p_i **in** $DataSet$)
  $\delta = \emptyset$;
  for each (p_j **in** $OutRefSet$)
    $Update(\delta, \mathrm{d}(p_i,p_j), p_j)$;
    $Update(\delta_j^t, \mathrm{d}(p_i,p_j), p_i)$;
    for each (p_j **in** $InRefSet$) **if**($\|\delta\|_r > \theta$) $Update(\delta, \mathrm{d}(p_i,p_j), p_j)$;
  if ($\|\delta\|_r > \theta$)
    for each (p_j **in** $OutRefSet$)
      $Update(\delta_j, \mathrm{d}(p_i,p_j), p_i)$;
      if ($\|\delta_j\|_r \leq \theta$)
        $OutRefSet = OutRefSet - \{p_j\}$;
        $InRefSet = InRefSet \cup \{p_j\}$;
    $Update(\delta, 0, p_i)$;
    if ($\|\delta\|_r > \theta$) $OutRefSet = OutRefSet \cup \{p_i\}$
    else $InRefSet = InRefSet \cup \{p_i\}$;
$RefSet = InRefSet \cup OutRefSet$; // — Second data set pass —
for each (p_i **in** ($DataSet - RefSet$))
  for each (p_j **in** $OutRefSet$) **if** ($\|\delta_j^t\|_r > \theta$)
    if ($i < j$) $Update(\delta_j^t, \mathrm{d}(p_i,p_j), p_i)$;
$IncrRefSet = \emptyset$; // — Reference set augmentation —
for each (p_j **in** $OutRefSet$) **if** ($\|\delta_j^t\|_r \leq \theta$)
  for each (p_i **in** $IncrRefSet$) $Update(\delta_j, \mathrm{d}(p_j,p_i), p_i)$;
  while ($\|\delta_j\|_r > \theta$)
    Let p_i be the object of ($\delta_j^t - \delta_j$) with the minimum value of $\mathrm{d}(p_i,p_j)$;
    $Update(\delta_j, d, p_i)$;
    $IncrRefSet = IncrRefSet \cup \{p_i\}$;
$RefSet = RefSet \cup IncrRefSet$;
return($RefSet$);

Fig. 1. The CNNDD rule

p_i is inserted in $RefSet$. In this case, the heap δ is updated with the object p_i, and the heaps δ_i and δ_i^t associated to p_i are set equal to δ. Furthermore, the heaps δ_j of the objects already contained in $OutRefSet$ are updated with p_i: if the value $\|\delta_j\|_r$ becomes less or equal than θ, then the object p_j is removed from $OutRefSet$ and inserted into $InRefSet$. We note that the heaps associated to these objects are no longer useful and can be discarded. Being $RefSet = OutRefSet \cup InRefSet$ a subset of $DataSet$, then it is the case that $\|\delta_{DataSet,\mathrm{d},k}(p)\|_r \leq \|\delta_{RefSet,\mathrm{d},k}(p)\|_r$. Thus, the points p of $DataSet$ not stored in $RefSet$, are such that $\|\delta_{DataSet,\mathrm{d},k}(p)\|_r \leq \|\delta_{RefSet,\mathrm{d},k}(p)\|_r \leq \theta$, and Property (1) is guaranteed for these objects. Furthermore, for each $p \in \overline{\mathcal{R}}(f)$, $\theta < \|\delta_{DataSet,\mathrm{d},k}(p)\|_r \leq \|\delta_{RefSet,\mathrm{d},k}(p)\|_r$, and, hence, $RefSet$ contains the set $\overline{\mathcal{R}}(f)$.

2nd phase: second dataset pass. Let p_j stored in $OutRefSet$ at the end of the first scan. Unfortunately, $\|\delta_j^t\|_r > \theta$ does not imply that $p_j \in \overline{\mathcal{R}}(f)$, as p_j was not compared with all the data set objects during the first data set pass. Thus, in order to establish whether $\|\delta_{DataSet,d,k}(p_j)\|_r$ is greater than θ, a second data set scan is performed. For each $p_j \in OutRefSet$, the heap δ_j^t is updated in order to compute exact value of $\delta_{DataSet,d,k}(p_j)$, by comparing p_j with all the objects p_i in $(DataSet - RefSet)$ such that $i < j$, i.e. with the objects preceding p_j that are not stored in $RefSet$. Indeed, a generic object p_j of $OutRefSet$ was compared, during the first scan, exactly with all the objects p_i of $DataSet$, $j < i$, and with all the objects $\{p_i \in RefSet \mid i < j\}$.

3rd phase: reference set augmentation: The third phase of the algorithm is introduced to guarantee Property (1) for the objects stored in $OutRefSet$. To this aim, the set $RefSet$ is augmented with the set $IncrRefSet$. In particular, for each $p_j \in OutRefSet$ such that $\|\delta_{DataSet,d,k}(p_j)\|_r \leq \|\delta_j^t\|_r \leq \theta$, $IncrRefSet$ is augmented with some nearest neighbors of p_j, until $\|\delta_{RefSet \cup IncrRefSet,d,k}(p_j)\|_r$ goes down the threshold θ. To conclude, we note that at the end of the algorithm, the objects of $RefSet$ having $\|\delta_j^t\|_r > \theta$ are the outliers of $DataSet$.

Theorem 4. The CNNDD rule computes a reference consistent subset for the NNDD rule.

The CNNDD rule is suitable for disk-resident data sets, as it tries to minimize the number of I/O operations performing only two data set passes. As for the spatial cost of the CNNDD rule, it is $\mathcal{O}(|RefSet| \cdot k)$, due to space needed to store heaps associated to objects in $OutRefSet$. The temporal cost is $\mathcal{O}(|RefSet| \cdot |DataSet| \cdot (d + \log k))$, where d is the cost of computing the distance between two objects, and $\log k$ is the cost of updating an heap of k elements. The cost above stated is a worst case, but usually each data set object is compared only with a fraction of the objects in the reference subset. Thus, the temporal cost of the CNNDD rule depends on the size of the computed reference subset and it is subquadratic in general, while it becomes quadratic when the reference subset consists of all the data set objects, i.e. if we set $\theta = 0$. As shown in the following section, for values of θ of interest, the reference consistent subset is composed by a fraction of the data set objects.

5   Experimental Results

In this section we describe experiments executed on the following data sets[1]: *Image segmentation* (19 attributes, 330 normal objects, 1,980 abnormal objects), *Ionosphere* (34 attributes, 225 normal objects, 126 abnormal objects), *Iris* (4 attributes, 50 normal objects, 100 abnormal objects)), *Letter recognition* (16 attributes, 789 normal objects representing the letter "a", 19,211 abnormal objects representing the other letters), *Satellite image* (36 attributes, 1,533 normal objects, 4,902 abnormal objects), *Shuttle* (9 attributes, 34,108 normal objects, 3,022

---

[1] See the UCI Machine Learning Repository for more information.

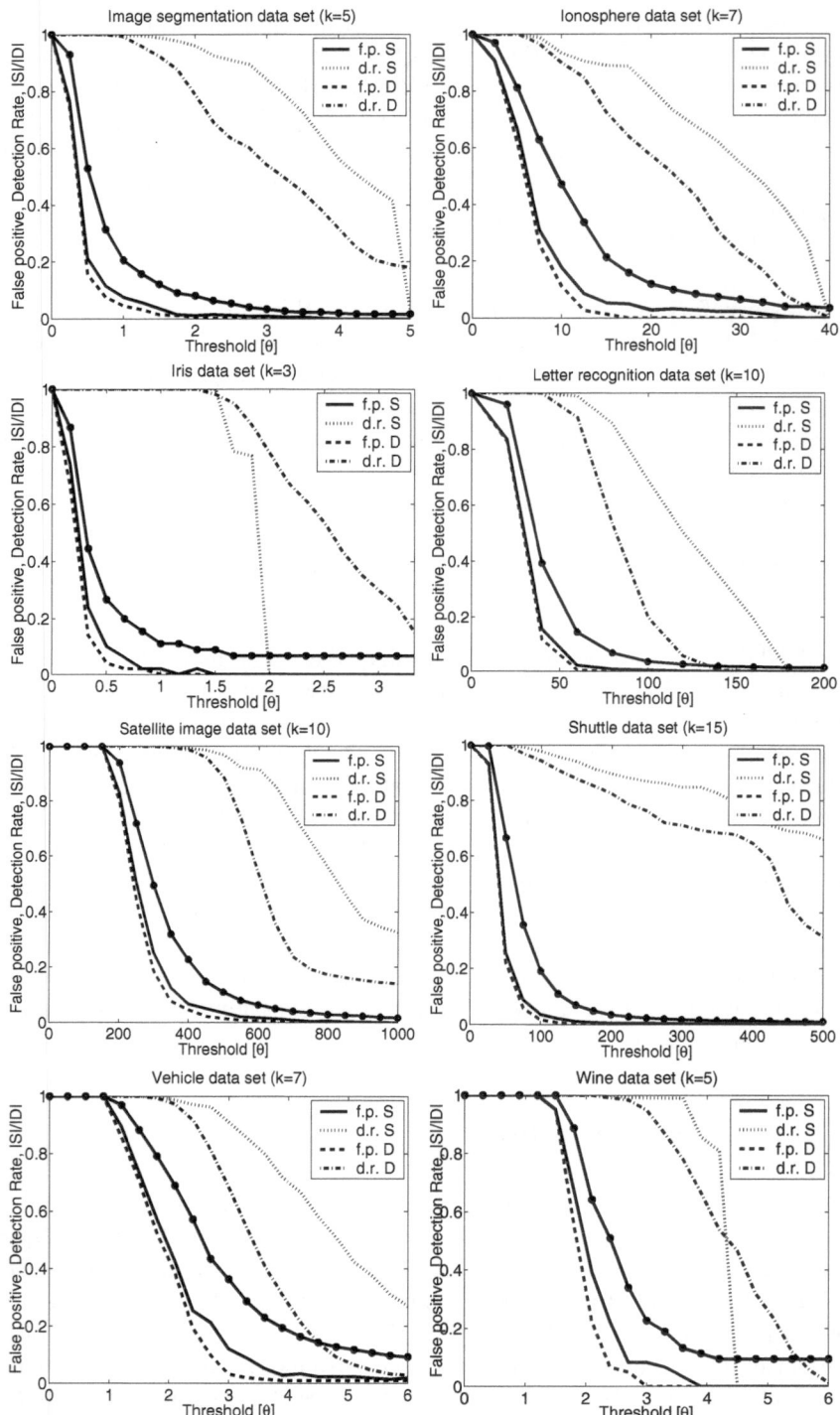

Fig. 2. Comparison between the NNDD and the CNNDD rules

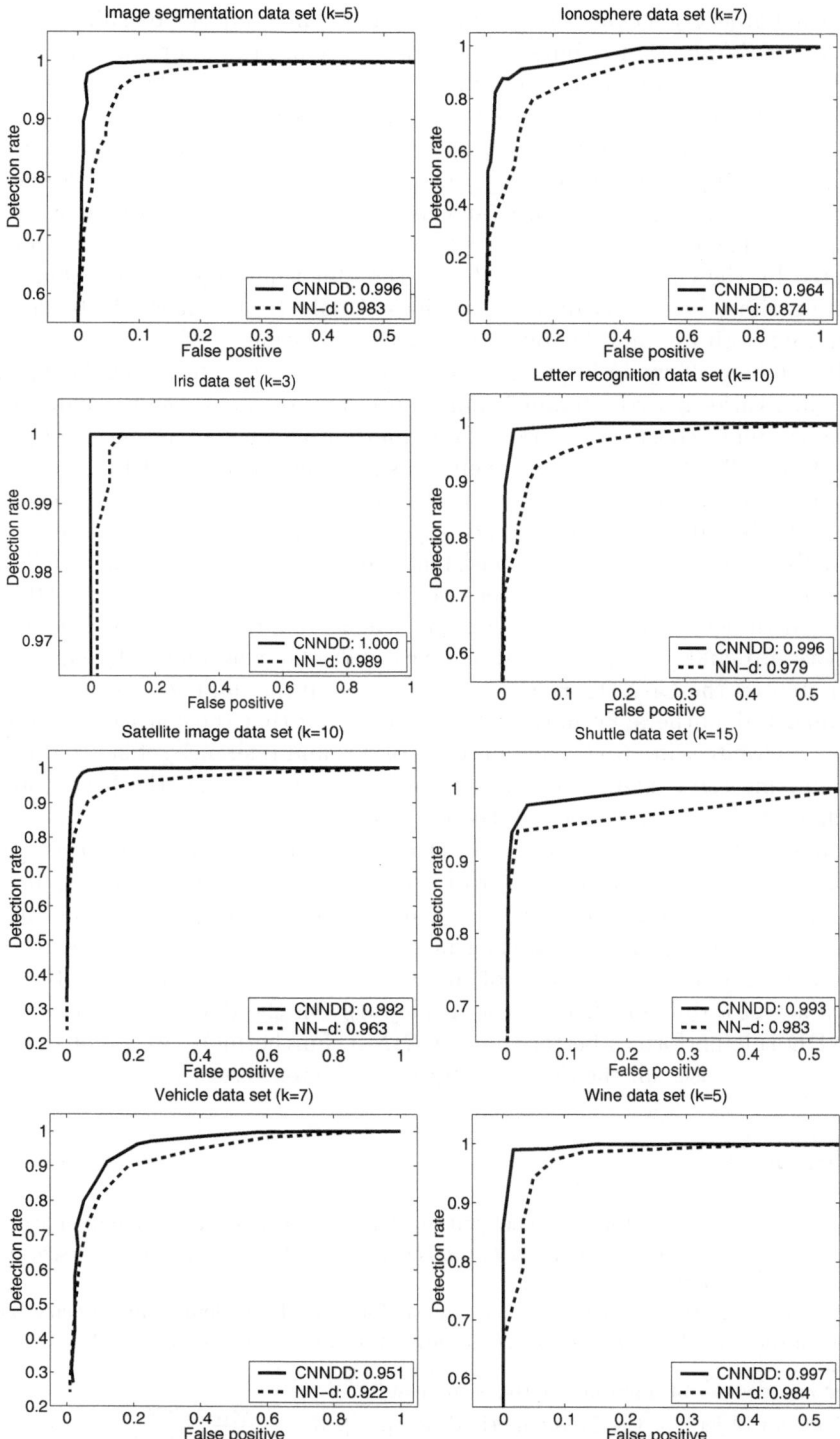

Fig. 3. ROC curves of the CNNDD and NN-d methods

abnormal objects), *Vehicle* (18 attributes, 218 normal objects, 628 abnormal objects), and *Wine* (13 attributes, 59 normal objects, 119 abnormal objects). We used the Euclidean distance, and set the parameter r to 1^2 in all the experiments.

First, we compared the NNDD and CNNDD rules. For a fixed value of k, we varied θ in the range $[0, \theta_{\max}]$, and measured the empirical error, the *false positive rate* (f.p. in the following) and the *detection rate* (d.r. in the following) of the NNDD and CNNDD rules, and the size of the consistent reference subset computed. The f.p. (d.r. resp.) is the fraction of normal (abnormal resp.) objects rejected by the classifier. We recall that the abnormal objects are unknown at learning time, being the data set composed only by the normal objects. We computed both f.p. and d.r. by 10-fold cross-validation.

Results are shown in Figure 2. The x axe reports the threshold value θ, while the y axe varies between 0 and 1 and reports f.p., d.r., and the normalized size of the reference consistent subsets. Solid and dotted lines represent the f.p. and d.r. of the CNNDD rule respectively. Dashed and dash-dotted lines represent the f.p. and d.r. of the NNDD rule. The empirical error is very close to the f.p. of the NNDD rule, thus it is not reported. Finally, pointed line reports the normalized size of the reference consistent subset computed by the CNNDD rule. We note that the CNNDD rule sensibly improves the d.r. over the NNDD rule with a little loss of f.p. when θ is high. When θ approaches the value zero, in any case f.p. and d.r. approach value one, while the consistent reference subset computed by the CNNDD rule tend to contain all the data set objects, as they are almost all outliers. From these figures it is clear that the consistent reference subset guarantees improvements in terms of d.r. and reference set size reduction, thus making the NNDD rule efficient and effective, and that the best trade-off is achieved in the curve elbow of the false positive rate.

Finally, we compared the CNNDD rule with a different nearest neighbor method, the NN-d method[3] [16] through ROC analysis. ROC curves are the plot of f.p. versus d.r., and the area under the curve gives an estimate of the ability of the method in separating inliers from outliers. Figure 3 reports the ROC curves and the ROC areas of the two methods (we obtained similar areas for values of k different from those displayed). On the data set considered the CNNDD rule performes better than the NN-d. Furthermore, we point out that the NN-d rule uses all the data set objects as reference set.

References

1. F. Angiulli and C. Pizzuti. Fast outlier detection in high-dimensional spaces. In *Proc. of the European Conf. on Principles and Practice of Knowledge Discovery in Databases*, pages 15–26, 2002.
2. M. Breunig, H.P. Kriegel, R. Ng, and J. Sander. Lof: identifying density-based local outliers. In *Proc. of the ACM Int. Conf. on Management of Data*, 2000.

---

[2] For $r = \infty$ we obtained about the same results.

[3] Given an object p, the NN-d method accepts p if $\frac{\|\delta_{D,\mathrm{d},k}(p)\|_\infty}{\|\delta_{D,\mathrm{d},k+1}(nn_k(p,D))\|_\infty} \leq \theta$, and rejects it otherwise ($k = 1$ and $\theta = 1$ are usually employed).

3. T.M. Cover and P.E. Hart. Nearest neighbor pattern classification. *IEEE Trans. on Information Theory*, 13:21–27, 1967.
4. L. Devroye. On the inequality of cover and hart. *IEEE Trans. on Pattern Analysis and Machine Intelligence*, 3:75–78, 1981.
5. L. Devroye, L. Gyorfi, and G. Lugosi. *A Probabilistic Theory of Pattern Recognition*. Springer-Verlag, New York, 1996.
6. E. Eskin, A. Arnold, M. Prerau, L. Portnoy, and S. Stolfo. A geometric framework for unsupervised anomaly detection: detecting intrusions in unlabeled data. *Applications of Data Mining in Computer Security*, 2002.
7. E. Fix and J. Hodges. Discriminatory analysis. non parametric discrimination: Consistency properties. In *Tech. Report 4*, USAF School of Aviation Medicine, Randolph Field, Texas, 1951.
8. M.R. Garey and D.S. Johnson. *Computer and Intractability*. W. H. Freeman and Company, New York, 1979.
9. P.E. Hart. The condensed nearest neighbor rule. *IEEE Trans. on Information Theory*, 14:515–516, 1968.
10. B. Karaçali and H. Krim. Fast minimization of structural risk by nearest neighbor rule. *IEEE Trans. on Neural Networks*, 14(1):127–137, 2003.
11. E. Knorr and R. Ng. Algorithms for mining distance-based outliers in large datasets. In *Proc. of the Int. conf. on Very Large Databases*, pages 392–403, 1998.
12. S. Ramaswamy, R. Rastogi, and K. Shim. Efficient algorithms for mining outliers from large data sets. In *Proc. ACM Int. Conf. on Managment of Data*, pages 427–438, 2000.
13. B. Schölkopf, C. Burges, and V. Vapnik. Extracting support data for a given task. In *Proc. of the Int. Conf. on Knowledge Discovery & Data Mining*, pages 251–256, Menlo Park, CA, 1995.
14. C. Stone. Consistent nonparametric regression. *Annals of Statistics*, 8:1348–1360, 1977.
15. D. Tax and R. Duin. Data domain description using support vectors. In *Proc. of the European Symp. on Artificial Neural Networks*, pages 251–256, Bruges (Belgium), April 1999.
16. D. Tax and R. Duin. Data descriptions in subspaces. In *Proc. of Int. Conf. on Pattern Recognition*, pages 672–675, 2000.
17. G. Toussaint. Proximity graphs for nearest neighbor decision rules: Recent progress. In *Tech. Report SOCS-02.5*, School of Computer Science, McGill University, Montréal, Québec, Canada, 2002.
18. V. Vapnik and A. Chervonenkis. On the uniform convergence of relative frequencies of events to their probabilities. *Theory of Probability and its Applications*, 16(2):264–280, 1971.
19. V. N. Vapnik. *Statistical learning theory*. S. Haykin Ed., Wiley, New York, 1998.
20. A. Ypma and R. Duin. Support objects for domain approximation. In *Proc of the ICANN*, 1998.

Balancing Strategies and Class Overlapping[*]

Gustavo E.A.P.A. Batista[1,2], Ronaldo C. Prati[1], and Maria C. Monard[1]

[1] Institute of Mathematics and Computer Science at University of São Paulo,
P. O. Box 668, ZIP Code 13560-970, São Carlos (SP), Brazil
[2] Faculty of Computer Engineering at Pontifical Catholic University of Campinas,
Rodovia D. Pedro I, Km 136, ZIP Code 13086-900, Campinas (SP), Brazil
{gbatista, prati, mcmonard}@icmc.usp.br

Abstract. Several studies have pointed out that class imbalance is a bottleneck in the performance achieved by standard supervised learning systems. However, a complete understanding of how this problem affects the performance of learning is still lacking. In previous work we identified that performance degradation is not solely caused by class imbalances, but is also related to the degree of class overlapping. In this work, we conduct our research a step further by investigating sampling strategies which aim to balance the training set. Our results show that these sampling strategies usually lead to a performance improvement for highly imbalanced data sets having highly overlapped classes. In addition, over-sampling methods seem to outperform under-sampling methods.

1 Introduction

Supervised Machine Learning – ML – systems aim to automatically create a classification model from a set of labeled training examples. Once the model is created, it can be used to automatically predict the class label of unlabeled examples. In many real-world applications, it is common to have a huge intrinsic disproportion in the number of examples in each class. This fact is known as the class imbalance problem and occurs whenever examples of one class heavily outnumber examples of the other class[1]. Generally, the minority class represents a circumscribed concept, while the other class represents the counterpart of that concept.

Several studies have pointed out that domains with a high class imbalance might cause a significant bottleneck in the performance achieved by standard ML systems. Even though class imbalance is a problem of great importance in ML, a complete understanding of how this problem affects the performance of learning systems is not clear yet. In spite of poor performances of standard learning systems in many imbalanced domains, this does not necessarily mean that class imbalance is solely responsible for the decrease in performance. Rather,

---

[*] This research is partly supported by Brazilian Research Councils CAPES and FAPESP.

[1] Although in this work we deal with two-class problems, this discussion also applies to multi-class problems. Furthermore, **positive** and **negative** labels are used to denominate the minority and majority classes, respectively.

A.F. Famili et al. (Eds.): IDA 2005, LNCS 3646, pp. 24–35, 2005.

it is quite possible that beyond class imbalance yields certain conditions that make the induction of good classifiers difficult. For instance, even for highly imbalanced domains, standard ML systems are able to create accurate classifiers when classes are linearly separable.

All matters considered, it is crucial to identify in which situations a skewed dataset might lead to performance degradation in order to develop new tools and/or to (re)design learning algorithms to cope with this problem. To accomplish this task, artificial data sets may provide a useful framework, since their parameters can be fully and easily controlled. For instance, using artificial data sets Japkowicz [3] showed that class imbalance is a relative problem depending on both the complexity of the concept and the overall size of the training set. Futhermore, in previous work [6] using artificial datasets, we showed that performance degradation of imbalanced domains is related to the degree of data overlapping between classes.

In this work, we broaden this research by applying several under and over-sampling methods to balance the training data. Under-sampling methods balance the training set by reducing the number of majority class examples, while over-sampling methods balance the training set by increasing the number of minority class examples. Our objective is to verify whether balancing training data is an effective approach to deal with the class imbalance problem, and how the controlled parameters, namely class overlapping and class imbalance, affect each balancing method. Our experimental results in artificial datasets show that balancing training data usually leads to a performance improvement for highly imbalanced data sets with highly overlapped classes. In addition, over-sampling methods usually outperform under-sampling methods.

This work is organized as follows: Section 2 presents some notes related to evaluating the performance of classifiers in imbalanced domains. Section 3 introduces our hypothesis regarding class imbalances and class overlapping. Section 4 discusses our experimental results. Finally, Section 5 presents some concluding remarks and suggestions for future work.

2   Evaluating Classifiers in Imbalanced Domains

As a rule, error rate (or accuracy) considers misclassification of examples equally important. However, in most real-world applications this is an unrealistic scenario since certain types of misclassification are likely to be more serious than others. Unfortunately, misclassification costs are often difficult to estimate. Moreover, when prior class probabilities are very different, the use of error rate or accuracy might lead to misleading conclusions, since there is a strong bias to favour the majority class. For instance, it is straightforward to create a classifier having an error rate of 1% (or accuracy of 99%) in a domain where the majority class holds 99% of the examples, by simply forecasting every new example as belonging to the majority class. Another point that should be considered when studying the effect of class distribution on learning systems is that misclassification costs and class distribution may not be static.

When the operating characteristics, *i.e.* class distribution and cost parameters, are not known at training time, other measures that disassociate errors (or hits) that occurred in each class should be used to evaluate classifiers, such as the ROC (Receiving Operating Characteristic) curve. A ROC curve is a plot of the estimated proportion of positive examples correctly classified as positive — the sensitive or true-positive rate (tpr) — against the estimated proportion of negative examples incorrectly classified as positive — the false alarm or false-positive rate (fpr) — for all possible trade-offs between the classifier sensitivity and false alarms. ROC graphs are consistent for a given problem even if the distribution of positive and negative examples is highly skewed or the misclassification costs change. ROC analysis also allows performance of multiple classification functions to be visualised and compared simultaneously. The area under the ROC curve (AUC) represents the expected performance as a single scalar. The AUC has a known statistical meaning: it is equivalent to the Wilconxon test of ranks, and is equivalent to several other statistical measures for evaluating classification and ranking models [2].

3   The Effect of Class Overlapping in Imbalanced Data

There seems to be an agreement in the ML community with the statement that imbalance between classes is the major obstacle when inducing good classifiers in imbalanced domains. However, we believe that class imbalance is not always the problem. In order to illustrate our conjecture, consider the two decision problem shown in Figure 1. The problem is related to building a classifier for a simple-single attribute problem that should be classified into two classes, positive and negative. The conditional probabilities of both classes are given by a one-dimensional unit variance Gaussian function, but the negative class centre is one standard deviation apart from the positive class centre in the first problem – Figures 1(a) and 1(b) – and four (instead of one) standard deviations apart from the positive class centre in the second problem– Figures 1(c) and 1(d).

In Figure 1(a), the aim is to build a (optimal) Bayes classifier, and perfect knowledge regarding probabilitiy distributions is assumed. The vertical line represents the optimal Bayes split. In such conditions, the optimal Bayes split should be the same however skewed the dataset is. On the other hand, Figure 1(b) depicts the same problem, but now no prior knowledge is assumed regarding probability distributions, and the aim is to build a Naïve-Bayes classifier only with the data at hand. If there were a huge disproportion of examples between classes, the algorithm is likely to produce poorer estimates for the class with fewer examples, rather than for the majority class. Particularly, in this figure, the variance is over-estimated at 1.5 (continuous line) instead of the truly variance 1 (dashed line). In other words, if we know beforehand the conditional probabilities (a constraint seldom applicable for most real-world problems) which makes the construction of a true Bayes classifier possible, class distribution should not be a problem at all. Conversely, a Naïve-Bayes classifier is likely to suffer from poor estimates due to few data available for the minority class.

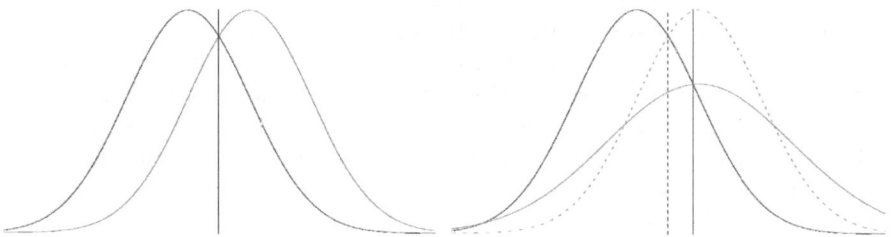

(a) The learner has perfect knowledge about the domain

(b) The learner only uses the data at hand

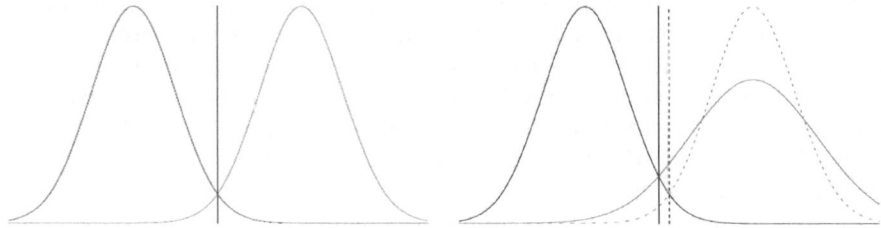

(c) The learner has perfect knowledge about the domain

(d) The learner only uses the data at hand

Fig. 1. Two different decision problems (vide text)

Consider now the second decision problem. As in Figure 1(a), Figure 1(c) represents the scenario where full knowledge regarding probabilities distribution is assumed, while Figure 1(d) represents the scenario where the learning algorithm must induce the classifier only with the data at hand. For the same reasons stated before, when perfect knowledge is assumed, the optimal Bayes classifier should not be affected by the class distribution. However, if this is not the case, the final classifier is likely to be affected. Nevertheless, due to low overlapping between the classes, the effect of class imbalance in this case is lower than when there is a high overlapping. This is to say that the number of examples misclassified in the former scenario is, therefore, higher than the number of examples misclassified in the latter. This might indicate that class probabilities are not solely responsible for hindering the classification performance, but instead the degree of overlapping between classes.

4 Experiments

The main goal of this work is to gain some insight on how balancing strategies may aid classifier's induction in the presence of class imbalance and class overlapping. In former work [6], we performed a study aimed to understand when class imbalance causes performance degradation on learning algorithms when applied to class overlapped datasets. In this work we broaden this research by investigating how several balancing methods affect the performance of learning in such conditions. We start describing the experimental set up conducted to perform our analysis, followed by a description of the balancing methods used in the experiments. To make this work more self-contained, we continue by summing up the findings reported in [6] and conclude the section with an analysis of the obtained results.

4.1   Experimental Set Up

To perform our analysis, we generated 10 artificial domains. Two clusters compose these domains: one representing the majority class and the other one representing the minority class. The data used in the experiments have two major controlled parameters. The first one is the distance between both cluster centroids, and the second one is the imbalance degree. The distance between centroids enable us to control the "level of difficulty" of correctly classifying the two classes. The grade of imbalance let us analyse if imbalance is a factor by itself for degrading performance.

Each domain is described by a 5-dimensional unit-variance Gaussian variable. Jointly, each domain has 2 classes: positive and negative. For the first domain, the mean of the Gaussian function for both classes is the same. For the following domains, we stepwise add 1 standard deviation to the mean of the positive class, up to 9 standard deviations. For each domain, we generated 14 data sets. Each data set has 10,000 examples with different proportions of examples belonging to each class, ranging from 1% up to 45% in the positive class, and the remainder in the negative class as follows: 1%, 2.5%, 5%, 7.5% 10%, 12.5%, 15%, 17.5%, 20%, 25%, 30%, 35%, 40% and 45%. We also included a control data set, which has a balanced class distribution.

Although the class complexity is quite simple (we generated data sets with only 5 attributes, two classes, and each class is grouped in only one cluster), this situation is often faced by supervised learning systems since most of them follow the so-called divide-and-conquer (or separate-and-conquer) strategy, which recursively divides (or separates) and solves smaller problems in order to induce the whole concept. Furthermore, Gaussian distribution might be used as an approximation of several statistical distributions. To run the experiments, we chose the C4.5 [8] algorithm for inducing decision trees. The reason for choosing C4.5 is twofold. Firstly, tree induction is one of the most effective and widely used methods for building classification models. Secondly, C4.5 is quickly becoming the community standard algorithm when evaluating learning algorithms in imbalanced domains. In this work, the induced decision trees were modified to produce probability decision trees (PET) [7], instead of only forecasting a class. We also use the AUC as the main method for assessing our experiments. All experiments were evaluated using 10-fold stratified cross validation.

Moreover, the choice of two Gaussian distributed classes enables us to easily compute the theoretical AUC values for the optimal Bayes classifier. The AUC can be computed using Equation 1 [5], where $\Phi(.)$ is the standard normal cumulative distribution, δ is the Euclidean distance between the centroids of the two distributions and ϕ_{pos} as well as ϕ_{neg} are, respectively, the standard deviation of positive and negative distribution.

$$AUC = \Phi\left(\frac{\delta}{\sqrt{\phi_{pos} + \phi_{neg}}}\right) \tag{1}$$

4.2 Summary of Our Previous Findings

Figure 2 summarises results of our previous findings [6]. For a better visualization, we have omitted some proportions and distances, however the lines omitted are quite similar, respectively, to the curves with 9 standard deviations apart and 50% of examples in each class. Figure 2(a) plots the percentage of positive examples in the data sets *versus* the AUC of the classifiers induced by C4.5 for different centroids of the positive class (in standard deviations) from the negative class. Consider the curve of the positive class where the class centroids are 2 standard deviations apart. Observe that these classifiers have good performance, with AUC higher than 90%, even if the proportion of positive class is barely 1%.

Figure 2(b) plots the variation of centroid distances *versus* the AUC of classifiers induced by C4.5 for different class imbalances. In this graph, we can see that the main degradation in the classifiers performance occurs mainly when the difference between the centre of positive positive and negative classes is 1 standard deviation apart. In this case, the degradation is significantly higher for highly imbalanced data sets, but decreases when the distance between the centre of the positive and negative classes increases. The difference in the performance of the classifiers are statistically insignificant when the difference between centres goes up 4 standard deviations, independently of how many examples belong to the

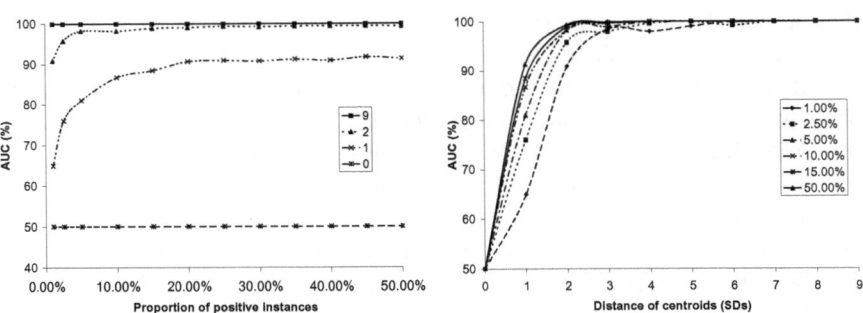

(a) Variation in the proportion of positive instances *versus* AUC.

(b) Variation in the centre of positive class *versus* AUC.

Fig. 2. Experimental results for C4.5 classifiers induced in data sets with several overlapping and imbalance rates [6]

positive class. Thus, these results suggest that datasets with linearly separable classes do not suffer from the class imbalance problem.

4.3   Balancing Methods

As summarized in Section 4.2, in [6] we analyzed the interaction between class imbalance and class overlapping. In this work we are interested in analysing the behaviour of methods that artificially balance the (training) dataset in the presence of class overlapping. Two out of five evaluated methods, described next, are non-heuristic methods, while the other three make use of heuristics to balance the training data. The non-heuristic methods are:

Random under-sampling is a method that aims to balance class distribution through random elimination of majority class examples.
Random over-sampling is a method that aims to balance class distribution through random replication of minority class examples.

Several authors agree that the major drawback of Random under-sampling is that this method can discard potentially useful data that could be important for the induction process. On the other hand, Random over-sampling can increase the likelihood of occurring overfitting, since it makes exact copies of the minority class examples. In this way, a symbolic classifier, for instance, might construct rules that are apparently accurate, but actually cover one replicated example. The remaining three balancing methods, which are described next, use heuristics in order to overcome the limitations of the non-heuristic methods:

NCL. Neighbourhood Cleaning Rule [4] uses the *Wilson's Edited Nearest Neighbour Rule* (ENN) [10] to remove majority class examples. ENN removes any example whose class label differs from the class of at least two of its three nearest neighbours. NCL modifies ENN in order to increase data cleaning. For a two-class problem the algorithm can be described in the following way: for each example E_i in the training set, its three nearest neighbours are found. If E_i belongs to the majority class and the classification given by its three nearest neighbours contradicts the original class of E_i, then E_i is removed. If E_i belongs to the minority class and its three nearest neighbours misclassify E_i, then the nearest neighbours that belong to the majority class are removed.
Smote. Synthetic Minority Over-sampling Technique [1] is an over-sampling method. Its main idea is to form new minority class examples by interpolating between several minority class examples that lie together. Thus, the overfitting that may occur with random over-sampling is avoided and causes the decision boundaries for the minority class to spread further into the majority class space.
Smote + ENN. Although over-sampling minority class examples can balance class distributions, some other problems usually present in data sets with skewed class distributions are not solved. Frequently, class clusters are not

well defined since some majority class examples might be invading the minority class space. The opposite can also be true, since interpolating minority class examples can expand the minority class clusters, introducing artificial minority class examples too deeply in the majority class space. Inducing a classifier under such a situation can lead to overfitting. In order to create better-defined class clusters, we propose applying ENN to the over-sampled training set as a data cleaning method. Differently from NCL, which is an under-sampling method, ENN is used to remove examples from both classes. Thus, any example that is misclassified by its three nearest neighbours is removed from the training set.

4.4 Experimental Results

From hereafter, we focus on data sets up to d = 3 standard deviations apart, since these data sets provided the most significant results. Furthermore, we generated new domains, also a 5-dimensional unit variance Gaussian variable having the same class distributions than the previous domains, every 0.5 standard deviation. Therefore, 7 domains were analysed in total, with the following distances in standard deviation between the centroids: 0, 0.5, 1, 1.5, 2, 2.5 and 3. The results of theoretical AUC values for these distances are shown in Table 1. As we want to gain some insight into the interaction between large class imbalances and class overlapping, we also constraint our analysis for domains up to 20% of examples in the positive class, and compared results with the naturally balanced dataset.

Table 1. Theoretical AUC values

δ	0	0.5	1	1.5	2	2.5	3
AUC	50.0 %	78.54 %	94.31 %	99.11 %	99.92 %	99.99 %	99.99 %

Smote, Random over-sampling and Random under-sampling methods have internal parameters that enable the user to set up the resulting class distribution obtained after the application of these methods. We decided to add/remove examples until a balanced distribution was reached. This decision is motivated by the results presented in [9], in which it is shown that when AUC is used as a performance measure, the best class distribution for learning tends to be near the balanced class distribution.

Figure 3 presents in graphs[2] the experimental results for distances between centroids of 0, 0.5, 1 and 1.5 standard deviations apart. Note that we have promoted a change in the scale of the AUC axis, in order to better present the results. These graphs show, for each distance between centroids, the mean AUC

---

[2] Due to lack of space, tables with numerical results were not included in this article. However, detailed results, including tables, graphs and the data sets used in the experiments can be found in http://www.icmc.usp.br/~gbatista/ida2005.

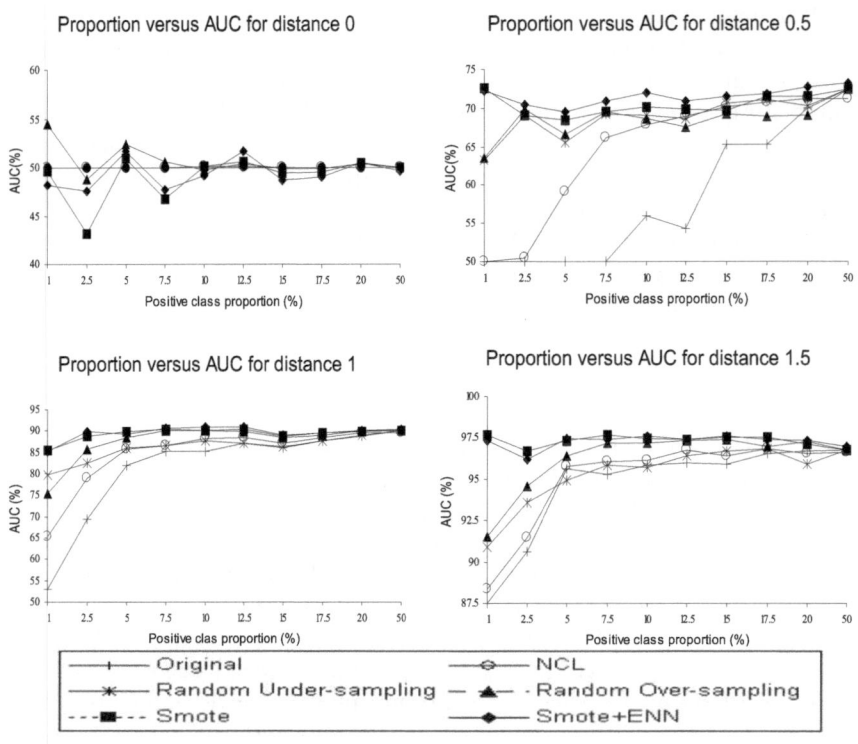

Fig. 3. Experimental results for distances 0, 0.5, 1 and 1.5.

measured over the 10 folds versus the number of positive (minority) class examples in percentage of the total number of examples in the training set. Distance 0 was introduced into the experiments for comparison purposes. As expected, AUC values for this distance oscillate (due to random variation) around random performance (AUC = 50%). In our experiments, the major influence of class skew occurs when the distance is 0.5 standard deviations. In this case, the theoretical AUC value is 78.54%, but the archived AUC values for the original data set is under 60% for proportions under 15% of positive instances.

In almost all cases, sampling methods were able to increase the AUC values for the induced classifiers. As can be observed, NCL shows some improvements over the original data, however these improvements are smaller than those obtained by Random under-sampling; the over-sampling methods usually provide the best results, with the Smote-based methods achieving almost a constant performance for all class distributions. Smote + ENN presents better results than other methods for almost all class distributions. We believe this is due to the data cleaning method, which seems to be more efficient in highly overlapped regions. The ENN data cleaning method starts to become less effective as the distance increases, since there are less data to be cleaned when the clusters are more distant from each other. Therefore, results obtained by Smote + ENN are

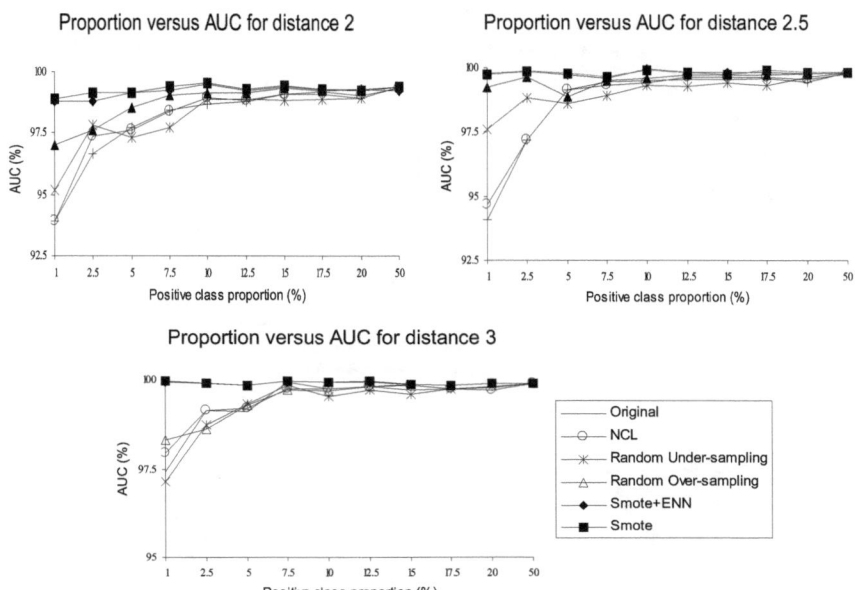

Fig. 4. Experimental results for distances 2, 2.5 and 3.

becoming more similar to the results obtained by Smote. From distance 1.5, almost all methods present good results, with most values greater than 90% AUC, and the over-sampling methods reaching almost 97% AUC. Nevertheless, the Smote-based methods produced better results and an almost constance AUC value in the most skewed region.

Figure 4 presents the experimental results for distances between centroids of 2, 2.5 and 3 standard deviations apart. For these distances, the over-sampling methods still provide the best results, especially for highly imbalanced data sets. Smote and Smote + ENN provide results that are slightly better than Random over-sampling, however the data cleaning provided by ENN becomes very ineffective. Observe that the Smote-based methods provide an almost constant, near 100% AUC for all class distributions. It is interesting to note that the performance decreases for the Random over-sampling method for distance 3 and highly imbalanced data sets. This might be indicative of overffiting, but more research is needed to confirm such a statement.

In a general way, we are interested in which methods provide the most accurate results for highly imbalanced data sets. In order to provide a more direct answer to this question, Figure 5 shows the results obtained for all distances for the most imbalanced proportions: 1%, 2.5% and 5% of positive examples. These graphs clearly show that the over-sampling methods in general, and Smote-based methods in particular, provide the most accurate results. They also show that, as the degree of class imbalance decreases, the methods tend to achieve similar performance.

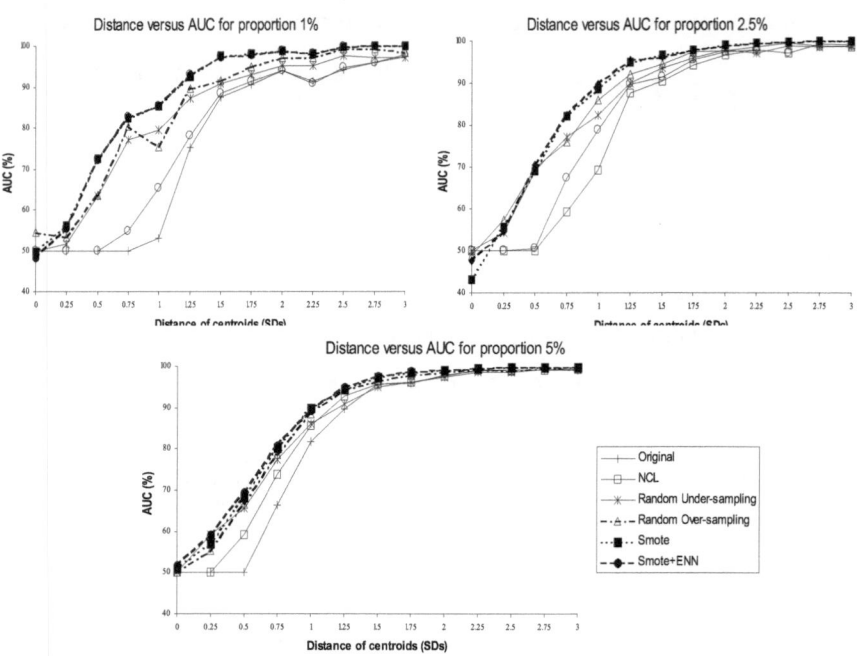

Fig. 5. Experimental results in graphs for proportions 1%, 2.5% and 5%.

5   Conclusion and Future Work

In this work, we analyse the behaviour of five methods to balance training data
in data sets with several degrees of class imbalance and overlapping. Results
show that over-sampling methods in general, and Smote-based methods in particular, are very effective even with highly imbalanced and overlapped data sets.
Moreover, the Smote-based methods were able to achieve a similar performance
as the naturally balanced distribution, even for the most skewed distributions.
The data cleaning step used in the Smote + ENN seems to be especially suitable
in situations having a high degree of overlapping.

In order to study this question in more depth, several further approaches can
be taken. For instance, it would be interesting to vary the standard deviations of
the Gaussian functions that generate the artificial data sets. It is also worthwhile
to consider the generation of data sets where the distribution of examples of the
minority class is separated into several small clusters. This approach can lead to
the study of the class imbalance problem together with the small disjunct problem, as proposed in [3]. Another point to explore is to analyse the ROC curves
obtained from the classifiers and simulate some misclassification cost scenarios.
This approach might produce some useful insights in order to develop or analyse
methods for dealing with class imbalance.

References

1. N. V. Chawla, K. W. Bowyer, L. O. Hall, and W. P. Kegelmeyer. SMOTE: Synthetic Minority Over-sampling Technique. *Journal of Artificial Intelligence Research*, 16:321–357, 2002.
2. D. J. Hand. *Construction and Assessment of Classification Rules*. John Wiley and Sons, 1997.
3. N. Japkowicz. Class Imbalances: Are We Focusing on the Right Issue? In *Proc. of the ICML'2003 Workshop on Learning from Imbalanced Data Sets (II)*, Washington, DC (USA), 2003.
4. J. Laurikkala. Improving Identification of Difficult Small Classes by Balancing Class Distribution. Technical Report A-2001-2, University of Tampere, 2001.
5. C. Marzban. The ROC Curve and the Area Under it as a Performance Measure. *Weather and Forecasting*, 19(6):1106–1114, 2004.
6. R. C. Prati, G. E. A. P. A. Batista, and M. C. Monard. Class Imbalances *versus* Class Overlapping: an Analysis of a Learning System Behavior. In *3rd Mexican International Conference on Artificial Intelligence (MICAI'2004)*, volume 2971 of *LNAI*, pages 312–321, Mexico City, 2004. Springer-Verlag.
7. F. Provost and P. Domingos. Tree Induction for Probability-Based Ranking. *Machine Learning*, 52:199–215, 2003.
8. J. R. Quinlan. *C4.5 Programs for Machine Learning*. Morgan Kaufmann, 1988.
9. G. M. Weiss and F. Provost. Learning When Training Data are Costly: The Effect of Class Distribution on Tree Induction. *Journal of Artificial Intelligence Research*, 19:315–354, 2003.
10. D. L. Wilson. Asymptotic Properties of Nearest Neighbor Rules Using Edited Data. *IEEE Trans. on Systems, Management, and Communications*, 2(3):408–421, 1972.

Modeling Conditional Distributions of Continuous Variables in Bayesian Networks*

Barry R. Cobb[1], Rafael Rumí[2], and Antonio Salmerón[2]

[1] Virginia Military Institute, Department of Economics and Business,
Lexington, VA 24450, USA
cobbbr@vmi.edu

[2] University of Almería, Department of Statistics and Applied Mathematics,
Ctra. Sacramento s/n, La Cañada de San Urbano, 04120 - Almería, (Spain)
{rrumi, Antonio.Salmeron}@ual.es

Abstract. The MTE (mixture of truncated exponentials) model was introduced as a general solution to the problem of specifying conditional distributions for continuous variables in Bayesian networks, especially as an alternative to discretization. In this paper we compare the behavior of two different approaches for constructing conditional MTE models in an example taken from Finance, which is a domain were uncertain variables commonly have continuous conditional distributions.

1 Introduction

A Bayesian network is a model of an uncertain domain which includes conditional probability distributions in its numerical representation. Methods of modeling the conditional density functions of continuous variables in Bayesian networks include discrete approximations and conditional linear Gaussian (CLG) models [5]. Modeling continuous probability densities in Bayesian networks using a method that allows a tractable, closed-form solution is an ongoing research problem.

Recently, mixtures of truncated exponentials (MTE) potentials [6] were introduced as an alternative to discretization for representing continuous variables in Bayesian networks. Moral *et al.* [8] suggest a mixed tree structure for learning and representing conditional MTE potentials. Cobb and Shenoy [2] propose operations for inference in continuous Bayesian networks where variables can be linear deterministic functions of their parents and probability densities are approximated by MTE potentials. This approach can also be implemented to represent the conditional density of a continuous variable, as we demonstrate in this paper.

This paper compares the results obtained using the four previously mentioned methods of modeling conditional densities of continuous variables in Bayesian networks. The remainder of the paper is outlined as follows. In Section 2 we

---

* This work was supported by the Spanish Ministry of Science and Technology, project TIC2001-2973-C05-02 and by FEDER funds.

A.F. Famili et al. (Eds.): IDA 2005, LNCS 3646, pp. 36–45, 2005.

establish the notation and give the definition of the MTE model. Section 3 contains the description of the models of conditional distributions used in this work, and Section 4 reports on the comparison of those models for an econometric example. The paper ends with conclusions in Section 5.

2   Notation and Definitions

Random variables will be denoted by capital letters, e.g., A, B, C. Sets of variables will be denoted by boldface capital letters, e.g., $\mathbf{X}$. All variables are assumed to take values in continuous state spaces. If $\mathbf{X}$ is a set of variables, $\mathbf{x}$ is a configuration of specific states of those variables. The continuous state space of $\mathbf{X}$ is denoted by $\Omega_{\mathbf{X}}$. MTE potentials are denoted by lower-case greek letters.

In graphical representations, continuous nodes are represented by double-border ovals and nodes that are deterministic functions of their parents are represented by triple-border ovals.

A mixture of truncated exponentials (MTE) [6,9] potential has the following definition.

Definition 1 (MTE potential). *Let $\mathbf{X} = (X_1, \ldots, X_n)$ be an n-dimensional random variable. A function $\phi : \Omega_{\mathbf{X}} \mapsto \mathbb{R}^+$ is an MTE potential if one of the next two conditions holds:*

1. The potential ϕ can be written as

$$\phi(\mathbf{x}) = a_0 + \sum_{i=1}^{m} a_i \; exp\left\{ \sum_{j=1}^{n} b_i^{(j)} x_j \right\} \tag{1}$$

for all $\mathbf{x} \in \Omega_{\mathbf{X}}$, where $a_i, i = 0, \ldots, m$ and $b_i^{(j)}$, $i = 1, \ldots, m$, $j = 1, \ldots, n$ are real numbers.

2. The domain of the variables, $\Omega_{\mathbf{X}}$, is partitioned into hypercubes $\{\Omega_{\mathbf{X}_1}, \ldots, \Omega_{\mathbf{X}_k}\}$ such that ϕ is defined as

$$\phi(\mathbf{x}) = \phi_i(\mathbf{x}) \qquad if \; \mathbf{x} \in \Omega_{\mathbf{X}_i} \; , \; i = 1, \ldots, k \; , \tag{2}$$

where each $\phi_i, i = 1, \ldots, k$ can be written in the form of equation (1) (i.e. each ϕ_i is an MTE potential on $\Omega_{\mathbf{X}_i}$).

In the definition above, k is the number of *pieces* and m is the number of exponential *terms* in each piece of the MTE potential. We will refer to ϕ_i as the i-th piece of the MTE potential ϕ and $\Omega_{\mathbf{X}_i}$ as the portion of the domain of $\mathbf{X}$ approximated by ϕ_i. In this paper, all MTE potentials are equal to zero in unspecified regions. Cobb and Shenoy [1] define a general formulation for a 2-piece, 3-term un-normalized MTE potential which approximates the normal PDF is

$$\psi'(x) = \begin{cases} \sigma^{-1}(-0.010564 + 197.055720 \ \exp\{2.2568434(\frac{x-\mu}{\sigma})\} \\ \qquad -461.439251 \ \exp\{2.3434117(\frac{x-\mu}{\sigma})\} \\ \qquad +264.793037 \ \exp\{2.4043270(\frac{x-\mu}{\sigma})\}) & \text{if } \mu - 3\sigma \leq x < \mu \\ \\ \sigma^{-1}(-0.010564 + 197.055720 \ \exp\{-2.2568434(\frac{x-\mu}{\sigma})\} \\ \qquad -461.439251 \ \exp\{-2.3434117(\frac{x-\mu}{\sigma})\} \\ \qquad +264.793037 \ \exp\{-2.4043270(\frac{x-\mu}{\sigma})\}) & \text{if } \mu \leq x \leq \mu + 3\sigma. \end{cases}$$
$$(3)$$

An MTE potential f is an *MTE density* for $\mathbf{X}$ if it integrates to one over the domain of $\mathbf{X}$. In a Bayesian network, two types of probability density functions can be found: marginal densities for the root nodes and conditional densities for the other nodes. A *conditional MTE density* $f(x|\mathbf{y})$ is an MTE potential $f(x, \mathbf{y})$ such that after fixing $\mathbf{y}$ to each of its possible values, the resulting function is a density for X.

3   Modeling Conditional Distributions

3.1   Discrete Approximations

Discretization of continuous distributions can allow approximate inference in a Bayesian network with continuous variables. Discretization of continuous chance variables is equivalent to approximating a probability density function (PDF) with mixtures of uniform distributions. Discretization with a small number of states can lead to poor accuracy, while discretization with a large number of states can lead to excessive computational effort. Kozlov and Koller [4] improve discretization accuracy by using a non-uniform partition across all variables represented by a distribution and adjusting the discretization for evidence. However, the increased accuracy requires an iterative algorithm and is still problematic for continuous variables whose posterior marginal PDF can vary widely depending on the evidence for other related variables.

Sun and Shenoy [10] study discretization in Bayesian networks where the tails of distributions are particularly important. They find that increasing the number of states during discretization always improves solution accuracy; however, they find that utilizing undiscretized continuous distributions in this context provides a better solution than the best discrete approximation.

3.2   Conditional Linear Gaussian (CLG) Models

Let X be a continuous node in a hybrid Bayesian network, $\mathbf{Y} = (Y_1, \ldots, Y_d)$ be its discrete parents, and $\mathbf{Z} = (Z_1, \ldots, Z_c)$ be its continuous parents. *Conditional linear Gaussian (CLG)* potentials [5] in hybrid Bayesian networks have the form

$$\mathcal{L}(X \mid \mathbf{y}, \mathbf{z}) \sim N(w_{\mathbf{y},0} + \sum_{i=1}^{c} w_{\mathbf{y},i} z_i, \sigma_{\mathbf{y}}^2), \qquad (4)$$

where **y** and **z** are a combination of discrete and continuous states of the parents of X. In this formula, $\sigma_{\mathbf{y}}^2 > 0$, $w_{\mathbf{y},0}$ and $w_{\mathbf{y},i}$ are real numbers, and $w_{\mathbf{y},i}$ is defined as the i-th component of a vector of the same dimension as the continuous part **Z** of the parent variables. This assumes that the mean of a potential depends linearly on the continuous parent variables and that the variance does not depend on the continuous parent variables. For each configuration of the discrete parents of a variable X, a linear function of the continuous parents is specified as the mean of the conditional distribution of X given its parents, and a positive real number is specified for the variance of the distribution of X given its parents. CLG models cannot accommodate continuous random variables whose distribution is not Gaussian unless each such distribution is approximated by a mixture of Gaussians.

3.3   Mixed Probability Trees

A conditional density can be approximated by an MTE potential using a *mixed probability tree* or mixed tree for short. The formal definition is as follows:

Definition 2. (Mixed tree) *We say that a tree $\mathcal{T}$ is a mixed tree if it meets the following conditions:*

i. *Every internal node represents a random variable (either discrete or continuous).*
ii. *Every arc outgoing from a continuous variable Z is labeled with an interval of values of Z, so that the domain of Z is the union of the intervals corresponding to the arcs Z-outgoing.*
iii. *Every discrete variable has a number of outgoing arcs equal to its number of states.*
iv. *Each leaf node contains an MTE potential defined on variables in the path from the root to that leaf.*

Mixed trees can represent MTE potentials that are defined by parts. Each entire branch in the tree determines one sub-region of the space where the potential is defined, and the function stored in the leaf of a branch is the definition of the potential in the corresponding sub-region. In [7], a method for approximating conditional densities by means of mixed trees was proposed. It is based

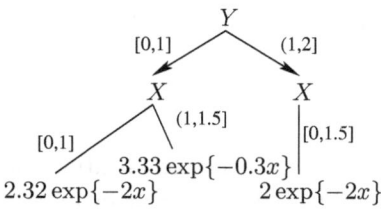

Fig. 1. A mixed probability tree representing the potential ϕ in equation (5)

on fitting a univariate MTE density in each leaf of the mixed tree. For instance, the mixed tree in Figure 1 represents the following conditional density:

$$\phi(x,y) = f(x|y) = \begin{cases} 2.32\exp\{-2x\} & \text{if } 0 \le y \le 1, \ 0 \le x \le 1 \\ 3.33\exp\{-0.3x\} & \text{if } 0 \le y \le 1, \ 1 < x \le 1.5 \\ 2\exp\{-2x\} & \text{if } 1 < y \le 2, \ 0 \le x \le 1.5 \end{cases} \quad (5)$$

3.4   Linear Deterministic Relationships

Cobb and Shenoy [2] describe operations for inference in continuous Bayesian networks with linear deterministic variables. Since the joint PDF for the variables in a continuous Bayesian network with deterministic variables does not exist, these operations are derived from the method of convolutions in probability theory.

Consider the Bayesian network in Figure 2. The variable X has a PDF represented by the MTE potential, $\phi(x) = 1.287760 - 0.116345\exp\{1.601731x\}$, where $\Omega_X = \{x : x \in [0,1]\}$. The variable Z is a standard normal random variable, i.e. $\mathcal{L}(Z) \sim N(0,1)$, which is represented by the 2-piece, 3-term MTE approximation to the normal PDF defined in (3) and denoted by φ. The variable Y is a deterministic function of X and Z, and this relationship is represented by the conditional mass function (CMF), $\alpha(x,y,z) = p_{Y|\{x,z\}}(y) = 1\{y = 3x + z + 2\}$, where $1\{A\}$ is the indicator of the event A.

Fig. 2. The Bayesian network used to demonstrate the operations for linear deterministic relationships

The joint PDF for $\{X, Z\}$ is a 2-piece MTE potential, defined as

$$\vartheta(x,z) = (\phi \otimes \varphi)(x,z) = \begin{cases} \phi(x) \cdot \varphi_1(z) \text{ if } (-3 \le z < 0) \cap (0 \le x \le 1) \\ \phi(x) \cdot \varphi_2(z) \text{ if } (0 \le z \le 3) \cap (0 \le x \le 1) \ , \end{cases}$$

where φ_1 and φ_2 are the first and second pieces of the MTE potential φ. The symbol '$\otimes$' denotes pointwise multiplication of functions. The un-normalized joint PDF for $\{Y, Z\}$ is obtained by transforming the PDF for $\{X, Z\}$ as follows:

$$\theta(y,z) = \begin{cases} \phi((y-z-2)/3) \cdot \varphi_1(z) \text{ if } (-3 \le z < 0) \cap (0 \le (y-z-2)/3 \le 1) \\ \phi((y-z-2)/3) \cdot \varphi_2(z) \text{ if } (0 \le z \le 3) \cap (0 \le (y-z-2)/3 \le 1) \ . \end{cases}$$

This transformation is a marginalization operation where X is removed from the combination of ϑ and α and is denoted by $\theta = (\vartheta \otimes \alpha)^{-X}$. The function θ remains an MTE potential because the function substituted for x in θ is linear in Y and Z. The un-normalized marginal PDF for Y is obtained by integrating the MTE potential for $\{Y, Z\}$ over the domain of Z as follows:

$$\eta(y) = \theta^{-Z}(y) = \begin{cases} \int_{-3}^{y-2} \theta_1(y,z)\,dz & \text{if } -1 \le y < 2 \\[2ex] \int_{y-5}^{0} \theta_1(y,z)\,dz + \int_{0}^{y-2} \theta_2(y,z)\,dz & \text{if } 2 \le y < 5 \\[2ex] \int_{y-5}^{3} \theta_2(y,z)\,dz & \text{if } 5 \le y \le 8 \ . \end{cases} \tag{6}$$

The variables Y and Z are dependent, so the limits of integration in (6) are defined so that the function is integrated over the joint domain of Y and Z. The result of the operation in (6) is an MTE potential, up to one linear term in the first and third pieces. This linear term is replaced by an MTE potential so the densities in the Bayesian network remain in the class of MTE potentials (for details, see [3]). The above operations are extended with new notation to model linear deterministic relationships in hybrid Bayesian networks in [3].

4   An Example and Comparison

In this section we compare the four methods described in Section 3 using an example taken from an econometric model. Consider the model in Figure 3 where the daily returns on Chevron-Texaco stock (Y) are dependent on the daily returns of the Standard & Poor's (S&P) 500 Stock Index (X).

Suppose S_n is the stock or index value at time n. The return on S between time n and time $n+1$ is a rate r defined such that $S_n \exp\{r\} = S_{n+1}$, assuming continuous compounding. Thus, we can calculate the daily returns as $r = \ln(S_{n+1}/S_n)$ if the time interval is assumed to be one day. If we assume that stock or index prices follow a geometric Brownian motion (GBM) stochastic process, the distribution of stock or index prices is lognormal with parameters determined by the drift and volatility of the GBM process. If stock prices are lognormal, stock returns are normally distributed because the log of stock prices are normally distributed and because $r = \ln(S_{n+1}) - \ln(S_n)$. Thus, stock returns are a linear combination of normal random variables, which is itself a normal random variable.

In this example, we use data on daily closing prices for the S&P 500 and Chevron-Texaco for each business day in 2004 to calculate daily returns. There are 251 observations in the sample. We randomly selected 50 as a holdout sample to test the marginal distributions created for Chevron-Texaco returns (Y) and used the remaining 201 to parameterize the various models.

A least-squares regression of Chevron-Texaco stock prices on S&P 500 index prices defines the linear equation $y_i = a + b \cdot x_i + \epsilon_i$, where a is an intercept, b

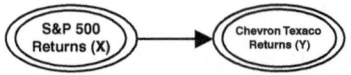

Fig. 3. The Bayesian network for the Chevron-Texaco stock example

is a slope coefficient, and ϵ_i is an error term for observation i. Estimating the parameters for this model from the data yields the equation $\hat{y}_i = \hat{a} + \hat{b} \cdot x_i$. The residuals, calculated as $e_i = y_i - \hat{y}_i$, are an estimate of the error term in the model, and are assumed to be normally distributed with a mean of zero and a variance denoted by σ_Z^2.

Using the 2004 data for Chevron-Texaco and the S&P 500 yields the linear model, $\hat{y}_i = 0.083749 + 0.305849 \cdot x_i$, with $\sigma_Z^2 = 1.118700$. For this model, the coefficient $\hat{b}$ is referred to as the *beta* of the stock, which is an index of the stock's systematic risk, or the sensitivity of returns on the stock to changes in returns on the market index. This coefficient is statistically significant with a t−score of 2.86 and a two-tailed p−value of 0.0043.

We use the parameters from the linear regression model and the data on daily returns for Chevron-Texaco and the S&P 500 to parameterize sixteen Bayesian network models and compare the results obtained with the actual distribution of Chevron-Texaco prices using the KS test statistic. Where applicable, the methods are tested using 2, 3, and 4-piece MTE approximations to the marginal distribution for S&P 500 returns (X) determined using the method in [7]. We will refer to these as *marginal approximations* (MAs).

4.1 Discrete Approximation

We have considered three discretizations by dividing the domain of the continuous variables into 6, 9 and 12 sub-intervals respectively. The intervals have been determined according to the data such that each contains the same number of sample points. The probability for each discretized split is calculated by maximum likelihood.

4.2 CLG Model

Using the CLG model for this example requires that we assume S&P 500 returns (X) are normally distributed, i.e. $\pounds(X) \sim N(\mu_X, \sigma_X^2)$, and Chevron-Texaco returns (Y) are normally distributed with a mean stated as a linear function of X and a variance independent of X, i.e. $\pounds(Y \mid x) \sim N(a\mu_X + b, \sigma_Z^2)$. The mean and variance of the S&P 500 returns calculated from the data are 0.028739 and 0.487697, respectively, i.e. $X \sim N(0.028739, 0.487697)$. We use the results of the regression model to define $Y \mid x \sim N(0.305849x + 0.083749, 1.18700)$. Since $E(Y) = 0.305849 \cdot \mu_X + 0.083749$ and $Var(Y) = (0.305849)^2 \cdot Var(X) + \sigma_Z^2$, the CLG model determines the marginal distribution of Y as $N(0.102797, 1.137741)$.

4.3 Mixed Tree Model

The three mixed tree models considered in this example are constructed according to the method proposed in [8], partitioning the domain of the variables into 2, 3 and 4 pieces respectively. In each piece, an MTE density with two exponential terms plus a constant is fitted, i.e., an MTE of the form $\phi(y) = a + b\exp\{cx\} + d\exp\{ex\}$. The marginals are computed using the Shenoy-Shafer propagation algorithm adapted to MTEs [9].

During the computation of the marginals, which involves multiplication of MTE potentials, the number of exponential terms in each potential increases. In order to keep the complexity of the resulting MTE potentials—measured in total number of terms used—equivalent to the number of splits in the discrete approximation, we employ an approximate version of the Shenoy-Shafer algorithm which restricts potentials to two exponential terms and a constant, with extra terms pruned as described in [9]. We refer to this model as *pruned mixed tree*.

4.4   Linear Deterministic Model

The linear deterministic model assumes that Chevron-Texaco returns (Y) are a linear deterministic function of S&P 500 returns (X) and a Gaussian noise term (Z), $Y = a + b \cdot X + Z$. A Bayesian network representation is shown in Figure 4.

Fig. 4. The Bayesian network for the linear deterministic model in the Chevron-Texaco stock example

To parameterize the model, we use the 2, 3, and 4-piece marginal MTE potentials for X obtained by the *marginal approximation* and denoted by ϕ. The variable Z is a Gaussian noise term which is modeled by the 2-piece, 3-term MTE approximation to the normal PDF defined in (3) with $\mu = 0$ and $\sigma^2 = 1.118700$ and denoted by φ . The CMF for Y given $\{X, Z\}$ is $\alpha(x, y, z) = p_{Y|\{x,z\}}(y) = 1\{y = 0.305849x + z + 0.083749\}$. In each test case, the joint PDF $\vartheta = (\phi \otimes \varphi)$ for $\{X, Z\}$ is an MTE potential. The un-normalized joint PDF $\theta(y, z) = (\vartheta \otimes \alpha)^{-X}$ is obtained by substituting $(y - z - 0.083749)/0.305849$ into the joint PDF for $\{X, Z\}$.

The marginal PDF for Y is obtained by integrating the joint PDF θ for $\{Y, Z\}$ over the domain of Y, as in (6). The marginal distributions for Y determined using the 2, 3 and 4-piece marginals for X have 8, 11, and 14 pieces, respectively.

4.5   Comparison

The methods in Sections 4.1 through 4.4 are compared using the Kolmogorov-Smirnov (KS) statistic, which is defined as

$$D(F, G) = \sup_{-\infty < x < \infty} |F(x) - G(x)| , \tag{7}$$

where F is a target distribution and G is the empirical distribution of the sample. This statistic can be used to construct a test for the hypothesis that the data actually come from the target distribution F. The results of the KS test (value of the D statistic and p-value of the test) are displayed in Table 1.

The results of the test for the example considered in this paper show that the methods based on MTE models provide better results than the discrete approximation. Even the case of pruned mixed trees outperforms the discrete approximation.

Table 1. KS test statistics comparing the holdout sample with the marginal distributions for each method

		KS Test		
Method	Intervals	D	p-value	*No. of terms*
	2	0.099	0.71	6
Marginal approx.	3	0.1013	0.6842	9
	4	0.102	0.6758	12
	2	0.1086	0.5969	16
Mixed tree	3	0.0912	0.7996	27
	4	0.0842	0.8702	76
Pruned	2	0.1098	0.5826	6
mixed	3	0.0974	0.73	9
tree	4	0.1031	0.6628	12
	6	0.1136	0.5389	6
Discrete approx.	9	0.1226	0.4399	9
	12	0.1352	0.3198	12
Linear	2	0.0871	0.8426	64
deterministic	3	0.0832	0.8794	85
model	4	0.0866	0.8472	110
CLG	1	0.0863	0.85	2

The theoretical model for this example is a CLG model. The results of the experiment show that the MTE models result in marginal distributions for the dependent variables that are very similar to the CLG model, with the best results obtained by the linear deterministic model. This result is not surprising since the relationship between the variables is assumed to be linear with Gaussian noise, as in the theoretical model. However, it must be pointed out that case of four intervals is favorable to the mixed tree model in terms of p-value, and using a lower number of terms.

Most importantly, the MTE models yield similar results (in terms of marginal distributions) to the CLG model and the discrete approximations. The MTE model is extended to the non-Gaussian case as easily as the discrete model, and marginals are defined in terms of densities rather than discrete probabilities. The discrete approximations tested in this paper are comparable in computational complexity to the MTE models, e.g., the 6-bin discrete approximation has the same number of parameters as the 2-piece MTE approximation.

5   Conclusions

In this paper we compared alternative methods of specifying conditional distributions in Bayesian networks with continuous variables. The results show that the MTE model is appropriate in problems where the theoretical model has Gaussian distributions, which is common in Econometric examples. We use an example where continuous variables are known to be normal to compare the results with CLG models; however, since the MTE model can be used to approximate any continuous probability distribution, the results extend to models with non-Gaussian densities. Using the results in this paper, we can deduct that the results obtained in larger models using the two approaches for estimating conditional MTE models are likely to be accurate; however, additional research is needed to compare the complexity of propagation in large models that employ the two approaches.

References

1. Cobb, B.R. and P.P. Shenoy: Inference in hybrid Bayesian networks with mixtures of truncated exponentials. WP#294, School of Business, University of Kansas, Lawrence, KS (2003). Available for download at: http://www.people.ku.edu/~brcobb/WP294.pdf
2. Cobb, B.R. and P.P. Shenoy: Inference in hybrid Bayesian networks with deterministic variables. In P. Lucas (ed.): *Proc. of the 2nd European Workshop on Probabilistic Graphical Models (PGM'04)* Leiden, Netherlands (2004) 57–64. Available for download at: http://hdl.handle.net/1808/176
3. Cobb, B.R. and P.P. Shenoy: Propagation in hybrid Bayesian networks with linear deterministic variables. WP#314, School of Business, University of Kansas, Lawrence, KS (2005). Available for download at: http://www.people.ku.edu/~brcobb/WP314.pdf
4. Kozlov, A.V. and D. Koller: Nonuniform dynamic discretization in hybrid networks. In D. Geiger and P.P. Shenoy (eds.): *Uncertainty in Artificial Intelligence* **13** (1997) 314–325, Morgan–Kaufman, San Francisco.
5. S.L. Lauritzen and F. Jensen: Stable local computation with conditional Gaussian distributions. *Statistics and Computing* **11** (2001) 191–203.
6. Moral, S., Rumí, R. and A. Salmerón: Mixtures of truncated exponentials in hybrid Bayesian networks. Lecture Notes in Artificial Intelligence **2143** (2001) 156–167.
7. Moral, S., Rumí, R. and A. Salmerón: Estimating mixtures of truncated exponentials from data. In J.A. Gámez and A. Salmerón (eds.): *Proc. of the 1st European Workshop on Probabilistic Graphical Models (PGM'02)* Cuenca, Spain (2002) 135–143.
8. Moral, S., Rumí, R. and A. Salmerón: Approximating conditional MTE distributions by means of mixed trees. Lecture Notes in Artificial Intelligence **2711** (2003) 173–183.
9. Rumí, R. and A. Salmerón: Penniless propagation with mixtures of truncated exponentials. ECSQARU'05.
10. Sun, L. and P.P. Shenoy: Using Bayesian networks for bankruptcy prediction. WP#302, School of Business, University of Kansas, Lawrence, KS (2003). Available for download at: http://lark.cc.ku.edu/ pshenoy/Papers/WP302.pdf

Kernel K-Means for Categorical Data

Julia Couto

James Madison University,
Harrisonburg VA 22807, USA
coutoji@jmu.edu

Abstract. Clustering categorical data is an important and challenging data analysis task. In this paper, we explore the use of kernel K-means to cluster categorical data. We propose a new kernel function based on Hamming distance to embed categorical data in a constructed feature space where the clustering is conducted. We experimentally evaluated the quality of the solutions produced by kernel K-means on real datasets. Results indicated the feasibility of kernel K-means using our proposed kernel function to discover clusters embedded in categorical data.

1 Introduction

Clustering is an important data analysis task aimed to partition data into groups such that objects in the same group are similar among themselves while objects in different clusters are different. Most of the clustering algorithms found in the literature seek to cluster numerical data. Typically, numerical clustering algorithms rely on a distance metric, such as Euclidean distance or Minkowski distance, to measure the dissimilarity among objects. Categorical data, data whose attributes are discrete and unordered, lack a natural metric to assess the dissimilarity among categorical objects [5]. As a consequence, clustering categorical data is a difficult and challenging problem. The discovery of natural groups embedded in categorical datasets is a relevant issue in several fields such as psychology and bioinformatics.

In recent years, several clustering algorithms for categorical data have been proposed [1,2,5,9,20]. In this paper, we present a novel approach for clustering categorical data by means of kernel methods. Kernel methods [17] focus on the application of standard machine learning algorithms to data embedded into an inner product feature space through kernel functions. In applying this approach, we propose to embed categorical objects into an inner product feature space of large dimensionality where the clustering is conducted. The embedding of the categorical data into the feature space is expected to exploit the intrinsic correlations of the groups in the data [17]. The inner product in the new space defines a distance metric among the embeddings of the objects. The computation of the inner product of the embedding of any pair of objects is performed through kernel functions. A standard clustering algorithm that relies on a distance metric can be applied to discover the clusters of the categorical data in the feature space. To the best of our knowledge, there are no other categorical clustering

A.F. Famili et al. (Eds.): IDA 2005, LNCS 3646, pp. 46–56, 2005.

algorithms that follow this approach. In this paper, we formulate a novel kernel function for categorical data based on Hamming distance to embed the data into a constructed inner product feature space. Due to its simplicity, we have chosen the popular clustering algorithm K-means as the clustering algorithm to carry out the discovering of the clusters in a feature space.

This paper is organized as follows. Section 2 briefly describes some related work in clustering algorithms for categorical data. Section 3 formulates a new kernel function based on Hamming distance to compare categorical objects and describes the family of diffusion kernel functions for categorical data. Section 4 describes kernel K-means, an extension of K-means for clustering data in a feature space. Section 5 compares and discusses the quality of the clustering produced by kernel K-means using the kernel functions discussed in Section 3 with respect to other clustering algorithms for categorical data.

2   Related Work

K-modes [11] is an extension of the K-means clustering algorithm for categorical data. K-modes uses the modes of the objects grouped in the same cluster as its representative. The algorithm minimizes the dissimilarity of the objects in a cluster with respect to its mode.

STIRR [7] is an iterative method based on non-linear dynamic systems on multiple instances of weighted hypergraphs (known as basins). Each attribute value is represented as a weighted vertex. Two vertices are connected when the attribute values they represent co-occur at least once in the dataset. The weights are propagated on each hypergraph until the configuration of weights in the main basin converges to a fixed point.

ROCK [9] is an agglomerative hierarchical clustering algorithm for categorical data that uses the concept of links between objects. A link between two categorical objects is defined as the number of common neighbors. Two objects are neighbors when their Jaccard coefficient exceeds a certain threshold θ defined by the user. ROCK proceeds in an agglomerative fashion to maximize its criterion function. The choice of threshold θ is critical to the quality of the clusters found by ROCK and seems to be dataset dependent. ROCK does not exploit correlations among attributes and it does not deal with noise or missing values in a dataset.

CACTUS [5] is a combinatorial search based algorithm that uses intra-attribute and inter-attribute summary information to discover clusters of attribute values. A cluster is defined as a maximal set of strongly connected attribute values. A set of attribute values are strongly connected if the number of tuples in the dataset containing the attribute values exceeds their expected co-occurrence by a user-defined threshold under the attribute independence assumption. CACTUS uses the intra-attribute and inter-attribute summaries to compute all the cluster-projections on each attribute. Then, CACTUS heuristically constructs a set of candidate clusters by combining cluster-projections to ensure that the attribute values in the candidate clusters are pairwise strongly connected. At last, CACTUS discards those candidate clusters whose attribute values are not strongly connected.

CLICK [20] is a graph approach for clustering categorical data sets that characterizes a cluster as a maximal k-partite clique. In CLICK, each attribute value is a vertex in a k-partite graph, two vertices in the k-partite graph are linked by an edge when they belong to different attributes and are strongly connected [5]. CLICK uses a heuristic approach to detect all the maximal k-partite strongly connected.

COOLCAT [2] and LIMBO [1] use information theory to discover clusters in categorical data. COOLCAT is an incremental partition clustering algorithm to minimize the expected entropy. COOLCAT uses a sample to identify k categorical objects with maximum pairwise entropy. Afterward, COOLCAT incrementally places each object in a cluster that achieves the minimum expected entropy. COOLCAT assumes the independence of the attributes to compute the entropy of each cluster. LIMBO is a scalable two-stage clustering algorithm for large categorical datasets based on the agglomerative information bottleneck algorithm (AIB) [18]. LIMBO starts partitioning the dataset into a set of initial clusters in such a way that the loss of information is minimized. Then, LIMBO applies AIB to the initial clusters until it obtains the desired number of clusters.

More recently, some clustering kernel methods have been proposed [3,6,17,21]. Ben et al.[3] formulate the clustering problem as a convex optimization problem that finds the smallest enclosing sphere of the embedding of the data in a feature space. The preimages of the smallest enclosing sphere define the contours of the clusters in the data. In [6], the author proposes an iterative procedure similar to expectation maximization to discover clusters in a feature space in such a way that the intra-cluster distance is minimized. Objects whose embeddings belong to the same cluster in the feature space are clustered together. A kernel clustering scheme based on K-means for large datasets is proposed in [21].

3   Kernel Functions for Categorical Data

Kernel methods applies standard learning machine algorithms that rely on distance metrics or inner products to data embedded into a feature space using kernel functions. The embedding of the data into a feature space is expected to capture and enhance the patterns and regularities in the data [17]. Kernel methods proceed in two steps. The first step embeds the data into a feature space of high or infinite dimension, while the second step uses standard algorithms for classification, clustering and principal component analysis to detect the regularities of the data in the feature space. The core of kernel methods relies on the use of kernel functions. A kernel function computes the inner product in a feature space of the embedding of two data points under a certain mapping ϕ. Formally speaking:

Definition 1. *Let X be an n-dimensional input space and F be a N-dimensional inner product feature space F, $N >> n$. A kernel function $K : X \times X \to \Re$ is a symmetric function such that for all $x, y \in X$ satisfies*

$$K(x, y) = < \phi(x), \phi(y) > . \tag{1}$$

where $\phi : X \to F$ is a mapping between X and F such that for all $x \in X$

$$\phi(x) \to (\phi_1(x), \phi_2(x), \ldots, \phi_N(x)). \tag{2}$$

$\phi_i(x)$, $i = 1, \ldots, N$, are the features of x in the feature space F.

Definition 2. *The normalised kernel $\widetilde{K}$ of a kernel function K is computed as follows:*

$$\widetilde{K}(x, y) = \frac{K(x, y)}{\sqrt{K(x, x) K(y, y)}}. \tag{3}$$

The square distance between the embeddings of two points $x, y \in X$, $\phi(x)$ and $\phi(y)$ respectively, is computed in terms of the kernel function $K(x, y)$:

$$d^2(\phi(x), \phi(y)) = \|\phi(x) - \phi(y)\|^2 = K(x, x) - 2K(x, y) + K(y, y) . \tag{4}$$

A kernel function can be formulated by defining the mapping between input space and some feature space where the inner product is computed [17]. A kernel function defined in this way requires the characterization of the feature space F, the specification of the embedding $\phi : X \to F$, and finally the computation of the inner product between the embedding of two points. However, the computation of the features and the evaluation of the inner product have high computational costs that depend on the dimension of the feature space. An efficient approach uses computational methods such as dynamic programming to compute the kernel function for any pair of points without explicitly embedding the points in the feature space and then computing their inner product [16,17], which is the approach applied in this paper. Alternatively, the characterisation of kernel functions as finitely positive semi-definite functions [17] allows determining whether a function is a kernel function without knowing the nature of the feature space and the specification of the mapping ϕ. Finally, kernel functions satisfy several closure properties that allows constructing complex kernel functions by manipulating and combining simpler ones[17].

3.1   Hamming Distance Kernel Function

We formulate a kernel function for categorical data that uses Hamming distance to embed categorical data into a constructed inner product feature space. Our proposed kernel function does not depend on a generative model or a priori information about the nature of the data. The construction of the feature space and the kernel function follows the same methodology proposed in [16].

Definition 3. *Let D_i be a finite domain of categorical values. Let $(a_1, \ldots, a_n)$ be a categorical object such that $a_i \in D_i$. Let $D^n = \prod_{i=1}^n D_i$ be the cross product over all the domains of the attributes such that for each $(u_1, \ldots, u_n) \in D^n$, $u_i \in D_i$. Given a categorical object $s = (s_1, \ldots, s_n)$, s_k denotes the value of the k-th attribute of s. The feature space F is a subspace of $\Re^{D^n}$.*

Definition 4. *The mapping of a categorical object s into the feature space F is defined by the u coordinate $\phi_u(s) = \lambda^{H(u,s)}$, for all $u \in D^n$, $\lambda \in (0,1)$. The Hamming distance $H(u,s)$ between s and u is defined as:*

$$H(u,s) = \sum_{i=1}^{n} \delta(u_i, s_i) \ . \tag{5}$$

where $\delta(x,y)$ is 0 when $x = y$ and 1 otherwise. The u coordinate of s according to the mapping ϕ can be rewritten as:

$$\phi_u(s) = \lambda^{H(u,s)} = \prod_{i=1}^{n} \lambda^{\delta(u_i, s_i)} \ . \tag{6}$$

Definition 5. *The kernel function $K_H(s,t)$ between two input categorical objects s and t is defined as:*

$$K_H(s,t) = \sum_{u \in D^n} \phi_u(s)\phi_u(t) = \sum_{u \in D^n} \prod_{i=1}^{n} \lambda^{\delta(u_i, s_i)} \lambda^{\delta(u_i, t_i)} \ . \tag{7}$$

It can be shown that the kernel function $K_H(s,t)$ can be computed recursively in the following manner:

$$K^0(s,t) = 1$$
$$K^j(s,t) = (\lambda^2(|D_j| - 1 - \delta(s_j, t_j)) + (2\lambda - 1)\delta(s_j, t_j) + 1)K^{j-1}(s,t) \quad 1 \le j \le n$$
$$K_H(s,t) = K^n(s,t) \ . \tag{8}$$

Due to lack of space, we omit the proof of the correctness of this recursion. Finally, $\widetilde{K}_H(s,t)$ denotes the normalised kernel of $K_H(s,t)$.

3.2   Diffusion Kernels

Kondor and Lafferty [14] proposed a family of kernel functions for categorical data based on an extension of hypercube diffusion kernels. The feature space is a graph induced by the set D^n. Each categorical object $s \in D^n$ is a vertex in the graph. Two vertices v_s and v_t are connected by an edge whenever their underlying categorical objects s and t differ only in the value of one attribute, i.e., $H(s,t) = 1$. Let β be a bandwidth parameter, the family of diffusion kernel functions $K_{DK}(\beta)$ for categorical data with n attributes is defined in the following way:

$$K_{DK}(\beta)(x,y) = \prod_{i=1}^{n} \left(\frac{1 - e^{-|D_i|\beta}}{1 + (|D_i| - 1)e^{-|D_i|\beta}} \right)^{\delta(x_i, y_i)} \ . \tag{9}$$

where $x = (x_1, \ldots, x_n)$ and $y = (y_1, \ldots, y_n)$ are categorical objects.

4 Kernel K-Means

Kernel K-means is an extension of the popular clustering algorithm K-means to discover clusters in a feature space in which the distance is calculated via kernel functions. Let $\phi : X \to F$ be an embedding of a set X into a feature space F and $K : X \times X \to \Re$ its associated kernel function. Let $z_1, \ldots, z_k$ be the centroids of clusters $C_1, \ldots, C_k$ respectively. Kernel K-means can be formulated in the following way:

1. Initialization Step: Select k data points and set their embeddings in feature space as the initial centroids $z_1, \ldots, z_k$.
2. Assignment Step: Assign each data point x_i to a cluster C_q such that:

$$q = \arg\min_{j} d^2(\phi(x_i), z_j)$$

$$z_j = \frac{1}{|C_j|} \sum_{x_p \in C_j} \phi(x_p)$$

$$d^2(\phi(x_i), z_j) = K(x_i, x_i) - \frac{2}{|C_j|} \sum_{x_p \in C_j} K(x_i, x_p) + \frac{1}{|C_j|^2} \sum_{x_p \in C_j} \sum_{x_m \in C_j} K(x_p, x_m) \; .$$

3. Repeat 2 until convergence.

We use the scheme for kernel K-means for large datasets proposed in [21]. Nevertheless, in our implementation of kernel K-means, the initialization step applies a heuristic similar to [13] to select the initial centroids. The heuristic selects k well-scattered points in the feature space by maximizing their minimum pairwise distance. The initialization heuristic proceeds as follows:

1. Selects the two most distant embeddings of the dataset in the feature space as the initial centroids of the first two clusters C_1 and C_2, namely z_1 and z_2.

2. Then, the selection of the initial centroid z_i, $i = 3, \ldots, k$, of the remaining clusters proceeds in an iterative fashion. First, it computes the minimum distance between the embeddings of the remaining points to the existing centroids. Then, the object with the largest minimum distance in the feature space to the existing centroids is selected as the initial centroid z_i of cluster C_i. This step is repeated until k initial centroids have been obtained.

Table 1

Dataset	N. Records	N.Classes	Attributes	Missing Values
Votes	435	2	16	288
Mushrooms	8125	2	22	2480
Soybean	47	4	35	0
Zoo	101	7	15	0

5 Experimental Evaluation

We conducted a series of experiments to evaluate and compare the quality of the clustering produced by kernel K-means using both the kernel functions $\widetilde{K}_H$ (KKM-NH) and diffusion kernels $K_{DK}(\beta)$ (KKM-DK) with ROCK, COOLCAT and K-modes. The experiments were run in a DELL computer equipped with a Pentium 4 running at 2.8 GHz, and 1 Gigabyte of main memory, running SUSE Linux 9.1. We assessed the quality of the clustering produced by aforementioned clustering algorithms on four real categorical datasets obtained from the UCI Machine Learning Repository [4]. Missing values were treated as another attribute value. The characteristics of the datasets are summarized in Table 1.

5.1 Clustering Quality Measures

Validation of the quality of the clustering produced by a clustering algorithm is one of the most important issues in clustering analysis. Several quantitative measures have been proposed to evaluate the quality of the clustering solution found by a clustering algorithm [12,10]. In our experiments, we used three quantitative measures, External Entropy, F-Measure and Category Utility, to assess the quality of the solutions produced by the aforementioned algorithms. When comparing several clustering algorithms, if a clustering algorithm outperforms the others in most of these measures for a given dataset, it is assumed to be the best clustering algorithm for that dataset [19].

External Entropy. The external entropy is a measure of the purity of the clusters found by a clustering algorithm. Let $D = \{x_1, \ldots, x_N\}$ be a dataset of categorical objects. Let $C = \{C_1, \ldots, C_k\}$ be a clustering and let $c = \{c_1, \ldots, c_k\}$ be the classes in the data. The expected entropy of the clustering C is the weighted external entropy of each cluster:

$$E(C) = \sum_{i=1}^{k} \frac{|C_i|}{N} \sum_{j=1}^{k} P(c_j|C_i) log(P(c_j|C_i)) \ . \tag{10}$$

F–Measure. The F–measure is a combination of precision and recall measurements from information retrieval [15]. Let $P(c_i, C_j)$ be the precision of a class c_i in a cluster C_j and $R(c_i, C_j)$ be the recall of a class c_i in a cluster C_j. The F-measure of a class c_i in a cluster C_j is defined as follows:

$$F(c_i, C_j) = \frac{2R(c_i, C_j)P(c_i, C_j)}{R(c_i, C_j) + P(c_i, C_j)} \ . \tag{11}$$

The overall F–measure of a clustering is given by [19]:

$$F = \sum_{i} \frac{|c_i|}{N} \max_{j}\{F(c_i, C_j)\} \ . \tag{12}$$

A larger F–measure indicates a better quality of the clustering.

Category Utility (CU). Category utility [8] measures the increase of the expected probability of attribute values of the objects in the same clusters over the expected probability of the attributes. Category Utility is computed as follows:

$$CU = \sum_{j=1}^{k} \frac{|C_j|}{N} \sum_{i=1}^{n} \sum_{v \in D_i} [P(A_i = v|C_j)^2 - P(A_i = v)^2] \ . \tag{13}$$

5.2   Experimental Details

We performed several experiments on synthetic datasets to empirically determine the parameters λ and β for KKM-NH and KKM-DK respectively that produce the best clustering of the data. Our experiments indicated that the parameters for KKM-NH and KKM-DK that achieve the best clustering are dataset dependent. Nevertheless, KKM-NH with λ between 0.6 and 0.8 produced a good clustering of the data with respect to External Entropy, F-Measure and Category Utility. In our experiments on the UCI datasets, we set the parameter λ to 0.6 for KKM-NH and the parameter β for KKM-DK was set between 0.1 and 2.0.

COOLCAT is sensitive to the size of the sample used to seed the initial clusters as well as the ordering of the data points in the datasets [1]. For each dataset, we ran COOLCAT on twenty random orderings. In each run, the whole dataset was set as the sample used to find the initial seeds of the clusters.

The quality of the clustering produced by ROCK is highly influenced by both the choice of the threshold θ as well as the ordering of the data. The threshold θ that results in the best performance is dataset dependent. For each dataset, we generated twenty random orderings and ran ROCK with threshold θ between 0.1 to 0.95.

Both K-modes and kernel K-means are sensitive to the initial centroids of the clusters. In our experiments, we ran K-modes with twenty random restarts. Kernel K-means KKM-NH and KKM-DK were run using twenty different initial centroids selected according to the initialization heuristic explained in section 4.

Finally, the quality measures reported for all the clustering algorithms are averages over all the runs. For ROCK and KKM-DK, we also report the algorithm's parameter that produced the best average results.

5.3   Results

The average quality measures produced by K-modes, COOLCAT, ROCK, KKM-HN and KKM-DK on the UCI datasets are shown in Table 2.

KKM-NH and KKM-DK achieved the best clustering for Congressional Votes and Soybean with respect to the three quality measures. On the ZOO dataset, KKM-NH and KKM-DK outperformed the other algorithms with respect to External Entropy and Category Utility. F-measure obtained by both KKM-DK and KKM-DK on this dataset were comparable to the F-Measure obtained by ROCK. On the Mushrooms dataset, K-modes produced the best results with

Table 2

Dataset	Clustering Algorithm	EE	F	CU
Congressional Vote	K-Modes	0.519	0.864	2.896
	ROCK (θ=0.73)	0.654	0.798	1.891
	COOLCAT	0.511	0.864	2.839
	KKM-NH	0.477	0.880	2.941
	KKM-DK (β=1.6)	0.475	0.880	2.941
Mushrooms	K-Modes	0.751	0.706	1.504
	ROCK (θ=0.8)	0.849	0.653	1.064
	COOLCAT	0.791	0.701	1.465
	KKM-NH	0.910	0.618	1.313
	KKM-DK (β=0.5)	0.811	0.634	1.510
	KKM-NH(*)	0.715	0.751	1.404
	KKM-DK(*) (β=0.3)	0.786	0.713	1.183
Soybean	K-Modes	1.229	0.560	2.950
	ROCK (θ=0.75)	0.021	0.996	5.493
	COOLCAT	0.033	0.986	5.489
	KKM-NH	0.000	1.000	5.558
	KKM-DK (β=0.6)	0.000	1.000	5.558
Zoo	K-Modes	1.229	0.560	2.950
	ROCK (θ=0.69)	0.294	0.898	4.127
	COOLCAT	0.376	0.793	4.320
	KKM-NH	0.272	0.803	4.454
	KKM-DK (β=1.2)	0.262	0.844	4.476

(*) Random centroids

respect to the three quality measures. However, the poor performance of KKM-NH and KKM-DK can be explained by an inadequate selection of the initial centroids using our initialization heuristic. To confirm this hypothesis, we ran 20 trials of KKM-NH and KKM-DK selecting the initial centroids at random (Table 2). Our experiments showed an overall improvement in the quality of the clustering produced by KKM-NH with respect to External Entropy, F-Measure and CU. The results for KKM-DK(β=0.3) showed an improvement of External Entropy and F-Measure. Nevertheless, its CU was significantly lower than the one obtained by KKM-DK(β=0.5) applying the initialization method explained in section 4.

6   Conclusions and Future Work

In this paper, we have proposed the use of kernel clustering methods to cluster categorical data in a constructed feature space via kernel functions. We have introduced a new kernel function for categorical data, $\widetilde{K}_H$, based on the Hamming distance. We have applied kernel K-means to cluster categorical data embedded in a feature space via the kernel functions $\widetilde{K}_H$ and diffusion kernels $K_{DK}(\beta)$. The results of our experiments indicate that the embedding of categorical data by means of the kernel functions $\widetilde{K}_H$ and diffusion kernels $K_{DK}(\beta)$ preserves the clusters in the data. Furthermore, our results demonstrate that the solutions produced by kernel K-means embedding categorical data through the new kernel function $\widetilde{K}_H$ (λ=0.6) are generally better than the other categorical clustering algorithms compared in this paper. With regard to KKM-DK, our experiments show that the choice of the parameter β is crucial for discovering the clusters in the data. As a consequence, the application of KKM-DK for clustering categorical data is deterred by the selection of the appropriate parameter β that fits the data.

In our future work, we will focus on an incremental approach for kernel K-means to overcome the disk-space and I|O requirements of the method when dealing with massive datasets. In addition, we plan to investigate the performance of KKM-NH on datasets containing noise and missing values. Finally, we will evaluate the sensitivity of KKM-HN to the number of classes and number of relevant attributes defining the classes of a dataset.

References

1. Andritsos, P., Tsaparas, P., Miller, R. J., Sevcik., K. C.: LIMBO: Scalable Clustering of Categorical Data. In Proceedings of the 9th International Conference on Extending Database Technology (EDBT 2004), Heraklion, Crete, Greece, March 2004.

2. Barbara, D., Couto, J., Li Y.: Coolcat: An Entropy-based algorithm for Categorical Clustering. In Proceedings of the 11th ACM Conference on Information and Knowledge Management (CIKM 02), McLean, Virginia, USA, November 2002, ACM Press, pp. 582–589.

3. Ben-hur, A., Horn, D., Siegelmann, H.T., Vapnik V.: Support Vector Clustering. Journal of Machine Learning Research 2, pp. 125–137.

4. Blake, C.L., Merz, C.J.: UCI Repository of Machine Learning Databases. http://www.ics.uci.edu/~mlearn/MLRepository.html. University of California, Department of Information and Computer Science, Irvine, CA.

5. V. Ganti, J. Gehrke, and R. Ramakrishnan.: CACTUS: Clustering Categorical Data using Summaries. In Proceedings of the 5th ACM SIGKDD International Conference on Knowledge Discovery and Data Mining (KDD), San Diego, CA, USA, August 1999, ACM Press, pp. 73–83.

6. Girolami, M.: Mercer Kernel Based Clustering in Feature Space. IEEE Transactions on Neural Networks, 13(4), pp. 780–784, 2002.

7. Gibson, D., Kleinberg, J., Raghavan, P.: Clustering Categorical Data: An Approach Based on Dynamical Systems. In Proceedings of the 24th International Conference on Very Large Data Bases, (VLDB), New York, NY, USA, August 1998, Morgan Kaufmann, pp. 311–322.

8. Gluck, A., Corter, J.: Information, Uncertainty, and the Utility of Categories. In Proceedings of the 7th Annual Conference of the Cognitive Science Society, Irvine, California, 1985, Laurence Erlbaum Associates, pp. 283–287.

9. Guha, S., Rastogi, R, Shim, K.: ROCK: A Robust Clustering Algorithm for Categorical Attributes. Journal of Information Systems, 25(5), pp. 345–366, 2000.

10. M. Halkidi ,Y. Batistakis, M. Vazirgiannis.: On Clustering Validation Techniques. Journal of Intelligent Information Systems, 17(2–3), pp. 107–145, 2001.

11. Huang, Z.: Extensions to the K-Means Algorithm for Clustering Large Data Sets with Categorical Values. Data Mining and Knowledge Discovery, 2(3), pp. 283 – 304, 1998.
12. Jain, A.K., Dubes, R.C.: Algorithms for Clustering Data. Prentice Hall, Englewood Cliffs, NJ, 1988.
13. Katsavounidis, I., Kuo, C., Zhang,Z.: A New Initialization Technique for Generalized Lloyd Iteration. IEEE Signal Processing Letters, 1(10), pp. 144–146, 1994.
14. Kondor, R.I.,Lafferty, J.: Diffusion Kernels on Graphs and Other Discrete Structures. In Sammut,C. and Hoffmann, A.G. (eds), Machine Learning. In Proceedings of the 19th International Conference on Machine Learning (ICML 2002), Morgan Kaufmann, pp. 315–322.
15. Larsen, B., Aone, C.: Fast and Effective Text Mining Using Linear-time Document Clustering. In Proceedings of the 5th ACM SIGKDD International Conference on Knowledge Discovery and Data Mining, San Diego, CA, August 1999, ACM Press, pp. 16–22.
16. H. Lodhi, J. Shawe-Taylor, N. Cristiani and C. Watkins.: Text Classification using String Kernels. Journal of Machine Learning Research, 2, pp. 419–444.
17. Shawe-Taylor, J., Cristiani, N.: Kernel Methods for Pattern Analysis. Cambridge University Press, 2003.
18. Slonim, N., Tibshy, N.: Agglomerative Information Bottleneck. In Proceedings of the Neural Information Processing Systems Conference 1999 (NIPS99), Beckenridge, 1999.
19. Steinbach, M., Karypis, G., Kumar, V.: A Comparison of Document Clustering Technique. Technical Report #00–034, University of Minnesota, Department of Computer Science and Egineering.
20. Zaki, M. J., Peters, M.: CLICK: Mining Subspaces Clusters in Categorical Data via K-partite Maximal Cliques. TR 04-11, CS Dept., RPI, 2004.
21. Zhang, R., Rudnicky, A.: A Large Scale Clustering Scheme for Kernel K-means. In Proceedings of the 16th International Conference on Pattern Recognition (ICPR 02), Quebec City, Canada, August 2002, pp. 289–292.

Using Genetic Algorithms to Improve Accuracy of Economical Indexes Prediction

Óscar Cubo[1], Víctor Robles[1], Javier Segovia[2], and Ernestina Menasalvas[2]

[1] Departamento de Arquitectura y Tecnología de Sistemas Informáticos,
Universidad Politécnica de Madrid, Madrid, Spain
{ocubo, vrobles}@fi.upm.es
[2] Departamento de Lenguajes y Sistemas,
Universidad Politécnica de Madrid, Madrid, Spain
{fsegovia, emenasalvas}@fi.upm.es

Abstract. All sort of organizations needs as many information about their target population. Public datasets provides one important source of this information. However, the use of these databases is very difficult due to the lack of cross-references.

In Spain, two main public databases are available: Population and Housing Censuses and Family Expenditure Surveys. Both of them are published by Spanish Statistical Institute. These two databases can not be joined due to the different aggregation level (FES contains information about families while PHC contains the same information but aggregated). Besides, national laws protects this information and makes difficult the use of the datasets.

This work defines a new methodology for join the two datasets based on Genetic Algorithms. The approach proposed could be used in any case where data with different aggregation level need to be joined.

1   Introduction

Nowadays marketing needs all possible social, demographic and economical information about its potential customers and target environments. Marketing uses the typologies provided by micromarketing tools such as MOSAIC [7]. These typologies must be updated periodically with actual data.

In Spain, the National Statistics Institute (INE) [3] publishes the most reliable source of this kind of information which is protected by Spanish privacy laws. This protection makes very difficult the use of this data.

INE publishes two different databases:

- **Population and Housing Censuses (PHC) [4].** Provides demographic data every 10 years. The last two census were done on 1991 (published on 1996) and 2001 (published on first quarter of 2005).
- **Family Expenditure Surveys (FES) [5].** Information about consumptions of a sample of 9000 individual families each year. INE selects the most representative families.

It is possible to assign consumption (economic) indexes to each *censal section* by crossing PHC and FES data sources but a problem arises: this joining is impossible due to the different aggregation level and the lack of cross-references.

A.F. Famili et al. (Eds.): IDA 2005, LNCS 3646, pp. 57–65, 2005.

PHC contains demographic data aggregated at censal section level. A *censal section* is a group of 500 households (Spain is covered by 32000 censal sections) without any kind of reference about its members.

On the other hand, FES contains a detailed description about families. The data of each family is referenced by economical status of its main earner. The selection algorithm assures that the chosen families represents the whole population: INE selects 8 families from the more important autonomous region and stratum.

A previous work [11] solved the same problem using the previous databases for PHC and FES of the year 1991. This work used one attribute (*pollster identifier*) to group near families. The new datasets erase the *pollster identifier* so the previous work can not be applied and other approach is needed.

This paper proposes a new approach for this issue and expose a method to assign economical indexes to each *censal section* based on the use of Genetic Algorithms [1]. These indexes could be used to estimate the economical situation of new customers.

This paper is organized as follows. Second section explains the format of the used data. Third section explains the new defined process and the results of the experimentation done for the region of Madrid. The last section presents the conclusions and further future work

2   Data Fusion

The work is related to Data Fusion, also know as micro data set merging. In short, we have two data sets that could not be joined but share some variables. The target of Data Fusion is add new variables from one dataset based on shared variables.

Nowadays, it is used to reduce the cost of surveys. For example, we could create two samples of the target population and use two small surveys to each sample. The results of the two surveys could be joined using Data Fusion techniques.

As said in [9], one of the most complex issues in data fusion is the measure of the quality. In our particular case, this issue has an important impact because this quality is the best measurement for fitness functions.

3   Data Formats

This work is heavily based on the available data. All steps must take into account data sources, their formats and aggregation levels.

Spanish Statistics Institute (INE) published two main sources of information. These two datasets are:

- **Population and Housing Censuses (PHC).** Contains all demographic data aggregated at *censal section* level.
  This data set contains attributes such as sex (male, female), age (ranges), professional occupation, income... as shown in Table 1.
- **Family Expenditure Surveys (FES).** This dataset contains mainly economic data with a few demographic variables aggregated at family level. Each family is represented by its main earner.

Table 1. Example of PHC data set

Censal section	Sex Male	Sex Female	Basic studies	Medium studies	High studies	...	Income
CS1	60	40	10	60	30	...	20,000
CS2	45	55	20	40	40	...	15,000
CS3	49	51	35	40	25	...	12,500
...	...	...	...	...	...	...	...

Table 2. Example of FES data set

Family Id.	Sex	Age	Study level	...	Income
1	Male	40	Basic	...	20000
2	Male	55	Medium	...	22000
3	Female	35	Medium	...	15000
4	Male	60	High	...	17000
5	Female	40	Medium	...	19000
...	...	...	...	...	...

Table 3. FES data set from Table 2 converted to unified format

Family Id.	Sex male	Sex female	...	Age 35-39	Age 40-44	Age 45-49	...	Income Index
1	100%	0%	...	0%	100%	0%	...	1.08
2	100%	0%	...	0%	0%	0%	...	1.18
3	0%	100%	...	100%	0%	0%	...	0.81
4	100%	0%	...	0%	0%	0%	...	0.91
5	0%	100%	...	0%	100%	0%	...	1.02
...	...	...	...	...	...	...	...	...

The dataset contains data such as age, studies, income, expenses... as shown in Table 2

These two datasets are converted to a unified data format. This unified format, shown in Table 3, is based on PHC format with few changes and defines two sets of variables:

- Demographic data attributes are converted to marks. For each value or range of values a new variable is created (i.e., *Studies Basic*, *Studies_Medium* and *Studies_High* are derived from *Studies* attribute and *Age_0-4*, *Age_5-9*, *Age_10-14*... are derived from *Age* variable).
  The data stored is the percent of the value over the total: $value_i = \frac{value_i}{\sum_{\forall j} value_j}$
- Economic data are converted to indexes relative to the mean of the values: $index_i = \frac{value_i}{mean}$

This proccess is also called binarization.

4   A New Proposal to Join PHC and FES Datasets

Previous work [11] solved this problem with the data of 1991. This study involved three steps:

1. The groups of families were generated. The generation took into account a geographical proximity criterion: the *pollster identifier*.
2. The groups created in the previous step were used to train models for economical indexes. These models had as input the variables shared between FES and PHC and as output the desired economic indexes.
3. The models were applied to the PHC dataset. This step assigned the economic indexes to each *censal section*.

This old approach can not be applied to the new dataset due to the lack of the *pollster identifier* used as proximity criterion. A new methodology must be defined based on the ideas of this previous work. Besides, we can use new knowledge related to the selection of families done by INE for the FES dataset.

The documentation of the FES [2], provided by INE, shows a stratified selection of families:

– First, the most important *censal sections* of each autonomic region are chosen.
– Then, on each censal section, the criteria of the importance of the town (the *stratum* attribute of the dataset) is used. Groups of eight families are randomly chosen for each stratum.

In conclusion, the National Statistics Institute chooses a number of groups taking into account the importance of the autonomous region and stratum. The number of groups can be calculated with the expression $\frac{households\_in\_FES}{8}$. For example, in Madrid, 776 families are chosen in 97 groups, as shown in Table 4.

The new methodology has the same steps that the old one but differs in the generation of the groups of families. The new methodology tries to create groups of 8 families as similar as possible to the real ones. Our process mimics the INE procedure and creates groups of 8 families from the same autonomic region and stratum. Two different approaches has been proposed and compared:

– Random selection
– Optimization based selection. This approach uses Genetic Algorithms as optimizer.

Table 4. Number of groups of families in each stratum (Madrid)

Autonomous Region	Stratum (importance of the town)	Groups	Families
Madrid	1	57	456
Madrid	2	22	176
Madrid	3	6	48
Madrid	4	5	40
Madrid	5	3	24
Madrid	6	4	32
Total	-	97	776

4.1   Random Approach

The first idea used to solve the problem is the random approach. The detailed steps are shown in Figure 1.

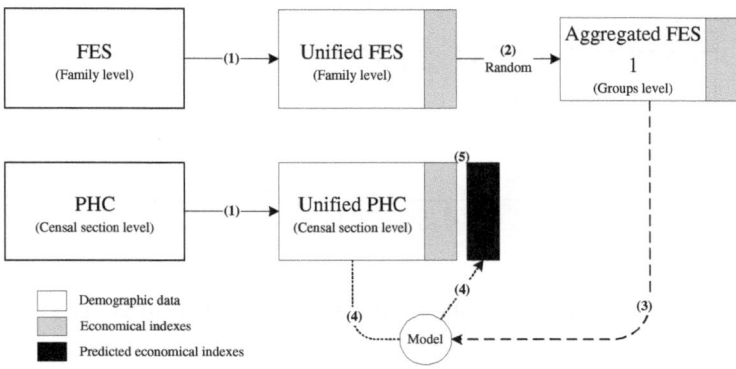

Fig. 1. Random process

The main steps are:

1. PHC and FES dataset are converted to the unified format (described on Section 3). The result is two datasets with the same attributes. The difference between these datasets is the aggregation level:
   - PHC aggregated at *censal section* level.
   - FES aggregated at family level

   In both datasets, we could distinguish two sets of attributes:
   - Demographic data used as input.
   - Economic indexes used as output.

2. Assign random groups to each family of the unified FES and aggregate the resulting groups. The resulting dataset is aggregated at group level.
   The algorithm generates groups of 8 families, randomly chosen without repetition, from each autonomous region and *stratum.*

3. Economic indexes models are trained using the previous groups. In this case, we train lineal models using as input the demographic attributes and as output the income index.

4. Trained models are applied to PHC dataset (it has the same unified format). This will assign the predicted economical indexes to each *censal section.*

5. Thus, if the real index is available it is possible to measure the accuracy of the prediction.

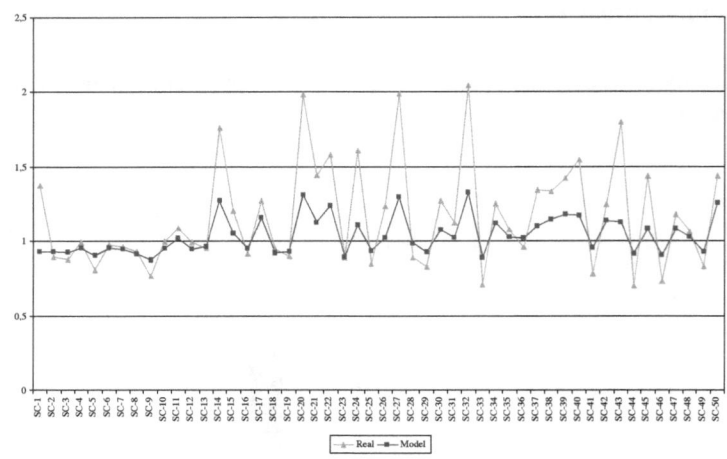

Fig. 2. Random grouping algorithm for Madrid (Experiment 1)

4.1.1 Results

The random approach is tested with two set of groups. Each group is treated separately to analyze how the initial random initialization of groups affects the final results.

The Figure 2 and Figure 3 shows the comparison between real income index (published by the INE) and the index predicted by the model for 50 *censal sections* of Madrid.

The results of each experiment depend on the random set of groups generated. This random generation introduces noise in the models so the final results have high errors.

The cause of these bad results is the poor similarity between generated groups. As an example, Figure 4 shows the income index distributions used for training. The random selection creates random distributions so the results depend on this selection so the generated distributions differ from the real ones.

4.2 Genetic Algorithm (GA) Approach

Due to the bad results obtained by the random approach, other approach is proposed. On this second approach, we use a grouping algorithm that takes into account the final objective. This method could improve the results of the trained models.

The proposed method is based on genetic algorithms. The group generation and selection are controlled by a genetic algorithm which evaluation function is the accuracy of the trained models.

In this sense, the evolution of the genetic algorithm will create the best set of groups of families.

The global view of the algorithm:

1. As in the random approach, both PHC and FES are converted to the unified format. The result is two datasets with the same set of attributes but aggregated at different levels:

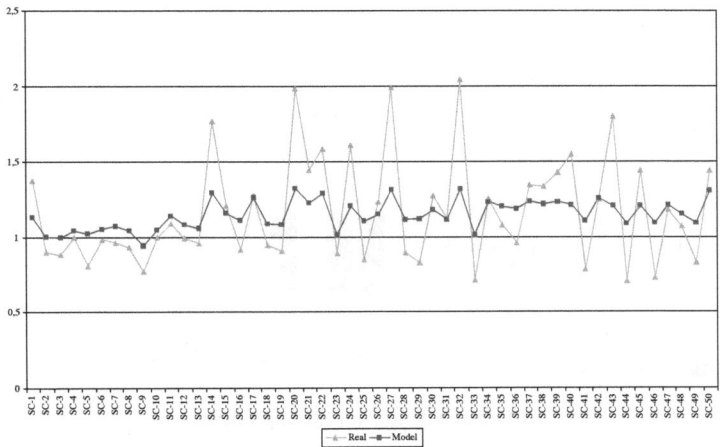

Fig. 3. Random grouping algorithm for Madrid (Experiment 2)

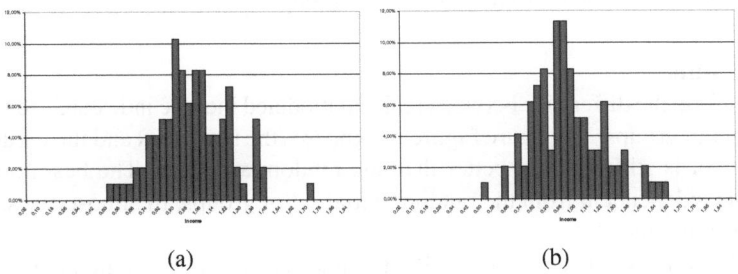

(a)   (b)

Fig. 4. Distributions of the Income Index used for training

- PHC aggregated at *censal section* level
- FES aggregated at family level
2. The genetic algorithm will create "random" groups derived from its individuals. The data in FES dataset is aggregated according with the generated groups. At this point both PHC and FES has the same format and both are aggregated.
3. Then the fitness of each group is calculated. These main steps are involved:
   (a) Each set of groups (individual) of the previous step is used to train a model for economical indexes. This model uses the demographic data as input and income index as output.
   (b) Each previous model is applied to the PHC dataset. This step assigns the predicted income index to each *censal section*.
   (c) Using the prediction of income index and the real index (available in PHC) is possible to measure the accuracy of the set. In this case the sum of the differences is used as the fitness value of each set of groups.

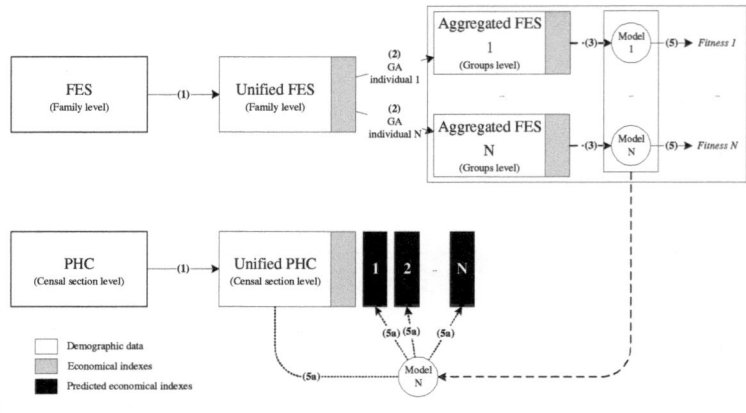

Fig. 5. Genetic Algorithm process

In this case, a Steady-State genetic algorithm is used. The individuals are a array of real numbers as long as the number of families. This array is sorted and divided in groups of 8 (according with the rank) to generate the groups.

4.2.1   Results

As said before, the difference between real and modeled income index are used to evaluate each set of groups. The figure Figure 6 compares the real index and the modeled one.

The error is still high but better than the random approach. The best point is that the error is independent from the random initial set of groups. In almost cases, the final solution is the same.

The differences between families in the same *censal section* could be the reason of the high error. *Censal sections* with average income have the lowest error while those with lowest of highest income have the highest errors.

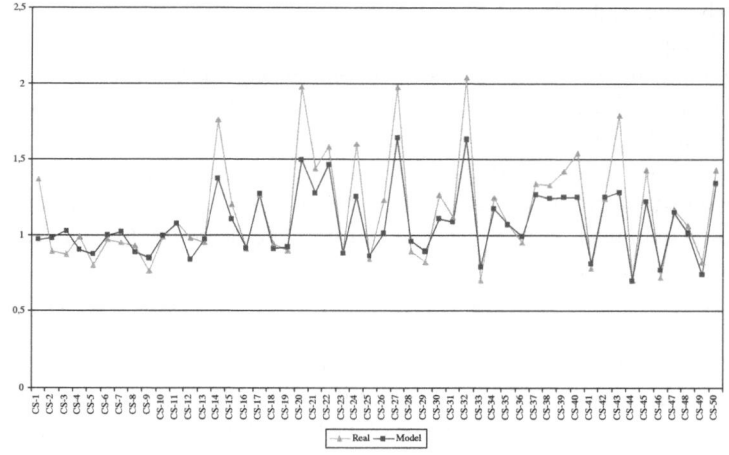

Fig. 6. Optimizer Algorithm for Madrid

5   Conclusions and Future Work

This paper proposes a new workaround to join Spanish Population and Housing Censuses and Families Expenditure Surveys datasets. The work studies two different approaches to the problem: random and Genetic Algorithms.

The results show that the Genetic Algorithm approach obtains better results that the random approach due to the noise introduces by the random selection of families. But these are only the first steps to be done to resolve this issue. A lot of work are still undone.

First of all, the accuracy of the results used could be improved. This could be achieved changing the model that is learned from data or changing the optimization algorithm.

The learned models could be improved using the most significant variables as inputs. This involves the feature subset selection (FSS) of the used variables.

Multiple optimization algorithms could be used. This work uses only Steady State Genetic Algorithm but it is possible to use other implementations of GA [1], Estimation Distribution Algorithms (EDA) [6], Local Search [8], mixed hybrid approach, etc.

Only the income index is used to evaluate the groups generated by the optimization algorithm. If more indexes were used the results could be more accurate and the possible over-fitting of the groups will be avoided.

References

1. J.H. Holland. *Adaption in natural and artificial systems*. The University of Michigan Press, Ann Harbor, MI, 1975.
2. INE. http://www.ine.es/daco/daco43/metodo_ecpf_trimestral.doc.
3. INE. http://www.ine.es, 2005.
4. INE. http://www.ine.es/censo2001/censo2001.htm, 2005.
5. INE. http://www.ine.es/daco/daco43/notecpf8597.htm, 2005.
6. P. Larrañaga and J.A. Lozano. *Estimation of Distribution Algorithms. A New Tool for Evolutionary Computation*. Kluwer Academic Publisher, 2001.
7. MOSAIC.   http://www.business-strategies.co.uk/Content.asp?ArticleID=629, 1999.
8. C.H. Papadimitriou and K. Steiglitz. *Combinatorial Optimization: Algorithms and Complexity*. Prentice-Hall, 1982.
9. Joost N. Kok Peter van der Putten and Amar Gupta. Data fusion through statistical matching. *Center for eBussiness@MIT*, 2002.
10. R Development Core Team. *R: A language and environment for statistical computing*. R Foundation for Statistical Computing, Vienna, Austria, 2004. ISBN 3-900051-07-0.
11. Cesar Montes Sonia Frutos, Ernestina Menasalvas and Javier Segovia. Calculating economic indexes per household and censal section from official spanish databases. *ECML/PKDD*, 2002.

A Distance-Based Method for Preference Information Retrieval in Paired Comparisons

Esther Dopazo, Jacinto González-Pachón, and Juan Robles

Facultad de Informática, Universidad Politécnica de Madrid, Campus de Montegancedo,
28660-Boadilla del Monte (Madrid), Spain
{edopazo, jgpachon, jrobles}@fi.upm.es

Abstract. The pairwise comparison method is an interesting technique for assessing priority weights for a finite set of objects. In fact, some web search engines use this inference tool to quantify the importance of a set of web sites. In this paper we deal with the problem of incomplete paired comparisons. Specifically, we focus on the problem of retrieving preference information (as priority weights) from incomplete pairwise comparison matrices generated during a group decision-making process. The proposed methodology solves two problems simultaneously: the problem of deriving preference weights when not all data are available and the implicit consensus problem. We consider an approximation methodology within a flexible and general distance framework for this purpose.

1 Introduction

The pairwise comparison method is a useful tool for assessing the relative importance of several objects, when this can not be done by direct rating. In fact, some web search engines use this inference tool to quantify the importance of a set of web sites ([9]).

Formally, the problem we are interested in can be formulated as follows. Suppose there is a finite set of objects $X = \{x_1, ..., x_n\}$ which are compared by an expert in the form of paired comparisons, i.e. he assigns the value $m_{ij} > 0$ to the answer to the question "of object x_i and x_j, which is more important and by what ratio?". Then, an $n \times n$ pairwise comparison (pc) matrix $M = (m_{ij})_{i,j}$ is defined.

A wide range of techniques have been developed to deriverelative importance as a priority vector from a pc matrix. In a multicriteria decision-making context, where objects are criteria, Saaty [11] proposed a well-known solution in his Analytical Hierarchical Process (AHP) method. This solution is based on searching the principal eigenvector associated with a "rational" pc matrix.

In practice, noise and/or imperfect judgements lead to non "rational" pc matrices. In this case, the challenge is to obtain a priority vector from non-ideal matrices. To do this, a distance-based point of view may be adopted. The problem should be stated as follows. How can a "rational" matrix, B, which is "as close as possible" to M be found?. Priority weights associated with M are obtained from B. Most of the papers

A.F. Famili et al. (Eds.): IDA 2005, LNCS 3646, pp. 66–73, 2005.

using this approach has been based on Euclidean distance (see [2] and [8]). A more general l_p-distance framework was stated in [4].

Another problem associated with pairwise comparison methods is missing information. There are several possible sources for incomplete information (see [13]): too many paired comparisons to process, time pressure, lack of knowledge and a DM's limited expertise related to the problem domain...

There are some studies in the literature on how to deal with incomplete information. In an AHP context, some methods for estimating unknown data are introduced in [7], [3] and [14]. For specific distance-based frameworks, [12] and [5] solve the problem using mathematical programming.

In this paper, we are interested in dealing with a particular case that leads to incomplete matrices: a group of experts with limited expertise give their preference information on subsets of X in which their domains of expertise are defined. In this case, two different problems appear simultaneously: the problem of deriving preference weights when not all data are available and the implicit consensus problem. We propose a general l_p-distance framework, where the p-parameter can be interpreted as having a preference meaning. A study for the complete case appears in [6].

The paper is organized as follows. The main definitions are introduced in section 2. Section 3 focuses on the formulation of the problem and the proposed problem-solving method. The above ideas are illustrated with the help of a numerical example in section 4. Finally, the main conclusions derived from this research and some applications of preference information retrieval are included in section 5.

2 Preliminaries

Let $X = \{x_1,...,x_n\}$ be a finite set of $n\,(n \geq 2)$ objects. Let $E = \{E_1,...,E_m\}$ be a group of m experts. Let $X_k \subseteq X$ be the cluster (subset) of n_k objects, in which expert E_k is considered to be qualified to express precise preference information. Some points should be made about these subsets:

1. $X = \bigcup_{k=1}^{m} X_k$

2. $\forall i\ \exists j \neq i$ such that $X_i \cap X_j \neq \emptyset\quad i,j = 1,...,m$.

We assume that expert E_k only reports his preferences on the elements of X_k. This is the reason why incomplete pairwise comparison (pc) matrices are considered in this paper. Thus expert E_k gives an incomplete pc matrix on X, $M^k = (m_{ij})_{i,j}$ as follows: m_{ij}^k represents the estimation of the preference ratio between elements x_i and x_j if $x_i, x_j \in X_k$. Otherwise, the comparison value is underdetermined, there is no information, and we will denote the missing comparison by zero .

The problem is how to retrieve global preference information from incomplete pc matrices M^k in order to obtain the consensus preference weights for the elements of

X. In the underlying estimation problem, there are two issues about rationality for consideration:

1. The preference information of each expert E_k is expected to verify normative properties ("local rationality") as defined by Saaty, i.e. for each $k \in \{1,...,m\}$

$$m_{ij}^k m_{ji}^k = 1 \quad \forall i,j \text{ such that } x_i,x_j \in X_k \qquad \text{(local reciprocity)} \qquad (1)$$

$$m_{ij}^k = m_{il}^k m_{lj}^k \quad \forall i,j,l \text{ such that } x_i,x_j,x_l \in X_k \qquad \text{(local consistency)} \qquad (2)$$

2. In the ideal case, interactions of individual preferences should verify normative properties ("global rationality"), i.e. for all $k_1,k_2,k_3 \in \{1,...,m\}$

$$m_{ij}^{k_1} m_{ji}^{k_2} = 1 \quad \forall i,j \text{ such that } x_i,x_j \in X_{k_1} \cap X_{k_2} \qquad \text{(global reciprocity)} \qquad (3)$$

$$m_{ij}^{k_1} = m_{il}^{k_2} m_{lj}^{k_3} \quad \forall i,j,l \text{ such that } x_i \in X_{k_1} \cap X_{k_2},$$
$$x_j \in X_{k_1} \cap X_{k_3}, x_l \in X_{k_2} \cap X_{k_3}. \qquad \text{(global consistency)} \qquad (4)$$

We find that the second issue deals implicitly with a consensus pc matrix that verifies rational properties as described by Saaty. On the other hand, it is quite clear that global properties imply local ones. Therefore, we will refer to global rationality in the rest of the paper.

In practice, the complexity of the decision-making problem, the existence of imperfect and subjective judgements, independent evaluations... lead to pairwise information without rational properties. Moreover, different values could be assigned by different experts to the same objects. In this context, the challenge is to derive a consensus preference weight vector $(w_1,...,w_n)$ for the elements of X.

3   Problem Formulation and Problem-Solving Method

Our objective is to derive a consensus priority vector for the problem stated above. The main idea is to deal simultaneously with an estimation problem and a consensus problem. A general l_p - distance framework is proposed for this purpose.

Let $w = (w_1,...,w_n)$ be the positive vector that we are looking for, whose components are normalized, i.e. $\sum_{i=1}^n w_i = 1$. In the ideal case, the relations between non-null elements of pc matrices and the components of w are as follows

$$m_{ij}^k - \frac{w_i}{w_j} = 0 \qquad (5)$$

$\forall i, j \in \{1,...,n\}$ and $k \in \{1,...,m\}$ such that $m_{ij}^k > 0$.

The above equations are non linear. To take advantage of linear algebra tools, the following equivalent linear equations are considered for $m_{ij} > 0$

$$m_{ij}^k w_j - w_i = 0$$

(6)

$\forall i, j \in \{1,...,n\}$ and $k \in \{1,...,m\}$ such that $m_{ij}^k > 0$.

This constitutes an overdetermined homogeneous linear system. Generally, the above linear system does not have an exact non-trivial solution because rationality conditions are not verified in practice. Hence we shall look for a solution that gets compatibility as best as possible between the above equations in an l_p - metric space. Formally, this is equivalent to searching approximated solutions that minimize the aggregation of residuals $r_{ij}^k = \left| m_{ij}^k w_j - w_i \right|$ for $m_{ij}^k > 0$ in an l_p - metric, $1 \le p \le \infty$. Notice that the solution for the case $p = 2$ is the well-known linear least square solution of the over-determined linear system (6) (see [1]).

Therefore, the following optimization problem is obtained for metric $1 \le p < \infty$

$$\min \left[\sum_{k=1}^m \sum_{\{i,j:m_{ij}^k \neq 0\}} \left| m_{ij}^k w_j - w_i \right|^p \right]^{1/p}$$

(7)

subject to

$$\sum_{i=1}^n w_i = 1$$

(8)

$w_i > 0 \quad \forall i = 1,...,n$.

The feasible set is defined by normalization and positivity conditions for weights. For $p = \infty$, the objective function is expressed as follows

$$\min \left[\max_{\{i,j,k;m_{ij}^k \neq 0\}} \left| m_{ij}^k w_j - w_i \right| \right]$$

$$\sum_{i=1}^n w_i = 1$$

(9)

$w_i > 0 \quad \forall i = 1,...,n$.

In the above problems, the residual aggregation is affected by the p-parameter. Accordingly, as $p \in [1,\infty)$ increases, more importance is given to the larger residual values. Hence, the case $p = 1$ leads to a more robust estimation, whereas the estimation for $p = \infty$ is more sensitive to extreme residual values.

Once the analytical framework has been established, we focus on computing the approximated weights for different p-metrics. The above optimization problems can be formulated as mathematical programming problems. The residuals r_{ij}^k are replaced in this formulation by $r_{ij}^k = n_{ij}^k + p_{ij}^k$ with

$$n_{ij}^k = \frac{1}{2}\left[\left|m_{ij}^k w_j - w_i\right| + (m_{ij}^k w_j - w_i)\right]$$

$$p_{ij}^k = \frac{1}{2}\left[\left|m_{ij}^k w_j - w_i\right| - (m_{ij}^k w_j - w_i)\right].$$

(10)

Thus, for $p \in [1, \infty)$, the optimization problem is equivalent to the following mathematical programming problem

$$\min\left[\sum_{k=1}^m \sum_{\{i,j:m_{ij}^k \neq 0\}} \left(n_{ij}^k + p_{ij}^k\right)^p\right]^{1/p}$$

subject to

(11)

$$m_{ij}^k w_j - w_i - n_{ij}^k + p_{ij}^k = 0, \quad \forall i,j,k: m_{ij}^k \neq 0$$

$$n_{ij}^k, p_{ij}^k \geq 0, \quad \forall i,j,k: m_{ij}^k \neq 0$$

$$w_i > 0 \quad \forall i = 1,...,n$$

$$\sum_{i=1}^n w_i = 1.$$

For $p = \infty$, it can also be shown that the optimization problem is equivalent to the following linear programming problem

$$\min D$$

$$n_{ij}^k + p_{ij}^k \leq D \quad \forall i,j,k: m_{ij}^k \neq 0$$

$$m_{ij}^k w_j - w_i - n_{ij}^k + p_{ij}^k = 0, \quad \forall i,j,k: m_{ij}^k \neq 0$$

$$n_{ij}^k, p_{ij}^k \geq 0, \quad \forall i,j,k: m_{ij}^k \neq 0$$

$$\sum_{i=1}^n w_i = 1, \quad w_i > 0 \quad \forall i = 1,...,n.$$

(12)

where D is an extra positive variable that quantifies the maximum deviation.

We should make some important points from a computational point of view for the most used commonly values of p. For $p = 1$ and $p = \infty$, the above formulations are reduced to linear programming problems that can be solved using the simplex method. The case $p = 2$ corresponds to the least square solution of an overdetermined linear for which several numerical tools are available (see[1]).

4 A Numerical Example

We present a numerical example to illustrate the proposed method. We consider a set of four elements $X = \{x_1, x_2, x_3, x_4\}$ and a set of two experts $\{E_1, E_2\}$. We assume that experts E_1 and E_2 give their opinions on elements of $X_1 = \{x_1, x_2, x_3\}$ and $X_2 = \{x_2, x_3, x_4\}$, respectively, in terms of the incomplete pc matrices M^1 and M^2, respectively:

$$M^1 = \begin{pmatrix} 1 & 2 & 0.5 & 0 \\ 0.5 & 1 & 1/3 & 0 \\ 2 & 3 & 1 & 0 \\ 0 & 0 & 0 & 0 \end{pmatrix} \qquad M^2 = \begin{pmatrix} 0 & 0 & 0 & 0 \\ 0 & 1 & 0.25 & 2 \\ 0 & 4 & 1 & 5 \\ 0 & 0.5 & 0.2 & 1 \end{pmatrix} \tag{13}$$

Note that the zero entries in the above matrices represent missing of data. In one case, expert E_1 is not able to give information about alternative x_4. On the other hand, E_2 gives no information concerning the element x_1. In this example, we find that information provided by the two experts verifies the local reciprocity property but does not verify local consistency. Moreover, the estimations given by experts on the common subset $\{x_2, x_3\}$ are not compatible: A consensus between their preference information is required.

Now we translate the numerical estimations given in (13) by the experts in linear equations as in (6). This constitutes a linear system with more equations than unknown variables (overdetermined system). Because the matrices in question are not consistent, there is no exact positive solution. Hence we look for the l_p - solution with $p = 1, 2, \infty$, as described in the last section. For the case $p = 2$, the least square solution of the linear system (6), with positiveness and normalization restrictions, is computed. For the cases $p = 1$ and $p = \infty$, the solutions are computed by solving the linear programming problems (11) and (12), respectively. MATLAB numerical software was used to do the computations.

The preference weights and the associated rankings obtained by applying our methods for $p = 1, 2, \infty$ are as follows

Table 1. Priority vectors and their associated rankings for $p = 1, 2, \infty$

	Priority vector	Ranking
$p = 1$	$(0.256 \quad 0.128 \quad 0.512 \quad 0.102)$	$x_3 \succ x_1 \succ x_2 \succ x_4$
$p = 2$	$(0.256 \quad 0.139 \quad 0.494 \quad 0.094)$	$x_3 \succ x_1 \succ x_2 \succ x_4$
$p = \infty$	$(0.275 \quad 0.137 \quad 0.482 \quad 0.103)$	$x_3 \succ x_1 \succ x_2 \succ x_4$

In the above example, the original pc matrices are near to locally consistent matrices. Therefore, the priority vectors for different metrics are close and their associated rankings are the same.

To illustrate the effect of the p-parameter in the method, we assume that expert E_2 changes the comparison between objects x_2 and x_3. Now, M^2 is not consistent and not even reciprocal. This change leads matrix M^2 to be "more inconsistent". On the other hand, "the level of consensus" between experts is lower. In this case, we get the following results.

Table 2. Priority vectors and rankings for the perturbed example

	Priority vector	Ranking
$p = 1$	$(0.303 \quad 0.151 \quad 0.454 \quad 0.090)$	$x_3 \succ x_1 \succ x_2 \succ x_4$
$p = 2$	$(0.175 \quad 0.108 \quad 0.101 \quad 0.044)$	$x_1 \succ x_2 \succ x_3 \succ x_4$
$p = \infty$	$(0.384 \quad 0.256 \quad 0.205 \quad 0.153)$	$x_1 \succ x_2 \succ x_3 \succ x_4$

We can see that the effect of the perturbation on the results differs depending on which p-parameter is used. The perturbation causes a rank reversal in the case of $p = 2, \infty$, whereas the method with $p = 1$ is more conservative.

5   Conclusions

We have provided a flexible method for retrieving preference information (as priority weights) from incomplete pc matrices. We have focused on a particular problem where a set of incomplete pc matrices is obtained during a group decision-making process. The proposed methodology simultaneously solves two different problems: the problem of deriving preference weights when not all data are available and the problem of searching a consensus priority vector. These two problems have been solved using a unified l_p- distance framework. This is equivalent to adopting the same point of view for the deviations of both local and global rationality.

Moreover, the methodology proposed here can be applied to other related scenarios. For instance, the proposed method could manage pc matrices with missing data in a general framework. Thus, our techniques do not force the decision maker to manage unknown or imprecise data. Following the procedure, point/local comparisons are smoothed out by the other seeking global rationality.

On the other hand, our approach can be used to deal with pc problems on a set of objects with high cardinality. Thus, the whole set X could be split into several smaller sets $X_k \subseteq X$, based on an "idea of similarity". This leads to a set of incomplete pc matrices. In this context, more accurate decision maker estimations can be obtained.

Finally, the l_p-distance framework offers a new preference interpretation based on the value of the p –parameter. Thus, the case $p = 1$ leads to a more robust estimation, whereas the estimation for $p = \infty$ is more sensitive to extreme residual values.

Acknowledgements

The authors would like to thank the reviewers for their suggestions and comments. Thanks go to Rachel Elliot for checking the English.

References

1. Björck, A.: Solution of Equations in R^n. Least Squares Methods. In: Ciarlet, P. G., Lions, J. L. (eds.): Handbook of Numerical Analysis. Volume I. North-Holland, The Netherlands (1990)
2. Chu, M. T.: On The Optimal Consistent Approximation to Pairwise Comparison Matrices. Linear Algebra and its Applications, Vol. 272 (1998) 155-168
3. Carmone, F. J., Kara, A., Zanakis, S. H.: A Montecarlo Investigation of Incomplete Pairwise Comparison Matrices in AHP. Eur. J. Opl. Res., Vol. 102 (1997) 538-553
4. Dopazo, E., González-Pachón, J.: Computational Distance-Based Approximation to a Pairwise Comparison Matrix. Kybernetika, Vol. 39 (2003) 561-568
5. Fan, Z., Ma, J.: An Approach to Multiple Attribute Decision Making Based on Incomplete Information on Alternatives. In: Sprague, R. H. Jr. (ed.): Proceedings of the 32nd Annual Hawaii International Conference on Systems Sciences. IEEE Comput. Soc., Los Alamitos, CA, USA (1999)
6. González-Pachón, J., Romero, C.: Inferring Consensus Weights from Pairwise Comparison Matrices without Suitable Properties (submitted) (2004)
7. Harker, P. T.: Incomplete Pairwise Comparisons in the Analytic Hierarchy Process. Mathl. Modelling, Vol. 9 (1987) 837-848
8. Koczkodaj, W., Orlowski, M.: Computing a Consistent Approximation to a Generalized Pairwise Comparisons Matrix. Computers and Mathematics with Applications, Vol. 37 (1999) 79-85
9. Langville, A.N., Meyer, C. D : A Survey of Eigenvector Methods of Web Information Retrieval. *The SIAM Review,* Vol. 47(1) (2005) 135-161
10. Romero, C.: Handbook of Critical Issues in Goal Programming. Pergamon Press (1991)
11. Saaty, T. L.: The Analytic Hierarchy Process. McGraw-Hill (1980)
12. Triantaphyllou, E.: Multi-Criteria Decision Making Methods: A Comparative Study. Kluwer Academic Publishers, The Netherlands (2000)
13. Xu, Z. S.: Goal Programming Models for Obtaining the Priority Vector of Incomplete Fuzzy Preference Relation. International Journal of Approximate Reasoning, Vol. 36 (2004) 261-270
14. Yoon, M. S., Whang, K. S.: Pooling Partial Pairwise Comparisons in the AHP. Int. J. Mngt. Sci., Vol. 4 (1998) 35-57

Knowledge Discovery in the Identification of Differentially Expressed Genes

A. Fazel Famili[1], Ziying Liu[1], Pedro Carmona-Saez[2], and Alaka Mullick[3]

[1] Institute for Information Technology National Research Council of Canada,
Ottawa, ON, K1A 0R6, Canada
{fazel.famili, ziying.liu}@nrc-cnrc.gc.ca
[2] Centro Nacional de Biotecnología (CNB - CSIC), Madrid 28049, Spain
pcarmona@cnb.uam.es
[3] Biotechnology Research Institute, National Research Council of Canada,
Montreal, QC. H4P 2R2, Canada
alaka.mullick@nrc-cnrc.gc.ca

Abstract. High-throughput microarray data are extensively produced to study the effects of different treatments on cells and their behaviours. Understanding this data and identifying patterns of groups of genes that behave differently or similarly under a set of experimental conditions is a major challenge. This has motivated researchers to consider multiple methods to identify patterns in the data and study the behaviour of hundreds of genes. This paper introduces three methods, one of which is a new technique and two are from the literature. The three methods are cluster mapping, Rank Products and SAM. Using real data from a number of microarray experiments comparing the effects of two very different products we have identified groups of genes that share interesting expression patterns. These methods have helped us to gain an insight into the biological problem under study.

1 Introduction

Over the last few years we have seen an explosion of high throughput microarray data being produced by biologists and other researchers, studying the behaviour of multiple genes at the same time. These experiments, mostly related to gene response analysis, have been applied to several biological processes. One of the most popular applications is to detect the differences of gene expressions between two or more conditions. Each condition may be related to a treatment, physiological state or other type of study. Each experiment normally involves some biological replicates. When conditions or treatments are studied, two hypotheses may exist:

(i) there is no difference in gene expressions between two or more conditions, when conditions or treatments are compared directly. This implies that the true ratio between the expression of each gene in the comparing samples is one,

(ii) there is a significant difference in gene expressions between two or more conditions, when conditions or treatments are compared. This implies that the ra-

A.F. Famili et al. (Eds.): IDA 2005, LNCS 3646, pp. 74–85, 2005.

tios between the two conditions is not the same and the goal is to identify group of genes that behaved differently and look for patterns that indicate their differences.

The problem studied here was gene response analysis of microarray data from multiple biological experiments that involve using various treatments. The overall goal of this investigation was to identify the effects of these treatments on a particular problem under consideration.

To achieve our data mining objectives, three issues were important: (i) selecting the right method, (ii) applying the correct data analysis strategy, and (iii) providing a certainty factor for each identified gene. Here we applied three methods, two of which are listed in the literature and one that has been introduced as part of our research. No a-priori information about attributes of interest or their behaviour was used in these studies. However, extensive validation techniques were used to evaluate the set of identified attributes.The paper continues as follows. We first provide a brief overview of related work and introduce methods applied. We then follow with a detailed section on experimental analysis that consists of description of the data, our data preprocessing, results and validation. In the last section we present our conclusions.

2 Related Work

Accurate identification of differentially expressed genes and their related patterns using high throughput data has been investigated by many researchers. Here we report most of the research related to the knowledge discovery aspect of this paper. Considering gene expression data as a matrix (the rows are genes and the columns the results of each experiment), identifying differentially expressed genes can be done by comparing rows or analyzing experiments. While most researchers investigate either gene dimensions or experiments, a few investigations combine both [1]. Getz et al [6] proposed a complex, two-way clustering method with the idea of identifying subsets of the genes and samples so that when one group is used to cluster the others, stable and significant partitions are identified. Tang et al [8] also investigated a two-way clustering method in which relationships between genes and experiments are dynamically taken into account. The method iteratively clusters through both gene dimensions and experiments. Troyanskaya et al [9] compare three model-free approaches, to identify differentially expressed genes. These are: non-parametric t-test, Wilcoxon Rank Test, and a heuristic method based on high Pearson correlation. Their results using simulated and real data showed very low false positive rates. Cui and Churchill [4] applied modified t-test and ANOVA to detect differential expressed genes in microarray experiments. Similarly, Tsai et al [10] used a combination of type-I error, power of one- and two-sample t-tests and one- and two-sample permutation tests for detecting differentially expressed genes. Their results showed the two-sample t-test to be more powerful than others. Of other comparative studies to be listed is the research on feature selection and classification by Li et al [7] where multi-class classification of samples based on gene expressions is investigated.

Among related work on methods directly related to our research are: (i) Rank Products [2] and (ii) Significance Analysis of Microarrays-SAM [11]. The Rank Products method is based on biological reasoning and has been evaluated on biological data and shown to perform better than a *t*-test and SAM. SAM, on the other hand assigns a score to each gene on the basis of change in gene expression, relative to the standard deviation of all measurements. Performance of SAM was reported in the same paper to be better than conventional methods, in terms of false discovery rates. These methods are explained in the next section.

3   Methods

This section provides an overview of the three methods applied in this research. We start with Cluster Mapping, which is introduced in this paper, and continue with a brief description of the other two methods that are listed in the literature.

3.1   Cluster Mapping

This method was originally introduced to search for interesting patterns in time series data [5]. It consists of a combination of unsupervised and supervised learning techniques. Unsupervised learning does not need any user's involvement or interference during the entire data mining process (e.g. clustering). Supervised learning requires some forms of user's participation along the line of data analysis process. The first step is to apply a sliding window of size x for partitioning experiments (e.g. time points) and move the sliding window by a step of one. Therefore, for a data set consisting of n experiments (n attribute vectors containing gene expression data), the total number of windows to analyze, S (or number of combined data points selected), is (n-x) + 1. For example, for a data set with 5 experiments (n=5) and a window size of 2 (x=2) with a step of one, we will have S=4.

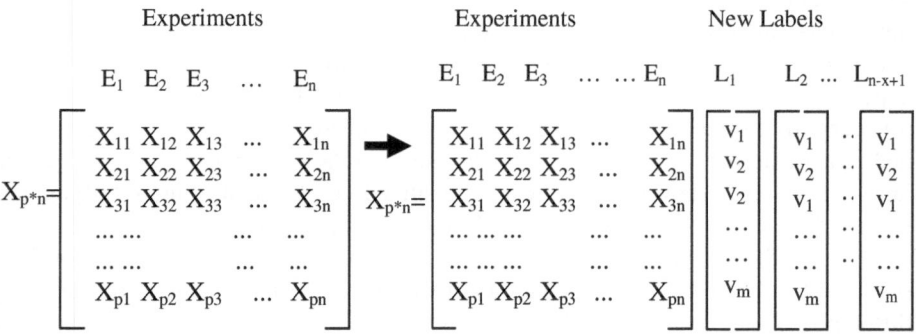

Fig. 1. Left side of the figure shows the initial structure of the data matrix and right side of the figure shows the structure of the data matrix with the list of all clusters obtained with the assigned labels

In the second step, an unsupervised learning process, a clustering method, is applied to each window to identify group of genes that, based on a measure of similarity, belong to a particular group. The unsupervised method selected for this step will depend on the characteristics of the application for which the data is generated. The gene expression data matrix is then labeled with cluster assignments (Fig. 1).

We then group together genes that always remain in the same cluster in the sequences of clustering on each window. Following is the pseudo code of the algorithm which recursively splits the data matrix based on the labels

```
Procedure SplitData (DataMatrix, StartLabelIndex)
     Attribute at StartLabelIndex with outcomes v₁, v₂ …, vₘ;
     m is number of the clusters at the StartLabel (ini-
     tially, L₁ is the start label);
     #Split DataMatrix D into subsets Dᵥ₁,…, Dᵥⱼ, …, Dᵥₘ;
     count = 0; # for new labels
     For i=0; i<m; i⁺⁺ ;
          If (StartLabelIndex of Dᵥᵢ + 1 != n - x + 1)
               Then SplitData (Dᵥᵢ, (StartLabelIndex + 1))
          Else
               count++; {Dᵥᵢ ∈ D, Label Dᵥᵢ with Lcount};
          End if
     End for
End
```

As an example, if we use K-Means for clustering with K=k, the total number of new attribute vectors S=s, and then the maximum number of new clusters could be k^s. The patterns in clusters would then be evaluated based on some domain knowledge and three main properties of cluster centroid information: (i) properties of individual experiments (e.g. mean, median, etc.), (ii) properties of each experiment with respect to comparing experiments (e.g. dimensionless terms such as forward-centroid ratio, backward-centroid ratio, etc.), and (iii) properties of all or a sub-set of experiments (e.g. partitioned slope).

In this study, instead of clustering every two or three adjacent experiments or conditions, we applied K-Means clustering method, with k=8 to cluster all the genes in each individual experimental condition, which was the average of all biological replicates under that condition. The value of k=8 was chosen based on a set of experiments, in which we tried to minimize the number of genes belonging to more than one cluster. The results showed visually good separation that the highly over- and under- expressed genes were clearly distinguished from other genes under each individual experimental condition. Due to the characteristic of the data, the way of choosing k could be priori, which usually requires a good understanding of the characteristic of the data and the background knowledge of the data. After choosing the value k, we then applied the algorithm described above to generate a set of new clusters.

3.2 Rank Products

This method has been recently introduced by Breitling *et al* [2], and is based on rank-ing of genes across different experiments or replicates. The rank of up-regulation (denoted as r^{up}) for each gene in each experiment is defined as its position on the list after sorting all genes by decreasing expression values. Using these rank values across experiments, the combined probability of observing a certain rank pattern in random lists of genes can be estimated as $RP_g^{up} = \prod_{i=1}^{K}(r_{i,g}^{up}/n_i)$, where $r_{i,g}^{up}$ is the position of gene g in the list of genes sorted by decreasing expression values in the *ith* experiment and n_i is the total number of genes. In this way, lower RP values indicate a lower likeli-hood of observing a gene on the top of the list of differentially expressed genes (up-regulated genes) just by chance. The same procedure is carried out to detect down-regulated genes, but sorting them by increased expression values. Breitling *et al* [2] also proposed a simple procedure to measure the statistical significance of observed differentially expressed genes based on the likelihood of observing a given RP value or better in a random set of experiments. The procedure is based on generating a number of random experiments by randomly shifting ranks of genes from the original dataset. Then, for each gene, RP values are calculated in each random dataset and the number of simulated RP values smaller than or equal to a given experimental RP value are counted. We can then calculate the average expected value, E(RP), just dividing by the number of random experiments. For each gene g, the percentage of false-positives if this gene (and all genes with RP values smaller than this cutoff) would be considered as significantly differentially expressed can be also estimated as $q_g = E(RP_g)/rank(g)$, where rank (g) denotes the position of gene g in a list of all genes sorted by increasing RP value. This estimates the false discovery rate and pro-vides a way to assign a significance level to each gene.

3.3 SAM (Significance Analysis of Microarrays)

This statistical technique was introduced by Tusher *et al* [11] to identify differentially expressed genes under different experimental conditions. The method assigns a statis-tics score to each gene by considering the relative change of each gene expression level with respect to the standard deviation of repeated measurements. The relative difference is calculated as following:

$$d(i) = \frac{\overline{x}_a(i) - \overline{x}_b(i)}{s(i) + s_0} \tag{1}$$

where $\overline{x}_a(i)$ and $\overline{x}_b(i)$ are defined as the average levels of expression for gene i in condition a and b, respectively, and $s(i)$ is the standard deviation of the repeated experiments:

$$s(i) = \sqrt{q\left\{ \sum_m \left[x_m(i) - \bar{x}_a(i) \right]^2 + \sum_n \left[x_n(i) - \bar{x}_b(i) \right]^2 \right\}} \qquad (2)$$

where $\sum_m$ and $\sum_n$ are summations of the expression measurements in condition a and b, respectively. In this equation, $q = (1/n_1 + 1/n_2)/(n_1 + n_2 - 2)$, where n_1 and n_2 are the numbers of measurements in condition a and b. s_0 is a small constant which is chosen to minimize the coefficient of variation. The genes with scores greater than a threshold are deemed potentially significant. A false discovery rate, which is the percentage of genes identified by chance, is also estimated by performing permutation. The number of falsely discovered genes corresponding to each permutation is computed by counting the number of genes that exceed a user defined cutoff for the induced and repressed genes.

4 Experimental Analysis

We performed a series of experiments to analyze the data, discover the most useful knowledge related to these experiments, and also evaluate the usefulness of CM in multi-experiment comparison. The following sections provide some details on these studies.

4.1 The Data Sets

The data used in this study were a large data set representing a set of attributes for multiple biological experiments. Each biological experiment had 2-4 replicates, with 4 treatments of substance A and B. Each data set contained 31200 data points, with two measurements for each attribute of interest. Therefore, each experiment represented a log-ratio of biological stimulate and control for 15600 values. Missing data were flagged and the entire data was normalized using the Lowess method [3].

4.2 Data Preprocessing

The preliminary investigation on the data characteristics showed no particular anomalies, and there were only 0.29% of the values that were found to be missing. According to the correlation of gene expression among the replicates under the same treatment, four biological samples did not correlate with others and therefore were removed. In addition, 67 clones were removed due to a high standard deviation (threshold used was 1, empirically determined) in duplicated data points. Then the gene expression values of intra duplicated clones were averaged. 53 clones were also filtered out due to a high standard deviation (threshold used was 1, empirically determined) among replicates. Since some of our analysis methods did not accept data with missing values, we removed 40 data points which contained missing data. Finally, there were 15440 clones across the 8 experiments left for data analysis. They were: 3 replicates of Substance A, 3 replicates of one kind of Substance B and 2 replicates of another Substance B. We note that when CM was applied, the average of the biological replicates under each experimental condition was used.

4.3 Search for Patterns

To identify the most informative genes and to discover all associated patterns in the data, we defined a data analysis strategy that is shown in figure 2. We applied the three methods that were described earlier and selected a common strategy to validate the significance of these genes. The main biological objective was to identify the most informative genes that showed a marked:

(i) over- or under-expression in response to two different preparations of Substance B (common genes among Substance B);

(ii) over- or under-expression in response to Substance B and Substance A compared to untreated cells (common genes among Substance B and Substance A);

(iii) difference in their expression behavior in response to Substance A compared to Substance B treatments (different genes among Substance B and Substance A).

To this end we evaluated the results reported by: (i) CM, SAM and RP, (ii) SAM and RP and not CM. We were further interested to learn about all the genes that were validated using one of the acceptable techniques.

We applied CM, SAM and RP to detect genes that were significantly over- or under- expressed in response to Substance A and Substance B treatments as well as genes that showed differences in their expression patterns between both treatments. Specifically, to identify meaningful clusters applying CM, we obtained new features (e.g. forward centroid slope) from the centroids of the new clusters generated (as described in section 3.1). Two criteria are used to determine whether a gene cluster is differently expressed or similarly expressed under two conditions, the absolute value of centroid and the slope of the centroid under the two experimental conditions.

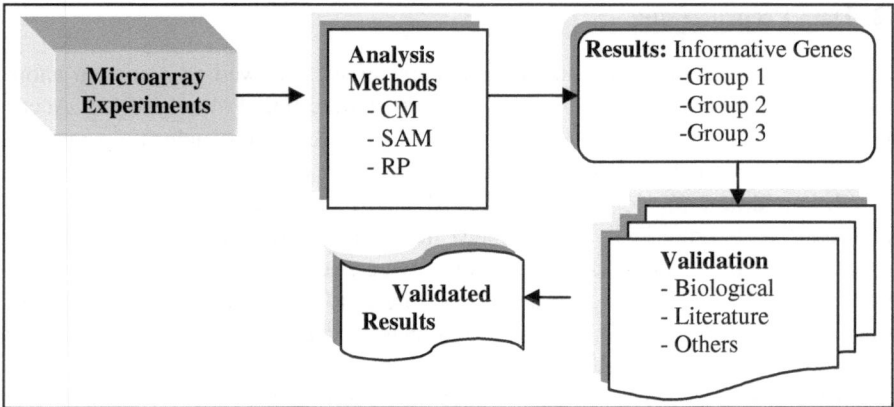

Fig. 2. Data analysis process

If the absolute value of the slope was greater than or equal to certain threshold (1 was used in our case, which was determined by a domain expert), and the absolute value(s) of the centroid under either of the two conditions was greater than certain threshold (0.8 was used (in log 2 ratio), which was also determined by a domain expert), then we considered the cluster of genes as differently expressed under the two conditions. Otherwise, if the absolute value of the slope was less than certain threshold (e.g. 1), and the absolute values of the centroid under both conditions were greater than certain threshold (e.g. 0.8), then we considered the cluster of genes similarly expressed under the two conditions. In our experiments for RP, the expected RP-values and False Discovery Rate (FDR) were calculated using 100 random experiments (number of permutations) of the same size of the original dataset. We selected genes based on the zero false discovery rate. As for SAM, a one-class response was applied to identify the genes which were highly over- or under-expressed in Substance B (similarly expressed genes among Substance B), and also applied to determine the genes which were highly over- or under-expressed in Substance B and Substance A (similarly expressed among Substance B and Substance A). Two-class unpaired analysis was applied to identify genes which were similarly expressed among Substance B but different with respect to Substance A. In order to make proper comparison between the genes discovered by SAM and RP, we applied the following strategy: based on the X number of genes identified by RP, we selected approximately the same number of genes from SAM. We should mention that the false discovery rate for SAM was between 0.38 and 10.00 and the analysis was based on 100 random permutations.

4.4 Results

Our first attempt was to list all the genes identified by the three methods for all biological problems (groups). Table 1 shows the number of genes identified by all three methods. The numbers in brackets represent unique genes and do not include the unknown ones.

Table 1. Number of genes discovered by different methods

	SAM	RP	CM	SAM & RP	CM & SAM & RP
Group1*	127	104	83	86 (60 known)	69 (46 known)
Group2*	190	216	74	150 (106 known)	71 (48 known)
Group3*	56	45	30	41 (13 known)	25 (9 known)

Group1*: Highly over- and under-expressed genes in Substance B (similarly expressed genes among all treatments of Substance B). **Group2***: Highly over- and under- expressed genes in Substance B and Substance A (similarly expressed genes among Substance B and Substance A). **Group3***: Similarly expressed genes among Substance B, but differently with respect to Substance A.

The very first observation in this study was that the number of genes reported by CM method, especially in the case of genes that were differentially expressed in the treatments with respect to the control, was less than the other two methods (e.g. 83 for CM, vs. 127 and 104 for the other two, in group 1, in Table 1). SAM and RP methods tend to detect genes that are highly over- or under-expressed based on fold-changes in each condition compared to the control. CM aims to detect genes that show high absolute ratios of treatment/control, but also show similarities in their expression patterns across treatments.

To evaluate the usefulness of a complementary method, we defined two main properties for the list of genes in a Venn diagram, as listed in Figure 3. The usefulness of these genes was evaluated at a later step.

4.5 Validation: Biological, Literature and Others

To verify the biologically relevant gene expression changes, a series of literature and biological experimental validations were performed based on the random selection of the known genes from each group (unknown genes and replicates were not considered). In this study, we compared the ratios of the positive discovery of the number of genes identified by all three methods and by two only (SAM and RP).

The three methods combined in this study, were able to take into account the statistical significance of the genes, and also the gene expression patterns. Tables 2 and 3 show that the true discovery rate of genes (which were calculated based on biological experimental validation and literature validation) related to the problem under study is increased when CM is involved for each biological problem (groups 1-2). Table 2 shows the genes found in the literature. Table 3 contains results of biological experimental validation.

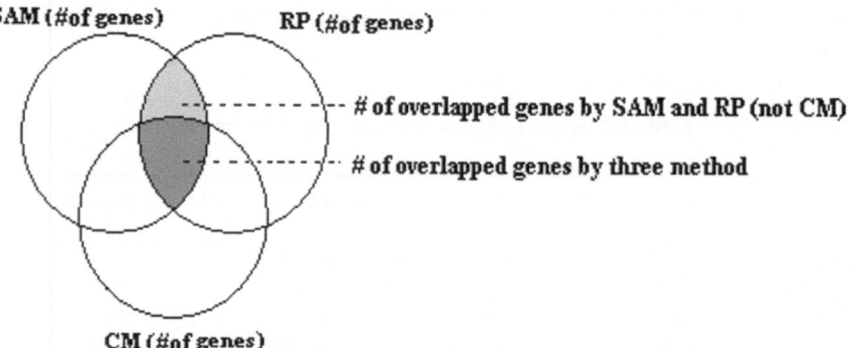

Fig. 3. The shadow in dark gray represents the number of genes identified by the three methods. The shadow in light gray is for the number of genes identified by SAM and RP only, and not CM

Table 2. Number and percentage of literature validated genes from the known gene lists discovered by all three methods and the genes discovered by SAM and RP (refer to fig. 3)

	SAM and RP discovery rate (dark gray + light gray)	SAM and RP and CM discovery rate (dark gray)	SAM and RP (No CM) discovery rate (light gray)
Group1*	22/60 known genes = 36.6%	18/46 known genes = 39%	4/14 known genes = 28.6%
Group2*	34/106 known genes = 32%	22/48 known genes = 45.8%	12/58 known genes = 20.1%
Group3*	6/13 know genes = 46%	2/9 known genes = 22%	4/4 known genes = 100%

Table 3. Number and percentage of biological experimental (RT-PCR) validated genes from randomly selected gene lists (refer to fig. 3)

	SAM and RP discovery rate (dark gray + light gray)	SAM and RP and CM discovery rate (dark gray)	SAM and RP (No CM) discovery rate (light gray)
Group1*	8/60 known genes = 13%	7/46 known genes = 15%	1/14 known genes = 7%
Group2*	7/106 known genes = 6.6%	6//48 known genes = 12.5%	1/58 known genes =1.7%

CM has obviously been able to reduce the false discovery rate of the other two methods. This is evident from literature and biological experimental validation. For example, in table 2, for group 1 and 2, when CM was applied, the discovery rate increased from 36.6% to 39% and 32% to 45.8%, respectively. However, for group 3 CM did not perform this way. It is important to note that the validated results are based on some arbitrary selection of genes and did not follow any particular selection process. For example, in group3, for SAM and RP (not CM), all 4 genes were evaluated; however for SAM, RP and CM, only 2 out of 9 genes were evaluated. This was due to the amount of time that was required for validation. Table 3 also shows that the CM involvement reduced the false discovery rate for the list of genes only listed by the other two (SAM and RP). For groups 1 and 2, the discovery rate increased from 13% to 15% and 6.6% to 12.5%, respectively, when CM was applied.

Overall the discovered patterns were very interesting and most of them had not been reported or validated before.

5 Conclusion

This paper deals with analyzing data from multiple biological experiments to identify gene responses to different experimental conditions. The main motivation for this research was to complement existing methods to achieve the best discovery rate when one needs to study the behaviour of hundreds of genes using an unsupervised approach. Two of the methods applied are from literature and one is a new approach.

These methods have been applied to analyze data from a number of microarray experiments comparing the effects of two very different products. We have identified groups of genes that share interesting expression patterns. Through random selection, we have further validated certain genes from the list of genes identified by these methods. The approach has demonstrated (i) the strength and weakness of the three methods applied to genomics and (ii) that a single method may not be able to identify all gene responses under different experimental conditions, let alone that most methods by themselves provide a large list of genes.

Overall, these methods have helped us to gain insight into the biological problem under study. The results also show that over-fitting may be resolved when multiple methods are applied. In addition to the methods presented here, other methods such as Wolpert's stacked generalization [12], boosting and bagging also could be suitable. In the future research we will explore the possibility of using different k value for K-Mean clustering, and also applying other clustering techniques such as SOM and Hierarchical clustering. We may also evaluate these methods to other data sets and consider other approaches for gene validation. This would be valuable support for gene identification and gene response analysis using microarray data and many other genomics data mining tasks that require a complex data analysis process.

Acknowledgement

The authors would like to acknowledge the contributions of all members of the BioMine project and a number of former students. Special thanks to Brandon Smith and Rita Lo for their help in data preparation. Thanks to Junjun Ouyang and Bob Orchard for reviewing an earlier version of this paper. Mr. Carmona-Saez is the recipient of a fellowship from Comunidad de Madrid.

References

1. Brazma A and Vilo J., (2000), Gene expression data analysis, Federation of European BiochemicalSociety, 480, 17-24.
2. Breitling R., Armengaud P. Amtmann A., and Herzyk P., (2004) Rank products: a simple, yet powerful, new method to detect differentially regulated genes in replicated microarray experiments, FEBS letters 573, pp 83-92.
3. Cleveland, W.S. (1979) Robust locally weighted regression and smoothing scatter plots, *J. Amer. Stat. Assoc.* 74, 829–836.
4. Cui X. and Churchill G., (2003) Statistical tests for differential expression in cDNA microarray experiments, *Genome Biology*, 4:210
5. Famili, A, Liu Z, Ouyang J, Walker R., Smith B. O'Connor, M, and Lenferink A. (2003) A novel data mining technique for gene identification in time-series gene expression data. ECAI Workshop on Data Mining in Genomics and Proteomics pp. 25–34.
6. Getz G., Levine E. and Domany E. (2000), Coupled two-way clustering analysis of gene microarray data, PNAS 97 (22)12079-12084.

7. Li T., Zhang C., Ogihara M, (2004) A comparative study of feature selection and multi-class classification methods for tissue classification based on gene expression, Journal of Bioinformatics, 20(15) 2429-2437.
8. Tang C., Zhang Li, Zhang A. and Ramanathan M., (2001) Interrelated two-way clustering: an unsupervised approach for gene expression data analysis, Proceedings of the 2nd International Symposium on Bioinformatics and Biocomputing, , 41-48.
9. Troyanskaya O.G., Garber M.E., and Brown, P.O., (2002) Nonparametric methods for identifying differentially expressed genes in microarray data, Journal of Bioinformatics, 18(11) 1454-1461.
10. Tsai C-A, Chen Y-J, and Chen J., (2003) Testing for differentially expressed genes with microarray data, Journal of Nucleic Acids Research, 31(9) e52.
11. Tusher V. G., Tibshirani R. and Chu G. (2001) Significance analysis of microarrays applied to the ionizing radiation response, PNAS, Vol.98 (9), 5116-5121.
12. Wolpert, D.H. (1992), Stacked Generalization, Neural Networks, Vol. 5, pp. 241-259, Pergamon Press.

Searching for Meaningful Feature Interactions with Backward-Chaining Rule Induction

Doug Fisher[1], Mary Edgerton[2,3], Lianhong Tang[4],
Lewis Frey[3], and Zhihua Chen[1]

[1] Department of Electrical Engineering and Computer Science, Vanderbilt University,
Nashville, TN, USA 37235
{douglas.h.fisher, mary.edgerton, l.tang, lewis.j.frey,
zhihua.chen} @vanderbilt.edu
http://www.vuse.vanderbilt.edu/~dfisher/
[2] Department of Pathology, Vanderbilt University Medical Center
[3] Department of Biomedical Informatics, Vanderbilt University Medical Center
[4] Vanderbilt Ingram Cancer Center, Vanderbilt University Medical Center

Abstract. Exploring the vast number of possible feature interactions in domains such as gene expression microarray data is an onerous task. We propose Backward-Chaining Rule Induction (BCRI) as a semi-supervised mechanism for biasing the search for plausible feature interactions. BCRI adds to a relatively limited tool-chest of *hypothesis generation* software, and it can be viewed as an alternative to purely unsupervised association rule learning. We illustrate BCRI by using it to search for gene-to-gene causal mechanisms. Mapping hypothesized gene interactions against a domain theory of prior knowledge offers support and explanations for hypothesized interactions, and suggests gaps in the current domain theory, which induction might help fill.

1 Introduction

With the increasing investment in gene expression microarray technology, there has been a move toward a "systems biology" approach to understanding the coupling of gene networks and signaling cascades that describe the phenotypes of living matter (e.g., [1],[2],[3]). This has led to a call for tools to (semi-)automatically explore the space of genomic interactions (e.g., [4]) in order to reduce the set of interactions to a manageable set for examination. The goal of this exploration is to focus analysts on plausible interactions, pathways, and markers, which can then be scrutinized further with hypothesis testing methods.

Consistent with research on other exploratory strategies (e.g., [5],[6],[7],[8]), we describe an investigation of *backward-chaining rule induction* (BCRI) for hypothesizing molecular causality and functional interactions from gene expression microarray data. BCRI is a novel strategy for restricting the search through a rule-space to those rules with traceable influence on a given top-level target class. Put simply, BCRI is given a top-level classification with labeled data, and rule induction is performed to find rules that predict the specified class. Antecedent conditions found in discovered rules then become "sub-goals", and rule induction is repeated on the data using these

A.F. Famili et al. (Eds.): IDA 2005, LNCS 3646, pp. 86–96, 2005.

sub-goal conditions as classes. The process of backward-chaining on rule antecedent conditions is repeated until a termination condition is satisfied.

BCRI is intermediate between supervised rule induction and unsupervised rule induction (e.g., association rule learning). Rather than an unconstrained exploration of the space of associations between variables, as would occur in association rule learning [9], only associations that are weakly tied to a top-level class are examined. While BCRI's search through association space will miss many associations (with any given top-level class), we expect that the density of "interesting" rules that it discovers will be higher than if uncovered by standard association rule learning, though this paper does not test this hypothesis directly.

BCRI can be viewed as one component in a process of *iterative exploration*. Induction from data (e.g., BCRI) can be used to find plausible interactions, which are then compared against prior knowledge. Prior knowledge can be used to (1) explain plausible interactions found through induction, (2) filter or rank these possibilities for an analyst (e.g., interactions that are already well-established in the literature might be ranked low, as might be those in which prior knowledge offers too few constraints on possible explanations), (3) implicate additional features or suggest pruning "redundant" features for subsequent induction (e.g., feature selection), (4) reveal gaps in current knowledge that induction may help fill. We look at examples of this last case in **Section 3**.

Our contributions are (1) the definition of the BCRI task abstraction, (2) the implementation of an initial prototype of BCRI, which we call C45-BCRI, (3) an illustration of BCRI in the domain of cancer prognosis, and (4) a demonstration of how BCRI generated hypotheses (e.g., gene interactions) may help fill gaps in prior knowledge. In **Section 2** we describe our implementation of BCRI and report our results in the domain lung cancer prognosis from clinical and gene expression data. In **Section 3** we match selected rules against prior knowledge in the form of an established gene interaction network. Inductively derived rules suggest values for gaps in current knowledge and suggest other plausible hypotheses. **Section 4** closes with a discussion on automating aspects of iterative exploration by coordinating the application and derivation of domain and induced knowledge.

2   Backward-Chaining Rule Induction (BCRI)

To summarize, the initial step of BCRI builds decision rules for predicting a user-specified class or outcome. The antecedents of rules discovered in this first step then become outcomes for which decision rule models are constructed in the second step. Antecedents of rules found in this second step, then become outcomes for decision rule learning in the third step, and so on.

As an illustrative example, in this paper we apply BCRI to published gene-expression and clinical data from lung cancer patients [10]. The data contains 61 instances defined over 4,996 gene attributes and eleven clinical attributes (5007 total). Classification as *High versus Low* risk is the as a top-level task that "kick-starts" BCRI in our application. For our analysis, patients who died at 30.1 months or less following diagnosis are high risk, and others are low risk.

We distinguish the general BCRI task abstraction from our initial implementation of BCRI. We implement BCRI as a wrapper around a rule-induction engine, which is illustrated in **Table 1** with pseudo-C code (local variable declarations excluded). BCRI is passed the labeled data, a set of the target classes used to label the data, and three functions: ***RuleInducer, PriorityFn, and TerminateFn***.

Table 1. Pseudo-C for Backward-Chaining Rule Induction

```
RuleSet BCRI (DataSet Data,

                 TargetSet Classes,

                 RuleSet (* RuleInducer) (DataSet, TargetCondition),

                 float (* PriorityFn)(Rule),

                 int (* TerminateFn) (Rule)) {

    PQ = InitializePriorityQueue(PriorityFn);

    FOR each class in Classes, Enqueue(PQ, [class → ___ ]);

    WHILE (NOT Empty(PQ)) {

        R = Dequeue(PQ);    /* and place R in Results SET*/

        IF (NOT (* TerminateFn)(R) {

            FOR each a IN ANTECEDENTS(R)  {

                Children =  (* RuleInducer) (Data, a);

                FOR each c IN Children  Enqueue(PQ, c)

            }

        }

    } /* end WHILE */

} /* end BCRI */
```

RuleInducer can, in principle, be any supervised rule discovery system that, given a class, will return rules that predict that class (i.e., ***RuleInducer*** is not a classifier per se as no rules predicting the complement of class are explicitly returned). Parameters shown for ***RuleInducer*** might be changed in minor ways to support differing induction engines. Our current implementation uses C4.5-rules [11], which first builds a decision tree to discriminate the values of a dependent attribute (i.e., C4.5-rules builds the classifier), then converts the tree to a set of rules. We do not detail the process here, as it is well established in the literature. We use the standard defaults for C4.5-rules. Moreover, as stated, the wrapper model that we have implemented assumes that RuleInducer discovers rules whose consequents are all of the same class. Thus, while C4.5-rules would discover rules for the complement of a class as well, we filter these out. The BCRI prototype is not optimal in terms of cost, but it allows us to investigate the BCRI methodology by exploiting a well-established rule learning algorithm. In the remainder of this section, we will refer to this C4.5-rules procedure, with filtering of complement rules, as simply C4.5.

PriorityFn is applied to a rule and returns a score. This score is used to store the rule on a priority queue of other scored rules. In our current implementation, the coverage of the rule (i.e., the number of instances in the data that satisfy the rule's antecedent), is used to organize the priority queue. Other possibilities include the rule's accuracy or the like. Our choice of coverage, versus accuracy or a like measure, is motivated by the observation that rule-learning systems tend to produce accurate rules (relative to a data specific upper bound), but that these rules vary significantly in coverage. We prefer to favor rules that cover a large proportion of data.

TerminateFn returns 0/1, indicating whether a rule should be further expanded (i.e., continued backward chaining on its antecedents). Currently, we implement a depth bound and only backward chain a specified number of levels. Other strategies include specifying a minimal coverage or confidence bound.

C45-BCRI, which includes a wrapper around C4.5 as just described, is what we call this paper's implementation of BCRI. Using High and Low Risk as the top-level classification, C45-BCRI begins with (Risk=Low) (cov 42/61) and (Risk=High) (cov 19/61) placed on the priority queue (i.e., passed as Classes to BCRI). The term "cov" is an abbreviation for data coverage (described above in PriorityFn) for the condition just described.

(Risk=Low) is dequeued. Application of C4.5-rules yields a single rule, which is placed on the queue:

[(Stage=1) → (Risk=Low) (cov 48/61) ‖ (Risk=High) → (cov 19/61)].

(Stage=1) → (Risk=Low) is dequeued and C4.5-rules yields a rule, which is added to the queue:

[(ELA2 > 163.3) → (Stage 1) (cov 45/61)   ‖   (Risk = High → (cov 19/61)]

The first of these rules is dequeued. A new rule is learned:

(MRPL19<= 161.4) & (EIF2S1 > 52) & (KRT15 <= 616.8) → (ELA2 >163.3) (cov: 45/61)

This rule is queued, resulting in the following priority queue:

[(MRPL19<= 161.4) & ... → (ELA2 >163.3) (cov: 45/61) ‖ (Risk = High → (cov 19/61)]

Having the highest priority (coverage), this same rule is immediately dequeued. Each individual antecedent serves in turn as a class for rule induction. A simple depth bound is used to terminate backward chaining, and these labeled rules are terminal.

Table 2 shows the 19 rules learned from the lung cancer data by backward chaining with C45-BCRI to a depth of three, beginning with an initial queue of

[(Risk = Low) → ‖ (Risk = High) →]

The ordering of rules is not strictly indicative of the order in which they were discovered. Rule number is given with its associated depth in the backward chaining process. Indentation indicates a parent child relationship. "acc" denotes accuracy (percent correct predictions) and "cov" denotes coverage, which is the number of cases satisfying the antecedents over the total number of samples.

The network of rules learned by BCRI (C45-BCRI and otherwise) is an AND/OR graph, much like the rule bases of expert systems such as Mycin [12]. We do *not* discuss the inference possibilities of such networks from an expert-system perspective. Rather, our primary goal here is a limited, focused exploration of the associations between variables, which is directly (initially) or indirectly (as backward chaining proceeds) tied to top-level class(es).

3 Hypothesized Pathways from BCRI and Prior Knowledge

BCRI rules can be used to find plausible interactions, which can then be compared against prior knowledge. Prior knowledge can be used to (1) explain plausible interactions found through induction, (2) filter or rank these possibilities for an analyst (e.g., interactions that are already well-established in the literature might be ranked low, as might be those in which prior knowledge offers too few constraints on possible explanations), (3) implicate additional features or suggest pruning "redundant" features for subsequent induction (e.g., feature selection), (4) reveal gaps in current knowledge that induction may help fill. We look at examples of this last case.

We will focus here on C45-BCRI rules that are reflected in existing knowledge, or pathways where we can make inferences from a combination of induced and existing knowledge. For each C45-BCRI rule we used Pathway Assist™ [13] to build the shortest pathway known from prior knowledge (as encoded in Pathway Assist™) between gene expression attributes of the rule. We have also used PubMED [14], LocusLink ([15],[16]), and GeneCards [17] as sources for peer-reviewed literature, chromosomal location, and functional annotation, respectively.

The type of interaction found in Pathway Assist™ pathways may involve gene expressions, which are the measured quantities in the data we used, or it may involve the protein product of gene expression, which is not measured but *might* be inferred from the gene expression level. We say "might" because gene expression quantities are not always directly proportional to protein product concentrations secondary to other factors affecting protein concentration (e.g. degradation, export, etc.)

Example 1: Our first example is one where a C45-BCRI-*discovered rule is reflected by existing knowledge.* In Rule 8./1, we have

8. /1   (ELA2 $\leq$ 163.3) & (SERPINA1 > 65) $\rightarrow$ (Stage=3) [acc: 89.1% cov: 12/61]

Existing knowledge from Pathway Assist™, illustrated in Figure 1, gives us relationships between ELA2 and SERPINA1, where the protein products are indicated by the large ovals, a binding interaction is indicated by the dot relationship between the ovals, and regulation is indicated by a square along a dotted line. Table 3 describes the type of reaction between specific nodes, the nodes themselves, and the effect of one node upon the other using the direction indicated in the nodes list.

Table 2. Rules induced by C45-BCRI

(Risk=Low) →

Rule # /Depth
1. /0   (Stage=1) → (Risk=Low)

2. /1   (ELA2> 163.3) → (Stage=1) [acc: 94.4% cov: 45/61]

3. /2   (MRPL19≤161.4) & (EIF2S1 > 52) & (KRT15 ≤ 616.8)
                → (ELA2 >163.3) [acc: 97.0% cov: 45/61]

4. /3   (TRIP12≤ 1176) & (NAPG <= 243) → (MRPL19 ≤161.4) [acc: 97.4% cov: 53/61]

5. /3   (FXN > 37.8) → (EIF2S1 >52) [acc: 97.6% cov: 57/61]

6. /3   (CTRL > 194.4) & (IDS ≤ 163.3) → (KRT15 <=616.8) [acc: 97.5% cov: 54/61]

(Risk=High) →

Rule # /Depth
7./0   (Stage=3) → (Risk=High)

8. /1   (ELA2 ≤ 163.3) & (SERPINA1> 65) → (Stage=3) [acc: 89.1% cov: 12/61]
9./2   (MRPL19 > 161.4) → (ELA2 ≤ 163.3) [acc: 84.1% cov: 8/61]

10./3   (TRIP12 >1176) → (MRPL19 >161.4) [acc: 75.8%cov: 5/61]

11./3   (NAPG> 243) → (MRPL19 >161.4) [acc: 70.7% cov: 3/61]

12./2   (KRT15>616.8)→(ELA2≤163.3) [acc: 79.4% cov: 5/61]

13./3   (CTRL ≤ 194.4) → (KRT15 >616.8) [acc: 70.7% cov: 4/61]

14./3   (IDS > 163.3) → (KRT15 >616.8) [acc: 45.3% cov: 3/61]

15./2   (PLAB ≤ 3703.9) & (H3FD ≤ 167.6) & (ANXA5 > 750) & (DDX5 ≤2804.7)
                → (SERPINA1 >65) [acc: 96.9% cov : 44/61]

16./3   (KIAA0618 > 27.2) → (PLAB ≤ 3703.9) [acc: 95.5% Data cov: 54/61]

17. /3   (SC4MOL > 32) → (H3FD ≤ 167.6) [acc: 97.5%cov: 54/61]

18./3   (AKAP > 496) & (SLC14A2 ≤ 397.1) → (ANXA5 >750) [acc: 97.4% cov: 53/61]

19./3   (KRT13 ≤ 262.9) → (DDX5 ≤ 2804.7) [acc: 97.6% cov: 57/61]

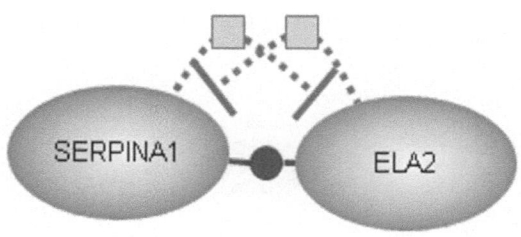

Fig. 1. Pathway Assist™ diagram showing SERPINA1 and ELA2 relationships of Example 1

Table 3. Details of Example 1 relationships given by Pathway Assist™

Type	Nodes	Effect
Binding	ELA2 ---- SERPINA1	
Regulation	ELA2 ---l SERPINA1	negative
Regulation	SERPINA1 ---l ELA2	negative

There are three known interactions between ELA2 and SERPINA1. Their protein products bind together, the protein product of ELA2 gene expression inhibits (i.e., Effect is negative) the gene expression of SERPINA1, and the protein product of SERPINA1 inhibits the gene expression of ELA2 (reciprocal down regulation).

The C45-BCRI rule suggests that ELA2 and SERPINA1 are also coupled by reciprocal down regulation of gene expression. The reciprocal negative regulation effect (down regulation) is reflected in the opposing relative quantities of the attributes in the antecedent of the rule, i.e., that ELA2 is depressed below a value and SERPINA1 is elevated above a value.

The regulatory relationship indicates that SERPINA1 will be highly expressed, ELA2 expression will be depressed. We can hypothesize that the activity of the protein product of ELA2, which is inhibited by binding with the protein product of SERPINA1, will be low as a condition for a high risk tumor. In a 1992 study of adenocarcinomas, high levels of alpha-1-antitrypsin, the protein product of SERPINA1 were found to be associated with higher stage disease [18]. However, high levels of elastase, the protein product of ELA2, in lung tumor tissue has also been correlated with higher stage tumors and poor survival in patients with lung cancer ([19],[20]).

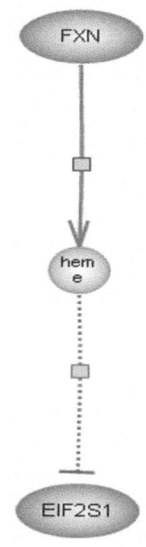

Fig. 2. Pathway Assist™ diagram illustrating FXN and EIF2S1 relationships of Example 2

Table 4. Details of Example 2 relationships given by Pathway Assist™

Type	Nodes	Effect
Regulation	heme ---l EIF2S1	negative
MolSynthesis	FXN ---> heme	unknown

We learn from BCRI in combination with existing knowledge that we may need to study the interaction of elastase and alpha-1-antitrypsin, and *not either of these in isolation*, to understand their role in lung cancer survival.

Example 2: As a second example, consider Rule 5./3,

5. /3   (FXN > 37.8) → (EIF2S1 >52)   [acc: 97.6% cov: 57/61]

Pathway Assist™ shows that FXN has an "unknown" effect on the molecular synthesis of heme, the interaction represented as a solid line with a square in Figure 2, and that heme, a small molecule depicted by the small, central oval, inhibits the gene expression of EIFS2. Relational details are listed in Table 4.

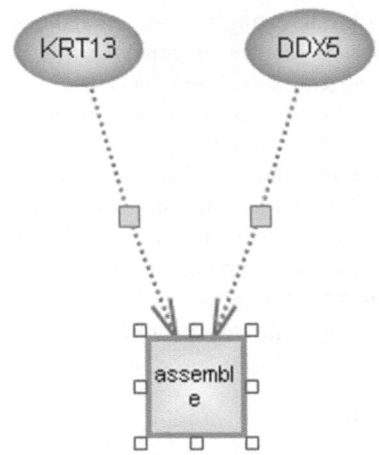

Fig. 3. Pathway Assist™ diagram of KRT13 and DDX5 relationships of Example 3

From our C45-BCRI rule, if we accept that elevated gene expression of FXN leads to elevated levels of its protein product frataxin, then we can *infer* that frataxin *blocks* the molecular synthesis of heme to results in elevated expression of EIFS2. Thus, the inductively derived rule, which might be tentatively abstracted as (FXN → EIF2S1, effect positive), together with (heme--lEIF2S1, effect negative) from prior knowledge, suggests that (FXN→ heme, effect *negative*, in place of unknown). This example illustrates where *induction can suggest fillers for gaps in background knowledge*.

Example 3: In Rule 19./3, 23 have from C45-BCRI

19./3   (KRT13 ≤ 262.9) → (DDX5 ≤ 2804.7)   [acc: 97.6% cov: 57/61]

Table 5. Details of relationships of Example 3 given by Pathway Assist™

Type	Nodes	Effect
Regulation	KRT13 ---> assemble	unknown
Regulation	DDX5 ---> assemble	unknown

and from Pathway Assist™ we have the diagram of Figure 3 and description of inter-actions in Table 5. The square labeled "assemble" represents a cell function of assembly, for example assembling a scaffold of filamentous proteins either into a structure for cell shape, or a scaffold upon which catalyzed reactions can take place.

From GeneCards and Locus Link, we learn that the protein product for KRT13 is keratin 13, a cytoskeletal protein that functions in maintaining the integrity of the cell shape and may also function as a support structure in cell reactions. p68 RNA helicase is the protein product for DDX5 and functions as an RNA-dependent ATPase (provides energy by breaking down ATP). Its presence in the nucleus is an indicator of proliferation, which is an important process in cancer.

The gene for KRT13 is located at chromosome position 17q12 to 17q21.2, while DDX5 is located at 17q21. This suggests that the transcription of DDX5 is associated with transcription of KRT13. Interestingly, Massion and Carbone [21] describe amplifications (increased numbers of copies of genes) in the 17q region of the genome (chromosome 17) that are associated with lung cancer.

From Pathway Assist™, we see that both the protein product of KRT13 and of DDX5 have an unknown role in assembly. Given the proximity of their locations on chromosome 17q, their common, though with unknown effect, role in assembly, and the correlation of amplification of 17q with lung cancer, we infer that 1) the genes on 17q that have a role in lung cancer include KRT13 and DDX5, 2) KRT13 and DDX5 are regulated by a common factor which controls transcription of the two together, 3) *the effect of KRT13 and DDX5 on assembly is the same* (either both positive or both negative), and 4) that the assembly process promotes proliferation.

4 Concluding Remarks

BCRI has been suggested as a means of biasing the search for gene interactions, and feature interactions generally. In particular, our contributions are (1) the definition of the BCRI task abstraction, (2) the implementation of an initial prototype of BCRI, which we call C45-BCRI, (3) an illustration of BCRI in the domain of cancer prognosis, and (4) an illustration of how (C45-)BCRI generated rules (e.g., gene interactions), coupled with prior knowledge, suggest hypotheses about the ways that genes interact that is not yet established in the literature.

Example 2 (Figure 2, Table 4), in particular, is a good example of a domain-independent hypothesis generation strategy. In this example, constraints from background knowledge and a C45-BCRI rule were sufficient to suggest, through qualitative reasoning, the value of an unknown effect. Of course, this reasoning only suggests hypotheses, but at least one other example of this possibility is found in our data. One direction of future research into iterative exploration is (semi-)automate the process of (1) examining a BCRI rule, (2) bolstering confidence in a suggested effect/correlation (positive, negative) through additional inductive means, (3) matching this rule against prior knowledge, (4) qualitatively reasoning about what can be inferred from the prior and inductive knowledge. This direction of research is related to work in scientific discovery (e.g., [22]) and theory revision (e.g., [23]).

A second direction is to use what is learned by what is found in prior knowledge to bias subsequent induction through feature selection with theory-derived features (e.g., [24]). We are pursuing theory driven feature selection strategies elsewhere ([25],[26]).

Finally, our only implementation of BCRI uses C4.5 as the core rule induction engine. C4.5 is biased to greedily find a minimal number of rules. A better choice as the base rule-induction engine may be a method such as Brute [27], which more extensively searches the space of rules, and generally returns many more rules. This latter characteristic will further motivate and necessitate work into using background knowledge to filter/rank hypothesized interactions for expert commentary.

Acknowledgements

We thank the reviewers for helpful and corrective comments. We also thank Mark Ross for his help in building and maintaining hardware on which BCRI was implemented. The research efforts of D.F., M.E., L.T., and Z.C. are supported in part by a National Institute of Health (NLM) grant (1R01LM008000) to M.E. and by funds from the Office of Research at Vanderbilt University Medical Center. L.F. is supported by a fellowship from the National Library of Medicine (5T15LM007450).

References

1. Guffanti, A. (2002). Modeling molecular networks: a systems biology approach to gene function. *Genome Biol* 3: reports4031.
2. Weston, A., & Hood, L. (2004) Systems biology, proteomics, and the future of health care: toward predictive, preventative, and personalized medicine. *J Proteome Res 3:179-96.*
3. Provart, N., & McCourt, P. (2004). Systems approaches to understanding cell signaling and gene regulation. *Curr Opin Plant Biol* 7:605-9.
4. Huels, C., Muellner, S., Meyer, H., et al. (2002). The impact of protein biochips and microarrays on the drug development process *Drug Discov Today* 7(18 Suppl):S119-24.
5. Evans, B., & Fisher, D. (1994). Overcoming process delays with decision tree induction. *IEEE Expert* 9: 60-66.
6. Evans B., & Fisher, D. (2002). Decision tree induction to minimize process delays. In Handbook of Data Mining and Knowledge Discovery, W. Klosgen & J. Zytkow (Eds). Oxford, UK. Oxford University Press. pp 874-881.
7. Waitman, L.R., Fisher, D., & King, P. (2003). Bootstrapping rule induction. In Proceedings of the IEEE International Conference on Data Mining. IEEE Computer Society Publications Office, Los Alamitos CA, pp. 677-680.
8. Waitman, L.R., Fisher, D., & King, P. (in press) Bootstrapping rule induction to achieve and increase rule stability. *Journal of Intelligent Information Systems.*
9. Mannila, H. (2002). Association rules. In Handbook of Data Mining and Knowledge Discovery, W. Klosgen & J. Zytkow (Eds). Oxford, UK. Oxford University Press. pp 344-348.

10. Beer, D, Kardia, S, Huang, C, et al. (2002). Gene-expression profiles predict survival of patients with lung adenocarcinoma. *Nature Medicine* 8: 816-824.
11. Quinlan, J.R. (1993). C4.5: Programs for Machine Learning. San Francisco. Morgan Kaufmann.URL: http://quinlan.com
12. Shortliffe, E., Davis, R., Axline, S., et al (1975). Computer –based consultations in clinical therapeutics: explanation and rule acquisition capabilities of the MYCIN system. *Comput. Biomed. Res* 8: 303-320.
13. Nikitin, A., Egorov, S., Daraselia, N., & Mazo, I. (2003). Pathway studio – the analysis and navigation of molecular networks. *Bioinformatics, 19*: 2155-2157.
14. PubMED Central, a free archive of life sciences journals.URL: http://www.pubmedcentral.nih.gov/
15. Pruitt, K., Katz, K., Sicotte, H. et al. (2000). Introducing RefSeq and LocusLink: curated human genome resources at the NCBI. *Trends Genet* 16(1):44-47.
16. URL: http://www.ncbi.nlm.nih.gov/projects/LocusLink/
17. Pruitt, K., & Maglott, D. (2001). RefSeq and LocusLink: NCBI gene-centered resources. *Nucleic Acids Res* 29(1):137-140.
18. Rebhan, M., Chalifa-Caspi, V., Prilusky, J., et al. (1997). GeneCards: encyclopedia for genes, proteins and diseases. Weizmann Institute of Science, Bioinformatics Unit and Genome Center (Rehovot, Israel)URL: http://bioinformatics.weizmann.ac.il/cards
19. Higashiyama, M., Doi, O., Kodama, K., et al. (1992). An evaluation of the prognostic significance of alpha-1-antitrypsin expression in adenocarcinomas of the lung: an immunohistochemical analysis. *Br J Cancer* 65: 300-302.
20. Yamashita, J., Tashiro, K., Yoneda, S., et al. (1996). Local increase in polymorphonuclear leukocute elastase is associated with tumor invasiveness in non-small cell lung cancer. *Chest* 109: 1328-1334.
21. Yamashita, J., Ogawa, M., Abe, M., et al (1997) Tumor neutrophil elastase is closely associated with the direct extension of non-small cell lung cancer into the aorta. *Chest* 111:885-90.
22. Massion, P., & Carbone, D. (2003). The molecular basis of lung cancer: molecular abnormalities and therapeutic implications. *Respiratory Research* 4: 12.
23. Langley, P., Shrager, J., & Saito, K. (2002). Computational discovery of communicable scientific knowledge. In "Logical and Computational Aspects of Model-Based Reasoning" L. Magnani, N.J. Nersessian, & C. Pizzi (Eds). Kluwer.
24. Mooney, R. (1993). Induction over the unexplained: Using overly-general theories to aid concept learning. *Machine Learning* 10: 79-110.
25. Ortega, J., & Fisher, D. (1995). Flexibly exploiting prior knowledge in empirical learning. In Proceedings of the International Joint Conference on Artificial Intelligence, Morgan Kaufmann, San Francisco, pp. 1041-1047.
26. Frey, L., Edgerton, M., Fisher, D., Tang, L., & Chen, Z. (2005). Discovery of molecular markers of poor prognosis from rule induction methods. Poster presented at the American Association for Cancer Research (AACR) Conference on Molecular Pathogenesis of Lung Cancer: Opportunities for Translation to the Clinic (San Diego, CA).
27. Frey, L., Edgerton, M., Fisher, D., Tang, L., & Chen, Z. (under review). Using prior knowledge and rule induction methods to discover molecular markers of prognosis in lung cancer. American Medical Informatics Association Symposium 2005 (Washington DC).
28. Riddle, P., Segal, R., & Etzioni, O. (1994). Representation Design and Brute-force induction in the Boeing Manufacturing Domain. *Applied Artificial Intelligence* 8: 125-147.

Exploring Hierarchical Rule Systems in Parallel Coordinates

Thomas R. Gabriel, A. Simona Pintilie, and Michael R. Berthold

ALTANA Chair for Bioinformatics and Information Mining,
Department of Computer and Information Science,
Konstanz University, Box M 712, 78457 Konstanz, Germany
{gabriel, pintilie, berthold}@inf.uni-konstanz.de

Abstract. Rule systems have failed to attract much interest in large data analysis problems because they tend to be too simplistic to be useful or consist of too many rules for human interpretation. We recently presented a method that constructs a hierarchical rule system, with only a small number of rules at each level of the hierarchy. Lower levels in this hierarchy focus on outliers or areas of the feature space where only weak evidence for a rule was found in the data. Rules further up, at higher levels of the hierarchy, describe increasingly general and strongly supported aspects of the data. In this paper we show how a connected set of parallel coordinate displays can be used to visually explore this hierarchy of rule systems and allows an intuitive mechanism to zoom in and out of the underlying model.

1 Introduction

Extracting rule models from data is not a new area of research. In [1] and [2], to name just two examples, algorithms were described that construct hyperrectangles in feature space. The resulting set of rules encapsulates regions in feature space that contain patterns of the same class. Other approaches, which construct fuzzy rules instead of crisp rules, were presented, for example, in [3,4,5] and [6]. What all of these approaches have in common is that they tend to build very complex rule systems for large data sets originating from a complicated underlying system. In addition, high-dimensional feature spaces result in complex rules relying on many attributes and increase the number of required rules to cover the solution space even further. An approach that aims to reduce the number of constraints on each rule individually was recently presented in [7]. The generated fuzzy rules only constrain few of the available attributes and hence remain readable even in the case of high-dimensional spaces. However, this algorithm also tends to produce many rules for large, complicated data sets.

In [8] we described a method that attempts to tackle this inherent problem of interpretability in large rule models. We achieve this by constructing a hierarchy of rules with varying degrees of complexity. The method builds a rule hierarchy for a given data set. The rules are arranged in a hierarchy of different

A.F. Famili et al. (Eds.): IDA 2005, LNCS 3646, pp. 97–108, 2005.

levels of precision; each rule only depends on few, relevant attributes thus making this approach also feasible for high-dimensional feature spaces. Lower levels of the hierarchy describe regions in input space with low evidence in the given data, whereas rules at higher levels describe more strongly supported concepts of the underlying data. The method is based on the fuzzy rule learning algorithm mentioned above [7,9], which builds a single layer of rules autonomously. We recursively use the resulting rule system to determine rules of low relevance, which are then used as a filter for the next training phase. The result is a hierarchy of rule systems with the desired properties of simplicity and interpretability on each level of the resulting rule hierarchy. Experimental results demonstrated that fuzzy models at higher hierarchical levels indeed show a dramatic decrease in number of rules while still achieving better or similar generalization performance than the fuzzy rule system generated by the original, non-hierarchical algorithm.

In this paper we show how an accompanying system of inter-connected rule visualizations in parallel coordinates can be used to intuitively explore the rule systems at each level of granularity while at the same time enabling the user to easily zoom in and out of the model, effectively changing to other levels of the hierarchy while maintaining the focus of analysis. The approach is based on recent work on visualization of fuzzy rules in parallel coordinates [10] and extends it using ideas from the information visualization community, so-called structure-based brushing techniques [11]. This method, however, cannot be used intuitively and hence is only useful for an expert user. Here we go beyond solely tying points in each view together, by allowing elements that are connected across different levels of abstraction to be highlighted, i.e. hierarchy layers in the case discussed here. The ability to highlight rule(s) in one layer of the hierarchy and immediately see related rules is a powerful way to quickly increase or reduce the level of detail in an inuitive manner.

The paper is organized as follows: In the next section we briefly describe the used hierarchical rule learning method, followed by an introduction to parallel coordinates, and how normal rule systems can be visualized in the section thereafter. We then describe how hierarchies of rules can be explored in parallel coordinates and illustrate the proposed method using the Iris data set, before we show how larger hierarchical rule sets can be visualized and explored for a number of real world data sets.

2   Hierarchical Rule System Formation

The rule induction algorithm used here is based on a method described in [7], which builds on an iterative algorithm. During each learning epoch, i.e. presentation of all training patterns, new fuzzy rules are introduced when necessary and existing ones are adjusted whenever a conflict occurs. For each pattern three main steps are executed. Firstly, if a new training pattern lies inside the support-region of an existing fuzzy rule of the correct class, its core-region is extended in order to cover the new pattern. Secondly, if the new pattern is not yet covered,

a new fuzzy rule of the correct class is introduced. The new example is assigned
to its core, whereas the support-region is initialized "infinite", that is, the new
fuzzy rule covers the entire domain. Lastly, if a new pattern is incorrectly covered
by an existing fuzzy rule, the fuzzy points' support-region is reduced so that the
conflict is avoided. This heuristic for conflict avoidance aims to minimize the
loss in volume. In [9], three different heuristics to determine the loss in volume
were compared in more detail. As discussed in [7], the algorithm terminates after only a few iterations over the set of example patterns. The resulting set of
fuzzy rules can then be used to classify new patterns by computing the overall
degree of membership for each class. The accumulated membership degrees over
all input dimensions and across multiple rules are calculated using fuzzy t-norm
and t-conorm respectively. For the purpose of this paper, we concentrate on the
rules' core only, that is, we consider only the part of each rule where the degree
of membership is equal to 1 – resulting in crisp rules [1].

In [8], an extension of this algorithm was proposed that allows the generation
of an entire hierarchy of such rules. The rule layers are arranged in a hierarchy
of different levels of precision. Lower levels of the hierarchy describe regions in
input space with low evidence in the given data, whereas rules at higher levels
describe more strongly supported concepts of the underlying data. We recursively
use the above-mentioned classical fuzzy rule induction algorithm to determine
rules of low relevance, which are then used as a filter for the next training
phase. Training examples that resulted in creation of small, less important rules
are therefore excluded from the training phase of the next layer, resulting in a
more general rule system, ignoring the withheld, small details in the training
data. The result is the desired hierarchy of rule systems with an increasing
generality towards higher levels. In [9] it was shown that the accuracy of these
hierarchies is comparable to the non-hierarchical algorithm. Additionally, it was
shown that the general rule system towards the top of the hierarchy alone often
also show comparable performance, sometimes even outperforming the classical
non-hierarchical system.

3   Rule Systems in Parallel Coordinates

Parallel coordinates [12,13] allow n-dimensional data to be visualized in 2D by
transforming multi-dimensional problems into 2D patterns without loss of information. Visualization is facilitated by viewing the 2D representation of the
n-dimensional data. Each of the n coordinate axes is taken and lined up in parallel, resulting in the basis for parallel coordinates. The distance between each
adjacent axis is assumed to be equal to 1. A point in n-dimensional space becomes a series of $n-1$ connected lines in parallel coordinates that intersect each
axis at the appropriate value for that dimension. A parallel coordinates example

---

[1]   Obviously, the extensions for visulizations of fuzzy rules described in [10] can also be
used but as this is not the central focus of this paper, it has therefore been omitted
for reasons of space.

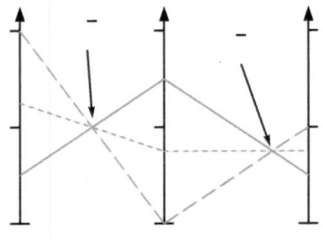   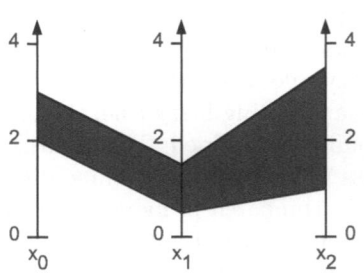

Fig. 1. Left: A parallel coordinate depiction of 3 points on a line in 3D. Right: A rule in parallel coordinates, expressing a disjunctive constraint on all three features.

of 3 points in 3D, $a = (1, 3, 1)$, $b = (4, 0, 2)$, and $c = (2.5, 1.5, 1.5)$, from a line is shown in Figure 1.

The dual of an n-dimensional line in Cartesian coordinates is a set of $n-1$ points in parallel coordinates [14,15]. For the example in Figure 1 (left), these are indicated by $\bar{l}_{0,1} = (0.5, 2)$ and $\bar{l}_{1,2} = (0.75, 1.5)$, which uniquely describe a line in 3 dimensions.

The n-dimensional line in Cartesian coordinates can be represented by $(n-1)$ linearly-independent equations each of which results from equating a different pair of the following fractions [12]:

$$\frac{x_0 - a_0}{u_0} = \frac{x_1 - a_1}{u_1} = \ldots = \frac{x_{n-1} - a_{n-1}}{u_{n-1}}. \tag{1}$$

Now it may be assumed that the $n-1$ linearly independent equations are obtained from pairing the $n-1$ adjacent fractions, with no loss in generality. This yields

$$x_{i+1} = m_i x_i + b_i, \ i = 0, 1, \ldots, n - 2, \tag{2}$$

where $m_i = u_{i+1}/u_i$ represents the slope and $b_i = (a_{i+1} - m_i a_i)$ the intercept of the x_{i+1}-axis of the projected line on the x_i/x_{i+1}-plane. The dual point of the n-dimensional line in parallel coordinates therefore corresponds to the set of $n-1$ indexed points:

$$\left(\frac{i}{1 - m_i}, \frac{b_i}{1 - m_i} \right), \text{ for } i = 0, 1, \ldots, n - 2. \tag{3}$$

In [16], an extension of parallel coordinates was presented that allows not only points to be visualized but also crisp and fuzzy rules. Crisp rules result in "bands" going through the parallel coordinates, visualizing the intervals representing the constraints on each axes. In Figure 1 (right) an example in 3D is shown, depicting the rule:

$$\text{IF } \ x_0 \in [2, 3] \wedge x_1 \in [0.5, 1.5] \wedge x_2 \in [1, 3.5] \ldots$$

The inherent imprecision of fuzzy rules was depicted using degrees of shading to visualize the degree of membership at each level, however, for the purpose of this paper we concentrate on crisp rules. The extension to the fuzzy case is straightforward. We will see examples of such visualizations in the next section.

4   Exploring Hierarchical Rule Systems: An Example

To illustrate the proposed hierarchical rule visualization scheme, the well-known Iris data [17] was used. The Iris data consists of 150 four-dimensional patterns describing three classes of Iris plants: Iris-setosa, Iris-versicolor, and Iris-virginica. The four dimensions consist of measurements for the petal and sepal, length and width.

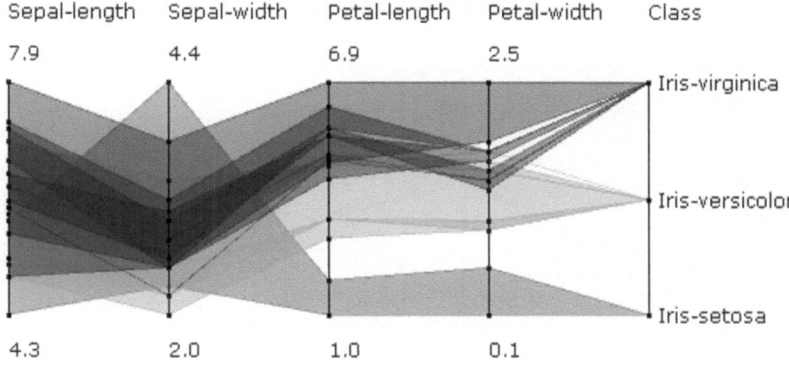

Fig. 2. The flat rule set for the Iris data

Figure 2 shows the flat non-hierarchical rule system as it would be generated by the original rule induction algorithm described in [7]. In Figure 3, the hierarchical rule learner produced three levels of rule systems. The top level has three rules, one for each class, which nicely describe the general trend in the data. At subsequent, lower levels, the granularity increases and finer details of the data are visually depicted. One can clearly see, how four isolated patterns of two classes were filtered out during the first stage of the hierarchy induction.

Obviously, such an easy example is only suitable to demonstrate the algorithm's operation. In the following section, we show how it also works on two real world data sets, discovering interesting structures in the data.

5   Application to a Real World Problem

5.1   Ocean Satellite Images

The first data set stems from a satellite used primarily to examine the ocean. The images are from the Coastal Zone Color Scanner (CZCS) and are of the West Florida shelf [18,19]. The CZCS was a scanning radiometer aboard the Nimbus-7 satellite, which viewed the ocean in six co-registered spectral bands 443, 520, 550, 670, 750 nm, and a thermal IR band. It operated from 1979-1986.

The features used were the 443, 520, 550, 670 nm bands; the pigment concentration value was derived from the lowest 3 bands. Atmospheric correction was

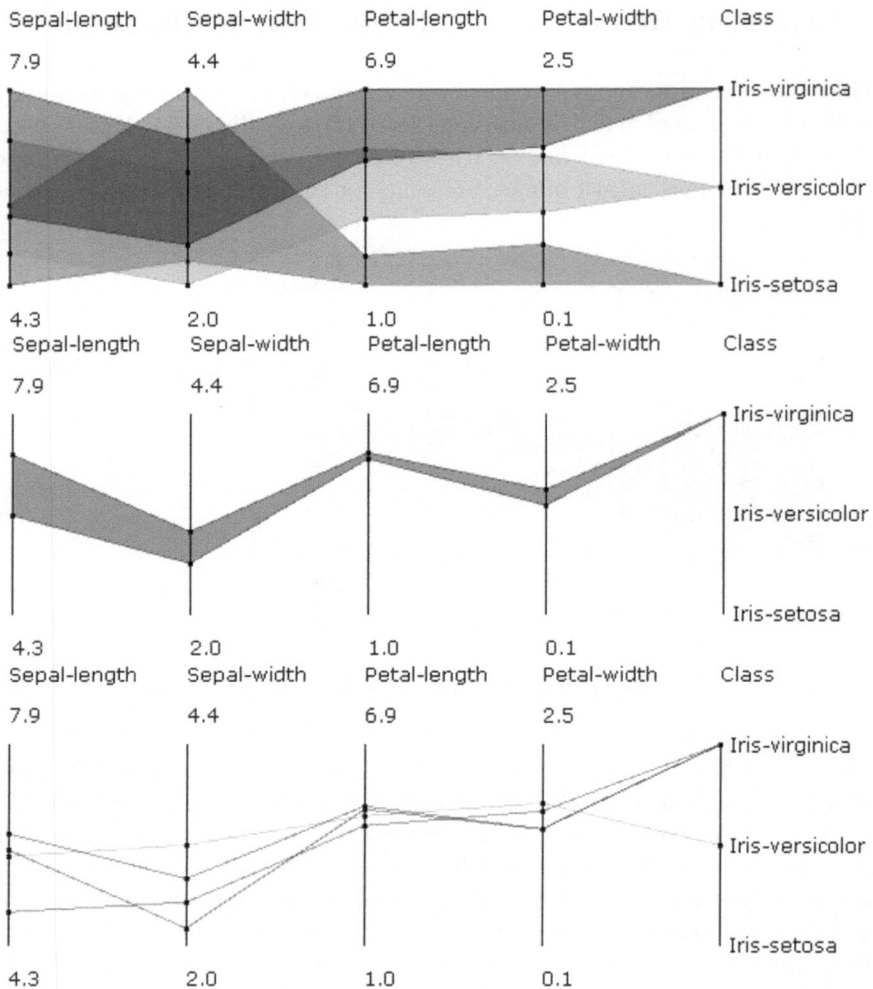

Fig. 3. The 3-level hierarchy for the Iris data. Bottom: the lowest level, showing rules for four isolated patterns which are in conflict with some of the rules of the higher levels. Middle: the next level, here only containing one rule for class Iris-virginica. Top: the top level, showing the three most general rules, one for each class.

applied to each image [20] before the features were extracted. A fast fuzzy clustering algorithm, mrFCM [21], was applied to obtain 12 clusters per image. There were five regions of interest in each image. These consist of red tide, green river, other phytoplankton blooms, case I (deep) water and case II (shallow) water. Twenty-five images were ground-truthed by oceanographers [22] and eighteen of these were used for training. The eighteen training images were clustered into 12 classes. Each class or cluster was labeled by the ground truth image as its majority class.

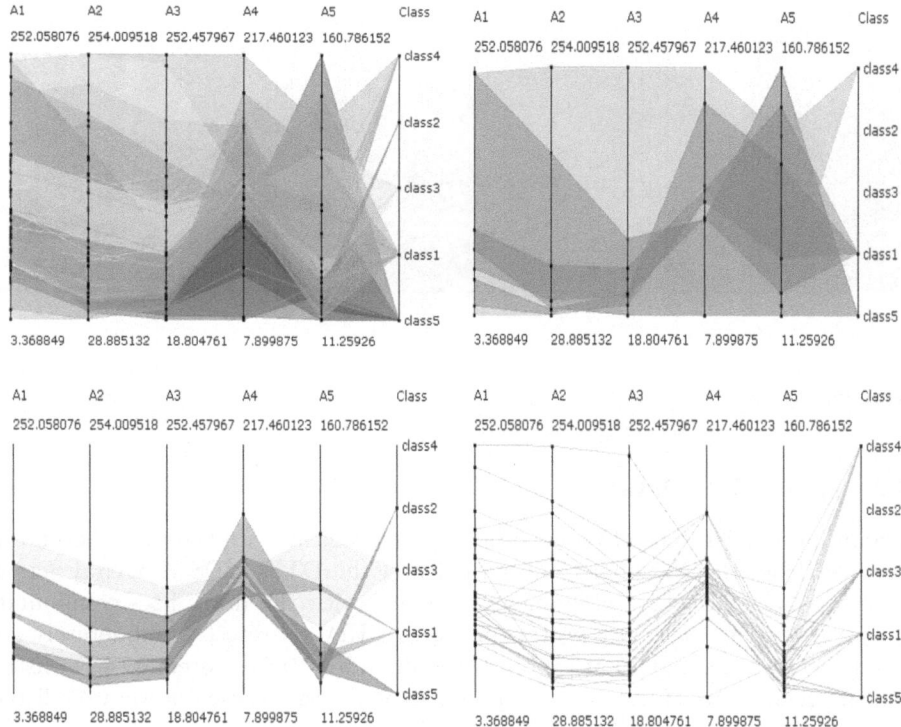

Fig. 4. Top left shows the flat rule system for the ocean satellite image data followed by the 3-level hierarchical rule system. Top right: the highest layer of the hierarchy showing four rules for three of the 5 classes. Bottom left: the middle layer, modeling less important classes and trends. Bottom right: the bottom layer of the hierarchy, modeling outliers, and rare cases.

The labeled cluster centers from the training images were then given to the rule induction tool used also with the Iris data. It generated a set of fuzzy rules, which are shown in Figure 4 (top left). Note how the parallel coordinate display is completely overloaded and essentially useless.

The hierarchical rule induction method generates three layers of hierarchy, which are shown again in Figure 4. Note how the top layer displays only the four most important rules, and even skips rules for two less frequent classes. Those classes are modeled by two extra rules on the middle layer of the hierarchy, which again only displays a few rules (five in this case), hence allowing interpretability even at this level. The bottom layer finally shows rules modeling outliers in the data, indicated by lines. Even this layer of the rule hierarchy still provides an interpretable overview of the structure of the remaining data. It is interesting to note that the complete hierarchy contains less rules than the original flat model itself. Therefore, not only does the hierarchical representation allow better interpretation of the resulting rule models, it is also a more compact representation of the data itself.

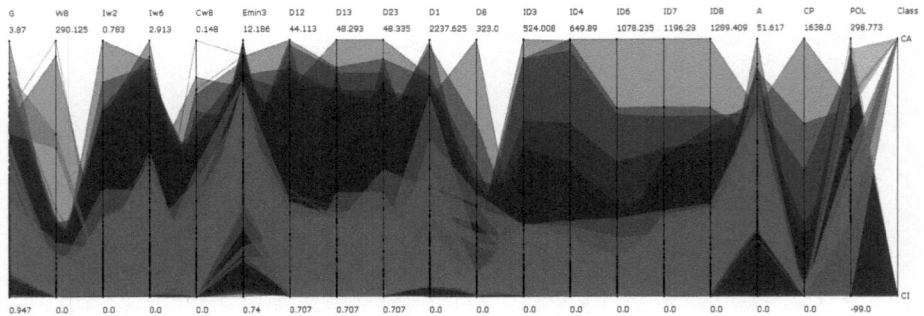

G	W8	Iw2	Iw6	Cw8	Emin3	D12	D13	D23	D1	D8	ID3	ID4	ID6	ID7	ID8	A	CP	POL	Class
3.87	290.125	0.783	2.913	0.148	12.186	44.113	48.293	48.335	2237.625	323.0	524.008	649.89	1078.235	1196.28	1289.409	51.617	1638.0	298.773	CA
0.947	0.0	0.0	0.0	0.0	0.74	0.707	0.707	0.707	0.0	0.0	0.0	0.0	0.0	0.0	0.0	0.0	0.0	-99.0	CI

Fig. 5. The bottom layer of the rule hierarchy for NCI's HIV dataset using VolSurf features

5.2   NCI's HIV Data

The proposed hierarchical visualization method was also applied to a well-known data set from the National Cancer Institute, the DTP AIDS Antiviral Screen data set [23]. The class assignment, provided with the data, lists compounds that provided at least 50% protection against HIV on retest as moderately active (**CM**), compounds that reproducibly provided 100% protection were listed as confirmed active (**CA**), and compounds not meeting these criteria were listed as confirmed inactive (**CI**). Available online [2] are screening results and chemical structural data on compounds that are not covered by a confidentiality agreement. We have generated VolSurf descriptors for these compounds [24], resulting in

- 325 compounds of class **CA**,
- 877 compounds of class **CM**, and
- 34, 881 compounds of class **CI**.

VolSurf computes 2D molecular descriptors based on grid maps modeling interaction energies at a molecular level. The used distance metric was the usual Euclidean distance, computed on a subset of 15 of the available descriptors. Patterns of class **CM** were not used in the following experiments.

Figure 5 shows the bottom level of the resulting hierarchy of rule models. Due to heavy overlap of many rules (210 in this case), not much useful information can be derived from this picture.

Figure 6 shows the top level of the model consisting of 30 rules (7 for class **CA**). Note how, especially for class **CA**, which is the class of interest in this application, a number of interesting observations can be made [3]. For instance, two main clusters can be distinguished that are clearly divided along dimensions G, D1, ID7, ID8, and POL. Also, a correlation across several attributes for rules of class **CA** is visible: ID7, ID8, and D1. In addition, it is interesting to note that

---

[2]   http://dtp.nci.nih.gov/docs/aids/aids_data.html
[3]   Naturally, the results can be seen more clearly on the screen.

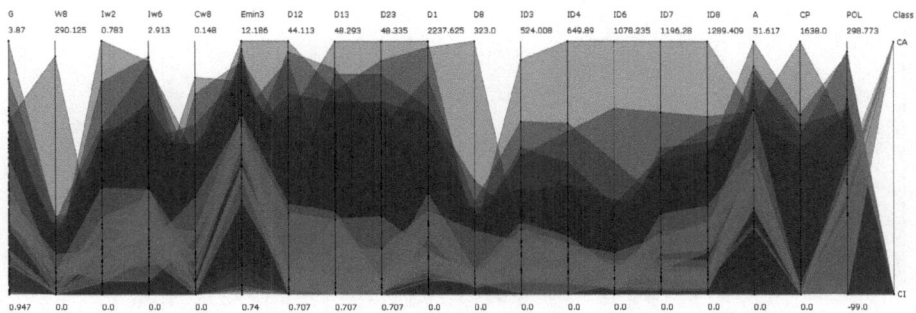

Fig. 6. The top layer of the rule hierarchy for NCI's HIV dataset, containing 30 rules of which only 7 belong to class **CA**, the class of interest

along attribute Iw2 only rules of class **CA** occupy a middle area, where no rules of class **CI** interfere.

Since these were early experiments, extensive evaluation with expert feedback was not able to be conducted. It would be interesting to find out if any of the above observations are correlated to information contained in the VolSurf descriptors.

6   Interactive Rule Exploration and Zooming

To demonstrate the power of interactive, visual brushing across different views in the parallel coordinate hierarchy, we trained a three-level fuzzy rule hierarchy on the vehicle silhouette dataset from the European StatLog–Project [25]. This 18-dimensional dataset consists of 846 samples belonging to 4 classes. The three levels of the fuzzy rule hierarchy contain 21 rules in the top, 47 in the middle, and 256 at the bottom-most level. Figure 7, 8, and 9 show two of three hierarchy models in parallel coordinates demonstrating the highlighting property of the views. The first picture displays all 256 rules of the bottom level — clearly no exploration is possible. Selecting one of the rules in the top level is shown in Figure 8, here all other non-selected rules are faded and moved to the background. This selection is automatically propagated to the other layers and highlights related rules in these views. As can be seen in Figure 9, in the bottom layer only 16 rules are related and hence highlighted. The user can easily identify these small rules, which explain outliers or artifacts in the data that are related to the rules selected in the top layer.

7   Conclusions

We have presented an approach to visualize hierarchical rule systems using a series of parallel coordinate displays. Experiments on three real world data sets show how complicated rule systems, which would otherwise be uninterpretable in a visual display, show interesting insights when displayed at different levels of

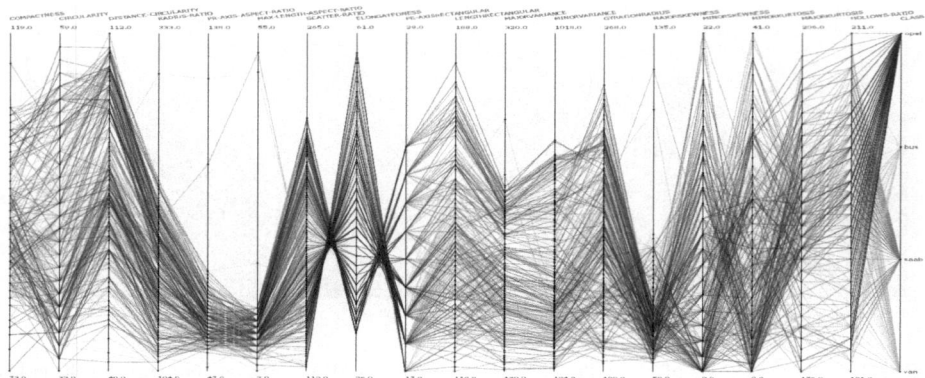

Fig. 7. Bottom level of the three-level hierarchy, which shows 256 rules for 4 classes

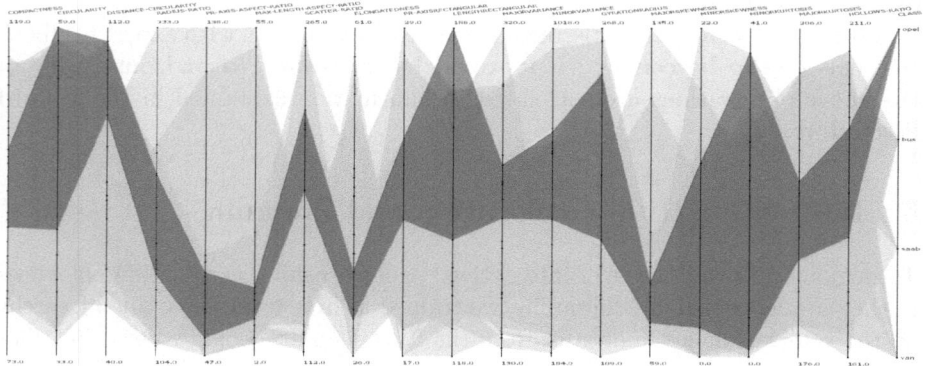

Fig. 8. Top level of the hierarchy, where one rule has been selected and the other 20 rules are faded

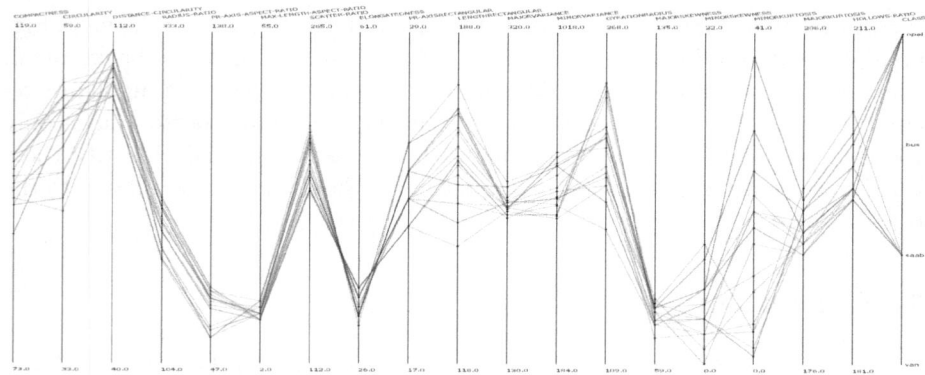

Fig. 9. Again the bottom level, showing 16 of the 256 rules that are related with the rule selected in the top level

abstraction. The ability to interact with the hierarchical rule system at different levels of detail shows promise for the analysis of large, complicated data sets. We are currently working on extending this tool to allow real visual zooming operations within the same view, which will make this type of hierarchical rule system visualization even more powerful for truly exploratory information mining.

Acknowledgments

This work was supported by the DFG Research Training Group GK–1042 "Explorative Analysis and Visualization of Large Information Spaces".

References

1. Salzberg, S.: A nearest hyperrectangle learning method. In: Machine Learning. 6 (1991) 251–276
2. Wettschereck, D.: A hybrid nearest-neighbour and nearest-hyperrectangle learning algorithm. In: Proceedings of the European Conference on Machine Learning. (1994) 323–335
3. Abe, S., Lan, M.S.: A method for fuzzy rules extraction directly from numerical data and its application to pattern classification. IEEE Transactions on Fuzzy Systems **3** (1995) 18–28
4. Higgins, C.M., Goodman, R.M.: Learning fuzzy rule-based neural networks for control. In: Advances in Neural Information Processing Systems. 5, California, Morgan Kaufmann (1993) 350–357
5. Simpson, P.K.: Fuzzy min-max neural networks – part 1: Classification. IEEE Transactions on Neural Networks **3** (1992) 776–786
6. Wang, L.X., Mendel, J.M.: Generating fuzzy rules by learning from examples. IEEE Transactions on Systems, Man, and Cybernetics **22** (1992) 1313–1427
7. Berthold, M.R.: Mixed fuzzy rule formation. International Journal of Approximate Reasoning (IJAR) **32** (2003) 67–84
8. Gabriel, T.R., Berthold, M.R.: Constructing hierarchical rule systems. In Berthold, M.R., Lenz, H.J., Bradley, E., Kruse, R., Borgelt, C., eds.: Proc. 5th International Symposium on Intelligent Data Analysis (IDA 2003). Lecture Notes in Computer Science (LNCS), Springer Verlag (2003) 76–87
9. Gabriel, T.R., Berthold, M.R.: Influence of fuzzy norms and other heuristics on "mixed fuzzy rule formation". International Journal of Approximate Reasoning (IJAR) **35** (2004) 195–202
10. Berthold, M.R., Hall, L.O.: Visualizing fuzzy points in parallel coordinates. IEEE Transactions on Fuzzy Systems **11** (2003) 369–374
11. Fua, Y.H., Ward, M., Rundensteiner, E.A.: Hierarchical parallel coordinates for exploration of large datasets. In: IEEE Conference on Visualization. (1999) 43–50
12. Inselberg, A., Dimsdale, B.: Multidimensional lines I: representation. SIAM J. Applied Math **54** (1994) 559–577
13. Inselberg, A., Dimsdale, B.: Multidimensional lines II: proximity and applications. SIAM J. Applied Math **54** (1994) 578–596
14. Inselberg, A.: Multidimensional detective. In: IEEE Symposium on Information Visualization, InfoVis, IEEE Press (1997) 100–107

15. Chou, S.Y., Lin, S.W., Yeh, C.S.: Cluster identification with parallel coordinates. Pattern Recognition Letters **20** (1999) 565–572
16. Berthold, M., Hand, D.J., eds.: Intelligent Data Analysis: An Introduction. 2nd edn. Springer Verlag (2003)
17. Fisher, R.A.: The use of multiple measurements in taxonomic problems. In: Annual Eugenics, II. 7, John Wiley, NY (1950) 179–188
18. Zhang, M., Hall, L., Goldgof, D.: Knowledge-based classification of czcs images and monitoring of red tides off the west florida shelf. In: The 13th International Conference on Pattern Recognition. Volume B. (1996) 452–456
19. Zhang, M., Hall, L.O., Goldgof, D.B., Muller-Karger, F.E.: Fuzzy analysis of satellite images to find phytoplankton blooms. In: IEEE International Conference on Systems Man and Cybernetics. (1997)
20. Gordon, H.R., Clark, D.K., Mueller, J.L., Hovis, W.A.: Phytoplankton pigments derived from the nimbus-7 czcs: comparisons with surface measurements. Science **210** (1980) 63–66
21. Cheng, T.W., Goldgof, D.B., Hall, L.: Fast fuzzy clustering. Fuzzy Sets and Systems **93** (1998) 49–56
22. Zhang, M., Hall, L., Goldgof, D.: Knowledge guided classification of coastal zone color images off the west florida shelf. Technical Report ISL-99-11, University of South Florida, Dept. of CSE, USF, Tampa, FL. (1999) Under review in International Journal of Pattern Recognition and AI.
23. Weislow, O., Kiser, R., Fine, D., Bader, J., Shoemaker, R., Boyd, M.: New soluble formazan assay for HIV-1 cytopathic effects: application to high flux screening of synthetic and natural products for AIDS antiviral activity. Journal National Cancer Institute **81** (1989) 577–586
24. Cruciani, G., Crivori, P., Carrupt, P.A., Testa, B.: Molecular fields in quantitative structure-permeation relationships: the VolSurf approach. Journal of Molecular Structure **503** (2000) 17–30
25. Michie, D., Spiegelhalter, D.J., Taylor, C.C., eds.: Machine Learning, Neural and Statistical Classification. Ellis Horwood Limited (1994)

Bayesian Networks Learning
for Gene Expression Datasets

Giacomo Gamberoni[1], Evelina Lamma[1], Fabrizio Riguzzi[1],
Sergio Storari[1], and Stefano Volinia[2]

[1] ENDIF-Dipartimento di Ingegneria, Università di Ferrara, Ferrara, Italy
{ggamberoni, elamma, friguzzi, sstorari}@ing.unife.it
[2] Dipartimento di Biologia, Università di Ferrara, Ferrara, Italy
s.volinia@unife.it

Abstract. DNA arrays yield a global view of gene expression and can
be used to build genetic networks models, in order to study relations
between genes. Literature proposes Bayesian network as an appropriate
tool for develop similar models. In this paper, we exploit the contribute of
two Bayesian network learning algorithms to generate genetic networks
from microarray datasets of experiments performed on Acute Myeloid
Leukemia (AML).

In the results, we present an analysis protocol used to synthesize
knowledge about the most interesting gene interactions and compare
the networks learned by the two algorithms. We also evaluated relations
found in these models with the ones found by biological studies performed
on AML.

1 Introduction

From DNA microarray experiments, we can obtain a huge amount of data about
gene expression of different cell populations. An intelligent analysis of these
results can be very useful and important for cancer research.

An important field of interest in this area is the discovering of genetic networks, intended as a synthetic representation of genetic interactions. Similar
problems are often studied in other fields, and several techniques were developed in order to learn interaction networks from examples. One of the most
used approach for this type of problems is the Bayesian Network one.

Bayesian networks are suitable for working with the uncertainty that is typical of real-life applications. These are robust models and usually maintain good
performance also with missing or wrong values.

A Bayesian network is a directed, acyclic graph (DAG) whose nodes represent
random variables. In Bayesian networks, each node is conditionally independent
of any subset of the nodes that are not its descendants, given its parent nodes.

By means of Bayesian networks, we can use information about the values of
some variables to obtain probabilities for the values of others. A probabilistic
inference takes place once the probability of the values of each node conditioned

A.F. Famili et al. (Eds.): IDA 2005, LNCS 3646, pp. 109–120, 2005.
© Springer-Verlag Berlin Heidelberg 2005

to just its parents are given. These are usually represented in a tabled form, called Conditional Probability Tables (CPTs).

Applying this theory to the Bioinformatic field, we aim to build a network in which nodes represent different genes or attributes of a biological sample. Studying several samples related to a particular pathology, we build up a network that represents the probabilistic relations between genes and attributes. This model may be useful for biologists, because it highlights these interactions in a synthetic representation.

Techniques for learning Bayesian networks have been extensively investigated (see, for instance [12]). Given a training set of examples, learning such a network is the problem of finding the structure of the direct acyclic graph and the CPTs associated with each node in the DAG that best match (according to some scoring metric) the dataset. Optimality is evaluated with respect to a given scoring metric (for example, description length or posterior probability [6,12,20]). A procedure for searching among possible structures is needed. However, the search space is so vast that any kind of exhaustive search cannot be considered, and a greedy approach is followed.

In the literature, we find two different approaches for learning Bayesian networks: the first one is based on information theory [5], while the second one is based on the search and score methodology [6,12,20].

In this paper we use the K2 and the K2-lift algorithm. The K2 algorithm [6] is one of the best known algorithms among those that follows the search and score methodology. K2-lift algorithm is a modified version of K2 algorithm that uses the lift parameter (defined in association rules theory [1]), in order to improve the quality of learned networks and to reduce the computational resources needed .

The paper is structured as follows. Section 2 provides an introduction to Bayesian networks and to algorithms for learning them. In Section 3 we present the experimental domain of the dataset used, presenting the Leukemia dataset used for testing and validation. In Section 3.3 we present the results, comparing K2 algorithm with the K2-lift one; then we report some considerations on the results, from the biological point of view. Related work is mentioned in Section 4. Finally, in Section 5, we conclude and present future work.

2 Bayesian Networks Theory and Learning

A Bayesian network $\mathcal{B}$ is defined as a pair $\mathcal{B} = (\mathcal{G}, \mathcal{T})$, where $\mathcal{G}$ is a directed, acyclic graph and $\mathcal{T}$ is a set of conditional probability tables. $\mathcal{G}$ is defined as a couple $\mathcal{G} = (\mathcal{V}, \mathcal{A})$, where $\mathcal{V}$ is a set of nodes $\mathcal{V} = \{V_1, \ldots, V_n\}$, representing a set of stochastic variables, and $\mathcal{A}$ is a set of arcs $\mathcal{A} \subseteq \mathcal{V} \times \mathcal{V}$ representing conditional and unconditional stochastic independences among the variables [14,16]. In the following, variables will be denoted by upper-case letters, for example V, whereas a variable V which takes on a value v , that is $V = v$, will be abbreviated to V.

The basic property of a Bayesian network is that any variable corresponding to a node in the graph $\mathcal{G}$ is conditionally independent of its non-descendants given its parents; this is called the *local Markov property*. A joint probability distribution $Pr(V_1, \ldots, V_n)$ is defined on the variables. As a consequence of the local Markov property, the following decomposition property holds:

$$Pr(V_1, \ldots, V_n) = \prod_{i=1}^{n} Pr(V_i | \pi(V_i)) \tag{1}$$

where $\pi(V_i)$ denotes the set of variables corresponding to the parents of V_i, for $i = 1, \ldots, n$.

Once the network is built, probabilistic statements can be derived from it by probabilistic inference, using one of the inference algorithms described in the literature (for example [14,16]).

Given a training set of examples, learning a Bayesian network is the problem of finding the structure of the direct acyclic graph and the Conditional Probability Tables (CPTs) associated with each node that best match (according to some scoring metrics) the dataset.

2.1   Learning Algorithms Used

A frequently used procedure for Bayesian network structure construction from data is the K2 algorithm [6]. Given a database D, this algorithm searches for the Bayesian network structure $\mathcal{G}$ with maximal $Pr(\mathcal{G}, D)$, where $Pr(\mathcal{G}, D)$ is determined as described below. Let D be a database of m cases, where each case contains a value assignment for each variable in $\mathcal{V}$. Let $\mathcal{T}$ be the associated set of conditional probability distributions. Each node $V_i \in \mathcal{V}$ has a set of parents $\pi(V_i)$.

The K2 algorithm assumes that an ordering on the variables is available, and that all structures are a priori equally likely. For every node V_i, it searches for the set of parent nodes $\pi(V_i)$ that maximizes a function $g(V_i, \pi(V_i))$.

K2 adopts a greedy heuristic method. It starts by assuming that a node lacks parents, and then, at every step, it adds the parent whose addition mostly increases the function $g(V_i, \pi(V_i))$. K2 stops adding parents to the nodes when the addition of a single parent does no longer increase $g(V_i, \pi(V_i)))$.

K2 is characterized by the insertion of a large number of extra arcs. The extra arc problem of K2 arises especially when the network is characterized by a lot of root nodes (nodes without parents). During network learning, the algorithm tries to add parents to each of these nodes until it maximizes function $g(V_i, \pi(V_i))$. The algorithm will add at least one arc to root nodes because the value of the heuristic for this new structure is always better than the value of the previous structure.

K2-lift [13] is an extension of the K2 algorithm. It uses parameters normally defined in relation to association rules [1].

Association rules represent co-occurrence between events $ant \Rightarrow cons$ in which an event is defined as the association of a value to an attribute, while ant and $cons$ are set of events.

The support of a set of events is the fraction of records that contain all the events in the set.

In K2-lift we focused our attention on rules with one item in the antecedent and one item in the consequent (called one-to-one rules) and on the lift parameter [4] of a rule, a measure of rule interest, computed as follows: $lift = support(ant \cup cons)/(support(ant) \times support(cons))$. The knowledge represented by one-to-one association rules parameters is used to reduce the set of nodes from which the K2 algorithm tries to identify the best set of parents. This reduces the problem of extra arcs.

3   Experiments

In this section we present the experimental dataset (in section 3.1), then, in section 3.2 we decribe the analysis protocol used. Finally, in section 3.3 we present and evaluate the results of the performed experiments.

3.1   Dataset

Recent technical and analytical advances make it practical to evaluate quantitatively the expression of thousands of genes in parallel using microarrays (as described extensively in [2,11,7]). A microarray experiment consists of measurements of the relative representation of a large number of mRNA species in a set of biological samples. This mode of analysis has been used to observe gene expression variation in a variety of human tumors.

The analyzed dataset, available on-line in the ArrayExpress repository of the European Bioinformatics Institute[3], regroups the results of 20 microarray experiments, divided as follows:

10 Acute Myeloid Leukemia (AML) samples;
10 MyeloDysplastic Syndrome (MDS) samples.

Acute Myeloid leukemia (AML) may develop de novo or secondarily to Myelo-Dysplastic Syndrome (MDS). Although the clinical outcome of MDS-related AML is worse than that of de novo AML, it is not easy to differentiate between these two clinical courses without a record of prior MDS. Large-scale profiling of gene expression by DNA microarray analysis is a promising approach with which to identify molecular markers specific to de novo or MDS-related AML.

The experiments were performed using Affymetrix[4] Genechip Human Genome U95Av2 arrays. The Detection algorithm (included in the Affymetrix Microarray Suite Version 5.0) uses probe pair intensities to assign a Present, Marginal, or Absent call. This is a very reliable discretization method that reveal if a probe is expressed or not in the sample. Notice that Bayesian Networks can handle only discrete attributes, so we absolutely need such a discretized expression level.

The data from M experiments considering N probes, may be represented as a $M \times N$ detection matrix, in which each of the M rows consists of a N-element detection vector for a single sample.

---

[3] http://www.ebi.ac.uk/arrayexpress/, access code E-MEXP-25
[4] http://www.affymetrix.com

In order to obtain a more readable Bayesian Network, we need to reduce the number of attributes. So we conducted the experiments focusing the attention on specific aspects of the illnesses described in the dataset and, in particular, on the probes of our dataset related to each one of these aspects. Filtering the dataset probes in such way we created new smaller datasets to be analyzed.

In these dataset we kept only the probes related to an aspect. In order to get their names, we used the NetAffx Analysis Center[5], searching for a term and specifying the GeneChip Array name.

Following the indications of our biologist, we considered two different aspects and subsequently worked on two different datasets.

The first interesting aspect of the pathology under evaluation is the Negative Regulation of Cell Proliferation (GO:0008285): this GO term refers to any process that stops, prevents or reduces the rate or extent of cell proliferation. The initial dataset contains 32 probes related to this GO term so we create a new smaller dataset (named *NRCP_dataset*) composed by 20 rows (cases) and 33 columns (32 probes expression levels and the class attribute).

The second interesting aspect of the pathology under evaluation is the study of the biological process related to Hepatocyte Growth Factor or Substrate (HGF, HGS). The initial dataset contains 9 probes related to this aspect so we create a new smaller dataset (named *HGS/HGF_dataset*) composed by 20 rows (cases) and 10 columns (9 probes expression levels and the class attribute);

3.2 Analysis Protocol

Given a dataset, the analysis protocol followed in our experiments consists of 3 steps :

1. Generate a set of 20 random attribute orderings named SAO_i, with $i = 1, .., 20$. The attribute ordering is required by the Bayesian network learning algorithms described in Section 2.1. The generation of the set of SAO_i is necessary because the optimal attribute ordering is unknown in our experiments.

2. For each learning algorithm $La \in \{K2, K2 - lift\}$:
   (a) For i=1,..,20
       i. Learn the Bayesian network $BN_{La,i}$ by using La on SAO_i
       ii. Compute the Bayes score $BS_{La,i}$ of $BN_{La,i}$
   (b) Rank the learned network $BN_{La,i}$ according to their score $BS_{La,i}$
   (c) Analyze the first five learned networks $BN_{La,i}$ and identify:
       – frequent parent probes
       – probes with frequently the same subset of parent probes (found on more than 3 networks over 5)

3. Compare the results achieved by using $K2$ and $K2 - lift$

The analysis performed on the learned networks $BN_{La,i}$ is preliminary and the definition of a more complete methodology is required. For example, the

---

[5] https://www.affymetrix.com/analysis/netaffx/index.affx

probabilistic relations represented by a Bayesian network consist of both qualitative and quantitative probabilistic meanings, but in our preliminary analysis for each Bayesian probabilistic relation found we consider only the qualitative one.

3.3   Results and Discussion

Results of application of the analysis protocol on the *HGS/HGF_dataset* by using K2-lift are presented in Table 1. In each cell you can find the number of occurrence of a relation between a parent node (on the column) and a child node (in the row), frequent relations are in bold.

A graphical representation of the results is presented in Figure 1, notice that the width of the arcs are proportional to the number of relations found (we omitted relations found only once, in order to keep the graph simple).

The most frequent parents in the obtained networks are the probes *829_s_at (GSTP1)* with frequency 19.5%, *1095_s_at (HGF)* with frequency 29.3% and the *class* attribute with frequency 26.8%.

Results of application of the analysis protocol on the *HGS/HGF_dataset* by using K2 are presented in Table 2. In each cell you find the number of occurrence of a relation between a parent node (on the column) and a child node (in the row), frequent relations are in bold.

The most frequent parents are the probes *742_at (HABP2)* with frequency 25%, *1095_s_at (HGF)* with frequency 18.9% and the *class* attribute with frequency 20.8%.

Applying the analysis protocol on the *NRCP_dataset* (complete tables are omitted, due to lack of space) by using K2-lift algorithm, we observed that:

Table 1. HGS/HGF K2-lift results

	class	742_at HABP2	829_s_at GSTP1	1095_s_at HGF	1340_s_at HGF	33396_at GSTP1	33887_at HGS	35063_at HGFAC	36231_at MGC17330	40508_at GSTA4
class								**3**	1	
742_at										
829_s_at	2			**4**		1	1		1	
1095_s_at	3		1				1			
1340_s_at										
33396_at	1		2	2						
33887_at	2		**3**	1		1			1	
35063_at										
36231_at				**5**						
40508_at	3		2							

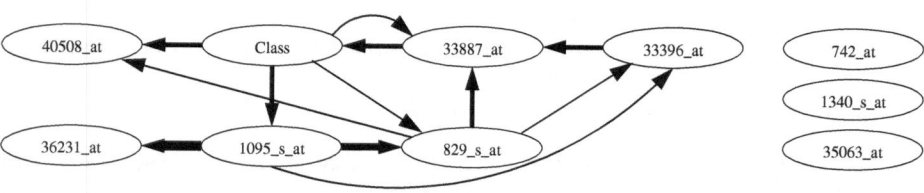

Fig. 1. K2-lift network

Table 2. HGS/HGF K2 results

	class	742_at HABP2	829_s_at GSTP1	1095_s_at HGF	1340_s_at HGF	33396_at GSTP1	33887_at HGS	35063_at HGFAC	36231_at MGC17330	40508_at GSTA4
class							3		1	
742_at	2			1		1				
829_s_at	2			4		1	1		1	
1095_s_at		3			1					
1340_s_at	2	1		1		1				
33396_at	1	1			2					
33887_at	2		3	1		1			1	
35063_at	2	1		1		1				
36231_at		4			1					
40508_at		3	1		1					

- the most frequent parent probes are *1880_at (MDM2)* with frequency 16.9%, *37107_at (PPM1D)* with frequency 11.3% and *40631_at (TOB1)* with frequency 10.5%;
- there are five frequent probabilistic probe relations composed by:
  - *36136_at (TP53I11)* with parents *1880_at (MDM2)* and *38379_at (GP-NMB)*;
  - *36479_at (GAS8)* with parent *1880_at (MDM2)*;
  - *38379_at (GPNMB)* with parent *1880_at (MDM2)*;
  - *38682_at (BAP1)* with parent *38379_at (GPNMB)*;
  - *41141 (PRKRIR)* with parent *37218_at (BTG3)*.

Applying the analysis protocol on the *NRCP_dataset* by using K2 algorithm we observed that:

- the most frequent parent probes are *1880_at (MDM2)* with frequency 27.1%, *34629_at (TP53I11)* with frequency 10%, *32568_at (BTG3)* with frequency 9.0% and the *class* attribute with frequency 10.5%;
- there are six frequent probabilistic probe relations composed by:
  - *36136_at (TP53I11)* with parent *1880_at (MDM2)*;
  - *36479_at (GAS8)* with parent *1880_at (MDM2)*;
  - *38379_at (GPNMB)* with parent *1880_at (MDM2)*;
  - *38639_at (MXD4)* with parent *1880_at (MDM2)*;
  - *38682_at (BAP1)* with parent *38379_at (GPNMB)*;
  - *41141_at (PRKRIR)* with parents *37218_at(BTG3)*and *38682_at (BAP1)*.

These results may be evaluated both regards the applied methods and the biological significance.

About the methods, some considerations arise about the difference between the results produced by K2 and K2-lift, and about the adopted analysis protocol.

Analyzing the Bayesian networks learned by K2, we can see that the first probe in the ordering is usually overrepresented as parent of the other probes. So it creates a more connected network that is difficult to evaluate by a biologist. K2-lift learns a more synthetic network which highlights the most interesting probes interactions. About the Bayes score of the learned networks, the score average of the best ones proposed by K2 and K2-lift are similar (-170.14 against -170.97 for the *NRCP_dataset* and -61.37 against -62.38 for the *HGS/HGF_dataset*).

About the adopted analysis protocol, the probabilistic relations found in the datasets consider only the qualitative meaning of such relations. The conditional probability table associated to each frequent relation found can be computed by standard statistic methodologies.

The results propose also some biological considerations. Acute myelogenous leukemia is a heterogeneous disease that appears to evade the normal regulatory controls of tumor suppressor genes (Stirewalt et al, 2000 [19]). Studies in AML have documented mutations in p53 (strictly related to *TP53I11* probe), but these mutations are relatively uncommon, especially compared to their mutational frequency in solid tumors. In addition, expression abnormalities have now been documented in several tumor suppressor genes or related genes including MDM2, p73, Rb, p14(ARF), p15(INK4B), and p16(INK4A). ERBB2 (strictly related to *TOB1* probe, found by K2-lift as a frequent parent) is a receptor protein tyrosine kinase frequently mutated in human cancer. The protein-kinase family is the most frequently mutated gene family found in human cancer and faulty kinase enzymes are being investigated as promising targets for the design of anti-tumour therapies.

Stephens et al (2004) [18] have sequenced the gene encoding the transmembrane protein tyrosine kinase ERBB2 from 120 primary lung tumours and identified 4% that have mutations within the kinase domain; in the adenocarcinoma subtype of lung cancer, 10% of cases had mutations. ERBB2 inhibitors, which have so far proved to be ineffective in treating lung cancer, should now be clinically re-evaluated in the specific subset of patients with lung cancer whose tumours carry ERBB2 mutations. MDM2 (found by K2-lift as a frequent parent) is a target gene of the transcription factor tumor protein p53. Overexpression of this gene can result in excessive inactivation of tumor protein p53, diminishing its tumor suppressor function.

Faderl et al (2000) [8] showed that overexpression of MDM-2 is common in AML and is associated with shorter complete remission duration and event free survival rate. It is striking to note that MDM2 and a TP53 induced protein are so tightly connected in the networks, since MDM2 is probably the most important protein for regulation of TP53 activity. The ERBB2 and MDM2 interaction is also very revealing and it should be noted that ERBB2 amplification or overexpression can make cancer cells resistant to apoptosis and promotes their growth.

Zhou et al (2001) [22] showed that ERBB2-mediated resistance to DNA-damaging agents requires the activation of Akt, which enhances MDM2-mediated ubiquitination and degradation of TP53.

We then compared our results to the original conclusions from the paper by Oshima and colleagues [15] who generated the datasets we used for analysis. They identified a set of genes associated to the course of the disease and to clinical classification. Furthermore they identified genes which are related to clinical outcome after induction chemotherapy. When compared to their gene lists, we identified novel gene interactions in our analysis, which might be very important for the clinical outcome. TP53 implications are outlined above while HGF has been widely implicated in tumor scattering and invasive growth and is of prognostic importance in AML (Verstovsek et al 2001 [21]). Our findings therefore represented novel potentially informative results related to the AML datasets. Thus it seems that by using the proposed method it is possible to mine useful data from microarray experiments in a previously undescribed way.

4 Related Works

A work related to ours is [9]. In it the authors learn causal networks from DNA microarray data with the aim of discovering causal relations between the different genes. A causal network is a network where the parent of a variable are its immediate causes. A causal network can be interpreted as a Bayesian network if we make the Causal Markov Assumption: given the values of a variable's immediate causes, it is independent of its earlier causes.

In order to learn causal networks, the authors make two assumptions: the first is that the unknown causal structure of the domain satisfies the Causal Markov Assumption, the second is that there are no latent or hidden variables. However, from these assumptions it is not possible to distinguish from observations alone between causal networks that specify the same independence properties. Therefore, what is learned is a partially directed acyclic graph (PDAG), where some of the edges can be undirected.

When data is sparse, a single PDAG can not be identified, rather a probability distribution over causal statements is induced. Moreover, the posterior probability is not dominated by a single model. In order to solve this problem, the authors try to identify *features*, i.e., relations between couples of variables. There are two types of features. The first is Markov relations: X is in a Markov relation with Y if Y is in the Markov blanket of X. The second is order relations: X is in an order relation with Y in a PDAG if there is a path between X and Y where all the edges are directed. The aim of the authors is to estimate the posterior probability of the features given the data. Ideally this should be done by sampling networks from the posterior to estimate this quantity. However, this is a hard problem. Therefore, they resort to a simpler analysis that consists of the bootstrap method: they generate perturbed versions of the data set and learn from them. In this way, they collect many networks that are reasonable models of the data. Then they compute the confidence of a feature as the fraction of networks containing the feature

In order to learn a PDAG from data they use the Sparse Candidate algorithm: a relatively small number of candidate parents for a variable can be identified by means of local statistics (such as correlation). Then the search is performed by picking parents for a variable only from the identified set.

They apply these techniques to DNA microarrays for S. cerevisiae. In particular they consider 800 genes whose expression varies over the different cell-cycle stages and 76 gene expression measurements. The learning experiment was conducted using 200-fold bootstrap. The results show that they were able to recover intricate structures even from such small data set. A biological analysis show that the results are well supported by current biological knowledge.

Our approach differs from the one of [9] because we learn a number of networks starting from different orders of the variables rather than from perturbed datasets. Moreover, we compute the confidence of the features only taking into account the best scoring networks according to a Bayesian metric.

Another work very related to ours is [10]. In it the author describe the state of the art in the use of probabilistic graphical models for inferring regulatory networks in cells. Besides the bootstrap approach of [9], the author describe two other studies that are relevant to ours. In the first, the authors examines only the networks in which a small number of regulators explain the expression of all other genes. This simplifies the learning procedure thus leading to statistical and computational advantages. They performed a systematic validation comparing the process and function annotation of the target set of each regulator with the known literature about the regulator. In most cases, they found a match between the annotation and the literature.

In the second study, the genes are divided into module that share a regulatory program. The learning procedure simultaneously identifies the composition of the modules and the regulators for each module. The module approach is in accordance with biological principles that suggest that a regulatory process usually involves many genes at the same time. Moreover, shared regulatory processes require less parameters thus leading to an improvement of the robustness of the model. Finally, the learned networks are easier to interpret, thanks to the module partition. The authors of this study confirmed the obtained results both by comparing the results with the literature and by examining gene expression of knockout strains.

These two studies are alternative to ours and exploit prior biological knowledge in order to target more effectively the gene expression domain, while we used general purpose techniques without exploiting other knowledge besides that contained in the microarray dataset.

5   Conclusions and Future Works

In this paper we describe the results of experiments conducted applying Bayesian network learning algorithm on microarray datasets. These preliminary results shows that in most of the cases, K2-lift creates a more synthetic network with respect to K2. It is also noteworthy that many relations found confirmation in biological literature.

The analysis protocol used for the result evaluation is very simple and need some enhancements both in complexity and in statistical significance. For these reasons, a bootstrap-based approach will be the subject of further studies. In a more complex protocol, we also need to consider the conditional probability tables and other information associated to the learned networks

In the future we also plan to empirically compare our approach to that of [9] in order to better compare the performances of the two methods. Moreover, we plan to improve our learning process by means of prior biological knowledge, as done in the studies described in [10].

References

1. Agrawal, R., Imielinski, T. and Swami, A.: Mining association rules between sets of items in large databases. Proceedings of the 1993 ACM SIGMOD International Conference on Management of Data (1993) 207–216
2. Alizadeh, A.A., Eisen, M.B., Davis, R.E., Ma ,C., Lossos, I.S., Rosenwald, A., Boldrick, J.C., Sabet, H., Tran, T., Yu, X., Powell, J.I., Yang, L., Marti, G.E., Moore, T., Hudson, J. Jr., Lu, L., Lewis, D.B., Tibshirani, R., Sherlock, G., Chan, W.C., Greiner, T.C., Weisenburger, D.D., Armitage, J.O., Warnke, R., Levy, R., Wilson, W., Grever, M.R., Byrd, J.C., Botstein, D., Brown, P.O., Staudt, L.M.: Distinct types of diffuse large B-cell lymphoma identified by gene expression profiling. Nature 403 (2000) 503–511
3. Anderson, D.R., Sweeney, D.J. and Williams, T.A. : Introduction to statistics concepts and applications, Third Edition. West Publishing Company (1994)
4. Berry, J.A. and Linoff, G.S.: Data Mining Techniques for Marketing, Sales and Customer Support. John Wiley & Sons Inc. (1997)
5. Cheng, J., Greiner, R., Kelly, J., Bell, D. and Liu, W.: Learning Bayesian networks from data: An information-theory based approach. Artificial Intelligence 137 (2002) 43–90
6. Cooper, G. and Herskovits, E.: A Bayesian method for the induction of probabilistic networks from data. Machine Learning 9 (1992) 309–347
7. Eisen, M.B., Spellman, P.T., Brown, P.O., and Botstein, D. : Cluster analysis and display of genome-wide expression patterns. Proc. Natl. Acad. Sci. USA 95 (1998) 14863–14688
8. Faderl, S., Kantarjian, H.M., Estey, E., Manshouri, T., Chan, C.Y., Rahman El-saied, A., Kornblau, S.M., Cortes, J., Thomas, D.A., Pierce, S., Keating, M.J., Estrov, Z., Albitar, M.: The prognostic significance of p16(INK4a)/p14(ARF) locus deletion and MDM-2 protein expression in adult acute myelogenous leukemia. Cancer 89(9) (2000) 1976–82
9. Friedman, N., Linial, M., Nachman, I., Pe'er, D.: Using Bayesian Networks to Analyze Expression Data. Journal of Computational Biology (7)3/4 (2000) 601–620
10. Friedman, N.: Inferring Cellular Networks Using Probabilistic Graphical Models. Science 303 (2004) 799–805
11. Golub, T.R., Slonim, D.K., Tamayo, P., Huard, C., Gaasenbeek, M., Mesirov, J.P., Coller, H., Loh, M.L., Downing, J.R., Caligiuri, M.A., Bloomfield, C.D., Lander, E.S.: Molecular classification of cancer: class discovery and class prediction by gene expression monitoring. Science 286 (1999) 531–537
12. Heckerman, D. and Geiger, D. and Chickering, D.: Learning Bayesian Networks: the combination of knowlegde and statistical data. Machine Learning 20 (1995) 197–243
13. Lamma, E., Riguzzi, F. and Storari, S.: Exploiting Association and Correlation Rules Parameters for Improving the K2 Algorithm. 16th European Conference on Artificial Intelligence (2004) 500–504
14. Lauritzen, S. L. and Spiegelhalter, D. J.: Local computations with probabilities on graphical structures and their application to expert systems. J. Royal Statistics Society B 50 (1988) 157–194
15. Oshima Y., Ueda M., Yamashita Y., Choi Y.L., Ota J., Ueno S., Ohki R., Koinuma R., Wada T., Ozawa K., Fujimura A., Mano H.: DNA microarray analysis of hematopoietic stem cell-like fractions from individuals with the M2 subtype of acute myeloid leukemia. Leukemia 17(10) (2003) 1900–1997

16. Pearl, J.: Probabilistic reasoning in intelligent systems: networks of plausible inference. Morgan Kaufmann (1988)
17. Ramoni, M. and Sebastiani, P.: Robust learning with missing data. Technical Report, Knowledge Media Institute, The Open University KMI-TR-28 (1996)
18. Stephens, P., Hunter, C., Bignell, G., Edkins, S., Davies, H., Teague, J., Stevens, C., O'Meara, S., Smith, R., Parker, A., Barthorpe, A., Blow, M., Brackenbury, L., Butler, A., Clarke, O., Cole, J., Dicks, E., Dike, A., Drozd, A., Edwards, K., Forbes, S., Foster, R., Gray, K., Greenman, C., Halliday, K., Hills, K., Kosmidou, V., Lugg, R., Menzies, A., Perry, J., Petty, R., Raine, K., Ratford, L., Shepherd, R., Small, A., Stephens, Y., Tofts, C., Varian, J., West, S., Widaa, S., Yates, A., Brasseur, F., Cooper, C.S., Flanagan, A.M., Knowles, M., Leung, S.Y., Louis, D.N., Looijenga, L.H., Malkowicz, B., Pierotti, M.A., Teh, B., Chenevix-Trench, G., Weber, B.L., Yuen, S.T., Harris, G., Goldstraw, P., Nicholson, A.G., Futreal, P.A., Wooster, R., Stratton, M.R.: Lung cancer: intragenic ERBB2 kinase mutations in tumours.Nature 431(7008) (2004) 525–6
19. Stirewalt, D.L., Radich, J.P.: Malignancy: Tumor Suppressor Gene Aberrations in Acute Myelogenous Leukemia. Hematology 5(1) (2000) 15–25
20. Suzuki, J.: Learning Bayesian Belief Networks Based on the MDL principle: An Efficient Algorithm Using the Branch and Bound Technique. IEICE Transactions on Communications Electronics Information and Systems (1999)
21. Verstovsek S., Kantarjian H., Estey E., Aguayo A., Giles F.J., Manshouri T., Koller C., Estrov Z., Freireich E., Keating M., Albitar M.: Plasma hepatocyte growth factor is a prognostic factor in patients with acute myeloid leukemia but not in patients with myelodysplastic syndrome. Leukemia 15(8) (2001) 1165–70
22. Zhou, B.P., Liao, Y., Xia, W., Zou, Y., Spohn, B., Hung ,M.C.: HER-2/neu induces p53 ubiquitination via Akt-mediated MDM2 phosphorylation. Nat Cell Biol. 3(11) (2001) 973–82

Pulse: Mining Customer Opinions from Free Text

Michael Gamon, Anthony Aue, Simon Corston-Oliver, and Eric Ringger

Natural Language Processing,
Microsoft Research, Redmond, WA 98052, USA
(mgamon, anthaue, simonco, ringger)@microsoft.com
http://research.microsoft.com/nlp/

Abstract. We present a prototype system, code-named *Pulse*, for mining topics and sentiment orientation jointly from free text customer feedback. We describe the application of the prototype system to a database of car reviews. Pulse enables the exploration of large quantities of customer free text. The user can examine customer opinion "at a glance" or explore the data at a finer level of detail. We describe a simple but effective technique for clustering sentences, the application of a bootstrapping approach to sentiment classification, and a novel user-interface.

1   Introduction

The goal of customer satisfaction studies in business intelligence is to discover opinions about a company's products, features, services, and businesses. Customer satisfaction information is often elicited in a structured form: surveys and focus group studies present customers with carefully constructed questions designed to gather particular pieces of information a company is interested in. The resulting set of structured, controlled data can easily be analyzed statistically and can be conveniently aggregated according to the specific dimensions of the survey questions or focus group setup. The drawbacks of structured studies are the expense associated with the design and administration of the survey, the limit that is necessarily imposed on the free expression of opinions by customers, and the corresponding risk of missing trends and opinions that are not expressed in the controlled situation. Additionally there is the risk of missing whole segments of the customer population that do not like to respond to a guided and structured set of questions.

Another potential source of information for business intelligence, which is becoming more and more pervasive and voluminous, is spontaneous customer feedback. This feedback can be gathered from blogs, newsgroups, feedback email from customers, and web sites that collect free-form product reviews. These can be rich sources of information, but these sources are much less structured than traditional surveys. The information is contained in free text, not in a set of answers elicited for a specific set of questions.

A.F. Famili et al. (Eds.): IDA 2005, LNCS 3646, pp. 121–132, 2005.

Paying people to mine this free-form information can be extremely expensive, and given the high volume of such free text is only feasible by careful sampling.[1]

With the advent of automatic techniques for text mining such as clustering and key term extraction, free-form customer opinions can be processed efficiently and distilled down to essential topics and recurring patterns of content. When trying to assess customer opinions, however, topic is only one of the dimensions that are of interest. As well as identifying what topics customers are talking about, it would be useful to characterize the opinions that they express about those topics.

Researchers have begun to focus on the analysis of opinion ('sentiment classification') typically using supervised machine learning techniques. [2] The project that we describe in this paper, code-named *Pulse*, combines the two dimensions of topic and sentiment and presents the results in an intuitive visualization. Pulse combines a clustering technique with a machine-learned sentiment classifier, allowing for a visualization of topic and associated customer sentiment. Pulse provides both a high-level overview of customer feedback and the ability to explore the data at a finer granularity. Pulse requires that only a small amount of data be annotated to train a domain-specific sentiment classifier.

Both sentiment detection and topic detection in Pulse are performed at the sentence level rather than at the document level. Document-level assessment, which is the focus of most sentiment classification studies, is too coarse for our purposes. In a review document, for example, we often find mixed positive and negative assessments such as: "OVERALL THE CAR IS A GOOD CAR. VERY FAST, THE ENGINE IS GREAT BUT FORD TRANSMISSIONS SUCK." Of course, even sentence-level granularity is too coarse in some instances, for example: "Its [sic] quick enough to get you and a few other people where you need to go although it isn't too flashy as far as looks go."[3] As we will discuss in further detail below, sentence-level granularity of analysis allows the discovery of new information even in those scenarios where an overall product rating is already provided at the document level.

We first describe the data to which Pulse has been applied (Section 2). We then describe the prototype system, consisting of a visualization component (Section 3.1), a simple but effective clustering algorithm (Section 3.2), and a machine-learned classifier that can be rapidly trained for a new domain (Section 3.3) by bootstrapping from a relatively small set of labeled data.

---

[1] It is worth noting that business intelligence is not the only scenario where customer satisfaction is of interest: individual customers often use resources on the web to find other people's reviews of products and companies to help them reach a decision on a purchase.

[2] Two notable exceptions are [1,2].

[3] In future work we intend to investigate sentences with mixed sentiment, analyzing them at the level of the clause or phrase.

2   Data

We applied Pulse to a sample of the car reviews database[3]. This sample contains 406,818 customer car reviews written over a four year period, with no editing beyond simple filtering for profanity. The comments range in length from a single sentence (56% of all comments) to 50 sentences (a single comment). Less than 1% of reviews contain ten or more sentences. There are almost 900,000 sentences in total.

When customers submitted reviews to the website, they were asked for a recommendation on a scale of 1 (negative) to 10 (positive). The average score was 8.3 suggesting that people are enamored of their cars, or that there is a self-selection in the reviewers. Even reviews with positive scores contain useful negative opinions: after all a less-than-perfect score often indicates that the car may have a few shortcomings, despite a relatively high score.

For this reason we ignore the document-level scores and annotated a randomly selected sample of 3,000 sentences for sentiment. Each sentence was viewed in isolation and classified as "positive", "negative" or "other". The "other" category was applied to sentences with no discernible sentiment, as well as to sentences that expressed both positive and negative sentiment and sentences with sentiment that cannot be deduced without taking context and/or world knowledge into account.

The annotated data was split: 2,500 sentences were used for the initial phase of training the sentiment classifier (Section 3.3); 500 sentences were used as a gold standard for evaluation. We measured pair-wise inter-annotator agreement on a separate randomly selected sample of 100 sentences using Cohen's Kappa score.[4] The three annotators had pair-wise agreement scores of 70.10%, 71.78% and 79.93%.This suggests that the task of sentiment classification is feasible but difficult even for people.

3   System Description

Pulse first extracts a taxonomy of major categories (makes) and minor categories (models) of cars by simply querying the car reviews database. The sentences are then extracted from the reviews of each make and model and processed according to the two dimensions of information we want to expose in the final visualization stage: sentiment and topic. To train the sentiment classifier, a small random selection of sentences is labeled by hand as expressing positive, "other", or negative sentiment. This small labeled set of data is used with the entirety of the unlabeled data to bootstrap a classifier (Section 3.3).

The clustering component forms clusters from the set of sentences that corresponds to a leaf node in the taxonomy (i.e. a specific model of car). The clusters are labeled with the most prominent key term. For our prototype we implemented a simple key-word-based soft clustering algorithm with tf·idf weighting and phrase identification (Section 3.2). Once the sentences for a make and model of car have been assigned to clusters and have received a sentiment score from

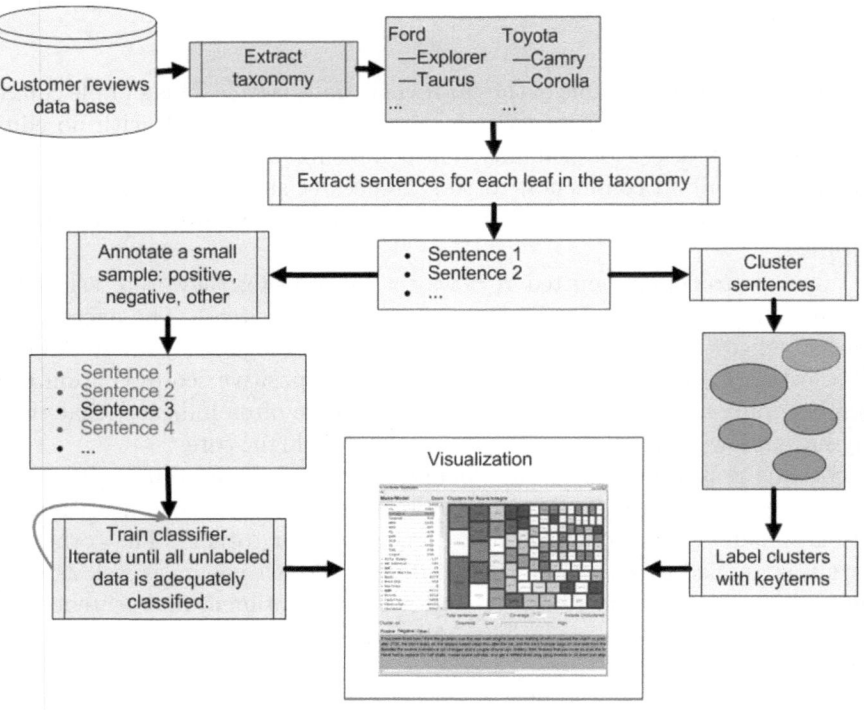

Fig. 1. Overview of the Pulse System Architecture

the sentiment classifier, the visualization component (Section 3.1) displays the clusters and the keyword labels that were produced for the sentences associated with that car. The sentences in a cluster can be displayed in a separate view. For each sentence in that view, the context (the original review text from which the sentence originated) can also be displayed. Figure 1 gives an overview of the system.

3.1   The Visualization Component

The visualization component needs to display the two dimensions of information, i.e. topic and sentiment, simultaneously. Another requirement is that it allow the user to easily access the specifics of a given topic. Pulse uses a Tree Map visualization [5] to display clusters and their associated sentiment. Each cluster is rendered as one box in the Tree Map. The size of the box indicates the number of sentences in the cluster, and the color indicates the average sentiment of the sentences in the box. The color ranges from red to green, with red indicating negative clusters and green indicating positive ones. Clusters containing an equal mixture of positive and negative sentiment or containing mostly sentences classified as belonging to the "other" category are colored white. Each box is also labeled with the key word for that particular cluster.

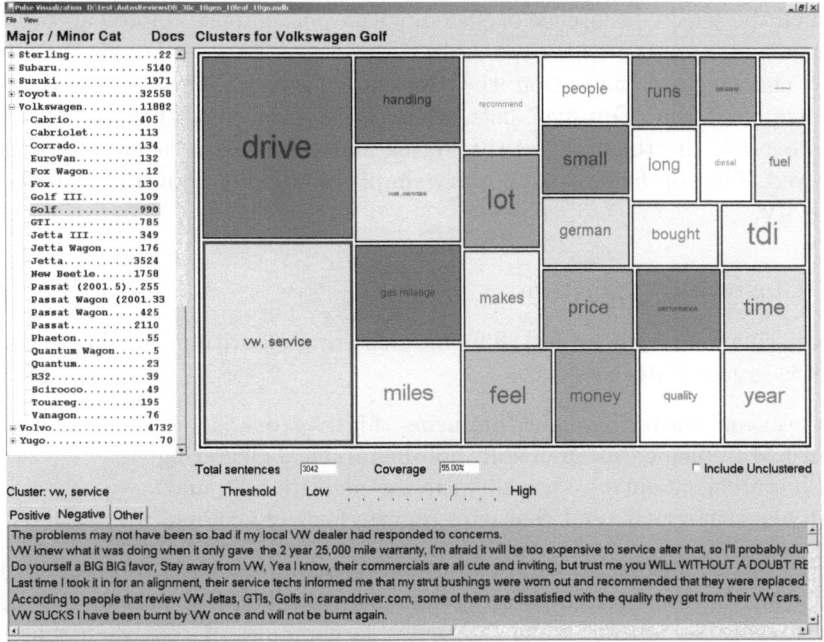

Fig. 2. Screenshot of the Pulse user interface showing the taxonomy of makes and models, the Tree Map with labeled clusters and sentiment coloring, and individual sentences from one cluster

The Tree Map visualization allows the identification of the following information about the sentences associated with a given make/model at a glance:

- the overall sentiment associated with the make/model (indicated by the relative area in the entire Tree Map colored red or green)
- the most common topics that customers mention in the reviews for the make/model as indicated by the larger boxes
- the most positive and the most negative topics, indicated by the darkest shades of green and red in the cluster boxes.

Figure 2 shows a screenshot of the visualization in the cluster view. The taxonomy of makes and models (i.e. major and minor category) is displayed in the left pane, the Tree Map to the right of it, and the sentences in the tabbed display at the bottom.

The user has selected the Volkswagen Golf. The two biggest clusters appear in the boxes at the left of the Tree Map: "drive", and "vw, service". The user has chosen to inspect the "vw, service" cluster by clicking on it and viewing the negative sentences in the tabbed display at the bottom of the screen. The threshold slider has been set approximately three quarters of the way along, restricting the display to only sentences with high class probability. This has the effect of increasing precision at the expense of recall. Clicking on a sentence in the tabbed display brings up a window (not shown) that displays the entire review in which the selected sentence occurred, with each sentence colored according to sentiment.

By choosing a menu option, the user can view a summary of the clusters in the form of simple "Top five" lists, where for a given make/model the top five terms overall, the top five positive terms and the top five negative terms are displayed. The top five display is very simple, and is not shown in the interests of brevity.

3.2   Clustering Algorithm

We experimented with several different clustering algorithms for finding salient patterns in the sentences:

- a k-means clustering algorithm using tf·idf vectors, as described in [6],
- an EM implementation of soft, non-hierarchical clustering[7],
- a hierarchical, entropy-based clustering algorithm[8], and
- an algorithm that used character n-gram feature vectors.

None of the approaches we tried produced clusters that we found satisfactory. Each algorithm was designed for a different task. The first two were designed for clustering documents which are much larger units of text than sentences. The third and fourth approaches were designed for clustering units of text that are much smaller than sentences, namely words and Internet search queries. We therefore formulated the following simple algorithm, which performs well.

The input to the clustering algorithm is the set of sentences S for which clusters are to be extracted, a stop-list W_{Stop} of words around which clusters ought not to be created, and (optionally) a "go list" W_{Go} of words known to be salient in the domain.

1. The sentences, as well as the stop and go lists, are stemmed using the Porter stemmer. [9]
2. Occurence counts C_W are collected for each stem not appearing in W_{Stop}.
3. The total count for stems occuring in W_{Go} is multiplied by a configurable parameter λ_1.
4. The total count for stems with a high tf·idf (calculated over the whole data set) is multiplied by a configurable parameter λ_2.
5. The total count for stems with a high tf·idf (calculated over the data in the given leaf node of the taxonomy) is multiplied by a configurable parameter λ_3.
6. The list of counts is sorted by size.
7. To create a set of N clusters, one cluster is created for each of the most frequent N stems, with all of the sentences containing the stem forming the cluster. The clusters are labeled with the corresponding stem St [4] An optional additional constraint is to require a minimum number M of sentences in each cluster.

---

[4] We experimented with N in the range 30–50. For larger values of N, the visualization became too cluttered to be useful.

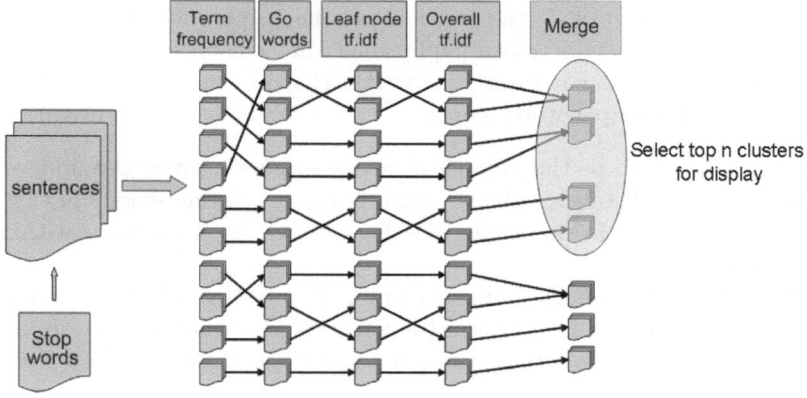

Fig. 3. Diagram of the clustering algorithm

8. Two clusters C_1 and C_2 are merged if the overlap of sentences S_{C1C2} contained in both C_1 and C_2 exceeds 50% of the set of sentences in C_1 or C_2. If the labels of C_1 and C_2 form a phrase in the sentences in S_{C1C2}, the new cluster C_{12} is labeled with that phrase, otherwise it is labeled with both labels, separated by a comma.

An overview of the clustering approach is presented in Figure 3. The initial set of clusters is determined by term frequency alone. Go words and the two tf·idf weighting schemes each re-rank the clusters, and finally some of the clusters are merged and a fixed number of clusters is selected off the top of the ranked list for display.

The stop word list consists of two components. The first is a manually specified set of function words and high frequency, semantically empty content words such as "put". The more interesting and essential part of the stop list, however, is the set of the top N features from the sentiment classifier, according to log likelihood ratio (LLR) with the target feature [10]. By disallowing words known to be highly correlated with positive or negative sentiment we ensure that the topics represented in the clusters are orthogonal to the sentiment of the feedback. Term frequency (tf)/inverse document frequency (idf) weighting is a common technique in clustering. Terms with high tf·idf scores are terms that have a high degree of semantic focus, i.e. that tend to occur frequently in specific subsets of documents. The tf·idf weighting scheme that we employed is formulated as

$$weight(i,j) = \left\{ \begin{array}{ll} (1 + log(tf_{i,j})log\frac{N}{df_i} & \text{if } tf_{i,j} \geq 1 \\ 0 & \text{otherwise} \end{array} \right\} \tag{1}$$

where $tf_{i,j}$ is the term frequency of a word w_i, and df_i is the document frequency of w_i, i.e. the number of documents containing w_i and N is the number of leaf nodes in the taxonomy ([6]).

Since we cluster sentences, i.e. sub-document units, we are not interested in using tf·idf for weight assignment in the sentence vectors themselves. We rather want to find out which of all the terms in all the reviews for one make/model leaf node should be given increased importance when clustering sentences in that leaf node. In order to assign a per-word weight that we can use in clustering, we calculate two different per-word scores:

1. We can take df_i to be the number of reviews under a given leaf node which contain w_i. $tf_{i,j}$ is taken to be the term frequency in the reviews in that leaf node. A high score in this scenario indicates high semantic focus within the specific leaf node.
2. If df_i is defined to be the number of reviews in the whole collection which contain w_i, and $tf_{i,j}$ is the term frequency in the whole collection, a high tf·idf score indicates a term with high semantic focus in the whole domain.

These two scores allow the customization of the weighting of terms according to their leaf-node specific salience or their domain-specific salience. The more uniform a collection of data is, the more the two measures will coincide. In addition to weighting the terms for clustering according to these two scores, Pulse also allows for the use of a go word list (i.e. a domain dictionary) where such a resource is available.[5] The go word list allows us to steer the clustering toward terms that we know to be salient in the domain, while at the same time still allowing us to discover new clusters automatically that do not appear in our domain dictionary. For example, for many makes and models of car, terms like "family" and "snow", which were not in the domain-specific go list, emerged as labels for clusters.

Finally, it must be noted that not all sentences are assigned to a cluster. Unassigned sentences are assigned to a nonce cluster, which is not displayed unless the user explicitly chooses to see it. Also, because more than one cluster keyword can appear in a given sentence, that sentence may correspondingly belong to more than one cluster (soft clustering).

3.3 Sentiment Analysis

As mentioned in the introduction, machine-learned approaches to sentiment analysis are a topic that has received considerable attention from researchers over the past few years. A number of different approaches have been applied to the problem. The annotated movie review data set made publicly available by Pang and Lee [11,12] has become a benchmark for many studies. The data consists of 2000 movie reviews, evenly split between positive and negative instances. The task is to determine which are positive and which are negative. Classification accuracies approaching 90% for this binary classification task are cited [11,12,13]. Features for sentiment classification typically consist of simple unigram (term) presence. However, the following characteristics of the car reviews data set rendered techniques previously cited in the literature unsuited to our task:

---

[5] For the autos domain, W_{Go} was created by extracting entry keywords from a freely-available online automotive dictionary.

1. Since we are aiming at sentence-level classification, we are dealing with much shorter textual units than the full movie reviews, which range from a few sentences to several paragraphs.
2. The car reviews are not annotated at the sentence level. Since one of the main purposes of Pulse is to avoid the cost associated with manual examination of data, we would like to be able make do with as little annotated data as possible.
3. The Movie Review data set is carefully selected to be balanced, and to contain only extremes, i.e. only very strong recommendations/disrecommendations. The car review data, on the other hand, are strongly imbalanced, with positive reviews predominating.
4. While the movie reviews are generally well-written, the car review sentences are frequently ungrammatical, fragmentary and idiosyncratic. They contain numerous misspellings, acronyms, and a more telegraphic style.

We ignored the recommendation scores at the review (document) level for two reasons. First, since we focus our classification on individual sentences, we cannot make the assumption that in a review all sentences express the same sentiment. If a reviewer decides to give 8 out of 10 stars, for example, the review is likely to contain a number of positive remarks about the car, with a few negative remarks–after all the reviewer had a reason to not assign a 10-out-of-10 score. Secondly, we wanted to investigate the feasibility of our approach in the absence of labeled data, which makes Pulse a much more generally applicable tool in other domains where customer feedback without any recommendations is common.

Because the sentences in the car reviews database are not annotated, we decided to implement a classification strategy that requires as little labeled data as possible. We implemented a modified version of Nigam et al.'s algorithm for training a Naive Bayes classifier using Expectation Maximization (EM) and bootstrapping from a small set of labeled data to a large set of unlabeled data [14]. The classification task in our domain is a three-way distinction between positive, negative, and "other". The latter category includes sentences with no discernible sentiment (a sentiment-neutral description of a model, for example), sentences with balanced sentiment (where both a positive and a negative opinion are expressed within the same sentence), and sentences with a sentiment that can only be detected by taking the review context and/or world knowledge into account. This bootstrapping allowed us to make use of the large amount of unlabeled data in the car reviews database, almost 900,000 sentences. The algorithm requires two data sets as input, one labeled (D_L), the other unlabeled (D_U).

1. An initial naive Bayes classifier with parameters θ is trained on the documents in D_L.
2. This initial classifier is used to estimate a probability distribution over all classes for each of the documents in D_U. (E-Step)
3. The labeled and unlabeled data are then used to estimate parameters for a new classifier. (M-Step)

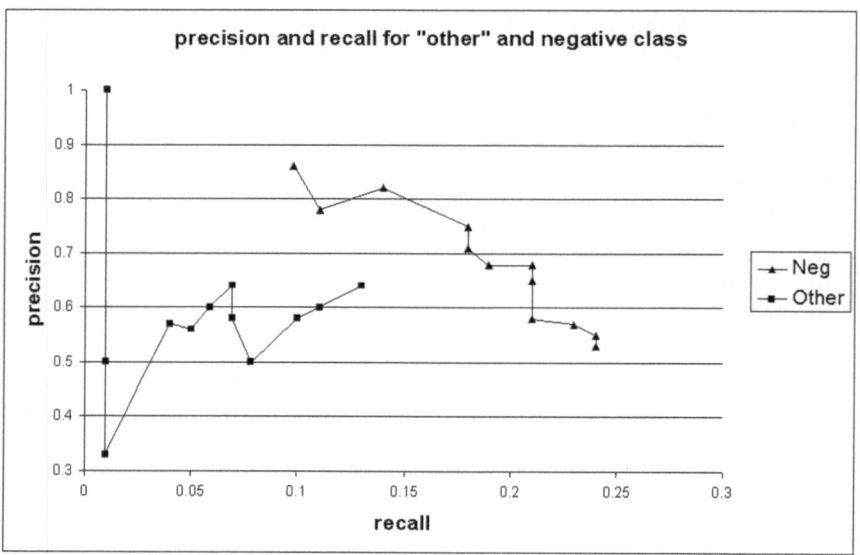

Fig. 4. Precision vs. Recall for Negative and OtherClass

Steps 2 and 3 are repeated until convergence is achieved when the difference in the joint probability of the data and the parameters falls below the configurable threshhold ϵ between iterations. We also implemented two additional modifications described by [14]:

1. A free parameter, δ, was used to vary the weight given to the unlabeled documents.
2. Mixtures were used to model each class.

In order to prepare the data for classification, we normalized each sentence using some simple filters. All words were converted to lower-case, and numbers were collapsed to a single token[6]. For each sentence, we produced a sparse binary feature vector, with one feature for each word or punctuation mark. Our labeled data were the hand-annotated sentences described in section 2. 2500 of these were used to train the classifier D_L, and the remaining 500 were reserved as a test set. The classifier was trained and then evaluated on the test set. The data set shows a clear skew towards positive reviews: in the annotated data set, positive sentences comprise 62.33% of the data, sentences of type "other" comprise 23.27%, and negative sentences 14.4%. Because of this skew toward a positive label in the data set, overall accuracy numbers are not very illuminating–naively classifying every sentence as positive will result in a 62.33% accuracy. Instead we evaluate

---

[6] We leave it for future research to also employ automatic spelling correction. We expect this to be useful in the car review domain, where misspellings are rather abundant (the word "transmission", for example, is spelled in 29 different ways in this data set).

the classifier by considering the precision vs. recall graph for the negative and "other" classes, which are the classes with the fewest occurrences in the training data. We achieved some of the best results on the negative and "other" classes by using a δ of 1.0.

Figure 4 shows that the classifier is able to achieve reasonable precision on the negative and "other" classes at the expense of recall. In domains with very large amounts of free-form customer feedback (typically so large that complete human analysis would not even be attempted) low recall is acceptable. The " other" category is clearly the hardest to identify, which is not surprising given its very heterogeneous nature. Recall on the positive class is nearly constant across precision values, ranging from 0.95 to 0.97.

4   Conclusion

Much has been written about the individual fields of clustering and sentiment analysis on their own. Combined, however, and paired with an appropriate visualization they provide a powerful tool for exploring customer feedback. In future work we intend to apply this combination of techniques to the analysis of a range of data, including blogs, newsgroups, email and different customer feedback sites. We are currently working with various end-users who are interested in using a practical tool for performing data analysis. The end-user feedback that we have received to date suggests the need for improved text normalization to handle tokenization issues, and the use of a speller tool to identify and normalize spelling variants and misspellings. Finally, our research will continue to focus on the identification of sentiment vocabulary and sentiment orientation with minimal customization cost for a new domain. We have begun experimenting with a variation of a technique for bootstrapping from seed words with known orientation [1,2] with promising initial results [15]. As opposed to the approach described here, the new approach only requires the user to identify a small (about ten item) seed word list with known strong and frequent sentiment terms and their orientation. The only additional task for the user would be to verify and edit an extended seed word list that the tool will automatically produce. Once this extended list has been verified, a sentiment classifier can be produced without further labeling of data.

References

1. Turney, P.D.: Thumbs up or thumbs down? semantic orientation applied to unsupervised classification of reviews. In: Proceedings of ACL 2002. (2002) 417–424
2. Turney, P.D., Littman, M.L.: Unsupervised learning of semantic orientation from a hundred-billion-word corpus. Technical Report ERC-1094 (NRC 44929), National Research Council of Canada (2002)
3. Microsoft Corporation: Msn autos (http://autos.msn.com/default.aspx) (2005)
4. Cohen, J.: A coefficient of agreement for nominal scales. Educational and Psychological measurements **20** (1960) 37–46

5. Smith, M.A., Fiore, A.T.: Visualization components for persistent conversations. In: CHI '01: Proceedings of the SIGCHI conference on Human factors in computing systems, ACM Press (2001) 136–143

6. Manning, C.D., Schütze, H.: Foundations of Statistical Natural Language Processing. The MIT Press, Cambridge, Massachusetts (1999)

7. Meila, M., Heckerman, D.: An experimental comparison of several clustering and initialization methods. Technical report, Microsoft Research (1998)

8. Goodman, J.: A bit of progress in language modeling. Technical report, Microsoft Research (2000)

9. Porter, M.: An algorithm for suffix stripping. Program **14** (1980) 130–137

10. Dunning, T.: Accurate methods for the statistics of surprise and coincidence. Computational Linguistics **19** (1993) 61–74

11. Pang, B., Lee, L., Vaithyanathan, S.: Thumbs up? sentiment classification using machine learning techniques. In: Proceedings of EMNLP 2002, EMNLP (2002) 79–86

12. Pang, B., Lee, L.: A sentimental education: Sentiment analysis using subjectivity summarization based on minimum cuts. In: Proceedings of ACL 2004, ACL (2004) 217–278

13. Bai, X., Padman, R., Airoldi, E.: Sentiment extraction from unstructured text using tabu search enhanced markov blanket. In: Proceedings of the International Workshop on Mining for and from the Semantic Web. (2004) 24–35

14. Nigam, K., McCallum, A., Thrun, S., Mitchell, T.: Text classification from labeled and unlabeled documents using em. Machine Learning **39(2/3)** (2000) 103–134

15. Gamon, M., Aue, A.: Automatic identification of sentiment vocabulary: exploiting low association with known sentiment terms. In: Proceedings of the ACL 2005 Workshop on Feature Engineering for Machine Learning in NLP, ACL (to appear)

Keystroke Analysis of Different Languages: A Case Study

Daniele Gunetti, Claudia Picardi, and Giancarlo Ruffo

Department of Informatics, University of Torino,
corso Svizzera 185, 10149 Torino, Italy
{gunetti, picardi, ruffo}@di.unito.it

Abstract. Typing rhythms are one of the rawest form of data stemming from the interaction between humans and computers. When properly analyzed, they may allow to ascertain personal identity. In this paper we provide experimental evidence that the typing dynamics of free text can be used for user identification and authentication even when typing samples are written in different languages. As a consequence, we argue that keystroke analysis can be useful even when people may use different languages, in those areas where ascertaining personal identity is important or crucial, such as within Computer Security.

1 Introduction to Keystroke Analysis

Keystroke Analysis is the biometric area concerned with the problem of ascertaining users' identity through the way they type on a computer keyboard [1]. As such, it is essentially a form of Pattern Recognition, as it involves *representation* of input data measures, *extraction* of characteristic features and *classification* or *identification* of patterns data so as to decide to which pattern class these data belong [9].

In the case of typing rhythms, input data is usually represented by a sequence of typed keys, together with appropriate timing information so that it is possible to compute the elapsed time between the release of the first key and the depression of the second (the so-called *digraph latency*) and the amount of time each key is held down (the *keystroke duration*). The extraction of such features turns a sequence of keystrokes into a *typing sample*. Appropriate algorithms are then used to classify a typing sample among a set of pattern classes, each one containing information about the typing habits of an individual. Pattern classes are often called *profiles* or *models*, and they are built using earlier typing information gathered from the involved individuals.

Within computer science, a biometric such as keystroke dynamics is particularly appealing, since it can be sampled without the aid of special tools, just the keyboard of the computer where the biometric analysis has to be performed. Keystroke analysis is however a difficult task, for several reasons: (1) keystrokes, unlike other biometric features, convey an unstructured and very small amount of information. Keystroke duration and digraph latency are in fact a pretty shallow kind of information. (2) Keystroke dynamics are a *behavioral* biometric, like voiceprints and handwritten signatures. As such, they are intrinsically unstable, and show a certain degree of variability even without any evident reason. After all, it is pretty difficult to control the number of milliseconds we hold

A.F. Famili et al. (Eds.): IDA 2005, LNCS 3646, pp. 133–144, 2005.

down a key when typing. (3) The variability of typing rhythms may be magnified by the fact that, of course, during the normal use of a computer, different texts are entered, possibly in different languages.

To deal with the instability of typing dynamics, most experiments within keystroke analysis have been limited to samples produced from a unique pre-defined text (e.g. [13,12,5,6,17,3]) or from a small number of different, but still pre-defined texts, (e.g. [14,10,4]). and we refer to [3] and [4] for a thorough descriptions of the various methods found in the literature. However, a large part of the interest in keystroke analysis lies in the possibility to use what stems from the normal use of a computer: the typing rhythms of free text. For example, Intrusion Detection techniques would benefit from such ability, as we discuss at the end of the paper. Unfortunately, when analyzing the typing dynamics of free text the variability of keystroke dynamics is akin to get worse, since the timings of a sequence of keystrokes may be influenced in different ways by the keystrokes occurring before and after the one currently issued. This is even more true if different languages are involved.

Analysis of "true" free text is attempted in [16], where the authors test different methods based on the Euclidean distance and on the mean typing speed and standard deviation of digraphs to measure similarities and differences among typing samples of 31 users, reaching a 23% of correct classification of the typing samples.

In [8] four users are monitored for some weeks during their normal activity on computers, so that thousands of digraphs latencies can be collected. Authors use both statistical analysis and different data mining algorithms on the users' data sets, and are able to reach an almost 60% of correct classification. Authors' approach is improved in [7], both in the outcomes and in the number of users (35) involved, collecting over three months of continuous monitoring more than 5 millions keystrokes.

In [4] we showed experimentally that, on the average, typing samples of different texts provided by the same individual are more similar than typing samples of the same text provided by different individuals. Thus, it was shown that keystroke analysis of free text, though more difficult than keystroke analysis of fixed text, can still be achieved.

In this paper we perform a further step, and show that it is possible to identify a user through the way he types on a keyboard, even when the user is entering free text in a language different from the one used to form his profile. Such ability is important since, for example, more and more people writing text with a computer may use their own language when communicating with others understanding the same language, but use English as the "Lingua Franca" to communicate with the rest of the world.

As we showed in [11], typing dynamics may provide meaningful information to improve the accuracy of an Intrusion Detection System, and may help to limit the number of false alarms. Thus, being able to deal with typing dynamics regardless of the language in use provides a double advantage. On the one hand, a legal user is free of entering text in the language she prefers, without particular risks of raising more alarms: the ability of the system to acknowledge her as the legal owner of the account under observation will not be affected by the use of a different language. On the other hand, intruders would not find any benefit by trying to disguise themselves using a language different from the one normally used by the intruded user: the system will not be fooled by the typing rhythms of a language different from the one of the user's profile.

As far as we know, this is the first work showing that keystroke analysis can be used to ascertain personal identity even when different languages are involved.

2   Computing the *Distance* Between Two Typing Samples

We will use the combination of two measures to evaluate the similarities and differences between the typing rhythms "recorded" in two samples we want to compare. We introduced the first measure, d_1, in [3]. The second measure, d_2, is described here for the first time. The only timing information we use in our experiments is the time elapsed between the depression of the first key and the depression of the second key of each digraph. We call such interval the *duration* of the digraph. If the typed text is sufficiently long, the same digraph may occur more than once. In such case, we report the digraph only once, and we use the mean of the duration of its occurrences.

Given any two typing samples **S1** and **S2**, each one turned into digraphs and sorted with respect to duration of such digraphs, we define the distance between **S1** and **S2**, $d_1(\mathbf{S1},\mathbf{S2})$, as the sum of the absolute values of the distances of each digraph of **S2** w.r.t. the position of the same digraph in **S1**. When computing $d_1(\mathbf{S1},\mathbf{S2})$, digraphs that are not shared between the two samples are simply removed. It is clear that, from the definition of d_1, we may compute the distance between any two typing samples, provided they have some digraphs in common, even if written in different languages. As an example, in the left part of the Table 1 we report typing samples **E1** and **E2** obtained typing, respectively the texts *mathematics* and *sympathetic*. Only digraphs shared between **E1** and **E2** are actually shown. Numbers beside digraphs are their typing speed in milliseconds. The right part of the table illustrates pictorially the computation of the distance between **E1** and **E2**. From the figure it is easy to see that: $d_1(\mathbf{E1},\mathbf{E2}) = 3+0+0+1+4 = 8$.

Given any two typing samples, the maximum distance they may have is when the shared digraphs, sorted by their typing speed, appear in reverse order in one sample w.r.t. the other sample. Hence, if two samples share N digraphs, the maximum distance they can have is given by: $N^2/2$ (if N is even); $(N^2-1)/2$ (if N is odd).

The above value can be used as a normalization factor of the distance between two typing samples sharing N digraphs, dividing their distance by the value of the maximum distance they may have. In this way it is possible to compare the distances of pairs of samples sharing a different number of digraphs: the normalized distance $d_1(\mathbf{S1},\mathbf{S2})$ between any two samples **S1** and **S2** is a real number between 0 and 1. Measure d_1 returns 0 when the digraphs shared by the two samples are exactly in the same order w.r.t. their duration, and returns 1 when the digraphs appear in reverse order ($d_1(\mathbf{S1},\mathbf{S2})$ is also set to 1 if S1 and S2 do not share any digraph). In our example, **E1** and **E2** share 5 digraphs. Thus, their normalized distance is $8/[(5^2-1)/2] = 0.66666$. From now on, in the paper we will always use the normalized version of d_1.

Distance d_1 performed very well to identify users through their typing rhythms on fixed text, and we refer to [3] for a thorough description of the measure and its properties. Readers may have noticed that d_1 completely overlooks any absolute value of the timings associated to the samples. Only the relative positions (which is a consequence of the typing speed) of the digraphs in the two samples are taken into consideration.

Table 1. Computation of the distance for typing samples **E1** and **E2**

E1	E2		E1			E2	
156 **ti** 270		ti	156	$d=3$		at	128
184 **ic** 136		ic	184	$d=0$		ic	136
195 **he** 201		he	195	$d=0$		he	201
197 **at** 128		at	197	$d=1$		th	250
207 **th** 250		th	207	$d=4$		ti	270

However, even the actual typing speed at which digraphs are entered can be useful to discriminate between different individuals. For example, users A and B may both type the word *on* more slowly than the word *of*, but if the average typing speed of the two words are, for user A, say: *on* = 127 millisec.; *of* = 115 millisec.; and for B: *on* = 239 millisec.; *of* = 231 millisec., than A and B can hardly be the same individual.

To take care of such situations, we introduce a second distance measure, d_2, based on the actual typing speeds of digraphs. We could just consider the average typing speed of samples entered by the user, but since we want to combine this new distance with d_1, we prefer to use a measure that considers the average typing speed of single digraphs, and that is normalized in the interval [0..1]. We define $d_2(\mathbf{S1,S2})$ be the number of digraphs shared by **S1** and **S2** whose typing speeds do not differ for more than 30%,[1] divided by the total number of digraphs shared by **S1** and **S2**. For example, in the case of samples **E1** and **E2**, it is easy to check that $d_2(\mathbf{E1,E2}) = 2/5 = 0.4$.

Finally, to combine together distances d_1 and d_2 we simply define $d(\mathbf{S1,S2})$, the distance between any two samples **S1** and **S2** that will be used in all the experiments described in this paper, as: $d(\mathbf{S1,S2}) = d_1(\mathbf{S1,S2}) + d_2(\mathbf{S1,S2})$.

3   Experiments in User Identification and User Authentication

To perform the experiments described in this paper, we asked 31 volunteers to provide two typing samples written in Italian and two typing samples written in English. All the people participating to the experiments are native speakers of Italian, and, though with varying typing skills, all of them are well used to type on normal computer keyboards. Moreover, all volunteers are more or less used to write in English, since they are colleagues and PhD students.

People provided the samples from their computer, through an HTML form with a text area of 780 characters to be filled by the users and submitted to the collecting server. A client side Javascript was used to record the time (in milliseconds) when a key was depressed, together with the ascii value of the key.

---

[1] In order to chose this "30% rule", at the same time trying to limit overfitting, we did the following. When the first five volunteers of our experiments had provided their samples, we performed the identification task described in Section 3, in order to test different percentages: 10%, 20%, 30% and 40%. The best outcomes were reached using a 30% rule, and thus this value is used in all the experiments of this paper. It is of course possible that better outcomes could be reached for some other values (say, 15% or 33%), but we did not bother to find such particular values, that would hardly perform in a similar way on a different set of users.

Volunteers were instructed to enter the samples in the most natural way, more or less as if they were writing an e-mail to someone. They were completely free to choose what to write, and the only limitations were of not typing the same word or phrase repeatedly in order to fill the form, and not to enter the same text in two different samples. People were free to make typos, and to correct them or not, using the backspace key or the mouse, as preferred. People were free to pause in every moment when producing a sample, for whatever reason and as long as they wanted. No sample provided by the volunteers was rejected, for any reason.

In our approach, a user's profile is simply made of a set of typing samples provided by that user. Hence, suppose we are given a set of users' profiles and a new typing sample from one of the users, so that we want to identify who actually provided the sample. If the measure d defined in Section 2 works well, we may expect the computed distance between two samples of the same user to be smaller than the distance between two samples coming from different users. As a consequence, we may expect the mean distance of a new sample X from (the samples in) the profile of user U to be smaller if X has been provided by U than if X has been entered by someone else.

Hence, suppose we have three users A, B and C, with, say, 3 typing samples each one in their profiles (so that, for example, A's profile contains typing samples A_1, A_2 and A_3). A new typing sample X has been provided by one of the users, and we have to decide who entered the sample. We may compute the mean distance (md for short) of X from each user's profile as the mean of the distances of X from each sample in the profile:

$$md(A,X) = (d(A_1,X) + d(A_2,X) + d(A_3,X))/3;$$
$$md(B,X) = (d(B_1,X) + d(B_2,X) + d(B_3,X))/3;$$
$$md(C,X) = (d(C_1,X) + d(C_2,X) + d(C_3,X))/3.$$

Then, we decide that X belongs to the user with the smallest mean distance among the three. This rule has been tested using all possible combinations of Italian and English samples in the profiles of the 31 volunteers, while one of the remaining samples is the one that must be identified. The outcomes of this experiment are reported in the "Identif. errors" columns of Table 2. Outcomes are grouped w.r.t. the number of samples in users' profiles, and are detailed w.r.t. the actual composition of the profiles. Right below each group we report the whole outcomes obtained for the corresponding group. Within brackets we indicate the numerical values that provide the corresponding percentages. For example, suppose there are 3 samples in users' profiles, two Italian samples and one English sample. In this case the system can be tested using the other English sample, for a total of 62 attempted classifications (since both English samples play, in turn, the role of testing sample). In this case all samples are correctly classified, with an identification error of 0.0%. When profiles contain one Italian sample and two English samples, the system makes 2 errors out of 62 attempts, for an identification error of 3.23%. On the whole, when there are 3 samples in users' profile, the system can be tested with 124 samples, and shows an error of 1.61%.

From the outcomes we see that the accuracy of the system increases with the number of samples in users' profiles. When profiles are made of just on sample, almost one out of three testing samples are not correctly classified, with an identification error of 29.57%. But such value quickly shrinks to 6.18% when users' profiles contain 2 samples, and to 1.61% with 3 samples in the profiles.

Quite obviously, when profiles contain exactly one sample in a given language, testing samples are more easily classified correctly if they are written in the same language. We detail more in depth this in the left part of the table. For example, when profiles contain only one Italian sample, we have 12 identification errors out of 62 attempts when trying to classify the other Italian sample, but 49 errors out of 124 attempts when trying to classify the two English samples.

When users' profiles contain two Italian samples, testing samples are all written in English, but less than one out of 15 are not correctly classified, for an identification error of 6.45%. The identification error is larger when users' profiles contain two English samples, and the Italian ones must be classified. Presumably, this is due to the fact that when users are writing in a language different from their own, their particular typing traits tend to remain more hidden. By putting together outcomes of these two identification tasks, we get 12 identification errors out of 124 attempts, that is, less than 10% of mistakes when attempting to identify a typing sample written in a language different from the one used for the two typing samples in the profiles. We get the best outcomes when profiles contains samples written in both languages. In this case it is easier to correctly identify the testing samples, regardless of the language used to write them.

Table 2. Results in user identification and authentication for different compositions of profiles

samples in profiles	Identif. errors	samples in profiles	Identif. errors	k = 0.9 IPR	k = 0.9 FAR	k = 0.8 IPR	k = 0.8 FAR
1 Italian sample	(*Ita.*) 19.35% (12/62)	2 Italian samples	6.45% (4/62)	2.07% (77/3720)	12.9% (8/62)	1.24% (46/3720)	16.13% (10/62)
	(*Eng.*) 39.51% (49/124)	2 English samples	12.9% (8/62)	2.07% (77/3720)	14.51% (9/62)	1.24% (46/3720)	17.74% (11/62)
- - -	- - -	1 Ita. + 1 Eng. sample	4.44% (11/248)	2.17% (323/14880)	5.65% (14/248)	1.44% (214/14880)	8.47% (21/248)
1 English sample	(*Ita.*) 32.26% (40/124)	2 samples	6.18%	2.14%	8.33%	1.37%	11.29%
		2 Ita.+1 Eng. samples	0.0% (0/62)	1.98% (147/7440)	0.0% (0/62)	1.07% (80/7440)	0.0% (0/62)
	(*Eng.*) 14.52% (9/62)	1 Ita.+2 Eng. samples	3.23% 2/62	2.02% (150/7440)	4.83% (3/62)	1.09% (81/7440)	6.45% (4/62)
1 sample	29.57%	3 samples	1.61%	1.99%	2.42%	1.08%	3.23%

The identification rule just described can be used to authenticate users simply by marking the samples with an identity: a new sample X claimed to come from user A is authenticated as belonging to A if $md(A,X)$ is the smallest among all known users. Now, the system can be evaluated w.r.t. two kinds of mistakes it can make: 1) the *Impostor Pass Rate (IPR)*, which is the percentage of cases in which a sample X from an unknown individual is erroneously attributed to one of the users of the system; 2) the *False Alarm Rate (FAR)*, which is the percentage of cases in which a sample belonging to some user is not identified correctly.

From the "Identif. errors" column of Table 2 it is easy to see that our system shows, e.g., an average FAR of 1.61% when users have in their profiles three samples: 2 samples out of 124 authentication attempts produce false alarms. But what about the IPR? If there are 31 users in the system, it is simply $(100/31)\% = 3.23\%$. In fact, an impostor unknown to the system, pretending to be a legal user U, has a chance out of 31 that the sample she provides is closer to U's profile than to any other profile known to the system. We may improve such *basic performance* by observing the following. Suppose again that we have 3 users A, B and C, with 3 samples in their profiles and a new sample X to be classified, so that we compute: $md(A,X)=0.419025$; $md(B,X)=0.420123$; $md(C,X)=0.423223$. As a consequence, X is classified as belonging to user A. However, suppose that the mean of the distances of the samples forming the model of A (denoted by $m(A)$) is:

$$d(A_1,A_2) = 0.312378; d(A_1,A_3) = 0.304381; d(A_2,A_3) = 0.326024.$$
$$m(A) = (0.312378 + 0.304381 + 0.326024)/3 = 0.314261.$$

Then, we may expect another sample of A to have a mean distance from the model of A similar to $m(A)$, which is not the case for X in the example above. Even if X is closer to A than to any other user's profile in the system, it should be rejected.

To deal with such situations, we restate the classification rule as follow: a new sample X claimed to belong to user A is classified as belonging to A if and only if:

1. $md(A,X)$ is the smallest w.r.t. any other user B and
2. $md(A,X)$ is *sufficiently* closer to $m(A)$ than to any other $md(B,X)$ computed by the system. Formally: $md(A,X) < m(A) + |k(md(B,X) - m(A))|$ for any user B, and for some k such that $0 < k \leq 1$.

If a user A meeting the above rules does not exist, X is rejected. Clearly, different values for k provide different trade-offs between IPR and FAR. Smaller values of k will allow to reject more samples from impostors, but could cause more false alarms. For $k = 1$, we fall back to the plain classification rule.

The IPR and FAR columns of Table 2 reports the outcomes of the experiments in user authentication for two different values for k. Again, in brackets are the numerical values from which we computed the corresponding percentage. For example, when profiles contain two samples, the system can be tested 22320 times for attacks from impostors: the profile of each user, in turn, is removed from the system,[2] and the Italian and English samples of that (now unknown) individual are used to attack all users in the systems.[3] Hopefully, the system should reject the attacking samples. Moreover,

---

[2] Otherwise, the attacking sample will be very likely attributed to the attacking user.

[3] Thus, we have (31 attacking users)·(4 attacking samples)·(30 attacked users)·(6 different pair of samples in a user's profile) = 22320 impostors' attacks.

the system is tested 372 times with legal samples claimed to belong to the users who actually provided them.[4]

The outcomes clearly show the effect of the authentication rule in use. For $k = 0.8$ and three samples in users' profiles, the system shows an IPR of 1.08%, that is, about one third of the IPR of the basic classification rule with 31 legal users. The cost is in the worsening of the ability to identify legal users, since the FAR = 3.23%, is now twice that of the basic classification method. Note also that, from the FAR columns we see that English samples appear easier to authenticate correctly using Italian samples in the profiles than vice versa. A result that we already noted in the experiments on identification. On the contrary, the corresponding IPRs do not change in both cases.

4 Discussion and Applications

Beside the outcomes of the previous section, an additional evidence of the fact that personal identity can be ascertained through the analysis of typing rhythms even when different languages are involved can be obtained by considering the mean distances (md for short in the table) reported in the last but one row of Table 3 for the samples gathered in our experiments.[5]

Table 3. Mean distances between different groups of samples

md between the Ita. samples provided by the same individual	md between the Eng. samples provided by the same individual	md between Ita. and Eng. samples provided by the same individual	md between any two Ita. samples provided by different individuals	md between any two Eng. samples provided by different individuals	md between any Ita. and Eng. samples provided by different individuals
md=1.11131 (31) [141]	md=1.12666 (31) [150]	md=1.15948 (124) [123]	md=1.36223 (1860) [140]	md=1.37821 (1860) [139]	md=1.38149 (3720) [122]

From the values in the table we see that typing samples of different text and language provided by the same individual (column 3) are, on the average, more similar than typing samples of different text but same language provided by different individuals (columns 4 and 5). Of course, even samples of different text and languages, coming from different individuals, have a larger distance between each other (column 6). Quite obviously, typing samples provided by the same individual in a certain language (columns 1 and 2), are more similar than typing samples provided by the same individual in different languages (column 3). But the mean of column 3 is only about

---

[4] In fact, we have (31 users)·(6 different pair of samples in a user's profile)·(2 testing samples) = 372 legal connections' attempts.

[5] Again, within round brackets we report the number of distances between samples used to compute the corresponding mean distance. For example, 62 English samples from different individuals allows to compute in (62·61)/2 - 31 = 1860 distances, where 31 is the number of comparisons between the two English samples provided by each volunteer.

4.24% greater than the mean value of column 1. On the contrary, the mean distance of typing samples written in the same language by different individuals (e.g., column 4) is about 16% greater that the mean distance between typing samples provided by the same individual in different languages (column 3). Thus, keystroke analysis involving different languages, though more difficult than when samples are all written in the same language, can still be achieved.

We also note that it is the combination of distances d_1 and d_2 that provides the good outcomes illustrated in the previous section. For example, when d_1 is used alone in the experiments in user identification, we get an identification error of 9.67% with 3 samples in users' profiles, and an error of 15.67% with 2 samples in users' profiles. When d_2 is used alone, the identification error is, respectively, 16.32% and 25.27%. The outcomes in user authentication worsen similarly when using only d_1 or d_2.

The accuracy of our method is related to the number of digraphs shared by the samples under comparison, as we showed in [3]. Samples written in different languages can be compared only if the two languages share some legal digraphs (That is, digraphs that occur in words belonging to the language). Within square brackets in Table 3 we report the average number of digraphs shared between any two samples of the corresponding columns. Samples of different languages (columns 3 and 6) share an average number of digraphs smaller than samples written in the same language. Note that English samples from the same user share a greater number of digraphs than Italian samples from the same user, probably because people tend to use a more restricted set of words when using a language different from their own. For a given length of the samples, the more similar the two languages, the larger the number of digraphs shared by the samples on the average, and the more accurate the distance between them returned by the distance measure used in this paper. Clearly, our method stops being useful when the languages involved (or just the samples under comparison) share a very small number of legal digraphs.

The outcomes of our experiments are among the best found in the literature about keystroke analysis of both free and fixed text, but one may wonder which is their statistical significance. A large amount of research on this issue, explicitly related to biometrics, is available, and we refer to [20] for a comprehensive treatment of the subject (or see [4] for a review of different available techniques). However, J. L. Wayman, Director of the U.S. National Biometric Test Center notes in [20] *our inability to predict even approximately how many tests will be required to have 'statistical confidence' in our results. We currently have no way of accurately estimating how large a test will be necessary to adequately characterize any biometric device in any application, even if error rates are known in advance.* In practice, the number of individuals and samples collected to test a system are not determined by pre-defined confidence intervals, but by the amount of time, budget and resources available [19]. Once test data has been collected and used on the system, it is then possible to estimate the uncertainty of the observed error rates with different methods, but such estimates will have to be taken with a grain of salt, due to the many sources of variability that affect biometric features [15]. We agree with the above view: especially in the case of an unstable biometric such as keystroke dynamics, the only way to evaluate a system is to test it in real conditions, with as many individuals as possible. The number of parameters that may influence keystroke rhythms is so high that any statistical evaluation of the system outcomes will very likely be of limited use.

We conclude this section by proposing possible applications of keystroke analysis of free text.

Intrusion detection. The generation of false alarms is an endemic problem within intrusion detection [2]. In principle, keystroke analysis can be used to notice possible anomalies in the typing dynamics of individuals connected to the system, that may be intruders. However, the inaccuracy of the analysis may itself be the source of false alarms or undetected intrusions. On the contrary, if keystroke analysis is used conjunction with other techniques, it may be useful to mitigate the problem of false alarms, by providing an additional evidence of identity, as we showed in [11]. A scenario where keystroke analysis can be useful even used alone is when it is performed off-line, on accounts monitored in the recent past, to look for possible anomalies that could be simply reported to the system administrator. At the very least, the legal user of the account could be suggested (possibly by an automatic procedure) to change his/her password. In such case, even a relatively high FAR of, say, 2% or 3% would not be a serious problem: false alarms will simply make users changing their passwords a bit more frequently than usual.

Intrusions are often successful because no monitoring procedure is active, and because different form of intrusions are used. Hence, it is important to "attack the attackers" with different and complementary techniques, in order to improve the chances to detect them reliably and quickly. Experiments in this paper show that keystroke analysis can be a valid aid to intrusion detection even when individuals under analysis are using different languages.

User identification over the Internet. The ability to identify users through their typing habits can be used to achieve some form of User and Usage Modeling, in order to be able to offer personalized graphical interfaces, services and advertising to users on their return on a Web site visited previously [18]. Keystroke analysis would in particular be of great help to identify returning users of web sites that provide mailing lists, forums, chat lines and newsgroups access. The use of such services produces a large amount of typed text, whose typing rhythms can be stored and used to identify people on their return to the site, especially when no form of registration is required to visit the site and use its services. User identification over the Internet through the analysis of typing rhythms would find an interesting application also within the investigation of illegal activities that use the web (e.g., newsgroups and anonymous mailing services) to exchange information. For example, the analysis of the typing rhythms coming from different anonymous accounts and web connections could be useful to restrict and direct investigations on a subset of the individuals under observation.

It is worth to note that the above use of keystroke analysis may raise some concern about user's privacy. As a consequence, users should at the very least be informed that some form of monitoring is going on. One may observe that if a typing sample is stored only in term of the digraphs it is made, it would in general be pretty difficult to recover the original text. However, various kind of digit sequences entered, such as phone numbers, numerical passwords and pins, could be easy to recover, thus undermining users' privacy.

5 Conclusion

In this paper we have shown that keystroke analysis of free text can be a useful tool for user identification and authentication even when the typing dynamics stem from the use of different languages. As far as we know, such a situation has never been investigated before in the literature. Our outcomes have been obtained without any particular form of overfitting or tailoring of the system on the given data set, and our technique does not rely on the classical training-testing approach that may require the system to be tuned anew when a different set of users' profiles is involved. We used in our experiments typing samples relatively long, but we believe that, at the current state of the art, keystroke analysis of free text cannot be performed with very short samples: timing analysis on such texts does not provide a sufficient amount of information to discriminate accurately among legal users. On the contrary, if relatively long sample texts are accepted, keystroke analysis can become a valid tool to ascertain personal identity.

The ability to deal with typing samples of different texts and languages improves the possibility of making computers safer and more able to fit personal needs and preferences. We believe keystroke analysis can be a practical tool to help implementing better systems able to ascertain personal identity, and our study represents a contribution to this aim.

Acknowledgements: We want to thank all the volunteers in our Department who contributed to our research.

References

1. J. Ashbourn. *Biometrics: Advanced Identity Verification. The Complete Guide.* Springer, London, GB, 2000.
2. S. Axelsson. The Base-rate Fallacy and the Difficulty of Intrusion Detection. *ACM Transactions on Information and System Security*, 3(3):186–205, 2000.
3. F. Bergadano, D. Gunetti and C. Picardi. User authentication through keystroke dynamics. *ACM Trans. on Information and System Security (ACM TISSEC)*, 5(4):1–31, 2002.
4. F. Bergadano, D. Gunetti and C. Picardi. Identity Verification through Dynamic Keystroke Analysis. *Journal of Intelligent Data Analysis*, 7(5), 2003.
5. S. Bleha, C. Slivinsky, and B. Hussein. Computer-access security systems using keystroke dynamics. *IEEE Trans. on Pattern Analysis and Machine Intelligence*, PAMI-12(12):1217–1222, 1990.
6. M. E. Brown and S. J. Rogers. User identification via keystroke characteristics of typed names using neural networks. *Int. J. of Man-Machine Studies*, 39, pages:999–1014. 1993.
7. P. Dowland, and S. Furnell. A Long-term Trial of Keystroke Profiling using Digraph, Trigraph and Keyword Latencies. In *Proc. of of IFIP/SEC 2004 - 19th Int. Conf. on Information Security*, Toulouse, France. Kluwer, 2004.
8. P. Dowland, S. Furnell and M. Papadaki. Keystroke Analysis as a Method of Advanced User Authentication and Response. In *Proc. of of IFIP/SEC 2002 - 17th Int. Conf. on Information Security*, Cairo, Egypt. Kluwer, 2002.
9. R. O. Duda, P. E. Hart and D. G. Stork. *Pattern Classification.* John Wiley and Sons, 2000.
10. S. Furnell, J. Morrissey, P. Sanders, and C. Stockel. Applications of keystroke analysis for improved login security and continuous user authentication. In *Proc. of the Information and System Security Conf.*, pages 283–294. 1996.

11. D. Gunetti and G. Ruffo. Intrusion Detection through Behavioural Data. In *Proc. of the Third Symp. on Intelligent Data Analysis*, LNCS 1642, Springer-Verlag, 1999.
12. R. Joyce and G. Gupta. newblock User authorization based on keystroke latencies. *Comm. of the ACM*, 33(2):168–176, 1990.
13. J. Leggett and G. Williams. Verifying identity via keystroke characteristics. *Int. J. of Man-Machine Studies*, 28(1):67–76, 1988.
14. J. Leggett, Gl. Williams and M. Usnick. Dynamic identity verification via keystroke characteristics. *Int. J. of Man-Machine Studies*, 35:859–870, 1991.
15. A. J. Mansfield and J. L. Wayman. Best Practices in Testing and Reporting Performances of Biometric Devices. Deliverable of the Biometric Working Group of the CESG Gov. Communication Headquarters of the United Kingdom. National Physical Laboratory, Report CMCS 14/02. *Teddington, United Kingdom*, 2002. Report available at www.cesg.gov.uk/technology/biometrics/media/Best%20Practice.pdf
16. F. Monrose and A. Rubin. Authentication via keystroke dynamics. In *Proc. of the 4th ACM Computer and Communications Security Conf.*, 1997. ACM Press.
17. M. S. Obaidat and B. Sadoun. A simulation evaluation study of neural network techniques to computer user identification. *Information Sciences*, 102:239–258, 1997.
18. M. Perkowitz and O. Etzioni. Adaptive Web Sites: Conceptual Framework and Case Study. *Artificial Intelligence*, 118(1,2):245–275, 2000.
19. J. L. Wayman. Fundamentals of Biometric Authentication Technologies. In *Proc. of CardTech/SecurTech Conference*, 1999.
    Available at www.engr.sjsu.edu/biometrics/nbtccw.pdf
20. J. L. Wayman (Editor). National Biometric Test Center: Collected Works 1997-2000. (Biometric Consortium of the U.S. Government interest group on biometric authentication) *San Jose State University, CA*. 2000.
    Report available at www.engr.sjsu.edu/biometrics/nbtccw.pdf

Combining Bayesian Networks with Higher-Order Data Representations

Elias Gyftodimos and Peter A. Flach

Computer Science Department,
University of Bristol, UK
{E.Gyftodimos, Peter.Flach}@bristol.ac.uk

Abstract. This paper introduces Higher-Order Bayesian Networks, a probabilistic reasoning formalism which combines the efficient reasoning mechanisms of Bayesian Networks with the expressive power of higher-order logics. We discuss how the proposed graphical model is used in order to define a probability distribution semantics over particular families of higher-order terms. We give an example of the application of our method on the Mutagenesis domain, a popular dataset from the Inductive Logic Programming community, showing how we employ probabilistic inference and model learning for the construction of a probabilistic classifier based on Higher-Order Bayesian Networks.

1 Introduction

In the past years there has been increasing interest in methods for learning and reasoning for structured data. Real-world problem domains often cannot be expressed with propositional, "single-table" relational representations. Probabilistic models, popular in propositional domains, have started being proposed for structured domains, giving rise to a new area of research referred to as probabilistic inductive logic programming or probabilistic/statistical relational learning [13,3]. At an earlier stage of our research [5,6] we have introduced Hierarchical Bayesian Networks, which define probability distributions over structured types consisting of nested tuples, lists and sets. In this paper we introduce Higher-Order Bayesian Networks (HOBNs), a probabilistic graphical model formalism which applies methods inspired by Bayesian Networks to complex data structures represented as terms in higher-order logics. We substantially expand our previous research, presenting a detailed formalism for dealing with a much broader family of higher-order terms. The novelty of our approach with respect to existing research on the field consists in the explicit handling of higher-order structures such as sets, rather than emulating these using first-order constructs.

The outline of the paper is as follows. The next section gives a brief overview of the higher-order logic we use for data representation. Section 3 contains the formal definitions of the proposed model, and section 4 defines the derived probability distribution over higher-order terms. Section 5 presents experimental results on a popular real-world benchmark dataset, briefly explaining how inference, model learning and classification is performed under the proposed framework. Section 6 gives a brief overview of existing related approaches. Finally, we summarise our main conclusions and discuss our perspective for further research.

A.F. Famili et al. (Eds.): IDA 2005, LNCS 3646, pp. 145–156, 2005.
© Springer-Verlag Berlin Heidelberg 2005

2 Representation of Individuals

Basic terms[10] are a family of typed higher-order terms that can be used for the intuitive representation of structured individuals. Constructs described by basic terms fall into three categories: The first is called *basic tuples* and includes individuals of the form $(t_1, ..., t_n)$, where each of the t_i is also a basic term. The second, *basic structures*, describes first-order structures such as lists or trees. A basic structure is a term of the form $(C\ t_1 ... t_n)$, where C is a *data constructor* (functor) of arity n and the t_i are basic terms. E.g. in order to define lists as in the LISP programming language, we need two data constructors, "*cons*" and "*nil*" of arity two and zero respectively; then the list with elements 1, 2, and 3 is written as $(cons\ 1\ (cons\ 2\ (cons\ 3\ nil)))$. Note that the enclosing parentheses are often dropped in the previous notation. The third category, *basic abstractions*, is suitable for the description of higher-order structures such as sets and multisets. A set of elements from a domain D can be viewed as a function of type $f : D \rightarrow \{\bot, \top\}$ where $f(x) = \top$ if and only if x is a member of the set. In general, a basic abstraction defines the characteristic function of an individual, and is a term t of the form

$$\lambda x.(if\ x = t_1\ then\ s_1\ else ... else\ if\ x = t_n\ then\ s_n\ else\ s_0)$$

where the t_i and s_i are basic terms and s_0 is a *default term*, i.e. a special term that is the default value of the characteristic function for each particular kind of basic abstractions (zero for multisets, $\bot$ for sets, etc). The set $supp(t) = \{s_1, ..., s_n\}$ is called the *support set* of t. The cardinality of $supp(t)$ will be called the *size* of the abstraction. The formal definition of basic terms also contains a definition of the class of default terms, as well as the definition of a total order on basic terms, so that a basic abstraction can be written in a unique manner with $t_1 < \cdots < t_n$.

Types are used to describe domains of basic terms. A basic tuple type is a Cartesian product $\tau_1 \times \cdots \times \tau_n$ of simpler types to which the elements of a tuple belong. A type of basic structures is defined by a *type constructor*, to which are associated a set of data constructors that define terms of that type. A type constructor takes some arguments, which are also types, and typically relate to the types of the components of the basic structures. E.g. we can define the type $L\ \tau$ of lists of elements from a type τ, with two associated data constructors *cons* and *nil*. Data constructors are also typed; for instance *cons* has a function type, accepting two arguments of types $\tau, L\ \tau$ respectively, and its value is of type $L\ \tau$. This is noted as $cons : \tau \rightarrow L\ \tau \rightarrow L\ \tau$. The type of a basic abstraction is a function type $\alpha \rightarrow \beta$, where α is the type of the argument and β is the type of the value domain of the abstraction. The formal definitions of higher-order types and basic terms can be sought at [10].

Additionally to standard basic terms, in the present paper we will refer to atomic terms and types. An *atomic type* is a domain of constants. It can be seen as a special case of a type constructor, to which all the associated data constructors have arity zero. An example of an atomic type is the type of booleans. An *atomic term* is a member of an atomic type. We also refer to a *canonical form* of types of basic structures: A type τ of basic structures, associated to a set of data constructors $C_i : \beta_{i,1} \rightarrow \cdots \rightarrow \beta_{i,k} \rightarrow \tau$, with $i = 1, ..., m$, has a canonical form $T\ \alpha_1 ... \alpha_n$ where T is a type constructor of arity n and $\bigcup_{i,j} \{b_{i,j}\} \setminus \{\tau\} = \{\alpha_1, ..., \alpha_n\}$. The reason for such a renaming is that we wish to be able to infer which are the types of the arguments of the basic structures in that type, simply by looking at the name of the type.

We will now define a type tree, which is a tree describing a higher-order type.

Definition 1 (Type tree). *The type tree corresponding to a type τ is a tree t such that: (a) If τ is an atomic type, t is a single leaf labelled τ. (b) If τ is a basic tuple type $(\tau_1 \times \cdots \times \tau_n)$, then t has root τ and as children the type trees that correspond to all the τ_i. (c) If τ is a basic structure type with a canonical form $T \alpha_1 \ldots \alpha_n$ then t has root τ and as children the type trees that correspond to all the α_i. (d) If τ is a basic abstraction type $\beta \rightarrow \gamma$, then t has root τ and as children the type trees corresponding to β and $\gamma' = \gamma \backslash \{s_0\}$, where s_0 is the default term associated to the particular type of basic abstractions.*

3 Higher-Order Bayesian Networks: Preliminaries

A standard Bayesian Network is a graphical model that encodes some conditional independence statements on a set of variables. It consists of two parts: the *structural* part, a directed acyclic graph in which nodes stand for random variables and edges for direct conditional dependence between them; and the *probabilistic* part that quantifies the conditional dependence. Higher-Order Bayesian Networks (HOBNs) are a generalisation of standard Bayesian Networks for basic terms. The structural part of an HOBN is a type tree over the domain, and a set of edges between nodes of the type tree that model correlations between them. The probabilistic part contains the parameters that quantify those correlations. We work under the assumption that domains are discrete. We will use a running example in order to explain the definitions and methodology as we proceed. The example domain concerns individuals corresponding to students. Each student has an "intelligence" property and is registered to a set of courses. Each course has a property "difficulty"; each registration has a property "grade"; finally, each student is associated to a property signifying an expected graduation mark. The structural part of the related HOBN is displayed in figure 1(a). The type of the basic terms which correspond to students is $(Int, Exp, ((Diff), Grade) \rightarrow \Omega)$ where $Int = \{i1, i2\}$, $Exp = \{1, 2i, 2ii, 3\}$, $Diff = \{d1, d2, d3\}$, $Grade = \{A, B, C\}$ and $\Omega = \{\top, \bot\}$. The type tree on which the HOBN structure of figure 1 is based represents exactly this type, where the names *Student*, *Regs*, *Reg* and *Course* are introduced for the types which lie in the internal nodes of the tree. We use $\{\tau\}$ as notational sugar for a type of sets over a domain τ. The type $RegMap = \{\top\}$ is introduced as a child of *Reg* in the type tree. Note that since *Regs* is a type of sets, *RegMap* is a trivial case of a domain with only one possible value and could easily be omitted, but it is used in order to demonstrate how general basic abstractions are handled.

We will refer to two distinct types of relationships between nodes of an HOBN. Firstly, relationships in the type tree called *t-relationships*. Secondly, relationships that are formed by the probabilistic dependence links (*p-relationships*). We will make use of everyday terminology for both kinds of relationships, and refer to *parents, ancestors, siblings, nephews* etc. with the obvious meaning. The t-parent and the p-parents of a node are subsequently used for defining the sets of *higher-level parents* (h-parents) and *leaf parents* (l-parents) for each node, which in turn are used for the definition of the

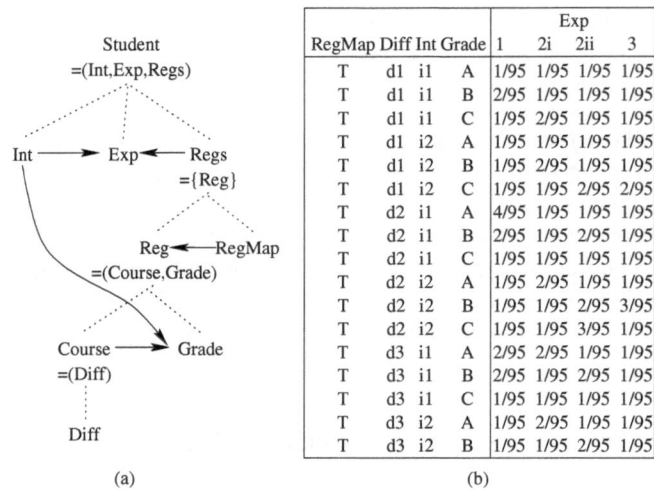

				Exp			
RegMap	Diff	Int	Grade	1	2i	2ii	3
T	d1	i1	A	1/95	1/95	1/95	1/95
T	d1	i1	B	2/95	1/95	1/95	1/95
T	d1	i1	C	1/95	2/95	1/95	1/95
T	d1	i2	A	1/95	1/95	1/95	1/95
T	d1	i2	B	1/95	2/95	1/95	1/95
T	d1	i2	C	1/95	1/95	2/95	2/95
T	d2	i1	A	4/95	1/95	1/95	1/95
T	d2	i1	B	2/95	1/95	2/95	1/95
T	d2	i1	C	1/95	1/95	1/95	1/95
T	d2	i2	A	1/95	2/95	1/95	1/95
T	d2	i2	B	1/95	1/95	2/95	3/95
T	d2	i2	C	1/95	1/95	3/95	1/95
T	d3	i1	A	2/95	2/95	1/95	1/95
T	d3	i1	B	2/95	1/95	2/95	1/95
T	d3	i1	C	1/95	1/95	1/95	1/95
T	d3	i2	A	1/95	2/95	1/95	1/95
T	d3	i2	B	1/95	1/95	2/95	1/95

(a)   (b)

Fig. 1. HOBN structure (a) and part of the parameters (b) for the "student-course" domain

probabilistic part of the model. E.g., in figure 1 node *Student* has three t-children *(Int, Exp, Regs)*; node *Grade* has two p-parents *(Course, Int)*.

We now define the structural part of an HOBN. Essentially this means adding the probabilistic dependence links between nodes of a type tree. Links are not allowed between every possible pair of nodes in the type tree. Intuitively, a p-relationship $A \to B$ is only allowed if B is either a "sibling" or a "nephew" node of A. This relates to some acyclicity condition which must be preserved (as in standard Bayesian Networks) and which will be clarified later.

Definition 2 (HOBN node and HOBN leaf). *An HOBN node associated to a type τ corresponds to a random variable of that type. We will occasionally use the HOBN node to refer to the corresponding variable or its domain; the meaning will always be clear from the context. If τ is atomic, the associated HOBN node is called an HOBN leaf.*

Definition 3 (HOBN structure). *Let τ be a type, and t its corresponding type tree. An HOBN structure T over the type tree t, is a triplet $\langle R, \mathcal{V}, \mathcal{E} \rangle$ where: (a) R is the root of the structure, and corresponds to a random variable of type τ. (b) $\mathcal{V}$ is a set of HOBN structures called the t-children of R. If τ is an atomic type then this set is empty, otherwise it is the set of HOBN structures over the children of τ in t. R is also called the t-parent of each element of $\mathcal{V}$. (c) Let $\mathcal{V}'$ be the set of t-descendants of R. $\mathcal{E} \subset \mathcal{V} \times \mathcal{V}'$ is a set of directed edges between nodes of the HOBN structure. For $(v, v') \in \mathcal{E}$ we say that v is a p-parent of v'.*

There are two additional constraints that a legal HOBN structure must satisfy. One is that if τ is a type $\beta \to \gamma$ corresponding to a basic abstraction, then the set of p-relationships $\mathcal{E}$ for the subtree under τ always contains the p-relationship (γ, β). The second constraint is that the structure needs to be acyclic. This is similar to the acyclicity property in Bayesian Networks, taking into account the propagation of the probabilistic dependence through the type tree. The formal definition of this property follows after the definition of the the notions of higher-level parents and leaf parents, which explain the propagation of the probabilistic dependence introduced by p-relationships down to the leaves of the HOBN structure.

Definition 4 (Higher-level parents and leaf parents). *The higher-level parents (h-parents) of a node t whose t-parent is t', are (i) its p-parents and (ii) the h-parents of t'. The leaf parents (l-parents) of a node are the HOBN leaves that are either (i) its h-parents or (ii) t-descendants of its h-parents.*

In our example, the h-parents of *Grade* are *Int*, *Course* and *RegMap*; the h-parents of *Exp* are *Int* and *Regs*, and its l-parents are *Int*, *RegMap*, *Grade* and *Diff*.
   It is now possible to define the acyclicity property that was mentioned above:

Definition 5 (Acyclic HOBN structure). *An HOBN structure is acyclic if no HOBN leaf in the structure is an l-ancestor of itself.*

This explains why a node may be the p-parent of its t-siblings and t-nephews, but not of itself or its t-children; in such a case some HOBN leaves under that node would become their own l-parents.
   The intended probabilistic semantics in an HOBN structure is that the value of a term corresponding to an HOBN node is independent of the nodes that are neither its t-descendants nor its p-descendants, given its p-parents. The probabilistic part of an HOBN contains the parameters that quantify the respective joint probabilities.

Definition 6. *The probabilistic part related to an HOBN structure T consists of a set of joint probability distributions, defined for some HOBN nodes in T, joint with the respective l-parents of each node. The following probability distributions are contained in the probabilistic part:*

1. *A joint probability distribution over each HOBN leaf X of type α, and its l-parents $X_1,\dots,X_n$, of types $\alpha_1,\dots,\alpha_n$ respectively. For every type $\xi = \alpha_{i_1} \times \cdots \times \alpha_{i_m}$, where all i_j are distinct, the conditional probability over the type α given the type ξ is derived by this joint probability and is denoted by $P_{\alpha|\xi}$.*
2. *For each node X associated to a type of basic abstractions τ, a joint probability distribution over the sizes of the basic abstractions that belong to τ, and the l-parents of X, namely $X_1,\dots,X_n$, of types $\alpha_1,\dots,\alpha_n$ respectively. For every type $\xi = \alpha_{i_1} \times \cdots \times \alpha_{i_m}$, where all i_j are distinct, the conditional probability over the size of the abstractions of type τ given the type ξ is derived by this joint probability and is denoted by $P_{size_\tau|\xi}$.*
3. *For each node X associated to a type of basic structures τ, a joint probability distribution over the domain $cons_\tau$ which contains the data constructors associated to that type, and the l-parents of X, namely $X_1,\dots,X_n$, of types $\alpha_1,\dots,\alpha_n$ respectively. For every type $\xi = \alpha_{i_1} \times \cdots \times \alpha_{i_m}$, where all i_j are distinct, the conditional probability over the size of the abstractions of type τ given the type ξ is derived by this joint probability and is denoted by $P_{cons_\tau|\xi}$.*

As a trivial case when a node has no l-parents the Cartesian product ξ corresponds to the nullary tuple type, and the conditional probability distribution reduces to an unconditional probability distribution.

An additional constraint in the probabilistic part is that the joint probabilities have to be consistent with each other, i.e. yielding the same marginals for the same groups of variables. In practise this happens naturally when the parameters are estimated from data observations, but particular attention is needed when the parameters are manually specified. Figure 1(b) shows the model parameters in our example which are associated to the HOBN node *Exp*. We can now give the definition of an HOBN:

Definition 7. *A Higher Order Bayesian Network is a triplet $\langle T, \Theta, t \rangle$ where t is a type tree, $T = \langle R, \mathcal{V}, \mathcal{E} \rangle$ is an HOBN structure over t and Θ is the probabilistic part related to T.*

4   Probability Distributions over Basic Terms

In this section a probability function over basic terms belonging to a type α is defined, using an HOBN over that type and exploiting the conditional independence assumptions that it introduces. As with standard Bayesian Networks, the joint probability is decomposed to a product of conditional probabilities using the chain rule, and the independent variables are eliminated from each posterior. In HOBNs the probability is defined recursively from the root to the leaves of the type tree, and the probability of a type in each node is expressed using the probabilities of its t-children. Before the formal definition, a short intuitive explanation of the distribution is given, describing how different cases of basic terms are treated.

We are using the notation $P_{\alpha|\xi}(t|c)$ as the probability of a term t of type α, conditional on some context c of type ξ. The definition has two parts: In the first part it is shown how to decompose the probability of t as a product of probabilities of simpler terms $P_{\alpha_i|\xi'}(t_i|c')$, which have types that correspond to the children of α in the type tree. At this stage the conditional part is augmented with additional knowledge on the p-parents of t_i. In the second part of the definition it is shown how in a similar way the probability under the conditional context c is expressed as a product of probabilities under simpler contexts whose types correspond to the t-descendants of ξ.

The first part of the definition has three different cases, according to t being a basic tuple, basic structure or basic abstraction. In the first case where $t = (t_1, \ldots, t_n)$, the probability of the term t is defined as the product of the probabilities of the terms t_i of type $\alpha_i, i = 1, \ldots, n$. The conditional independence statements that are derived from the HOBN structure are employed in order to simplify each posterior. In the second case where $t = C\, t_1 \ldots t_n$, a similar decomposition as with the tuple case is taking place. Each t_i is of a type $\alpha_i, i = 1, \ldots, n$, which is either one of the t-children of α or is α itself, for recursive data types. The probability of t is defined as the product of the probabilities of the t_i conditional on the values of the respective p-parents $P_{\alpha_i|\xi'}(t_i|c, \pi(t_i))$, also multiplied by the probability of the constructor $C, P_{cons_\alpha|\xi}(C|c)$. In the third case, where $t = \lambda x.(if\ x = t_1\ then\ v_1 \ldots\ else\ if\ x = t_n\ then\ v_n\ else\ v_0)$ the result is based on the product of the probabilities of each t_i conditional on the respective v_i.

The second part of the definition assumes that α is an atomic type. The conditional probability is recursively decomposed to a product where the context is replaced by its t-children, until the leaves of the type tree are reached. At each point when deriving the probability $P_{\alpha|\xi}(t|u, c)$, the context is a tuple of terms. The definition introduces a rotation of the tuple elements, by selecting the first element u in the tuple, and creating a new context which is the rest of the tuple with the t-children of the first element attached to the end. This gives a systematic way of reaching the point where the context is a tuple of atomic types. As in the first part of the definition, there are separate cases according to the type of the term in the context that is being decomposed, i.e. the element that is in the first position of the context tuple. If this term is a tuple $u = (u_1, \ldots, u_n)$, then the probability $P_{\alpha|\xi}(t|u, c)$ is simplified to $P_{\alpha|\xi'}(t|c, u_1, \ldots, u_n)$, where α, ξ, ξ' are the appropriate types. If the term is a basic structure $u = C\, u_1 \ldots u_n$, then the probability is defined using the probabilities $P_{\alpha|\xi_i}(t|c, u_i)$. If the term is a basic abstraction $u = \lambda x.(if\ x = u_1\ then\ v_1 \ldots\ else\ if\ x = u_\ell\ then\ v_\ell\ else\ v_0)$, then the probability is defined using the probabilities $P_{\alpha|\xi'}(t|c, u_i, v_i)$. In the course of applying the first part of the definition, a conditional context is introduced for the variable t. This conditional context contains the p-parents of the node corresponding to t. Subsequently t is decomposed to its t-descendants that lie on the leaves of the type tree. The p-parents of t are h-parents of those leaves. Finally, each of those h-parents is replaced by its t-descendants down to the leaves of the type tree. Therefore, after all the decomposition steps, the context is a tuple of atomic types which are the l-parents of τ, so the respective conditional probability is contained in the probabilistic part of the model, least a permutation of the l-parents in the context.

Definition 8. *Let t be a basic term whose type α is associated to a node A of the HOBN. By $\pi(t)$ we denote a basic tuple that contains the values of the p-parents of A. The conditional probability function $P_{\alpha|\xi}(t|c)$, where c is a term of type ξ (initially a nullary tuple, then determined by earlier recursive steps), is defined as follows:*

1 *If α is a non-atomic type, corresponding to either a basic tuple, basic structure or basic abstraction domain, then:*
1.a *If $\alpha = \alpha_1 \times \cdots \times \alpha_n$ and $t = (t_1, \ldots, t_n)$, then*

$$P_{\alpha|\xi}(t|c) = \prod_{i=1}^{n} P_{\alpha_i|\xi'}(t_i|c, \pi(t_i))$$

where ξ' is the type of the tuple $(c, \pi(t_i))$.
1.b *If t is a basic structure, $t = C\, t_1 \ldots t_n$, then*

$$P_{\alpha|\xi}(t|c) = P_{cons_\alpha|\xi}(C|c) \prod_{i=1}^{n} P_{\alpha_i|\xi'}(t_i|c, \pi(t_i))$$

where ξ' is the type of the tuple $(c, \pi(t_i))$ and α_i is the type associated to either the t-child of A corresponding to the type of the term t_i, or to A itself if t_i is of type α.
1.c *If t is a basic abstraction of type $\alpha = \beta \to \gamma$ with $t = \lambda x.(if\ x = t_1\ then\ v_1 \ldots\ else\ if\ x = t_\ell\ then\ v_\ell\ else\ v_0)$, then*

$$P_{\alpha|\xi}(t|c) = \sum_{\ell'=\ell}^{+\infty} \sum_{*} \ell'!\, P_{size_\alpha|\xi}(\ell'|c) \prod_{i=1}^{\ell} \frac{(P_{\beta|\xi'}(t_i|c, v_i) P_{\gamma|\xi}(v_i|c))^{x_i}}{x_i!} \qquad (1)$$

where $\gamma' = \gamma \setminus \{v_0\}$, ξ' is the type of the tuples (c, v_i) and the summation marked with $(*)$ is over all different integer solutions of the equation $x_1 + \cdots + x_\ell = \ell'$ under the constraints $x_i > 0, i = 1, \ldots, \ell$.

2 *If α is either atomic, abstraction size, or the domain of associated data constructors of a type, then $P_{\alpha|\xi}(t|c)$ is defined as follows (the type υ is associated to an HOBN node Y, and c' may trivially be a nullary tuple):*

2.a *If c is a tuple of atomic types or a nullary tuple, then $P_{\alpha|\xi}(t|c)$ is given in the HOBN probabilistic part associated to the HOBN node A.*

2.b *If $c = (u, c'), \xi = \upsilon \times \xi'$, where u is atomic but c' contains non-atomic terms, then*

$$P_{\alpha|\upsilon\times\xi'}(t|u,c') = P_{\alpha|\xi'\times\upsilon}(t|c',u)$$

2.c *If $c = (u, c'), \xi = \upsilon \times \xi'$, where u is a basic tuple $(u_1,\ldots,u_n)$ of type $\upsilon = \upsilon_1 \times \cdots \times \upsilon_n$, then*

$$P_{\alpha|\upsilon\times\xi'}(t|u,c') = P_{\alpha|\xi'\times\upsilon_1\times\cdots\times\upsilon_n}(t|c',u_1,\ldots,u_n)$$

2.d *If $c = (u, c'), \xi = \upsilon \times \xi'$, where u is a basic structure $Cu_1\ldots u_n$, then*

$$P_{\alpha|\upsilon\times\xi'}(t|u,c') = P_{\alpha|\xi'}(t|c')\prod_{i=1}^{n}\frac{P_{\alpha|\xi'\times\upsilon_i\times\upsilon_i'}(t|c',u_i,\pi(u_i))}{P_{\alpha|\xi'\times\upsilon_i'}(t|c',\pi(u_i))}$$

where υ_i' is the type of $\pi(u_i)$ and υ_i is associated to either the t-child of Y which corresponds to the type of the term u_i, or to Y itself if u_i is of type υ.

2.e *If $c = (u, c'), \xi = \upsilon \times \xi'$, where u is a basic abstraction of type $\upsilon = \beta \to \gamma$ with $u = \lambda x.(if\ x = u_1\ then\ v_1 \ldots else\ if\ x = u_\ell\ then\ v_\ell\ else\ v_0)$, then*

$$P_{\alpha|\upsilon\times\xi'}(t|u,c') = P_{\alpha|\xi'}(t|c')\sum_{\ell'=\ell}^{+\infty}\sum_{*}\prod_{i=1}^{\ell}\left(\frac{P_{\alpha|\xi'\times\beta\times\gamma}(t|c',u_i,v_i)}{P_{\alpha|\xi'}(t|c')}\right)^{x_i} \qquad (2)$$

where $\gamma' = \gamma\backslash\{v_0\}$, and the summation marked with $(*)$ is over all different integer solutions of the Diophantine equation $x_1 + \cdots + x_\ell = \ell'$ under the constraints $x_i > 0, i = 1,\ldots,\ell$.

This completes the definition for distributions over basic terms based on an HOBN.

The above definition is introducing some "naive" independence assumptions, which may not hold in the general case, in order to perform the decomposition of the probability into a product. E.g., the elements of a set are assumed to occur independently from each other. Such assumptions are needed for a tractable decomposition, and will render the computed function an approximation of the actual probability. Whether this approximation is meaningful will depend on the domain at hand. Under those assumptions, we can prove the following:

Proposition 1. *The function $P_{\alpha|\xi}$ given in definition 8 is a well-defined probability over basic terms of type α, under the assumption that the conditional independence statements employed hold given the relevant context.*

The proof of the above proposition cannot be presented here due to space constraints. It is derived in a straightforward way by applying in each case the chain rule of conditional probability and Bayes' theorem, and using standard combinatorics. Two additional results which facilitate in practise the computation of the probability in case of basic abstractions are given below:

Proposition 2. *When* $P_{\beta \times \gamma | \xi}(t_i, v_i | c) \ll 1$, *then the expression (1) is approximated by*

$$P_{\alpha | \xi}(t|c) = \ell! P_{size_\alpha | \xi}(\ell | c) \prod_{i=1}^{\ell} P_{\beta | \xi \times \gamma}(t_i | c, v_i) P_{\gamma | \xi}(v_i | c)$$

Proposition 3. *The expression (2) is approximated by:*

$$P_{\alpha | \upsilon \times \xi}(t | u, c) = P_{\alpha | \xi}(t|c) \prod_{i=1}^{\ell} \frac{P_{\alpha | \xi \times \beta \times \gamma}(t | c, u_i, v_i)}{P_{\alpha | \xi}(t|c)}$$

Looking back to our example, assume that we wish to calculate the probability for a student to have an expected final mark equal to 1, given that the value of the intelligence attribute is $i1$ and that the set of registrations is $\{((d3), A), ((d2), B), ((d1), C)\}$. We have (abbreviating the node names to their initials):

$$P_{E|R,I}(1|\{((d3), A), ((d2), B), ((d1), C)\}, i1) =$$

$$= P_{E|I}(1|i1) \frac{P_{E|I,R,RM}(1|i1, ((d3), A), \top)}{P_{E|I}(1|i1)} \cdot$$

$$\cdot \frac{P_{E|I,R,RM}(1|i1, ((d2), B), \top)}{P_{E|I}(1|i1)} \cdot \frac{P_{E|I,R,RM}(1|i1, ((d1), C), \top)}{P_{E|I}(1|i1)} = \ldots =$$

$$= \frac{P_{E|RM,D,I,G}(1|\top, d3, i1, A) P_{E|RM,D,I,G}(1|\top, d2, i1, B) P_{E|RM,D,I,G}(1|\top, d1, i1, C)}{P_{E|I}(1|i1)^2} =$$

$$= \frac{2/6 \cdot 2/6 \cdot 1/5}{(16/47)^2} = 0.1918$$

5   Experimental Evaluation

We present here the results of the application of our method on a real-world dataset, the Mutagenesis domain [15], consisting of a total of 188 instances. Instances in this domain are molecular structures classified as "mutagenic" or "non-mutagenic", and each one is described by four propositional attributes and a set of atoms. The atoms themselves are characterised by three propositional attributes and two sets of "incoming" and "outgoing" chemical bonds. Figure 2 shows an HOBN over that domain. Here *Atom-Map, FBM, and TBM* are associated to the singleton domain $\{\top\}$. The task is to predict whether particular molecules are mutagenic or not.

The first important issue concerning the application of HOBNs on data analysis is probabilistic inference, i.e. the calculation of a probability $P(Q|E)$ where Q and E ("query" and "evidence", respectively), are instantiated subsets of the problem domain. The method we are using in HOBNs is a straightforward extension of an approximate inference method for standard BNs: The graphical model is used as a generator of random instances on the domain. If we generate a sufficiently large number of such instantiations, the relative frequency of the cases where both Q and E hold divided by the relative frequency of the cases where E holds will converge to $P(Q|E)$. Probabilistic classification is a direct application of inference, where for each possible class C_i for a test instance T, the value of $P(C_i|T)$ is computed and the C_i which maximises that expression is the respective predicted value of the class. The second area of interest concerns the construction of an appropriate model given a set of training observations on the domain. In our analysis, we assume that the type tree is an inherent characteristic of the data and therefore is known. What we are learning is the p-relationships between the HOBN nodes, and the parameters of the model. When the HOBN structure is known (i.e. both the type tree and the p-relationships), training the parameters of the model is straightforward when there are no missing values in the data, using the relative frequencies of events in the database in order to estimate the values of the respective joint probabilities. Our approach for structure learning is based on a scoring function for candidate structures. Given such a function we employ a greedy best-first search method, starting from an initial structure (either empty or containing some p-links which are a priori known to be useful) and adding at a time the p-link which optimises the most the scoring function, until no further improvement occurs. The scoring function used in the present experiment was the accuracy of the model on the training data, using cross-validation to avoid over-fitting. The initial structure corresponds to a "naive Bayes" assumption, i.e. that all attributes are mutually independent given the class. This is established by the p-link *Class → Molecule* in the HOBN structure. Figure 2 shows the structure constructed by the learning algorithm for one of the training folds. Table 5 summarises the accuracies obtained in this dataset. The default performance is achieved by always predicting the majority class, and the best reported accuracy comes from [1]. We conclude that HOBNs approach the performance of state-of-the-art algorithms in the domain. It is also important that the learning algorithm employed gives a significant improvement compared to a naive Bayes approach.

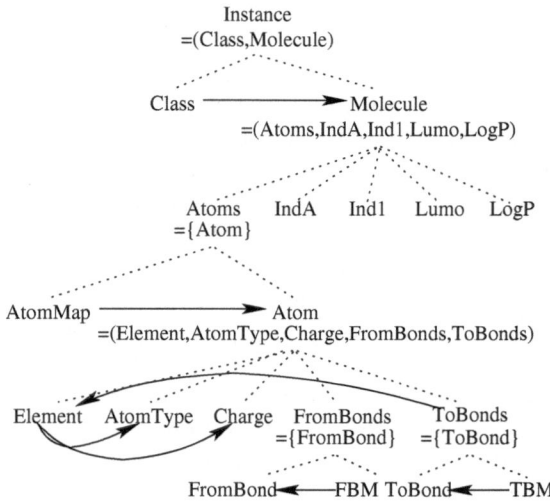

Fig. 2. HOBN structure for the mutagenesis domain

Table 1. Accuracy on the Mutagenesis dataset

Classifier	Accuracy
Default	66.5%
Best	89.9%
Naive HOBN	77.7%
Extended HOBN	88.8%

6   Related Research

Logic-based probabilistic models have been proposed in the past. Stochastic Logic Programs [2,11] use clausal logic, where clauses of a Prolog program are annotated with probabilities in order to define a distribution over the results of Prolog queries. Bayesian Logic Programs [7] are also based on clausal logic, and associate predicates to conditional probability distributions. Probabilistic Relational Models (PRMs) [8], which are a combination of Bayesian Networks and relational models, are also closely related to HOBNs. PRMs are based on an instantiation of a relational schema in order to create a multi-layered Bayesian Network, where layers are derived from different entries in a relational database, and use aggregation functions in order to model conditional probabilities between elements of different tables. Other related first-order probabilistic approaches include Independent Choice Logig (ICL) [12] and the probabilistic logic language PRISM (PRogramming In Statistical Modeling) [14]. Ceci et al. [1] have proposed a naive Bayes classification method for structured data in relational representations. Flach and Lachiche [4,9] have proposed the systems 1BC and 1BC2 which implement naive Bayes classification for first-order domains. To our knowledge, Higher-Order Bayesian Networks is the first attempt to build a general probabilistic reasoning model over higher-order logic-based representations.

7   Conclusions and Further Work

In this paper we have introduced Higher Order Bayesian Networks, a framework for inference and learning from structured data. We demonstrated how inference and learning methods can be employed for probabilistic classification under this framework.

Current results are encouraging for further development of HOBNs. More efficient methods for inference, inspired from methods applied to standard Bayesian Networks need to be researched, since these may boost the computational efficiency of the approach. Gradient methods for model training, generalising on the EM method that we are currently investigating, are likely to improve the performance of the model under the presence of missing values.

Acknowledgements

Part of this work was funded by the EPSRC project *Efficient probabilistic models for inference and learning* in the University of Bristol, and by a Marie Curie fellowship in the Institute for Computer Science at the University of Freiburg. The authors wish to thank the anonymous reviewers for helpful feedback, as well as Kristian Kersting for insightful discussions on earlier versions of this paper.

References

1. Michelangelo Ceci, Annalisa Appice, and Donato Malerba. Mr-sbc: A multi-relational naive bayes classifier. In Nada Lavrač, Dragan Gamberger, Ljupčo Todorovski, and Hendrik Blockeel, editors, *Proceedings of the 7th European Conference on Principles and Practice of Knowledge Discovery in Databases (PKDD 2003)*. Springer, 2003.
2. James Cussens. Parameter estimation in stochastic logic programs. *Machine Learning*, 44(3):245–271, 2001.
3. T. Dietterich, L. Getoor, and K. Murphy, editors. *Working Notes of the ICML-2004 Workshop on Statistical Relational Learning and Connections to Other Fields (SRL-2004)*, 2004.
4. Peter A. Flach and Nicolas Lachiche. 1BC: a first-order Bayesian classifier. In S. Džeroski and P. Flach, editors, *Proceedings of the 9th International Conference on Inductive Logic Programming*, pages 92–103. Springer-Verlag, 1999.
5. Elias Gyftodimos and Peter Flach. Learning hierarchical bayesian networks for human skill modelling. In Jonathan M. Rossiter and Trevor P. Martin, editors, *Proceedings of the 2003 UK workshop on Computational Intelligence (UKCI-2003)*. University of Bristol, 2003.
6. Elias Gyftodimos and Peter Flach. Hierarchical bayesian networks: An approach to classification and learning for structured data. In George A. Vouros and Themis Panayiotopoulos, editors, *Methods and Applications of Artificial Intelligence, Third Hellenic Conference on AI (SETN 2004), Proceedings*. Springer, 2004.
7. Kristian Kersting and Luc De Raedt. Bayesian logic programs. Technical report, Institute for Computer Science, Machine Learning Lab, University of Freiburg, Germany, 2000.
8. Daphne Koller. Probabilistic relational models. In Sašo Džeroski and Peter A. Flach, editors, *Inductive Logic Programming, 9th International Workshop (ILP-99)*. Springer Verlag, 1999.
9. Nicolas Lachiche and Peter A. Flach. 1BC2: a true first-order Bayesian classifier. In S. Matwin and C. Sammut, editors, *Proceedings of the 12th International Conference on Inductive Logic Programming*, pages 133–148. Springer-Verlag, 2002.
10. John W. Lloyd. *Logic for Learning: Learning Comprehensible theories from Structured Data*. Springer, 2003.
11. Stephen Muggleton. Stochastic logic programs. In Luc de Raedt, editor, *Advances in inductive logic programming*, pages 254–264. IOS press, 1996.
12. David Poole. Logic, knowledge representation, and bayesian decision theory. In John W. Lloyd, Verónica Dahl, Ulrich Furbach, Manfred Kerber, Kung-Kiu Lau, Catuscia Palamidessi, Luís Moniz Pereira, Yehoshua Sagiv, and Peter J. Stuckey, editors, *Computational Logic, First International Conference (CL-2000), Proceedings*. Springer, 2000.
13. Luc De Raedt and Kristian Kersting. Probabilistic inductive logic programming. In Shai Ben-David, John Case, and Akira Maruoka, editors, *Proceedings of the 15th International Conference in Algorithmic Learning Theory (ALT 2004)*, pages 19–36. Springer, 2004.
14. Taisuke Sato and Yoshitaka Kameya. Prism: A language for symbolic-statistical modeling. In *Proceedings of the 15th International Joint Conference on Artificial Intelligence (IJCAI 97)*, pages 1330–1339. Morgan Kaufmann, 1997.
15. A. Srinivasan, S. H. Muggleton, M. J. E. Sternberg, and R. D. King. Theories for mutagenicity: a study in first-order and feature-based induction. *Artificial Intelligence*, 85(1-2):277–299, August 1996.

Removing Statistical Biases in
Unsupervised Sequence Learning

Yoav Horman and Gal A. Kaminka

The MAVERICK Group,
Department of Computer Science, Bar-Ilan University, Israel
{hormany, galk}@cs.biu.ac.il

Abstract. Unsupervised sequence learning is important to many applications. A learner is presented with unlabeled sequential data, and must discover sequential patterns that characterize the data. Popular approaches to such learning include statistical analysis and frequency based methods. We empirically compare these approaches and find that both approaches suffer from biases toward shorter sequences, and from inability to group together multiple instances of the same pattern. We provide methods to address these deficiencies, and evaluate them extensively on several synthetic and real-world data sets. The results show significant improvements in all learning methods used.

1 Introduction

Unsupervised sequence learning is an important task in which a learner is presented with unlabeled sequential training data, and must discover sequential patterns that characterize the data. Applications include user modeling [1], anomaly detection [2], data-mining [3] and game analysis [4].

Two popular approaches to this task are frequency-based (*support*) methods (e.g., [3]), and statistical dependence methods (e.g., [5]—see Section 2 for background). We empirically compare these methods on several synthetic and real-world data sets, such as human-computer command-line interactions. The results show that statistical dependence methods typically fare significantly better than frequency-based ones in high-noise settings, but frequency-based methods do better in low-noise settings.

However, more importantly, the comparison uncovers several common deficiencies in the methods we tested. In particular, we show that they are (i) biased in preferring sequences based on their length; and (ii) are unable to differentiate between similar sequences that reflect the same general pattern.

We address these deficiencies. First, we show a length normalization method that leads to significant improvements in *all* sequence learning methods tested (up to 42% improvement in accuracy). We then show how to use clustering to group together similar sequences. We show that previously distinguished sub-patterns are now correctly identified as instances of the same general pattern, leading to additional significant accuracy improvements. The experiments show that the techniques are generic, in that they significantly improve all of the methods initially tested.

A.F. Famili et al. (Eds.): IDA 2005, LNCS 3646, pp. 157–167, 2005.

2 Background and Related Work

In unsupervised learning of sequences, the learner is given example streams, each a sequence $\alpha_1, \alpha_2, \ldots, \alpha_m$ of some atomic *events* (e.g., observed actions). The learner must extract sequential *segments* (also called *patterns*), consecutive subsequences with no intervening events, which characterize the example streams. Of course, not every segment is characteristic of the streams, as some of the segments reflect no more than a random co-occurrence of events. Moreover, each observation stream can contain multiple segments of varying length, possibly interrupted in some fashion. Thus it is important to only extract segments that signify invariants, made up of events that are predictive of each other. We provide the learner with as little assistance as possible.

The literature reports on several unsupervised sequence learning techniques. One major approach learns segments whose frequency (*support*) within the training data is sufficiently high [3]. To filter frequent segments that are due to chance—segments that emerge from the likely frequent co-occurrence of a frequent suffix and a frequent prefix—support-based techniques are usually combined with *confidence*, which measures the likelihood of a segment suffix given its prefix. In such combinations, the extracted segments are those that are more frequent than a user-specified *minimal support* threshold, and more predictive than a user-specified *minimal confidence*.

Another principal approach is statistical dependency detection (*DD*) [5]. DD methods test the statistical dependence of a sequence suffix on its prefix, taking into account the frequency of other prefixes and suffixes. To calculate the rank of a given segment S of size k, a 2×2 contingency table is built for its $(k-1)$-prefix p_r and suffix α_k (Table 2). In the top row, n_1 reflects the count of the segment S. n_2 is the number of times we saw a different event following the same prefix, i.e., $\sum_{i \neq k} count(p_r \alpha_i)$. In the second row, n_3 is the number of segments of length k in which α_k followed a prefix different than p_r ($\sum_{S \neq p_r, |S|=|p_r|} count(S\alpha_k)$). n_4 is the number of segments of length k in which a different prefix was followed by a different suffix ($\sum_{S \neq p_r, |S|=|p_r|} \sum_{i \neq k} count(S\alpha_i)$). The table margins are the sums of their respective rows or columns. A chi-square or G test [6] is then run on the contingency table to calculate how significant is the dependency of α_k on p_r. This is done by comparing the observed frequencies to the expected frequencies under the assumption of independence. DD methods have been utilized in several data analysis applications, including analysis of execution traces [5], time-series analysis [7], and RoboCup soccer coaching [4].

We focus in this paper on support/confidence and DD. However, other technique exist (see [8] for a survey), including statistical methods such as interest ([9]) and con-

Table 1. A statistical contingency table for segment S, composed of a prefix $p_r = \alpha_1, \alpha_2, \ldots, \alpha_{k-1}$ and a suffix α_k. In all cases, $|S| = |p_r|$.

	α_k	$\neg \alpha_k$	
p_r	n_1	n_2	$\sum_i count(p_r \alpha_i)$
$\neg p_r$	n_3	n_4	$\sum_{S \neq p_r, i} count(S\alpha_i)$
	$\sum_S count(S\alpha_k)$	$\sum_{S, i \neq k} count(S\alpha_i)$	

viction ([10]) that are often combined with support. We have successfully applied our methods to these, but do not discuss them here for lack of space. In addition, Likewise, we ignore here methods requiring human expert guidance or other pre-processing.

Also related to our work are [1] and [2], which use clustering techniques to learn the sequential behavior of users. A common theme is that clustering is done based on similarity between sequential pattern instances in the training data. Bauer then uses the resulting clusters as classes for a supervised learning algorithm. Lane and Brodley use the clusters to detect anomalous user behavior. Neither investigates possible statistical biases as we do.

3   A Comparison of Unsupervised Techniques

We conducted extensive experiments using synthetic data, comparing support, confidence, support/confidence and dependency-detection using a G-test. In each run, the techniques above were to discover five different re-occurring *true segments*, uniformly distributed within a file of 5000 streams. We refer to the percentage of the streams that contain true segments as *pattern rate*; thus low pattern rates indicate high levels of noise. The example streams might include additional random events before, after, or within a segment. We controlled *intra-pattern noise rate*: the probability of having noise inserted within a pattern.

In each experiment, each technique reported its best 10 segment candidates, and those were compared to the five true segments. The results were measured as the percentage of true segments that were correctly detected (the recall of the technique, hereinafter denoted *accuracy*). The support/confidence technique requires setting manual thresholds. To allow this method to compete, we set its thresholds such that no true pattern would be pruned prematurely. We refer to this technique as "Support/Confidence Optimal". We have also tested a more realistic version of the algorithm, using fixed min-

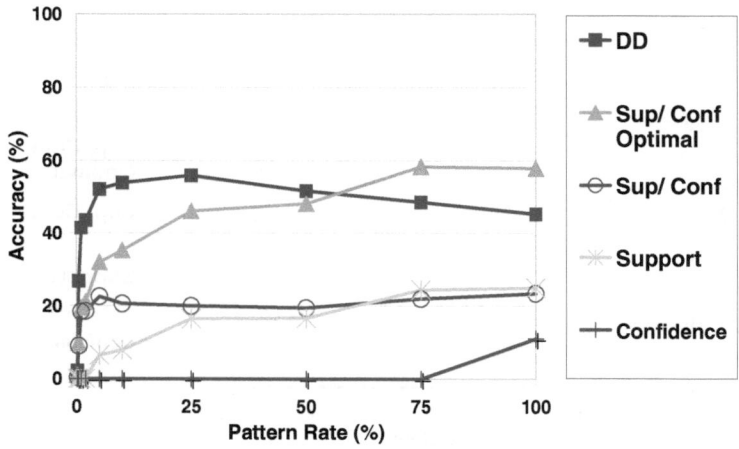

Fig. 1. Accuracy of unsupervised sequence learning methods

imal confidence of 20% ("Support/Confidence"). While the support/confidence method is meant to return all segments satisfying the thresholds, with no ordering, we approximated ranking of resulting segments by their support (after thresholding).

We varied several key parameters in order to verify the consistency of the results. For three different values of alphabet size, denoted T (5, 10 and 26) and three ranges of true-pattern sizes (2–3, 3–5 and 4–7) we have generated data sets of sequences with incrementing values of pattern rate. Intra-pattern-noise was fixed at 0%. For each pattern rate we have conducted 50 different tests. Overall, we ran a total of 4500 tests, each using different 5000 sequences and different sets of 5 true patterns.

The results are depicted in Figure 1. The X-axis measures the pattern rate from 0.2% to 100%. The Y-axis measures the average accuracy of the different techniques over the various combinations of T and pattern size. Each point in the figure reflects the average of *450* different tests. The "Optimal Support/Confidence" technique is denoted "Sup/Conf Optimal", where the standard method, using a fixed minimal confidence value, is denoted "Sup/Conf". The dependency-detection is denoted "DD".

The figure shows that dependency-detection (*DD*) outperforms all other methods for low and medium values of pattern rate. However, the results cross over and support/ confidence optimal outperforms DD at high pattern rates. The standard support/ confidence, as well as the simple support technique, provide relatively poor results. Finally, confidence essentially fails for most pattern rate values.

Figure 2 shows the results for the same experiment, focusing on pattern rates up to 5%. As can be clearly seen, DD quickly achieves relatively high accuracy, at least twice as accurate as the next best technique, support/confidence optimal. A paired one-tailed t-test comparing DD and support/confidence optimal for pattern rates of up to 5% shows that the difference is significant at the 0.05 significance level ($p < 1 \times 10^{-10}$).

For lack of space, we only briefly discuss the effect of the alphabet size T on the accuracy of the different algorithms. All methods achieve better results with greater alphabet sizes. However, when the alphabet is small ($T = 5$), the results of DD are up to 25 times more accurate than other methods, in high noise settings.

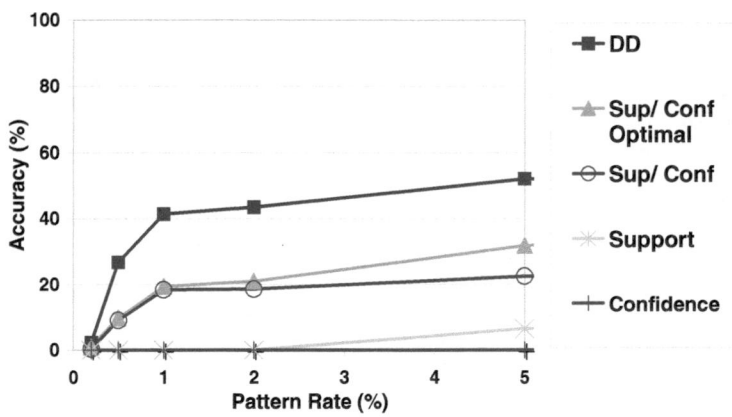

Fig. 2. Accuracy at low pattern rates (high noise)

4   Statistical Biases

We analyzed the results of the different techniques and found that all methods suffer from common limitations: (i) Bias with respect to the length of the segments (Section 4.1); and (ii) inability to group together multiple instances of the same pattern (4.2).

4.1   Removing the Length Bias

The first common limitation of the approaches described above is their bias with respect to the length of the segments. Figure 3 shows the average length of the segments returned by the learning algorithms, in a subset of the tests shown in Figure 1, for two different values of pattern rate (0.5% and 75%), where the length of the true patterns was set to 3–5 (average $\approx$ 4) and alphabet size was fixed at 10. The figure shows that the support algorithm prefers short segments. The optimal support/confidence algorithm behaves similarly, though it improves when pattern rate increases (75%). DD is slightly better, but also prefers shorter sequences at low noise (high pattern rate) settings. In contrast to all of these, Confidence prefers longer sequences.

Different methods have different reasons for these biases. Support-based methods have a bias towards shorter patterns, because there are more of them: Given a target pattern *ABCD*, the pattern *AB* will have all the support of *ABCD* with additional support from (random) appearances of *ABE,ABC,ABG,....* Confidence has a bias towards longer sequences, because their suffix can be easily predicted based on their prefix simply because both are very rare. Finally, DD methods prefer shorter segments at higher pattern rate settings. We found that this is due to DD favoring subsequences of true patterns to the patterns themselves. When pattern rate is high, significant patterns also have significant sub-patterns. Even more: The sub-patterns may have higher significance score because they are based on counts of shorter sequences—which are more frequent as we have seen. This explains the degradation in DD accuracy at higher pattern rates.

In order to overcome the length bias obstacle, we normalize candidate pattern ranks based on their length. The key to this method is to normalize all ranking based on units

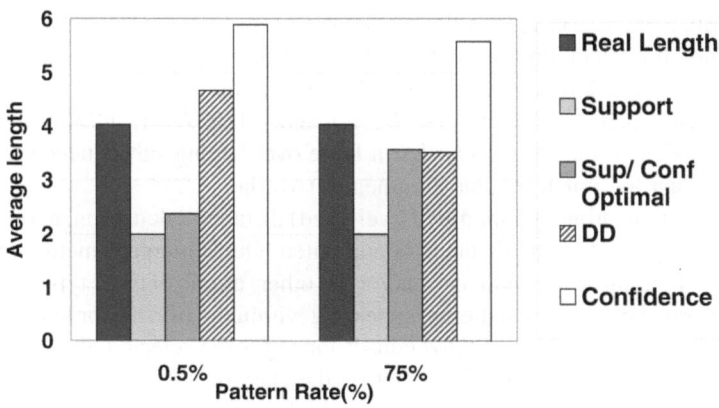

Fig. 3. Average segment length for two pattern rate values

of standard deviation, which can be computed for all lengths. Given the rank distribution for all candidates of length k, let $\bar{R}^k$ be the average rank, and $\hat{S}^k$ be the standard deviation of ranks. Then given a sequence of length k, with rank r, the normalized rank will be $\frac{r - \bar{R}^k}{\hat{S}^k}$. This translates the rank r into units of standard deviation, where positive values are above average. Using the normalized rank, one can compare pattern candidates of different lengths, since all normalized ranks are in units of standard deviation. This method was used in [7] for unsupervised segmentation of observation streams based on statistical dependence tests.

4.2 Generalizing from Similar Patterns

A second limitation we have found in existing methods is inability to generalize patterns, in the sense that sub-segments of frequent or significant patterns are often themselves frequent (or significant). Thus both segment and its subsegment receive high normalized ranks, yet are treated as completely different patterns by the learning methods. For instance, if a pattern $ABCD$ is ranked high, the algorithm is likely to also rank high the *shadow sub-patterns* ABC, BC, etc. Normalizing for length helps in establishing longer patterns as preferable to their shadows, but the shadows might still rank sufficiently high to take the place of other true patterns in the final pattern list.

We focus on a clustering approach, in which we group together pattern variations. We cluster candidates that are within a user-specified threshold of *edit distance* from each other. The procedure goes through the list of candidates top-down. The first candidate is selected as the representative of the first cluster. Each of the following candidates is compared against the representatives of each of the existing groups. If the candidate is within a user-provided edit-distance from a representative of a cluster, it is inserted into the representative's group. Otherwise, a new group is created, and the candidate is chosen as its representative. The result set is composed of all group representatives.

Generally, the edit-distance between two sequences is the minimal number of editing operations (insertion, deletion or replacement of a single event) that should be applied on one sequence in order to turn it into the other. For example, the editing distance between ABC and ACC is 1, as is the editing distance between AC and ABC. A well known method for calculating the edit distance between sequences is *global alignment* [11]. However, our task requires some modifications to the general method. For example, the sequence pairs $\{ABCDE, BCDEF\}$ and $\{ABCD, AEFD\}$ have an edit-distance of 2, though the former pair has a large overlapping subsequence ($BCDE$), and the latter pair has much smaller (fragmented) overlap $A??D$.

We use a combination of a modified (weighted) distance calculation, and heuristics which come to bear after the distance is computed. Our alignment method classifies each event (belonging to one sequence and/or the other) as one of three types: appearing before an overlap between the patterns, appearing within the overlap, or appearing after the overlap. It then assigns a *weighted* edit-distance for the selected alignment, where the edit operations have weights that differ by the class of the events they operate on. Edit operations within the overlap are given a high weight (called *mismatch weight*). Edit operations on events appearing before or after the overlap are given a low weight (*edge weight*). In our experiments we have used an infinite mismatch weight, meaning we did not allow any mismatch within the overlapping segment. However, both weight values are clearly domain-dependent.

In order to avoid false alignments where the overlapping segment is not a significant part of the overall unification, we set a minimal threshold upon the length of the overlapping segment. This threshold is set both as an absolute value and as a portion of the overall unification's length.

5   Experiments

To evaluate the techniques we presented, we conducted extensive experiments on synthetic (Section 5.1) and real data (5.2).

5.1   Synthetic Data Experiments

We repeated our experiments from Section 3, this time with the modified techniques. Figure 4 and Table 2 show the accuracy achieved at different pattern rates, paralleling Figures 1 and 2, respectively. Figure 4 shows all results, while Table 2 focuses on low pattern rates (high noise). Every point (table entry) is the average of *450* different tests, contrasting standard, normalized (marked *N*) and normalized-clustered (*NC*) versions of DD, Support, and Optimal Support/Confidence (marked simply as Sup/Conf).

The results show that length normalization improves *all* tested algorithms. For instance, the support technique has completely failed to detect true segments for a pattern rate of 1%, while its normalized version has achieved accuracy of 39% at this rate. Clustering the normalized results improved the results further, by notable margins.

The improvements derived from normalizing and clustering the results both proved to be statistically significant for all learning techniques. For instance, a paired one-tailed t-test shows that the normalized version of DD is significantly better than the standard version ($p < 1 \times 10^{-10}$) and that the clustered-normalized version of DD significantly outperforms the normalized version ($p < 1 \times 10^{-10}$).

Table 2. Accuracy at low pattern rates

Pattern Rate (%)	0.2	0.5	1	2	5	10
DD		2.7	26.8	41.1	45.9	51.9	54.4
N. DD	**13.6**	28.1	42.3	48.2	52.8	56.0
NC. DD	13.4	**28.3**	**45.6**	**53.0**	**60.8**	**66.5**
Sup/ Conf		1.3	10.2	20.3	19.6	33.0	34.9
N. Sup/ Conf	**16.6**	30.8	44.7	49.8	59.8	63.7
NC. Sup/ Conf	16.4	**32.6**	**51.0**	**54.9**	**70.4**	**73.0**
Sup		0.0	0.0	0.0	0.0	8.2	7.5
N. Sup	**16.6**	**29.9**	39.3	41.2	48.2	50
NC. Sup	16.4	29.7	**47.0**	**47.1**	**58.4**	**66.0**
Conf		0.0	0.0	0.0	0.0	0.0	0.0
N. Conf		0.0	1.0	13.4	21.3	31.0	37.1
NC. Conf		0.0	**1.2**	**14.4**	**22**	**34**	**42.0**

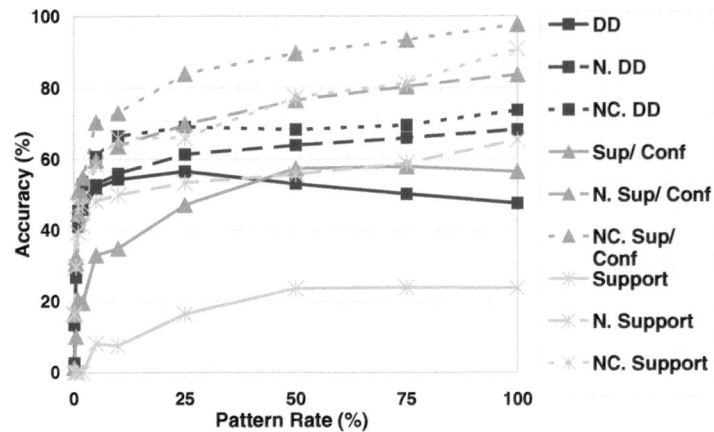

Fig. 4. Modified: All Pattern Rates, average results

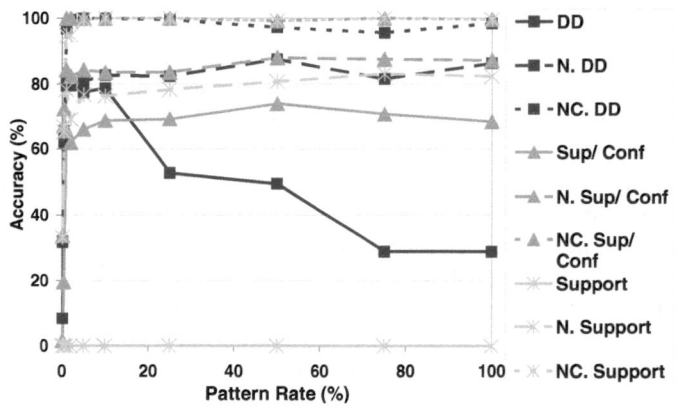

Fig. 5. Modified: $T = 26$, Patterns of size 3–5

The improvements are such that after normalization and clustering, the simple support technique outperforms the standard DD method for all pattern rate values, including 0.2%-0.5%. This is where standard DD performs significantly better than the other standard techniques. Indeed, after length-based standardization and clustering, DD may no longer be superior over the support/confidence approach.

Figure 5 shows the results from one specific setting, where both normalizing and normalizing-clustering proved particularly effective. Each point in the figure represents the average of 50 different tests, with an alphabet size 26, and true patterns composed of 3–5 events. In the figure, the normalized version of the support technique has achieved accuracy of 78% for a pattern rate of 1%, comparing to 0% accuracy of the standard version. The normalized clustered versions of all algorithms have achieved more than 95% accuracy for a pattern rate as low as 1%, where the accuracy of the standard techniques was 0% for support, 66% for support/confidence and 82% for DD.

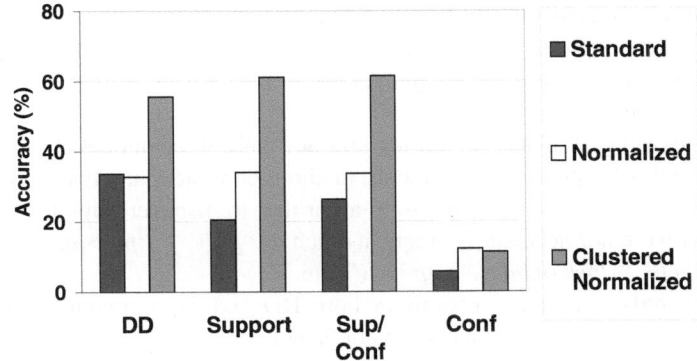

Fig. 6. Accuracy improvements: Orwell's 1984

We have also evaluated the effects of intra-pattern noise on the quality of results. In general, for all alphabet and pattern sizes, we have found that the Normalized Clustered versions offer consistent improvements to accuracy of Normalized methods, in the presence of up to 25% intra-pattern noise.

We evaluated our techniques with an additional dataset. We used the text of George Orwell's *1984* to test our modified techniques on data that was both more realistic, yet still allowed for controlled experiments. In one experiment, we changed the original text by introducing noise within the words and between them. For instance, the first sentence in the book - "It was a bright cold day in April" was replaced by "ItoH7l4H XywOct8M (...9 more noisy words) 6jOwas2x imfG8e1x (...2 more noisy words) nBaor1oL iWtHhTEq **bright**cT x**cold**Vuv vf**day**1Ap BsQG9pyK 8Nxf**in**XR 8TGmxcXO E1IenU2Q **April**ulxL". We inserted only fixed 8-character sequences, such that each actual word that is shorter than 8 characters was padded with noise, and words longer than 8 characters were cut. We set pattern rate to 40% by inserting 6 noisy streams, for each 4 containing actual words. Intra pattern noise was set at 10%. We then counted how many of the top 100 candidates returned by each technique are actual words appearing in the book. We hoped to find as many actual words in the results set as possible. The results, reflecting the average accuracy over the first 8 chapters of the book, are shown in Figure 6. Similar improvement results were achieved for other settings of pattern rate and intra pattern noise.

The results show that for each of the presented techniques the clustered normalized versions have significantly outperformed the standard versions, increasing accuracy by up to 41% for the support algorithm. The normalized versions have typically outperformed the standard versions, except for the case of DD, where the normalized results contained various sequences that reflected the same words (see Section 4.2), and were then significantly improved by our clustering approach. Note also that among the standard techniques, DD has once again outperformed the other methods.

5.2 Real World Experiments

We conducted real-world experiments on UNIX command line sequences. We utilized 9 data sets of UNIX command line histories, collected for 8 different users at Purdue university over the course of 2 years [12]. In this case, we do not know in advance what true patterns were included in the data, thus quantitative evaluation of accuracy is not possible. However, we hoped to qualitatively contrast the pattern candidates generated by the different methods.

The results suggest again that among non-normalized techniques, DD is superior. While the results of support and confidence methods consisted mainly of different variations of *ls*, *cd* and *vi*, DD was the only algorithm to discover obviously-sequential patterns (that were not necessarily frequent) such as "*g++ -g <file>;a.out*", "*| more*", "*ls -al*", "*ps -ef*", "*xlock -mode*", "*pgp -kvw*", etc.

The clustered-normalized versions of both DD and support/confidence detected more complex user patterns, which were not detected by the standard techniques. The results clearly show the ability of the improved techniques to discover valuable sequential patterns, which characterize interesting user behavior, and are overlooked by the standard methods. Among these sequential patterns are:

1. *ps -aux | grep <process>; kill -9*—a user looking for a certain process id to kill.
2. *tar <3 args>; cd; uuencode <2 args> > <file>; mailx*—a user packaging a directory tree, encoding it to a file, and sending it by mail.
3. *compress <arg>;quota;compress <arg>; quota*—a user trying to overcome quota problems by compressing files.
4. *latex <arg>; dvips <arg>; ghostview*—a latex *write → compile → view* cycle.
5. *vi <arg>; gcc <arg>; a.out; vi <arg>; gcc*—an *edit → compile → run* cycle.

6   Conclusions and Future Work

This paper tackles the problem of unsupervised sequence learning. The challenge is addressed by improving sequence learning algorithms, which extract meaningful patterns of sequential behavior from example streams. We empirically compared these algorithms, to determine their relative strengths. Based on the comparison, we noted several common deficiencies in all tested algorithms: All are susceptible to a bias in preferring pattern candidates based on length; and all fail to generalize patterns, often taking a high-ranked pattern candidate as distinct from its shorter sub-patterns.

We use a normalization method to effectively neutralize the length bias in all learning methods tested, by normalizing the frequency/significance rankings produced by the learning methods. Use of this method had improved accuracy by up to 42% in testing on synthetic data. We then use a clustering approach, based on a modified weighted edit-distance measure, to group together all patterns that are closely related. The use of clustering in addition to normalization had further improved accuracy by up to 22% in some cases. We also show that the techniques are robust to noise in and out of the patterns. Finally, the improved methods were run on two additional sets of data: sequences from Orwell's 1984, and UNIX real-world command-line data. The methods successfully detected many interesting patterns in both.

A weakness with the methods that we presented is their use with very large databases. For instance, normalization requires repeatedly counting all the patterns in the database, and would therefore be inefficient for large data-mining applications. However, the techniques we presented are well suited for typical agent-observation data (such as RoboCup soccer logs or UNIX command-line data). We plan to consider large data-mining applications in our future work.

References

1. Bauer, M.: From interaction data to plan libraries: A clustering approach. In: IJCAI-99. Volume 2., Stockholm, Sweden, Morgan-Kaufman Publishers, Inc (1999) 962–967
2. Lane, T., Brodley, C.E.: Temporal sequence learning and data reduction for anomaly detection. ACM Transactions on Information and System Security **2** (1999) 295–331
3. Agrawal, R., Srikant, R.: Mining sequential patterns. In Yu, P.S., Chen, A.S.P., eds.: Eleventh International Conference on Data Engineering, Taipei, Taiwan, IEEE Computer Society Press (1995) 3–14
4. Kaminka, G.A., Fidanboylu, M., Chang, A., Veloso, M.: Learning the sequential behavior of teams from observations. In: Proceedings of the 2002 RoboCup Symposium. (2002)
5. Howe, A.E., Cohen, P.R.: Understanding planner behavior. AIJ **76** (1995) 125–166
6. Sokal, R.R., Rohlf, F.J.: Biometry: The Principles and Practice of Statistics in Biological Research. W.H. Freeman and Co.,, New York (1981)
7. Cohen, P., Adams, N.: An algorithm for segmenting categorical time series into meaningful episodes. Lecture Notes in Computer Science **2189** (2001)
8. Tan, P.N., Kumar, V., Srivastava, J.: Selecting the right interestingness measure for association patterns. In: Proceedings of the eighth ACM SIGKDD international conference on Knowledge discovery and data mining, ACM Press (2002) 32–41
9. Silverstein, C., Brin, S., Motwani, R.: Beyond market baskets: Generalizing association rules to dependence rules. Data Mining and Knowledge Discovery **2** (1998) 39–68
10. Brin, S., Motwani, R., Ullman, J.D., Tsur, S.: Dynamic itemset counting and implication rules for market basket data. In Peckham, J., ed.: SIGMOD 1997, Proceedings ACM SIGMOD International Conference on Management of Data, May 13-15, 1997, ACM Press (1997) 255–264
11. Sellers, P.: The theory and computation of evolutionary distances: pattern recognition. Journal of Algorithms (1980) 1:359–373
12. Hettich, S., Bay, S.D.: The uci kdd archive. http://kdd.ics.uci.edu/ (1999)

Learning from Ambiguously Labeled Examples

Eyke Hüllermeier and Jürgen Beringer

Fakultät für Informatik,
Otto-von-Guericke-Universität Magdeburg, Germany
eyke.huellermeier@iti.cs.uni-magdeburg.de

Abstract. Inducing a classification function from a set of examples in
the form of labeled instances is a standard problem in supervised machine
learning. In this paper, we are concerned with *ambiguous label classification* (ALC), an extension of this setting in which several candidate labels
may be assigned to a single example. By extending three concrete classification methods to the ALC setting and evaluating their performance
on benchmark data sets, we show that appropriately designed learning
algorithms can successfully exploit the information contained in ambiguously labeled examples. Our results indicate that the fundamental idea
of the extended methods, namely to disambiguate the label information
by means of the inductive bias underlying (heuristic) machine learning
methods, works well in practice.

1 Introduction

One of the standard problems in (supervised) machine learning is inducing a
classification function from a set of training data. The latter usually consists of
a set of *labeled examples*, i.e., a set of objects (instances) whose correct classification is known. Over the last years, however, several variants of the standard
classification setting have been considered. For example, in *multi-label classification* a single object can have several labels (belong to several classes), that is,
the labels (classes) are not mutually exclusive [14]. In *semi-supervised learning*,
only a part of the objects in the training set is labeled [1]. In *multiple-instance
learning*, a positive or negative label is assigned to a so-called *bag* rather than to
an object directly [7]. A bag, which is a collection of several instances, is labeled
positive iff if contains at least one positive example. Given a set of labeled bags,
the task is to induce a model that will label unseen bags and instances correctly.

In this paper, we are concerned with another extension of the standard classification setting that has recently been introduced in [11,13], and that we shall
subsequently refer to as *ambiguous label classification* (ALC). In this setting, an
example might be labeled in a non-unique way by a *subset* of classes, just like in
multi-label classification. In ALC, however, the existence of a (unique) *correct*
classification is assumed, and the labels are simply considered as *candidates*.

In [11,13], the authors rely on probabilistic methods in order to learn a classifier in the ALC setting. The approach presented in this paper can be seen as
an alternative strategy which is more in line with standard (heuristic) machine

A.F. Famili et al. (Eds.): IDA 2005, LNCS 3646, pp. 168–179, 2005.

learning methods. Our idea is to exploit the inductive bias underlying these methods in order to disambiguate label information. This idea, as well as the relation between the two approaches, is discussed in more detail in Section 3. Before, the problem of ALC is introduced in a more formal way (Section 2). In Section 4, three concrete methods for ALC are proposed, namely extensions of nearest neighbor classification, decision tree learning, and rule induction. Experimental results are finally presented in Section 5.

2   Ambiguous Label Classification

Let $\mathcal{X}$ denote an instance space, where an instance corresponds to the attribute–value description x of an object: $\mathcal{X} = X_1 \times X_2 \times \ldots \times X_\ell$, with X_ι the domain of the ι-th attribute. Thus, an instance is represented as a vector $x = (x^1 \ldots x^\ell) \in \mathcal{X}$. Moreover, let $\mathcal{L} = \{\lambda_1 \ldots \lambda_m\}$ be a set of labels (classes). Training data shall be given in the form of a set $\mathcal{D}$ of examples (x_ι, L_{x_ι}), $\iota = 1 \ldots n$, where $x_\iota = (x_\iota^1 \ldots x_\iota^\ell) \in \mathcal{X}$ and $L_{x_\iota} \subseteq \mathcal{L}$ is a set of candidate labels associated with instance x_ι. L_{x_ι} is assumed to contain the true label λ_{x_ι}, and x_ι is called an *ambiguous* example if $|L_{x_\iota}| > 1$. Note that this includes the special case of a completely unknown label ($L_x = \mathcal{L}$), as considered in semi-supervised learning. Here, however, we usually have the case in mind where $1 \leq |L_x| < |\mathcal{L}|$. For example, in molecular biology the functional category of a protein is often not exactly known, even though some alternatives can definitely be excluded [2].

The learning task is to select, on the basis of $\mathcal{D}$, an optimal model (hypothesis) $h : \mathcal{X} \to \mathcal{L}$ from a hypothesis space $\mathcal{H}$. Such a model assigns a (unique) label $\lambda = h(x)$ to any instance $x \in \mathcal{X}$. Optimality usually refers to *predictive accuracy*, i.e., an optimal model is one that minimizes the expected loss (risk) with respect to a given loss function $\mathcal{L} \times \mathcal{L} \to \mathbb{R}$.

3   Learning from Ambiguous Examples

Ambiguous data may comprise important information. In fact, the benefit of this information might be especially high if it is considered, not as an isolated piece of knowledge, but in conjunction with the other data and the model assumptions underlying the hypothesis space $\mathcal{H}$. To illustrate this important point, consider a simple example in which the true label λ_{x_ι} of an instance x_ι is known to be either λ_1 or λ_2. Moreover, we seek to fit a classification tree to the data, which basically amounts to assuming that $\mathcal{X}$ can be partitioned by axis-parallel decision boundaries. Now, by setting $\lambda_{x_\iota} = \lambda_2$ we might find a very simple classification tree for the complete data, while $\lambda_{x_\iota} = \lambda_1$ requires a comparatively complex model (see Fig.1). Relying on the simplicity heuristic underlying most machine learning methods [8], this finding clearly suggests that $\lambda_{x_\iota} = \lambda_2$. Thus, looking at the original information $\lambda_{x_\iota} \in \{\lambda_1, \lambda_2\}$ with a view that is "biased" by the model assumptions, the benefit of this information has highly increased. As can be seen, the inductive bias underlying the learning process can help to *disambiguate* the label information given. This suggests that ambiguous label information might

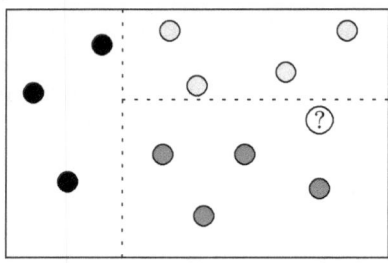

Fig. 1. Classification problem with three labels: black (λ_1), grey (λ_2), light (λ_3). The instance with a question mark is either black or grey. Assigning label grey allows one to fit a very simple decision tree (as represented by the axis-parallel decision boundaries). Note that this hypothetical labeling also provides important information on the decision boundary between the grey and light class.

indeed be useful and, in particular, that it might be easier for learning methods with a strong inductive bias to exploit such information than for methods with a weak bias. Both these conjectures will be supported by our experimental results in Section 5.

The above example has shown that candidate labels can appear more or less likely against the background of the underlying model assumptions. In fact, the insight that fitting a model to the data might change the likelihood of candidate labels can be formalized more rigorously in a probabilistic context. Assuming a parameterized model M_θ, the goal can roughly be stated as finding the parameter

$$\theta^* = \arg \max_\theta \prod_{i=1}^{n} \Pr(\lambda_{x_i} \in L_{x_i} \mid x_i, \theta).$$

This approach gives rise to an EM (expectation-maximization) approach in which model adaptation and modification of label information are performed alternately: Starting with a uniform distribution over each label set L_{x_i}, an optimal parameter θ^* is determined. Using this parameter resp. the associated model M_{θ^*}, the probabilities of the labels $\lambda \in L_{x_i}$ are then re-estimated. This process of estimating parameters and adjusting probabilities is iterated until convergence is eventually achieved [11,13].

On the one hand, this approach is rather elegant and first empirical evidence has been gathered for its practical effectiveness [11,13]. On the other hand, the assumption of a parameterized model basically restricts its applicability to statistical classification methods. Moreover, model optimization by means of EM can of course become quite costly from a computational point of view. Our idea of disambiguating label information by implementing a simplicity bias can be seen as an alternative strategy. As heuristic machine learning in general, this approach is of course theoretically not as well-founded as probabilistic methods. Still, heuristic methods have been shown to be often more effective and efficient in practical applications.

Unfortunately, standard classification methods generally cannot exploit the information provided by ambiguous data, simply because they cannot handle such data. This is one motivation underlying the development of methods for ALC (as will be done in Section 4). Note that a straightforward strategy for realizing ALC is a reduction to standard classification: Let the class of *selections*, $\mathcal{F}(\mathcal{D})$, of a set $\mathcal{D}$ of ambiguous data be given by the class of standard samples

$$\mathcal{S} = \{(x_1, \alpha_{x_1}), (x_2, \alpha_{x_2}), \dots, (x_n, \alpha_{x_n})\} \tag{1}$$

such that $\alpha_{x_i} \in L_{x_i}$ for all $1 \leq i \leq n$. In principle, a standard learning method could be applied to all samples $\mathcal{S} \in \mathcal{F}(\mathcal{D})$, and an apparently most favorable model could be selected among the models thus obtained. However, since the number of selections, $|\mathcal{F}(\mathcal{D})| = \prod_{i=1}^{n} |L_{x_i}|$, will usually be huge, this strategy is of course not practicable.

4 Methods for ALC

In this section, we present three relatively simple extensions of standard learning algorithms to the ALC setting, namely k-nearest neighbor classification, decision tree learning, and rule induction.

4.1 Nearest Neighbor Classification

In k-nearest neighbor (k-NN) classification [6], the label $\lambda_{x_0}^{est}$ hypothetically assigned to a query x_0 is given by the label that is most frequent among x_0's k nearest neighbors, where nearness is measured in terms of a similarity or distance function. In weighted k-NN, the neighbors are moreover weighted by their distance:

$$\lambda_{x_0}^{est} \stackrel{\mathrm{df}}{=} \arg\max_{\lambda \in \mathcal{L}} \sum_{i=1}^{k} \omega_i \, \mathbb{I}(\lambda = \lambda_{x_i}), \tag{2}$$

where x_i is the i-th nearest neighbor; λ_{x_i} and ω_i are, respectively, the label and the weight of x_i, and $\mathbb{I}(\cdot)$ is the standard $\{\mathrm{true}, \mathrm{false}\} \rightarrow \{0, 1\}$ mapping. A simple definition of the weights is $\omega_i = 1 - d_i \cdot (\sum_{j=1}^{k} d_j)^{-1}$, where the d_i are the corresponding distances.

Now, a relatively straightforward generalization of (2) to the ALC setting is to replace $\mathbb{I}(\lambda = \lambda_{x_i})$ by $\mathbb{I}(\lambda \in L_{x_i})$:

$$\lambda_{x_0}^{est} \stackrel{\mathrm{df}}{=} \arg\max_{\lambda \in \mathcal{L}} \sum_{i=1}^{k} \omega_i \, \mathbb{I}(\lambda \in L_{x_i}). \tag{3}$$

Thus, a neighbor x_i is allowed not one single vote only, but rather one vote for each its associated labels. If the maximum in (3) is not unique, one among the labels with highest score is simply chosen at random.[1]

---

[1] A reasonable alternative is to choose the prevalent class in the complete training set.

4.2   Decision Tree Induction

Another standard learning method whose extension to the ALC setting might be of interest, is decision tree induction [15]. Its basic strategy of partitioning the data in a recursive manner can of course be maintained for ALC. The main modification rather concerns the splitting measure. In fact, standard measures of the (im)purity of a set of examples, such as entropy, cannot be used, since these measures are well-defined only for a probability distribution over the label set.

As an extended measure of (im)purity, we propose the *potential entropy* of a set of examples $\mathcal{D}$, defined by

$$E^*(\mathcal{D}) \stackrel{\text{df}}{=} \min_{\mathcal{S} \in \mathcal{F}(\mathcal{D})} E(\mathcal{S}), \tag{4}$$

where $\mathcal{F}(\mathcal{D})$ is the set of selections (1) and $E(\mathcal{S})$ denotes the standard entropy: $E(\mathcal{S}) \stackrel{\text{df}}{=} - \sum_{i=1}^m p_i \log_2(p_i)$, with p_i the proportion of elements in $\mathcal{S}$ labeled by λ_i. As can be seen, (4) is the standard entropy obtained for the most favorable instantiation of the ALC-examples (x_i, L_{x_i}). It corresponds to the "true" entropy that would have been derived if this instantiation was compatible with the ultimate decision tree. Taking this optimistic attitude is justified since the tree is indeed hopefully constructed in an optimal manner.

Of course, computing the potential entropy comes down to solving a combinatorial optimization problem and becomes intractable for large samples. Therefore, we suggest the following heuristic approximation of (4):

$$E^+(\mathcal{D}) \stackrel{\text{df}}{=} E(\mathcal{S}^*), \tag{5}$$

where the selection $\mathcal{S}^*$ is defined as follows: Let q_i be the frequency of the label λ_i in the set of examples $\mathcal{D}$, i.e. the number of examples (x_j, L_{x_j}) such that $\lambda_i \in L_{x_j}$. The labels λ_i are first put in a (total) "preference" order according to their frequency: λ_i is preferred to λ_j if $q_i > q_j$ (ties are broken by coin flipping). Then, the most preferred label $\lambda_i \in L_{x_i}$ is chosen for each example x_i. Clearly, the idea underlying this selection is to make the distribution of labels as skewed (non-uniform) as possible, as distributions of this type are favored by the entropy measure. We found that the measure (5) yields very good results in practice and compares favorably with alternative extensions of splitting measures [12].

With regard to the stopping condition of the recursive partitioning scheme, note that a further splitting of a (sub)set of examples $\mathcal{D}$ is not necessary if $L(\mathcal{D}) \stackrel{\text{df}}{=} \bigcap_{x_i \in \mathcal{D}} L_{x_i} \neq \emptyset$. The corresponding node in the decision tree then becomes a leaf, and any label $\lambda \in L(\mathcal{D})$ can be chosen as the prescribed label associated with that node.

Pruning a fully grown tree can principally be done in the same way as pruning standard trees. We implemented the pruning technique that is used in C4.5 [15].

4.3   Rule Induction

An alternative to the *divide-and-conquer* strategy followed by decision tree learners is to induce rules in a more direct way, using a *separate-and-conquer* or *cover-*

ing strategy [10]. Concrete implementations of this approach include algorithms such as, e.g., CN2 [4,3] and Ripper [5].

In order to learn a concept, i.e., to separate positive from negative examples, covering algorithms learn one rule after another. Each rule *covers* a subset of (positive) examples, namely those that satisfy the condition part of the rule. The covered examples are then removed from the training set. This process is iterated until no positive examples remain. Covering algorithms can be extended to the m-class case ($m > 2$) in several ways. For example, following a one-versus-all strategy, CN2 learns rules for each class in turn, starting with the least frequent one. Since in the m-class case the order in which rules have been induced is important, the rules thus obtained have to be treated as a *decision list*.

A key component of all covering algorithms is a "find-best-rule" procedure for finding a good or even optimal rule that partly covers the current training data. Starting with a maximally general rule, CN2 follows a top-down approach in which the candidate rules are successively specialized (e.g. by adding conditions). The search procedure is implemented as a beam search, guided by the Laplace-estimate as a heuristic evaluation:

$$L(r) \stackrel{\mathrm{df}}{=} (p+1)(n+p+2)^{-1}, \tag{6}$$

where r is the rule to be evaluated, p is the number of positive examples covered by r, and n the number of negative examples. As a stopping criterion, CN2 employs a statistical significance test (likelihood ratio) that decides whether or not the distribution of positive and negative examples covered by the rule is significantly different from the overall distribution in the complete training set.

In order to adapt CN2 to the ALC setting, we have made the following modifications: Similarly to the generalization of the entropy measure, we have turned the Laplace-estimate into a "potential" Laplace-estimate: Considering label λ_j as the positive class, $p = p_j$ is given by the number of all examples x_i covered by the rule r and such that $\lambda_j \in L_{x_i}$. This way, (6) can be derived for each label, and the maximal value is adopted as an evaluation of the rule:

$$L(r) = \max_{1 \le j \le m} (p_j + 1)(|r| + 2)^{-1},$$

where $|r|$ is the number of examples covered by the rule. The consequent of r is then given by the label λ_j for which the maximum is attained.

As noted before, CN2 learns classes in succession, starting with the smallest (least frequent) one. As opposed to this, we learn rules without specifying a class in advance. Rather, the most suitable class is chosen depending on the condition part of a rule. In fact, the label predicted by a rule can even change during the search process. This modification is in agreement with our goal of to disambiguate by implementing a simplicity bias. Moreover, the focusing on one particular label is less useful in the ALC setting. In fact, in the presence of ambiguously labeled examples, it may easily happen that a rule r is dominated by a class λ_j while all of its direct specializations are dominated by other classes.

5   Experimental Results

The main purpose of our experimental study was to provide evidence for the conjecture that exploiting ambiguous data for model induction by using a suitable ALC-method is usually better than the obvious alternative, namely to ignore such data and learn with a standard algorithm from the remaining (exactly labeled) examples. We used the latter approach as a baseline method.

Note that this conjecture is by far not trivial. In fact, whether or not ambiguous data can be useful will strongly depend on the performance of the ALC-method. If this method is not able to exploit the information contained in that data, ambiguous examples might be misleading rather than helpful. In this connection, recall our supposition that the weaker the inductive bias of a learning method, the more likely that method might be misled by ambiguous examples.

5.1   Experimental Setup

We have worked with "contaminated" versions of standard benchmark data sets (in which each instance is assigned a unique label), which allowed us to conduct experiments in a controlled way. In order to contaminate a given data set, we have devised two different strategies:

Random model: For each example in the training set, a biased coin is flipped in order to decide whether or not this example will be contaminated; the probability of contamination is p. In case an example x_i is contaminated, the set L_{x_i} of candidate labels is initialized with the original label λ_{x_i}, and all other labels $\lambda \in \mathcal{L} \setminus \{\lambda_{x_i}\}$ are added with probability q, independently of each other. Thus, the contamination procedure is parameterized by the probabilities p and q, where p corresponds to the expected fraction of ambiguous examples in a data set. Moreover, q reflects the "average benefit" of a contaminated example x_i: The smaller q is, the smaller the (average) number of candidate labels becomes and, hence, the more informative such an example will be. In fact, note that the expected cardinality of L_{x_i}, in the case of contamination, is given by $1 + (m-1)q$.

Bayes model: The random model assumes that labels are added independently of each other. In practice, this idealized assumption will rarely be valid. For example, the probability that a label is added will usually depend on the true label. In order to take this type of dependency into account, our second approach to contamination works as follows: First, a Naive Bayes classifier is trained using the original data, and a probabilistic prediction is derived for each input x_i. Let $\Pr(\lambda \mid x_i)$ denote the probability of label λ as predicted by the classifier. Whether or not an example is contaminated is decided by flipping a biased coin as before. In the case of contamination, the true label λ_{x_i} is again retained. Moreover, the other $m-1$ labels $\lambda \in \mathcal{L} \setminus \{\lambda_{x_i}\}$ are arranged in an increasing order according to their probability $\Pr(\lambda \mid x_i)$. The k-th label, $\lambda^{(k)}$, is then added with probability $(2 \cdot k \cdot q)/m$. Thus, the expected cardinality of L_{x_i} is again $1 + (m-1)q$, but the probabilities of the individual labels are now biased in favor of the labels found to be likely by the Bayes classifier. Intuitively, the Bayes model should come along with a decrease in performance for the ALC approach because, roughly speaking, disambiguating the data might become more difficult in the case of a "systematic" contamination.

The experimental results have been obtained in the following way: In a single experiment, the data is randomly divided into a training set and a test set of the same size. The training set is contaminated as outlined above. From the contaminated data, a model is induced using an ALC-extension of a classification method (kNN, decision trees, rule induction). Moreover, using the classification method in its standard form, a model is learned from the reduced training set that consists of the non-contaminated examples. Then, the classification accuracy of the two models is determined by classifying the instances in the test set. The *expected* classification accuracy of a method – for the underlying data set and fixed parameters p, q – is approximated by averaging over 1,000 experiments.

For decision tree learning and rule induction, all numeric attributes have been discretized in advance using hierarchical entropy-based discretization [9]. We didn't try to optimize the performance of the three learning methods themselves, because this was not the goal of the experiments. Rather, the purpose of the study was to *compare* – under equal conditions – ALC learning with the baseline method.

5.2 Results

Due to reasons of space, results are presented for only five data sets from the UCI repository: (1) dermatology (385 instances, 34 attributes, 6 classes), (2) ecoli (336, 7, 8), (3) housing (506,13,10), (4) glass (214, 9, 6), (5) zoo (101, 16, 7) and a few combinations of (p, q)-parameters (more results can be found in an extended version, available as a technical report from the authors).

The results for k-NN classification with $k = 5$ are summarized in Table 1, where (r) stands for the random model and (b) for the Bayes model. As can be seen, the ALC version is generally superior to the standard 5-NN classifier. Exceptions (marked with a *) only occur in cases where both p and q are large, that is, where the data is strongly contaminated. Roughly speaking, the superiority of the ALC version shows that relying on a nearby ambiguous neighbor is usually better than looking at an exact example that is faraway (because the close, ambiguous ones have been removed). We obtained similar results for $k = 7, 9, 11$.

The results do not convincingly confirm the supposition that the ALC version will perform better for the random model than for the Bayes model. Even though it is true that the results for the former are better than for the latter in most cases, the corresponding difference in performance is only slight and much smaller than expected. In general, it can be said that the contamination model does hardly influence the performance of the classifier most of the time. In fact, there is only one noticeable exception: For the Zoo data, the performance for the random model is much better than for the Bayes model in the case of highly contaminated data.

For decision tree induction, the ALC-version consistently outperforms the standard version. As the results in Table 2 show, the gain in performance is

Table 1. Results for 5-NN classification (classification rate and standard deviation)

data	method	q	$p = .1$	$p = .5$	$p = .9$
derma	ALC (r)	.3	.959 (.013)	.955 (.014)	.943 (.017)
	ALC (b)	.3	.959 (.013)	.955 (.015)	.940 (.018)
	standard	.3	.958 (.014)	.948 (.018)	.910 (.039)
	ALC (r)	.5	.958 (.014)	.949 (.015)	.890 (.028)
	ALB (b)	.5	.958 (.014)	.946 (.016)	.874 (.031)
	standard	.5	.957 (.014)	.945 (.019)	.851 (.067)
	ALC (r)	.7	.959 (.014)	.938 (.019)	.746 (.050)
	ALC (b)	.7	.958 (.014)	.936 (.018)	.745 (.046)
	standard	.7	.958 (.014)	.945 (.020)*	.833 (.072)*
ecoli	ALC (r)	.3	.845 (.025)	.832 (.025)	.798 (.024)
	ALC (b)	.3	.846 (.025)	.830 (.028)	.798 (.029)
	standard	.3	.845 (.026)	.827 (.028)	.743 (.059)
	ALC (r)	.5	.844 (.027)	.815 (.023)	.715 (.039)
	ALC (b)	.5	.843 (.022)	.814 (.028)	.709 (.045)
	standard	.5	.843 (.027)	.815 (.030)	.691 (.099)
	ALC (r)	.7	.844 (.022)	.801 (.029)	.582 (.050)
	ALC (b)	.7	.841 (.024)	.802 (.032)	.593 (.052)
	standard	.7	.842 (.024)	.820 (.034)*	.699 (.087)*
glass	ALC (r)	.3	.634 (.041)	.620 (.043)	.592 (.045)
	ALC (b)	.3	.638 (.040)	.622 (.041)	.592 (.044)
	standard	.3	.630 (.041)	.604 (.048)	.510 (.070)
	ALC (r)	.5	.636 (.042)	.611 (.042)	.542 (.052)
	ALC (b)	.5	.635 (.042)	.607 (.043)	.529 (.051)
	standard	.5	.633 (.042)	.599 (.045)	.438 (.077)
	ALC (r)	.7	.633 (.042)	.602 (.045)	.463 (.061)
	ALC (b)	.7	.631 (.042)	.604 (.045)	.453 (.060)
	standard	.7	.631 (.041)	.595 (.051)	.408 (.077)
housing	ALC (r)	.3	.488 (.027)	.461 (.029)	.423 (.017)
	ALC (b)	.3	.476 (.026)	.457 (.029)	.412 (.032)
	standard	.3	.486 (.028)	.455 (.030)	.403 (.032)
	ALC (r)	.5	.476 (.028)	.431 (.030)	.320 (.034)
	ALC (b)	.5	.477 (.027)	.445 (.031)	.367 (.032)
	standard	.5	.474 (.028)	.444 (.030)	.362 (.046)*
	ALC (r)	.7	.486 (.027)	.440 (.033)	.271 (.035)
	ALC (b)	.7	.478 (.029)	.443 (.030)	.324 (.032)
	standard	.7	.484 (.027)	.454 (.031)*	.369 (.048)*
zoo	ALC (r)	.3	.926 (.038)	.912 (.041)	.887 (.054)
	ALC (b)	.3	.925 (.038)	.911 (.042)	.886 (.055)
	standard	.3	.925 (.039)	.896 (.053)	.782 (.104)
	ALC (r)	.5	.924 (.037)	.901 (.048)	.824 (.072)
	ALC (b)	.5	.925 (.039)	.895 (.048)	.777 (.091)
	standard	.5	.923 (.038)	.889 (.059)	.667 (.155)
	ALC (r)	.7	.922 (.038)	.885 (.058)	.673 (.110)
	ALC (b)	.7	.924 (.039)	.881 (.060)	.609 (.111)
	standard	.7	.921 (.038)	.884 (.061)	.655 (.162)

Table 2. Results for decision tree induction

data	method	q	p = .1	p = .5	p = .9
derma	ALC (r)	.3	.860 (.046)	.841 (.051)	.809 (.069)
	ALC (b)	.3	.861 (.047)	.839 (.055)	.802 (.069)
	standard	.3	.858 (.049)	.814 (.063)	.654 (.129)
	ALC (r)	.5	.858 (.047)	.818 (.057)	.742 (.098)
	ALB (b)	.5	.854 (.047)	.816 (.058)	.736 (.082)
	standard	.5	.858 (.049)	.807 (.069)	.488 (.155)
	ALC (r)	.7	.855 (.048)	.802 (.059)	.618 (.125)
	ALC (b)	.7	.854 (.048)	.801 (.063)	.659 (.092)
	standard	.7	.855 (.048)	.799 (.075)	.446 (.154)
ecoli	ALC (r)	.3	.705 (.039)	.676 (.041)	.645 (.043)
	ALC (b)	.3	.707 (.038)	.682 (.038)	.663 (.039)
	standard	.3	.704 (.043)	.655 (.058)	.543 (.115)
	ALC (r)	.5	.703 (.039)	.658 (.042)	.611 (.051)
	ALC (b)	.5	.703 (.038)	.669 (.039)	.639 (.045)
	standard	.5	.700 (.041)	.646 (.059)	.517 (.139)
	ALC (r)	.7	.700 (.039)	.648 (.043)	.567 (.065)
	ALC (b)	.7	.701 (.039)	.661 (.040)	.635 (.051)
	standard	.7	.699 (.041)	.643 (.064)	.509 (.149)
housing	ALC (r)	.3	.348 (.038)	.321 (.043)	.282 (.045)
	ALC (b)	.3	.348 (.038)	.333 (.042)	.311 (.044)
	standard	.3	.353 (.039)*	.334 (.051)*	.313 (.088)*
	ALC (r)	.5	.346 (.038)	.308 (.043)	.246 (.047)
	ALC (b)	.5	.348 (.038)	.331 (.044)	.294 (.043)
	standard	.5	.353 (.039)*	.336 (.051)*	.306 (.099)*
	ALC (r)	.7	.348 (.038)	.301 (.046)	.261 (.062)
	ALC (b)	.7	.352 (.036)	.321 (.044)	.286 (.078)
	standard	.7	.350 (.042)*	.337 (.052)*	.302 (.104)*
glass	ALC (r)	.3	.557 (.059)	.533 (.065)	.507 (.072)
	ALC (b)	.3	.559 (.059)	.534 (.069)	.496 (.080)
	standard	.3	.553 (.063)	.514 (.086)	.437 (.120)
	ALC (r)	.5	.556 (.055)	.525 (.066)	.460 (.085)
	ALC (b)	.5	.555 (.054)	.513 (.078)	.434 (.082)
	standard	.5	.551 (.064)	.497 (.091)	.395 (.152)
	ALC (r)	.7	.554 (.056)	.507 (.075)	.410 (.092)
	ALC (b)	.7	.557 (.057)	.504 (.079)	.382 (.065)
	standard	.7	.554 (.064)	.493 (.093)	.389 (.172)
zoo	ALC (r)	.3	.876 (.057)	.841 (.063)	.806 (.066)
	ALC (b)	.3	.876 (.058)	.843 (.060)	.807 (.063)
	standard	.3	.876 (.059)	.814 (.084)	.654 (.171)
	ALC (r)	.5	.877 (.057)	.827 (.067)	.765 (.079)
	ALC (b)	.5	.876 (.057)	.830 (.063)	.753 (.103)
	standard	.5	.873 (.060)	.811 (.091)	.552 (.245)
	ALC (r)	.7	.873 (.055)	.820 (.069)	.696 (.102)
	ALC (b)	.7	.874 (.054)	.825 (.066)	.553 (.206)
	standard	.7	.873 (.061)	.807 (.092)	.500 (.272)

even higher than for NN classification. Again, the results show that a systematic contamination of the data, using the Bayes instead of the random model, does hardly affect the performance of ALC. It is true that the classification performance deteriorates on average, but again only slightly and not in every case.

An interesting exception to the above findings is the Housing data (not only for decision tree learning but also for NN classification). First, for this data the standard version is down the line better than the ALC-version. Second, the ALC-version is visibly better in the case of the Bayesian model than in the case of the random model. A plausible explanation for this is the fact that for the Housing data the classes are price categories and hence do have a natural *order*. That is, we actually face a problem of *ordinal classification* rather than standard classification. (Consequently, ordinal classification methods should be applied, and the results for this data set should not be overrated in our context.) Moreover, the Bayesian model tends to add classes that are, in the sense of this ordering, neighbored to the true price category, thereby distorting the original class information but slightly. Compared to this, ambiguous information will be much more conflicting in the case of the random model.

Since the experimental results for rule induction are rather similar to those for decision tree learning, they are omitted here for reasons of space.

In summary, the experiments show that our ALC extensions of standard learning methods can successfully deal with ambiguous label information. In fact, except for some rare cases, these extensions yield better results than the baseline method (which ignores ambiguous examples and applies standard learning methods). A closer examination reveals two interesting points: Firstly, it seems that the gain in classification accuracy (of ALC compared with the baseline method) is a monotone increasing function of the parameter p (probability of contamination). With regard to the parameter q, however, the dependency appears to be non-monotone: The gain first increases but then decreases for large enough q-values. Intuitively, these findings can be explained as follows: Since q represents a kind of "expected benefit" of an ambiguous example, the utility of such an example is likely to become negative for large q-values. Consequently, it might then be better to simply ignore such examples, at least if enough other data is available. Secondly, the performance gain for decision tree learning seems to be slightly higher than the one for rule induction, at least on average, and considerably higher than the gain for NN classification. This ranking is in perfect agreement with our conjecture that the stronger the inductive bias of a learning method, the more useful ALC will be.

6   Concluding Remarks

In order to successfully learn a classification function in the ALC setting, where examples can be labeled in an ambiguous way, we proposed several extensions of standard machine learning methods. The idea is to exploit the inductive bias underlying these (heuristic) methods in order to disambiguate the label information. In fact, we argued that looking at the label information with a "biased view" may remove the ambiguity of that information to some extent. This idea gives rise to the conjecture that ALC learning methods with a strong (and of course approximately correct) bias can exploit the information provided by ambiguous examples better than methods with a weak bias. This conjecture has been supported empirically by experiments that have been carried out for three concrete learning techniques, namely ALC extensions of nearest neighbor classification, decision tree learning, and rule induction. The experiments also showed that applying our ALC methods to the complete data will usually yield better results than learning with a standard method from the subset of exactly labeled examples, at least if the expected benefit of the ambiguous examples is not too low. In any case, our approach can be seen as a simple yet effective alternative that complements the probabilistic approaches proposed in [11,13] in a reasonable way.

References

1. KP. Bennet and A. Demiriz. Semi-supervised support vector machines. In *Advances in Neural Information Processing 11*, pages 368–374. MIT Press, 1999.
2. J. Cavarelli et al. The structure of Staphylococcus aureus epidermolytic toxin A, an atypic serine protease, at 1.7 A resolution. *Structure* 5(6):813–24, 1997.
3. P Clark and R Boswell. Rule induction with CN2: Some recent improvements. In *Proc. 5th Europ. Working Session of Learning*, pages 151–163, Porto, 1991.
4. P. Clark and T. Niblett. The CN2 induction algorithm. *Machine Learning*, 3:261–283, 1989.
5. W.W. Cohen. Fast effective rule induction. *Proc. 12th ICML*, pages 115–123, Tahoe City, CA, 1995.
6. B.V. Dasarathy, editor. *Nearest Neighbor (NN) Norms: NN Pattern Classification Techniques*. IEEE Computer Society Press, Los Alamitos, California, 1991.
7. T.G. Dietterich, R.H. Lathrop, and T. Lozano-Perez. Solving the multiple-instance problem with axis-parallel rectangles. *Art. Intell. Journal*, 89, 1997.
8. P. Domingos. The role of Occam's razor in knowledge discovery. *Data Mining and Knowledge Discovery*, 3:409–425, 1999.
9. U. Fayyad and KB. Irani. Multi-interval discretization of continuos attributes as preprocessing for classification learning. Proc. *IJCAI–93*, pages 1022–1027, 1993.
10. J Fürnkranz. Separate-and-conquer rule learning. *AI Review*, 13(1):3–54, 1999.
11. Y Grandvalet. Logistic regression for partial labels. *IPMU–02*, pages 1935–1941, Annecy, France, 2002.
12. E. Hüllermeier and J. Beringer. Learning decision rules from positive and negative preferences. *IPMU-04*, Perugia, Italy, 2004.
13. R Jin and Z Ghahramani. Learning with multiple labels. *NIPS-02*, Vancouver, Canada, 2002.
14. A. McCallum. Multi-label text classification with a mixture model trained by EM. In *AAAI–99 Workshop on Text Learning*, 1999.
15. J.R. Quinlan. *C4.5: Programs for Machine Learning*. Morgan Kaufmann, San Mateo, CA, 1993.

Learning Label Preferences:
Ranking Error Versus Position Error

Eyke Hüllermeier[1] and Johannes Fürnkranz[2]

[1] Fakultät für Informatik,
Otto-von-Guericke-Universität Magdeburg, Germany
eyke.huellermeier@iti.cs.uni-magdeburg.de
[2] Fachbereich Informatik, TU Darmstadt, Germany
juffi@ke.informatik.tu-darmstadt.de

Abstract. We consider the problem of learning a *ranking function*, that is a mapping from instances to rankings over a finite number of labels. Our learning method, referred to as *ranking by pairwise comparison* (RPC), first induces pairwise order relations from suitable training data, using a natural extension of so-called pairwise classification. A ranking is then derived from a set of such relations by means of a *ranking procedure*. This paper elaborates on a key advantage of such a decomposition, namely the fact that our learner can be adapted to different loss functions by using different ranking procedures on the same underlying order relations. In particular, the Spearman rank correlation is minimized by using a simple weighted voting procedure. Moreover, we discuss a loss function suitable for settings where candidate labels must be tested successively until a target label is found, and propose a ranking procedure for minimizing the corresponding risk.

1 Introduction

Prediction problems involving complex outputs and structured output spaces have recently received a great deal of attention within the machine learning literature (e.g., [11]). Problems of that kind are particularly challenging, since the prediction of complex structures such as, say, graphs or trees, is more demanding than the prediction of single values as in classification and regression.

A common problem of this type is *preference learning*, the learning with or from preferences.[1] In the literature, we can identify two different learning scenarios for preference learning [8]: (i) learning from *object preferences*, where the task is to order a set of objects according to training information that specifies the preference relations between a set of training objects (see, e.g., [2]), and (ii) learning from *label preferences*, where the task is to learn a mapping from instances to rankings (total orders) over a finite number of class labels [7]. A

---

[1] Space restrictions prevent a thorough review of related work in this paper, but we refer to [6] and recent workshops in this area, e.g., those at NIPS-02, KI-03, SIGIR-03, NIPS-04, and GfKl-05 (the second and fifth organized by the authors).

A.F. Famili et al. (Eds.): IDA 2005, LNCS 3646, pp. 180–191, 2005.

corresponding *ranking function* can be seen as an extension of a standard classification function that maps instances to single class labels. In this paper, we focus on the second scenario, but our results can be carried over to the first scenario as well.

In [7], we have introduced a method for learning label preferences that we shall subsequently refer to as *ranking by pairwise comparison* (RPC). This method works in two phases. First, pairwise order relations (preferences) are learned from suitable training data, using a natural extension of so-called *pairwise classification*. Then, a ranking is derived from a set of such orders (preferences) by means of a *ranking procedure*.

The goal of this paper is to show that by using suitable ranking functions, our approach can easily be customized to different performance tasks, that is, to different loss functions for rankings. In fact, the need for a ranking of class labels may arise in different learning scenarios. In this work, we are particularly interested in two types of practically motivated learning problems, one in which the complete ranking is relevant and one in which the predicted ranking has the purpose of reducing the search effort for finding the single target label.

The remainder of the paper is organized as follows: The problem of preference learning is formally introduced in Section 2, and our pairwise approach is presented in Section 3. In Section 4, the aforementioned types of learning problems are discussed and compared in more detail. The ranking procedures suitable for the two types of problems are then discussed in Sections 5 and 6, respectively.

2   Learning from Label Preferences

We consider the following learning problem [8]:

Given:
  - a set of *labels* $\mathcal{L} = \{\, \lambda_i \,|\, i = 1 \ldots m \,\}$
  - a set of *examples* $\mathcal{S} = \{\, x_k \,|\, k = 1 \ldots n \,\}$
  - for each training example (instance) x_k:
    - a set of *preferences* $P_k \subseteq \mathcal{L} \times \mathcal{L}$, where $(\lambda_i, \lambda_j) \in P_k$ indicates that label λ_i is preferred over label λ_j for instance x_k.

Find: a function that orders the labels $\lambda \in \mathcal{L}$ for any given example.

We will abbreviate $(\lambda_i, \lambda_j) \in P_k$ with $\lambda_i \succ_{x_k} \lambda_j$ or simply $\lambda_i \succ \lambda_j$ if the particular example x_k doesn't matter or is clear from the context.

The above setting has recently been introduced as *constraint classification* in [9]. As shown in that paper, it is a generalization of several common learning settings, in particular

  - *ranking:* Each training example is associated with a total order of the labels.
  - *classification:* A single class label λ_x is assigned to each example x; implicitly, this defines the set of preferences $\{\, \lambda_x \succ \lambda \,|\, \lambda \in \mathcal{L} \setminus \{\lambda_x\} \,\}$.

- *multi-label classification:* Each example x is associated with a subset $L_x \subseteq \mathcal{L}$ of labels; implicitly, this defines the preferences $\{\lambda \succ \lambda' \mid \lambda \in L_x, \lambda' \in \mathcal{L} \backslash L_x\}$.

As mentioned above, we are mostly interested in the first problem, that is in predicting a ranking (complete, transitive, asymmetric relation) of the labels. The ranking $\succ_x$ of an instance x can be expressed in terms of a permutation τ_x of $\{1 \ldots m\}$ such that

$$\lambda_{\tau_x(1)} \succ_x \lambda_{\tau_x(2)} \succ_x \ldots \succ_x \lambda_{\tau_x(m)}. \tag{1}$$

Note that we make the simplifying assumption that all preferences are strict, i.e., we do not consider the case of indifference between labels.

An appealing property of this learning framework is that its input, consisting of *comparative* preference information of the form $\lambda_i \succ_x \lambda_j$ (x prefers λ_i to λ_j), is often easier to obtain than absolute ratings of single alternatives in terms of *utility degrees*. In this connection, note that knowledge about the complete ranking (1) can be expanded into $m(m-1)/2$ comparative preferences $\lambda_{\tau_x(i)} \succ \lambda_{\tau_x(j)}$, $1 \leq i < j \leq m$.

3   Learning Pairwise Preferences

The idea of pairwise learning is well-known in the context of classification [5], where it allows one to transform a multi-class classification problem, i.e., a problem involving $m > 2$ classes $\mathcal{L} = \{\lambda_1 \ldots \lambda_m\}$, into a number of *binary* problems. To this end, a separate model (base learner) $\mathcal{M}_{ij}$ is trained for each *pair* of labels $(\lambda_i, \lambda_j) \in \mathcal{L}$, $1 \leq i < j \leq m$; thus, a total number of $m(m-1)/2$ models is needed. $\mathcal{M}_{ij}$ is intended to separate the objects with label λ_i from those having label λ_j.

At classification time, a query x is submitted to all learners, and each prediction $\mathcal{M}_{ij}(x)$ is interpreted as a vote for a label. If classifier $\mathcal{M}_{ij}$ predicts λ_i, this is counted as a vote for λ_i. Conversely, the prediction λ_j would be considered as a vote for λ_j. The label with the highest number of votes is then proposed as a prediction.

The above procedure can be extended to the case of preference learning in a natural way [7]. A preference information of the form $\lambda_i \succ_x \lambda_j$ is turned into a training example (x, y) for the learner $\mathcal{M}_{ab}$, where $a = \min(i, j)$ and $b = \max(i, j)$. Moreover, $y = 1$ if $i < j$ and $y = 0$ otherwise. Thus, $\mathcal{M}_{ab}$ is intended to learn the mapping that outputs 1 if $\lambda_a \succ_x \lambda_b$ and 0 if $\lambda_b \succ_x \lambda_a$:

$$x \mapsto \begin{cases} 1 & \text{if} \quad \lambda_a \succ_x \lambda_b \\ 0 & \text{if} \quad \lambda_b \succ_x \lambda_a \end{cases}. \tag{2}$$

The mapping (2) can be realized by any binary classifier. Alternatively, one might of course employ a classifier that maps into the unit interval $[0, 1]$ instead of $\{0, 1\}$. The output of such a "soft" binary classifier can usually be interpreted as a probability or, more generally, a kind of confidence in the classification.

Thus, the closer the output of $\mathcal{M}_{ab}$ to 1, the stronger the preference $\lambda_a \succ_x \lambda_b$ is supported.

A preference learner composed of an ensemble of soft binary classifiers (which can be constructed on the basis of training data in the form of instances with associated partial preferences) assigns a *valued preference relation* $\mathcal{R}_x$ to any (query) instance $x \in \mathcal{X}$:

$$\mathcal{R}_x(\lambda_i, \lambda_j) = \begin{cases} \mathcal{M}_{ij}(x) & \text{if} \quad i < j \\ 1 - \mathcal{M}_{ij}(x) & \text{if} \quad i > j \end{cases}$$

for all $\lambda_i \neq \lambda_j \in \mathcal{L}$.

Given a preference relation $\mathcal{R}_x$ for an instance x, the next question is how to derive an associated ranking τ_x. This question is non-trivial, since a relation $\mathcal{R}_x$ does not always suggest a unique ranking in an unequivocal way. In fact, the problem of inducing a ranking from a (valued) preference relation has received a lot of attention in several research fields, e.g., in fuzzy preference modeling and (multi-attribute) decision making [4]. Besides, in the context of our application, it turned out that the *ranking procedure* used to transform a relation $\mathcal{R}_x$ into a ranking τ_x is closely related to the definition of the quality of a prediction and, hence, to the intended purpose of a ranking. In other words, risk minimization with respect to different loss functions might call for different ranking procedures.

4   Ranking Error Versus Position Error

In Section 2, we introduced the problem of predicting a ranking of class labels in a formal way, but did not discuss the semantics of a predicted ranking. In fact, one should realize that such a ranking can serve different purposes. Needless to say, this point is of major importance for the evaluation of a predicted ranking.

In this paper, we are especially interested in two types of practically motivated performance tasks. In the first setting, which is probably the most obvious one, the *complete ranking* is relevant, i.e., the positions assigned to all of the labels. As an example, consider the problem to order the questions in a questionnaire. Here, the goal is to maximize a particular respondents' motivation to complete the questionnaire. Another example is learning to predict the best order in which to supply a certain set of stores (route of a truck), depending on external conditions like traffic, weather, purchase order quantities, etc.

In case the complete ranking is relevant, the quality of a prediction should be quantified in terms of a distance measure between the predicted and the true ranking. We shall refer to any deviation of the predicted ranking from the true one as a *ranking error*.

To motivate the second setting, consider a fault detection problem which consists of identifying the cause for the malfunctioning of a technical system. If it turned out that a predicted cause is not correct, an alternative candidate must be tried. A ranking then suggests a simple (trial and error) search process, which successively tests the candidates, one by one, until the correct cause is found [1]. In this scenario, where labels correspond to causes, the existence of a single target label (instead of a target ranking) is assumed. Hence, an obvious measure of the quality of a predicted ranking is the number of futile trials made before that label is found. A deviation of the predicted target label's position from the top-rank will subsequently be called a *position error*.

The main difference between the two types of error is that an evaluation of a full ranking (ranking error) attends to all positions. For example, if the two highest ranks of the true ranking are swapped in the predicted ranking, this is as bad as the swapping of the two lowest ranks.

Note that the position error is closely related to the conventional (classification) error, i.e., the incorrect prediction of the top label. In both cases, we are eventually concerned with predictions for the top rank. In our setting, however, we not only try to maximize the number of correct predictions. Instead, in the case of a misclassification, we also look at the position of the target label. The higher this position, the better the prediction. In other words, we differentiate between "bad" predictions in a more subtle way.

Even though we shall not deepen this point in the current paper, we note that the idea of a position error can of course be generalized to multi-label (classification) problems which assume several instead of a single target label for each instance. There are different options for such a generalization. For example, it makes a great difference whether one is interested in having at least one of the targets on a top rank (e.g., since one solution is enough), or whether all of them should have high positions (resp. none of them should be ranked low). An application of the latter type has recently been studied in [3].

5   Minimizing the Ranking Error

The quality of a model $\mathcal{M}$ (induced by a learning algorithm) is commonly expressed in terms of its *expected loss* or *risk*

$$\mathbb{E}\left(D(y, \mathcal{M}(x))\right), \tag{3}$$

where $D(\cdot)$ is a loss or distance function, $\mathcal{M}(x)$ denotes the prediction made by the learning algorithm for the instance x, and y is the true outcome. The expectation $\mathbb{E}$ is taken over $\mathcal{X} \times \mathcal{Y}$, where $\mathcal{Y}$ is the output space (e.g., the set $\mathcal{L}$ of classes in classification).[2]

The simplest loss function, commonly employed in classification, is the 0/1–loss: $D(y, \hat{y}) = 0$ for $y = \hat{y}$ and $= 1$ otherwise. Given this loss function, the optimal (Bayes) prediction for a specific instance x is simply the most probable outcome y. In the classification setting, for example, where $\mathcal{Y} = \mathcal{L}$, this estimate is the class with maximum posterior probability $\mathsf{P}(\lambda_i \mid x)$.

A straightforward generalization of this principle to the ranking setting, where $\mathcal{Y}$ is the class of rankings over $\mathcal{L}$, leads to the prediction

$$\hat{\tau}_x = \arg \max_{\tau \in \mathcal{S}_m} \mathsf{P}(\tau \mid x),$$

---

[2] The existence of a probability measure over $\mathcal{X} \times \mathcal{Y}$ must of course be assumed.

where $P(\tau \mid x)$ is the conditional probability of a ranking (permutation) given an instance x, and $\mathcal{S}_m$ denotes the class of all permutations of $\{1 \ldots m\}$.

Obviously, the simple $0/1$–distance function is a rather crude evaluation measure for rankings, because it assigns the same loss to all rankings that differ from the correct ranking, and does not take into account that different rankings can have different degrees of similarity. For this reason, a number of more sophisticated distance measures for rankings have been proposed in literature.

In general, if $D(\tau, \tau')$ is a measure of the distance between two rankings τ and τ', the risk minimizing prediction is

$$\hat{\tau}_x = \arg \min_{\tau \in S_k} \sum_{\tau \in \mathcal{S}_m} D(\tau, \tau') \cdot P(\tau' \mid x). \tag{4}$$

A frequently used distance measure is the sum of squared rank distances

$$D(\tau', \tau) \stackrel{\text{df}}{=} \sum_{i=1}^{m} (\tau'(i) - \tau(i))^2 \tag{5}$$

which is equivalent to the *Spearman rank correlation*[3]

$$1 - \frac{6D(\tau, \tau')}{m(m^2 - 1)} \in [-1, 1].$$

RPC can yield a risk minimizing prediction for this loss function, if the predictions of the binary classifiers are combined by weighted voting, i.e., the alternatives λ_i are evaluated by means of the sum of weighted votes

$$S(\lambda_i) = \sum_{\lambda_j \neq \lambda_i} \mathcal{R}_x(\lambda_i, \lambda_j) \tag{6}$$

and ranked according to these evaluations:

$$\lambda_{\tau_x(1)} \succ_x \lambda_{\tau_x(2)} \succ_x \ldots \succ_x \lambda_{\tau_x(m)} \tag{7}$$

with τ_x satisfying $S(\lambda_{\tau_x(i)}) \geq S(\lambda_{\tau_x(i+1)})$, $i = 1 \ldots m - 1$.[4] This is a particular type of "ranking by scoring" strategy; here, the scoring function is given by (6).

Formally, we can show the following result, which provides a theoretical justification for the voting procedure (6). The proof of this theorem can be found in Appendix A.

Theorem 1. *Using the "ranking by scoring" procedure outlined above, RPC is a risk minimizer with respect to (5) as a loss function. More precisely, with*

$$\mathcal{M}_{ij}(x) = P(\lambda_i \succ_x \lambda_j) = \sum_{\tau : \tau(j) < \tau(i)} P(\tau \mid x),$$

the expected distance

---

[3] This is, of course, a similarity rather than a distance measure.
[4] Ties can be broken arbitrarily.

$$E(\tau') = \sum_{\tau} p(\tau) \cdot D(\tau', \tau) = \sum_{\tau} p(\tau) \sum_{\imath=1}^{m} (\tau'(\imath) - \tau(\imath))^2$$

becomes minimal by choosing τ' such that $\tau'(\imath) \leq \tau'(\jmath)$ whenever $S(\lambda_\imath) \geq S(\lambda_\jmath)$, where $S(\lambda_\imath)$ is given by (6).

6   Minimizing the Position Error

Despite the fact that (5) is a reasonable loss function for rankings, it is not always appropriate. In particular, it assumes that the *complete* ranking is relevant for the quality of a prediction, which is not the case in connection with the fault detection scenario outlined in the introduction. Here, only the prefix of a ranking τ_x is considered, up to the position of the target label λ_x, while the rest of the prediction is of no importance (since the search procedure stops if λ_x has been found). In this case, the loss function only depends on the rank of λ_x.

More specifically, we define the *position error* as $\tau_x^{-1}(\lambda_x)$, i.e., by the position of the target label λ_x in the ranking τ_x. To compare the quality of rankings of different problems, it is useful to normalize the position error for the number of labels. This *normalized position error* is defined as

$$\frac{\tau_x^{-1}(\lambda_x) - 1}{m - 1} \in \{0, 1/(m-1) \dots 1\}, \tag{8}$$

What kind of ranking procedure should be used in order to minimize the risk of a predicted ranking with respect to the position error as a loss function? Intuitively, the candidate labels λ should now be ordered according to their probability $P(\lambda = \lambda_x)$ of being the target label. Especially, the top-rank (first position) should be given to the label $\lambda_\top$ for which this probability is maximal. Regarding the second rank, recall the fault detection metaphor, where the second hypothesis for the cause of the fault is only tested in case the first one turned out to be wrong. In this setting, the second rank should not simply be given to the label with the second highest probability according to the measure $P_1(\cdot) = P(\cdot)$. Instead, it must be assigned to the label that maximizes the *conditional* probability $P_2(\cdot) = P(\cdot \mid \lambda_x \neq \lambda_\top)$, i.e., the probability of being the target label *given that the first proposal was incorrect.*

At first sight, passing from $P_1(\cdot)$ to $P_2(\cdot)$ might appear meaningless from a ranking point of view, since standard probabilistic conditioning (dividing all probabilities by $1 - P(\lambda_\top)$ and setting $P(\lambda_\top) = 0$) does not change the order of the remaining labels. One should realize, however, that standard conditioning is not an incontestable updating procedure in our context, simply because $P_1(\cdot)$ is not a "true" measure over the class labels. Rather, it is only an estimated measure coming from a learning algorithm. Thus, it seems sensible to perform "conditioning" not on the measure itself, but rather on the learner that produced the measure. By this we mean retraining the learner on the original data without the $\lambda_\top$-examples, something that could be paraphrased as "empirical conditioning". To emphasize that this type of conditioning depends on the data

$\mathcal{D}$ and the model assumptions (hypothesis space) $\mathcal{H}$ and, moreover, that it concerns an *estimated* ("hat") probability, the conditional measure $\mathsf{P}_2(\cdot)$ could be written more explicitly as

$$\mathsf{P}_2(\cdot) = \widehat{\mathsf{P}}(\cdot \,|\, \lambda_x \neq \lambda_\mathsf{T}, \mathcal{D}, \mathcal{M}).$$

To motivate the idea of empirical conditioning, suppose that the estimated probabilities come from a classification tree. Of course, the original tree trained with the complete data will be highly influenced by λ_T-examples, and the probabilities assigned by that tree to the alternatives $\lambda \neq \lambda_\mathsf{T}$ might be inaccurate. Retraining a classification tree on a reduced set of data might then lead to more accurate probabilities for the remaining labels, especially since the multi-class problem to be solved has now become simpler (as it involves fewer classes).

A problem of the above "ranking through iterated choice" procedure, that is, the successive selection of alternatives by estimating top-labels from (conditional) probability measures $\mathsf{P}_1(\cdot), \mathsf{P}_2(\cdot) \ldots \mathsf{P}_m(\cdot)$, concerns its computational complexity. In fact, realizing empirical conditioning by retraining a standard multi-class classifier comes down to training such a classifier for (potentially) each subset of the label set $\mathcal{L}$. Fortunately, empirical conditioning can be implemented much more efficiently by our pairwise approach, as will now be shown.

6.1 Implementing "Ranking Through Iterated Choice" by RPC

What kind of aggregation procedure is suitable for deriving an estimated probability distribution from pairwise classifications resp. valued preference $\mathcal{R}(\lambda_i, \lambda_j)$? Let E_i denote the event that $\lambda_i = \lambda_x$, i.e., that λ_i is the target label, and let $E_{ij} = E_i \vee E_j$ (either λ_i or λ_j is the target). Then,

$$(m-1)\,\mathsf{P}(E_i) = \sum_{j \neq i} \mathsf{P}(E_i) = \sum_{j \neq i} \mathsf{P}(E_i \,|\, E_{ij})\mathsf{P}(E_{ij}), \qquad (9)$$

where m is the number of labels. Considering the (pairwise) estimates $\mathcal{R}(\lambda_i, \lambda_j)$ as conditional probabilities $\mathsf{P}(E_i \,|\, E_{ij})$, we obtain a system of linear equations for the (unconditional) probabilities $\mathsf{P}(E_i)$:

$$\mathsf{P}(E_i) = \frac{1}{m-1} \sum_{j \neq i} \mathcal{R}(\lambda_i, \lambda_j)\mathsf{P}(E_{ij})$$

$$= \frac{1}{m-1} \sum_{j \neq i} \mathcal{R}(\lambda_i, \lambda_j)(\mathsf{P}(E_i) + \mathsf{P}(E_j)) \qquad (10)$$

In conjunction with the constraint $\sum_{i=1}^m \mathsf{P}(E_i) = 1$, this system has a unique solution provided that $\mathcal{R}(\lambda_i, \lambda_j) > 0$ for all $1 \leq i, j \leq m$ [12].

Based on this result, the "ranking through iterated choice" procedure suggested above can be realized as follows: First, the system of linear equations (10) is solved and the label λ_i with maximal probability $\mathsf{P}(E_i)$ is chosen as the top-label λ_T. This label is then removed, i.e., the corresponding row and column of the relation $\mathcal{R}$ is deleted. To find the second best label, the same procedure is then applied to the reduced relation, i.e., by solving a system of $m - 1$ linear equations. This process is iterated until a full ranking has been constructed.

Lemma 1. *In each iteration of the above "ranking through iterated choice" procedure, the correct conditional probabilities are derived.*

Proof. Without loss of generality, assume that λ_m has obtained the highest rank in the first iteration. The information that this label is incorrect, $\lambda_m \neq \lambda_x$, is equivalent to $P(E_m) = 0$, $P(E_m \mid E_{jm}) = 0$, and $P(E_j \mid E_{jm}) = 1$ for all $j \neq m$. Incorporating these probabilities in (10) yields, for all $i < m$,

$$(m-1)P(E_i) = \sum_{j=1...m, j \neq i} P(E_i \mid E_{ij})P(E_{ij})$$

$$= \sum_{j=1..m-1, j \neq i} P(E_i \mid E_{ij})P(E_{ij}) + 1P(E_{im})$$

and as $P(E_{im}) = P(E_i) + P(E_m) = P(E_i)$,

$$(m-2)P(E_i) = \sum_{j=1..m-1, j \neq i} P(E_i \mid E_{ij})P(E_{ij}).$$

Obviously, the last equation is equivalent to (10) for a system with $m-1$ labels, namely the system obtained by removing the m-th row and column of $\mathcal{R}$.   □

   As can be seen, the pairwise approach is particularly well-suited for the "ranking through iterated choice" procedure, as it allows for an easy incorporation of the information coming from futile trials. One just has to solve the system of linear equations (10) once more, with some of the pairwise probabilities set to 0 resp. 1 (or, equivalently, solve a smaller system of equations). No retraining of any classifier is required!

Theorem 2. *By ranking the alternative labels according to their (conditional) probabilities of being the top-label, RPC becomes a risk minimizer with respect to the position error (8) as a loss function. That is, the expected loss*

$$E(\tau) = \frac{1}{m-1} \sum_{i=1}^{m} (i-1) \cdot P\left(\lambda_{\tau(i)} = \lambda_x\right)$$

becomes minimal for the ranking predicted by RPC.

Proof. This result follows almost by definition. In fact, note that we have

$$E(\tau) \propto \sum_{i=1}^{m} P\left(\lambda_x \notin \{\lambda_{\tau(1)} \ldots \lambda_{\tau(i)}\}\right),$$

and that, for each position i, the probability to excess this position when searching for the target λ_x is obviously minimized when ordering the labels according to their (conditional) probabilities.   □

7   Concluding Remarks

By showing that RPC is a risk minimizer with respect to particular loss functions for rankings, this paper provides a sound theoretical foundation for our method of ranking by pairwise comparison. The interesting point is that RPC can easily be customized to different performance tasks, simply by changing the ranking procedure employed in the second step of the method. By modifying this procedure, the goal of RPC can be changed from minimizing the expected distance between the predicted and the true ranking to minimizing the expected number of futile trials in searching a target label. This can be done without retraining of the classifier ensemble.

Apart from these theoretical results, the practical validation of our method is of course an important issue. Regarding the ranking error, RPC has already been investigated empirically in [7,10], whereas empirical studies concerning the position error constitute a topic of still ongoing work. In this context, it is particularly interesting to compare the results obtained by the "ranking through iterated choice" procedure with predictions from standard ("non-iterated") probabilistic classification.

References

1. C. Alonso, JJ. Rodríguez, and B. Pulido. Enhancing consistency based diagnosis with machine learning techniques. In *Current Topics in AI*, vol. 3040 of LNAI, 312–321. Springer, 2004.
2. W.W. Cohen, R.E. Schapire, and Y. Singer. Learning to order things. *Journal of Artificial Intelligence Research*, 10:243–270, 1999.
3. K. Crammer and Y. Singer. A family of additive online algorithms for category ranking. *Journal of Machine Learning Research*, 3:1025–1058, 2003.
4. J. Fodor and M. Roubens. *Fuzzy Preference Modelling and Multicriteria Decision Support*. Kluwer, 1994.
5. J. Fürnkranz. Round robin classification. *Journal of Machine Learning Research*, 2:721–747, 2002.
6. J. Fürnkranz and E. Hüllermeier. Pairwise preference learning and ranking. Technical Report TR-2003-14, Österr. Forschungsinst. für Artif. Intell. Wien, 2003.
7. J. Fürnkranz and E. Hüllermeier. Pairwise preference learning and ranking. In *Proc. ECML-03*, Cavtat-Dubrovnik, Croatia, September 2003. Springer-Verlag.
8. J. Fürnkranz and E. Hüllermeier. Preference learning. *Künstliche Intelligenz*, 1/05:60–61, 2005.
9. S. Har-Peled, D. Roth, and D. Zimak. Constraint classification: a new approach to multiclass classification. In *Proc. ALT-02*, pp. 365–379, Lübeck, 2002. Springer.
10. E. Hüllermeier and J. Fürnkranz. Comparison of ranking procedures in pairwise preference learning. In *Proc. 10th Int. Conf. Information Processing and Management of Uncertainty in Knowledge-Based Systems (IPMU-04)*, Perugia, 2004.
11. I. Tsochantaridis, T. Hofmann, T. Joachims, and Y. Altun. Support vector machine learning for interdependent and structured output spaces. In *Proc. 21st Int. Conf. on Machine Learning (ICML-2004)*, pp. 823–830, Banff, Alberta, Canada, 2004.
12. T.F. Wu, C.J. Lin, and R.C. Weng. Probability estimates for multi-class classification by pairwise coupling. *J. of Machine Learning Research*, 5:975–1005, 2004.

A   Proof of Theorem 1

Lemma 1: Let s_i, $i = 1 \ldots m$, be real numbers such that $0 \leq s_1 \leq s_2 \ldots \leq s_m$. Then, for all permutations $\tau \in \mathcal{S}_m$,

$$\sum_{i=1}^{m}(i - s_i)^2 \leq \sum_{i=1}^{m}(i - s_{\tau(i)})^2 \tag{11}$$

Proof. We have

$$\sum_{i=1}^{m}(i - s_{\tau(i)})^2 = \sum_{i=1}^{m}(i - s_i + s_i - s_{\tau(i)})^2$$

$$= \sum_{i=1}^{m}(i - s_i)^2 + 2\sum_{i=1}^{m}(i - s_i)(s_i - s_{\tau(i)}) + \sum_{i=1}^{m}(s_i - s_{\tau(i)})^2.$$

Expanding the last equation and exploiting that $\sum_{i=1}^{m} s_i^2 = \sum_{i=1}^{m} s_{\tau(i)}^2$ yields

$$\sum_{i=1}^{m}(i - s_{\tau(i)})^2 = \sum_{i=1}^{m}(i - s_i)^2 + 2\sum_{i=1}^{m} i\, s_i - 2\sum_{i=1}^{m} i\, s_{\tau(i)}.$$

On the right-hand side of the last equation, only the last term $\sum_{i=1}^{m} i\, s_{\tau(i)}$ depends on τ. Since $s_i \leq s_j$ for $i < j$, this term becomes maximal for $\tau(i) = i$. Therefore, the right-hand side is larger than or equal to $\sum_{i=1}^{m}(i - s_i)^2$, which proves the lemma.   □

Lemma 2. *Let* $\mathsf{P}(\cdot \mid x)$ *be a probability distribution over* $\mathcal{S}_m$ *and let* $p(\tau) \stackrel{\mathrm{df}}{=} \mathsf{P}(\tau \mid x)$. *Moreover, let*

$$s_i \stackrel{\mathrm{df}}{=} m - \sum_{j \neq i} \mathsf{P}(\lambda_i \succ_x \lambda_j) \tag{12}$$

with

$$\mathsf{P}(\lambda_i \succ_x \lambda_j) = \sum_{\tau\,:\,\tau(j) < \tau(i)} \mathsf{P}(\tau \mid x). \tag{13}$$

Then, $s_i = \sum_{j \neq i} p(\tau)\, \tau(i)$.

Proof. We have

$$s_i = m - \sum_{j \neq i} \mathsf{P}(\lambda_i \succ_x \lambda_j) = 1 + \sum_{j \neq i}(1 - \mathsf{P}(\lambda_i \succ_x \lambda_j))$$

$$= 1 + \sum_{j \neq i} \mathsf{P}(\lambda_j \succ_x \lambda_i) = 1 + \sum_{j \neq i} \sum_{\tau\,:\,\tau(j) < \tau(i)} p(\tau)$$

$$= 1 + \sum_{\tau} p(\tau) \sum_{j \neq i} \begin{cases} 1 & \text{if} \quad \tau(i) > \tau(j) \\ 0 & \text{if} \quad \tau(i) < \tau(j) \end{cases}$$

$$= 1 + \sum_{\tau} p(\tau)(\tau(i) - 1) = \sum_{\tau} p(\tau)\, \tau(i)$$

Under the assumption that the base learners' estimates correspond exactly to the probabilities of pairwise preference, i.e.,

$$\mathcal{R}_x(\lambda_i, \lambda_j) = \mathcal{M}_{ij}(x) = \mathsf{P}(\lambda_i \succ_x \lambda_j), \tag{14}$$

$s_i \leq s_j$ is equivalent to $S(\lambda_i) \geq S(\lambda_j)$. Thus, ranking the alternatives according to $S(\lambda_i)$ (in decreasing order) is equivalent to ranking them according to s_i (in increasing order).

Theorem 1. *The expected distance*

$$E(\tau') = \sum_\tau p(\tau) \cdot D(\tau', \tau) = \sum_\tau p(\tau) \sum_{i=1}^m (\tau'(i) - \tau(i))^2$$

becomes minimal by choosing τ' such that $\tau'(i) \leq \tau'(j)$ whenever $s_i \leq s_j$, with s_i given by (12).

Proof. We have

$$E(\tau'_x) = \sum_\tau p(\tau) \sum_{i=1}^m (\tau'_x(i) - \tau(i))^2$$

$$= \sum_{i=1}^m \sum_\tau p(\tau)(\tau'_x(i) - \tau(i))^2$$

$$= \sum_{i=1}^m \sum_\tau p(\tau)(\tau'_x(i) - s_i + s_i - \tau(i))^2$$

$$= \sum_{i=1}^m \sum_\tau p(\tau) \left[(\tau(i) - s_i)^2 - 2(\tau(i) - s_i)(s_i - \tau'(i)) \right.$$

$$\left. + (s_i - \tau'(i))^2 \right]$$

$$= \sum_{i=1}^m \left[\sum_\tau p(\tau)(\tau(i) - s_i)^2 - 2(s_i - \tau'(i)) \cdot \right.$$

$$\left. \cdot \sum_\tau p(\tau)(\tau(i) - s_i) + \sum_\tau p(\tau)(s_i - \tau'(i))^2 \right]$$

In the last equation, the mid-term on the right-hand side becomes 0 according to Lemma 2. Moreover, the last term obviously simplifies to $(s_i - \tau'(i))$, and the first term is a constant $c = \sum_\tau p(\tau)(\tau(i) - s_i)^2$ that does not depend on τ'. Thus, we obtain $E(\tau'_x) = c + \sum_{i=1}^m (s_i - \tau'(i))^2$ and the theorem follows from Lemma 1. □

FCLib: A Library for Building Data Analysis and Data Discovery Tools

Wendy S. Koegler and W. Philip Kegelmeyer

Sandia National Laboratories, P.O. Box 969, Livermore CA 94551, USA
{wkoegle, wpk}@ca.sandia.gov

Abstract. In this paper we describe a data analysis toolkit constructed to meet the needs of data discovery in large scale spatio-temporal data. The toolkit is a C library of building blocks that can be assembled into data analyses. Our goals were to build a toolkit which is easy to use, is applicable to a wide variety of science domains, supports feature-based analysis, and minimizes low-level processing. The discussion centers on the design of a data model and interface that best supports these goals and we present three usage examples.

1   Introduction

In the past decade we have witnessed a mad race to create really big machines to run really big simulations [5]. Unfortunately, the creation of data analysis tools to handle really big data has lagged behind, particularly tools that support data discovery.

Data discovery is the iterative process of exploring data to extract information. It is by nature a human driven process, where an analyst uses a combination of data analysis, visualization, and other post-processing tools. Data analysis is an especially important tool when quantitative information is desired. As data increases in size and complexity, current methods (which depend on analysts examining all of the data and moving data between tools) become more cumbersome, slow, and error-prone. Better tools need to be developed that automate as much low-leveling processing as possible, yet still allow the analyst the freedom to flexibly compose their own chains of analysis. Further, a key feature of data analysis in large data lacking in most tools is the native ability to identify, manipulate and analyze small regions of interest (ROIs).

Our proposed solution is to provide the analyst with a toolkit of data analysis building blocks which can be used to assemble analyses for data discovery. Instead of worrying about low-level details, the analyst can compose data analyses at a higher level. We were not satisfied with existing tools (see Sec. 2) and so built our own as a library of C routines. Our focus was on developing an interface that better supports data analysis and discovery in large data, rather than performance, which can be improved later. We present here the basic structure of the library and evaluate whether we met our goals:

A.F. Famili et al. (Eds.): IDA 2005, LNCS 3646, pp. 192–203, 2005.

1. Minimize low-level processing.
2. Support feature-based analysis (where a feature is a region of interest).
3. Provide a variety of building blocks and make them easy to couple together.
4. Provide an 'elegant' interface – simple and easy to use, but still powerful.
5. Be general – applicable to a variety of science domains.

2   Background

The type of data that we address represents physically based phenomena, usually varying with time (i.e. spatio-temporal data). This type of data can be experimentally measured or created by simulation, and is generally represented as meshes with associated variable fields. A common mode of analysis in such data is to focus on regions of interest or *features*. Features are coherent structures that persist over some period of time. Some examples include vertex tubes in fluid dynamical systems, failure zones in mechanical systems, and hot spots in chemical systems. Although feature-based analysis is a common process for analysis, it is usually not supported by data analysis tools. Analysts typically hand-select regions of interest and export them for further analysis.

Our original plan was to build support for feature-based analysis into an existing data analysis system, but we could not find a satisfactory candidate and so ended up building our own: the Feature Characterization Library (FCLib). During evaluation we concluded that data access methods optimized for simulation are not optimized for post-processing [1]. For example, during a simulation it is necessary to write all variable fields at a single time step, but during analysis one usually wants access to a single variable field over all time steps.

We evaluated three potential systems in detail and found them unsatisfactory for the following reasons: TeraScale's Parallel Mesh Object (PMO) [4] because the data interface was focused on simulation; Sandia's internal Data Object Library (DOL) because its interface was too complicated; and Kitware's visualization toolkit (VTK) [3] because it does not inherently support time-varying data. In addition we felt that in all cases there was ample room to improve the interfaces for data discovery.

3   The Data Model

Because all data analyses must interact with the data, the data model and its interface strongly influence our usability goals for the toolkit. Our data model was evolved rather than designed; we started with a very simple model and adapted it as we built in the desired analysis capabilities. The data model design was also influenced by our experience with many systems including the Sets and Fields data model (SAF) [2] and those mentioned in Sec. 2. In this section we will first summarize the basic structure of the data model and then discuss how and why our model differs from the norm.

3.1   The Primary Data Objects

The data model consists of five core data objects, which are hierarchically organized as shown in Fig. 1.

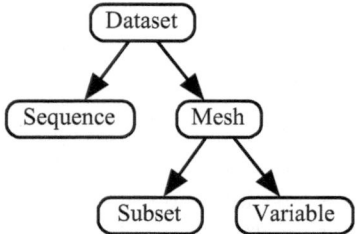

Fig. 1. The five primary data objects in the data model. The arrows indicate ownership relations (i.e. datasets own meshes and meshes own subsets).

1. **Datasets.** A *dataset* is a container for all data that correspond to a single simulation or experiment. A dataset can own multiple sequences and meshes.
2. **Sequences.** The data fields in a dataset often vary over some parameter space, most commonly time. A *sequence* describes the extent of such a parameter space (e.g. how many time steps and the time values at each step). A dataset can have multiple sequences but typically has only one.
3. **Meshes.** *Meshes* represent the spatial organization of the data. Each mesh is described by a set of vertices and a set of elements, which describe the mesh's geometry (spatial location) and topology, respectively. The topology of the edges and faces of the mesh are also available to the interface. Vertices, edges, faces and elements are all different types of *mesh entities*. The interface also has a special mesh entity type called *whole* which is used to represent the entire mesh.
4. **Subsets.** A *subset* describes an arbitrary collection of mesh entities in its parent mesh (e.g. the set of faces that describes the outside skin of the mesh).
5. **Variables.**
   (a) A *variable* contains a single data field associated with its parent mesh. The data field will be over all of one of the mesh entity types, e.g. temperature on the elements, or a velocity vector on the vertices.
   (b) Data fields that vary over the sequence parameter space are represented by an ordered array of variable objects, one for each step (e.g. a temperature variable for each time step of a time-varying temperature field). This ordered array of variables is called a *sequence variable*.

3.2   Design Discussion

Some of our most important design decisions were:
1. Have a minimal set of abstract data objects.
2. Subsets are full-fledged data objects.

3. Time-varying data is available per variable instead of per step.
4. Meshes have ownership of data (instead of global ownership).
5. Data objects have parent and child pointers.

The first design decision keeps the data model interface general yet small. The second provides support for feature-based analysis. The third and fourth design decisions support post-processing data access methods. And the last helps to keep the interface simple.

The break down of unstructured data into meshes and variables is nothing new (although it is not as common to support sequences and subsets as separate, full-fledged data objects). What we have done differently is to have a very small set of general data objects, and to move the handling of differences in the specifics of these data objects deeper into the interface. For example, many data models have separate types of variables depending on what mesh entity they are associated with, e.g. nodal variables and element variables. Instead, FCLib has a single variable object which you query for its type only when you need to know it. This allows us to design a smaller and simpler interface for data analysis.

We introduced the subset data object to represent regions of interest and features. It also turns out that being able to manipulate arbitrary subsets is extremely handy for supporting various characterization building blocks and creating data analyses. It is common for simulation codes to have specialized subsets for boundary conditions and other specifications, but neither simulation nor data analysis codes usually support arbitrary subsets.

One very difficult decision we faced was how to represent variable fields with a minimal number of data objects and still support both time-varying and non-time-varying variables. We choose to have the variable object encapsulate a single time step of variable data so that non-time-varying variables were a single variable and time-varying variables were a sequence of variables. An alternative would be to have a variable object encapsulate all time steps. This second option aligns better with our design goal of a single data object; but it ends up cluttering the interface for most variable routines because many operate per time step and the user would have to pass additional information about which time step is of interest. With the first choice we have some duplication of a few routines that need to handle time-varying and non-time-varying variables slightly differently, but overall the data model and data analysis interfaces are simpler.

As mentioned in Sec. 2, post-processing data access differs in a number of ways from simulation data access. For example, the analyst may only need to examine a few of the meshes in the dataset, or will examine the meshes one at a time in different ways. FCLib supports this by not having global subsets and variables. Instead, we choose to have subsets and variables be owned by meshes. This means, for example, that what one usually thinks of a single variable is split across meshes, e.g. there can be a temperature variable on each mesh and the temperature of the entire dataset is the collection of all of these variables. This design choice does make global operations over all meshes slightly more complicated, especially for datasets that have meshes that share vertices — a

Table 1. Partial listing of FCLib building blocks

	Mesh Topology
Get mesh entity children	E.g. get vertices that make up an element
Get mesh entity parents	E.g. get elements that contain a vertex
Get mesh entity neighbors	Get adjacent entities within a mesh
Segment	Separate mesh or subset into connected components
Get skin	Get the outside edges or faces of a mesh or subset
	Mesh Geometry (Spatial)
Find entities	Get mesh entities within bounding box or sphere
Get sizes	Determine edge lengths, surface areas, region volumes
Bounding box routines	Determine axis-aligned boundary of meshes and subset, can also combine and test for overlap of BBs
Centroid routines	Find center of mass of meshes and subsets
	Variable
Variable math	Create new variables as mathematic combinations of current variables (+, *, sqrt(), pow(), etc.)
Statistics routines	Determine min/max/mean/st.dev./sum
Decompose vectors	Decompose into normal and tangent components against an arbitrary vector
Kernel smooth variables	Replace variable field values with local averages
Threshold	Find subset of entities that meet threshold criteria
	Subsets
Set operations	Create new subsets using AND, OR, or XOR
Feature tracking	Track subsets over time (see Sec. 5.1)

naive global operation may double count those vertices. Another way that post-processing data access differs from simulation is that the analyst will access all time steps of a single variable while simulations access all variables at a single time step. Our data model favors the post-processing data access method, but does not preclude other access methods.

Another key design in our data objects is that they can be queried for their parent *and* child data objects. This helps to keep the interface simple because we can pass just one type of data object to a routine, but the routine can still access the object's parents or children if more information is needed.

4   Analysis Building Blocks

The bulk of the library that is not the data model is the routines for constructing data analyses. The data model and the analysis building blocks were co-evolved to allow easy chaining of building blocks into analyses, and much of the previous data model discussion applies here as well. The same philosophy of hiding details about the data objects applies and routines generally "do the right thing" for different types of an object.

A listing of some of the available building blocks and analyses is shown in Table 1. Building blocks range from simple helper routines like "get the Euclidean distance between two points" (not shown in table) to complete analyses like "determine spot weld breakage factors" (see Sec. 5.2). Because the library already supports a wide variety of building blocks, we are confident that new building blocks can be easily incorporated.

5   Data Analysis Examples

In this section we present three examples of using FCLib. The first is a simple feature-based analysis example intended to demonstrate the 'elegance' of the interface. The other two are real life examples and demonstrate using FCLib as an analysis development environment.

5.1   Feature Tracking and Feature-Based Analysis

This example is intended to show how easy it is to do feature-based analysis using FCLib. The problem is to find hot spots in a dataset and to report the maximum temperature of each hot spot. The complete code to accomplish this is shown in Fig. 2. For clarity, error checking and cleaning up of allocated memory has been left out. The procedure can be divided into three major steps: 1) setup (get handles to the appropriate data objects); 2) identify and track features; and 3) analyze features. Steps 2 and 3 are highlighted with boxes in the figure.

In FCLib, feature tracking consists of identifying regions of interest (ROIs) in each time step and then matching these regions up across time steps to create the features. In this example, ROIs are found by first finding all mesh entities that have a temperature greater than 30 using the library's thresholding routine, and then dividing these entities up into separate regions using the library's segmentation routine. Both the intermediate threshold subset and the final ROIs can be stored as subset data objects. As ROIs are found in each time step, they are passed to a feature tracking routine which decides how the ROIs are assembled into features. The default tracking algorithm decides this entirely on spatial overlap (i.e. if an ROI from the current time step occurs in the same spatial location as an ROI in a previous time step, they are the same feature). There are a variety of tracking algorithms and the interface permits the user to provide their own.

After feature tracking is performed, information about the system of features is available from the feature group object. The group knows which ROIs make up the features, and how features may split and merge to form new features. The feature graph shown in Fig. 3(a) (produced at line 26 in Fig. 2) represents how the hot spots interact. At $t0$ there are two features. At $t1$, Feature 0 splits into Features 2 and 3. At the next time step, Features 1 and 2 merge into Feature 4. Feature 3 ceases to exist at $t4$. Etc. Images of the features are provided in Fig. 4.

The feature group is also used to get the information needed to do feature-based analysis. The last part of the code in Fig. 3(b) iterates over each feature and uses a subset aware statistics routine to find the maximum temperature of

```
// This programs finds features--defined as regions where the variable
// is greater than 30--and then writes out the feature graph and prints
// the maximum value of the variable within the features.
01  #include <fc.h>
02
03  int main() {
04    int i, j, numStep, numROI, numFeature, *stepIDs;
06    FC_Dataset dataset;
07    FC_Mesh mesh;
08    FC_Variable *seqVar;
09    FC_Subset subset, *ROIs;
10    FC_FeatureGroup *featureGroup; // feature container
11    double min, max;
12
13    // initialize library and get data object handles
14    fc_initLibrary();
15    fc_loadDataset("gaussians.saf", &dataset);
16    fc_getMeshByName(datatset, "grid", &mesh);
17    fc_getSeqVariableByName(mesh, "temperature", &numStep, &seqVar);
18
19    // find and track features
20    fc_createFeatureGroup(&featureGroup);
21    for (i = 0; i < numStep; i++) {
22      fc_createThresholdSubset(seqVar[i], ">", 30, &subset);
23      fc_segment(subset, 0, &numROI, &ROIs);
24      fc_trackStep(i, numROI, ROIs, featureGroup);
25    }
26    fc_writeFeatureGraph(featureGroup, "graph.dot");
27
28    // report statistics of features
29    fc_featureGroup_getNumFeature(featureGroup, &numFeature);
30    for (i = 0; i < numFeature; i++) {
31      printf("Feature %d:\n", i);
32      fc_getFeatureROIs(featureGroup, i, &numROI, &stepIDs, &ROIs);
33      for (j = 0; j < numROI; j++) {
34        fc_getVariableSubsetMinMax(seqVar[stepIDs[j]], ROIs[i],
35                                   &min, &max);
36        printf("  time = %2d: max = %5.1f\n", stepIDs[j], max);
37      }
38    }
39
40    // clean up and exit
41    fc_finalLibrary();
42    exit(0);
43  }
```

Fig. 2. A complete program for performing feature tracking and report feature statistics. The first box highlights the core code for feature tracking and the second box the core code for feature analysis.

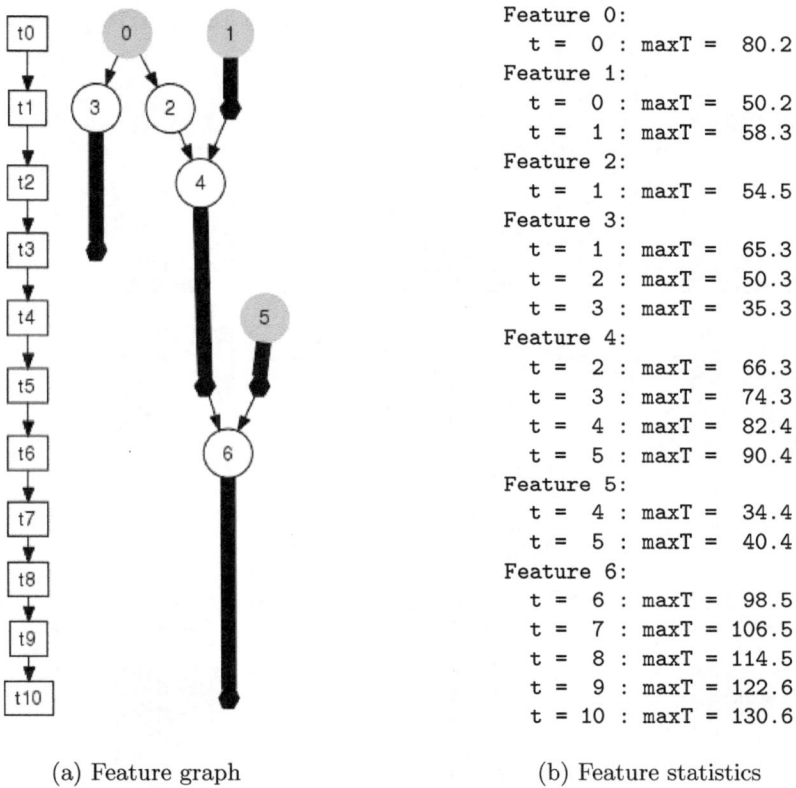

Feature 0:
    t =   0 : maxT =   80.2
Feature 1:
    t =   0 : maxT =   50.2
    t =   1 : maxT =   58.3
Feature 2:
    t =   1 : maxT =   54.5
Feature 3:
    t =   1 : maxT =   65.3
    t =   2 : maxT =   50.3
    t =   3 : maxT =   35.3
Feature 4:
    t =   2 : maxT =   66.3
    t =   3 : maxT =   74.3
    t =   4 : maxT =   82.4
    t =   5 : maxT =   90.4
Feature 5:
    t =   4 : maxT =   34.4
    t =   5 : maxT =   40.4
Feature 6:
    t =   6 : maxT =   98.5
    t =   7 : maxT = 106.5
    t =   8 : maxT = 114.5
    t =   9 : maxT = 122.6
    t = 10 : maxT = 130.6

(a) Feature graph   (b) Feature statistics

Fig. 3. Results of the feature tracking program in Fig. 2. Note that raw feature graph code was produced at line 26 and was post-processed with Graphviz's dot program to produce the graphic shown in (a). In (a), time runs from top to bottom; a labeled circle indicates the start of a feature (gray means no parents); and a thick black line indicates continuation of a feature.

the feature for each time step it exists. The results are shown in Fig. 3. After examining the results, feature tracking can let us conclude that most hot spots increased in temperature with time, except for Feature 3. The program can easily be expended to further analyze features by adding other functions, like getting the size of the ROI or statistics of other variables, into the analysis loop.

5.2 Evaluating Spot Weld Breakage

One of our first designs was to analyze spot weld breakage in mechanics simulations. The analyst's original method was to plot the force variable on each spot weld, and then to determine by visual inspection if and when the force dropped to zero, indicating that the spot weld had failed.

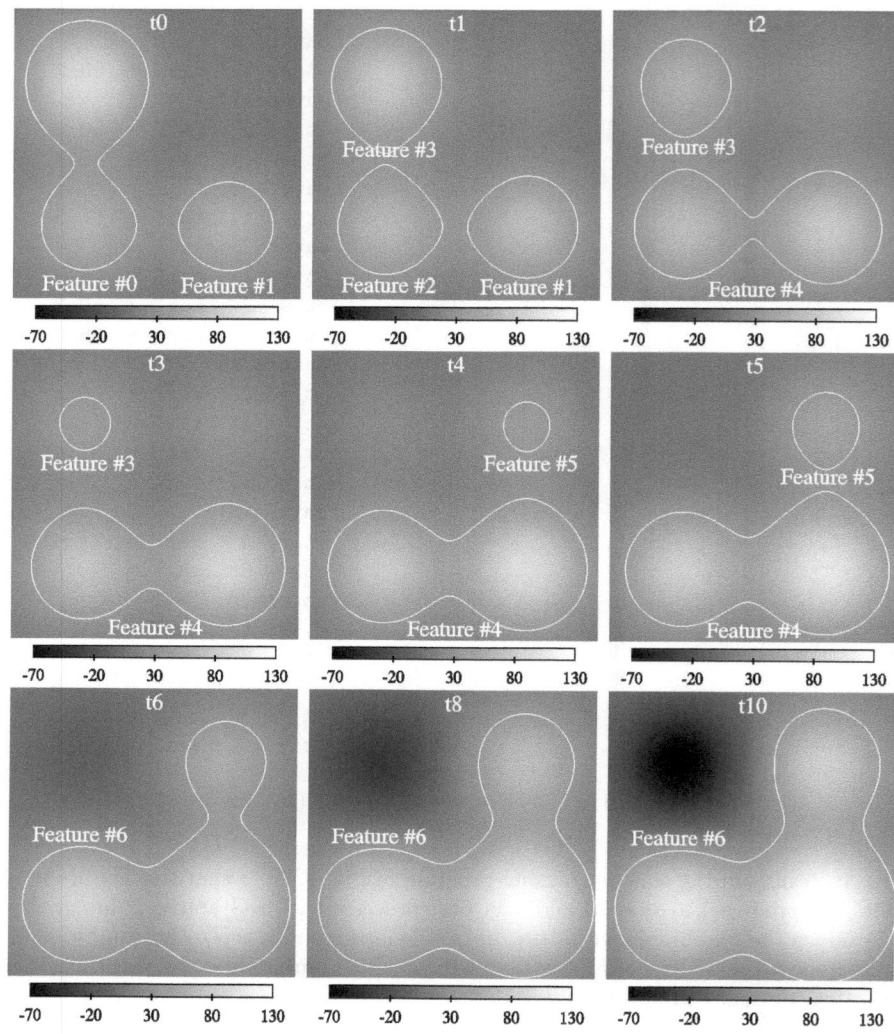

Fig. 4. Selected time steps of the dataset used in the feature tracking program in Fig. 2. The temperature variable was modeled with four Gaussians growing and shrinking at different rates. A contour line is drawn at 30 to visually identify the features corresponding to the results in Fig. 3.

In our first iteration of tool development we made two major improvements to this method. First, we completely automated the identification of failed spot welds and the collection and reporting of the results. We created a command line tool that took the input file used to run the simulation in addition to the resulting dataset, and used the input file to determine which subsets were spot welds and how to group spot welds by their specified parameters. The second improvement we made was to resolve the force vector into normal and tangential components relative to the instantaneous plane of the spot weld surface and to use the normal force for the calculation. This just consists of a library call to find the surface normal of the spot weld subset and then another library call to decompose the force vector into components relative to the surface normal. In the original analysis, the analysts used whichever axis-aligned component of the force vector was closest to normal to the starting orientation of the spot weld surface. The FCLib analysis was more correct and could handle situations where the spot weld surface changed orientation during the run.

In the second iteration of the tool, the analysts took advantage of our ability to easily access and use geometry, and asked for a more correct analysis of the spot welds. Instead of watching the force, which was a result of spot weld breakage, we reimplemented the same spot weld breakage calculation that the simulation code used to decide whether spot welds failed. This factor was based on the displacement of a spot weld's two attachment points and depends on resolving the displacement into normal and tangential components with reference to the plane of the spot weld. Besides giving more correct results for when spot welds failed, the spot weld failure factor can be used to judge how close spot welds are to failing. The analysts were extremely pleased with this additional measure that allowed them to more quantitatively assess their results.

2614 total lines of code were written for this analysis, including comments, a parser for the simulation input file which can be reused for other analyses, command-line argument handling, and error handling. The number of lines of core code for the subroutine that does each spot weld calculation was less than 100.

5.3   Developing a Gap Analysis

Often analysts don't know exactly how to analyze what they looking for and they need an environment where they can assemble and tweak analyses. For example, one of our analysts is doing mechanics simulations where a complex assembly of parts is exposed to shocks and stresses. He wants to know where gaps form between parts that should abut and he wants to be able to characterize these gaps (how big, etc).

Our initial iteration for this analysis is to create lines that join surfaces that started out near each other. Our next step will be to use these lines to quantify the gapping process. To create these *gap lines*, we first find surfaces that start out adjacent to each other. This is done by skinning each part to get a subset of all faces that are on the outside of it. Next we check the faces in pairs of skins to see if any are adjacent and collect those into a subset of shared faces. A line mesh is then created joining the shared surfaces. The lines are initially of zero length; as time progress, if any of these lines gets a non-zero length, then we know we have a gap forming. Fig. 5 shows this analysis applied to a gasket-like ring that falls out of the lip of a can.

When we take this initial result to the analysts they are very excited and have all sorts of ideas about what to do next. The analysts have lots of ideas about potential analyses, but lack the tools that will let them easily play with those ideas.

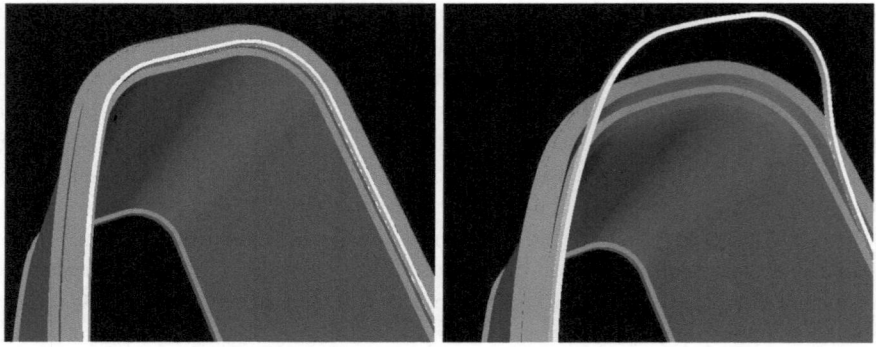

(a) At time zero (left), the white gasket-like ring rests inside the lip of the can. As time progresses (right), the ring falls out of the can (the orientation is upside down for better viewing).

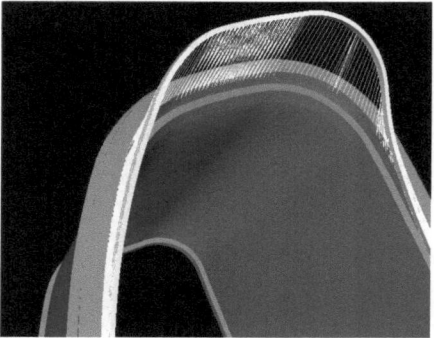

(b) The gap analysis creates line elements that join faces and vertices that were adjacent at time zero.

Fig. 5. Visualization of the gap analysis results

6   Conclusions and Future Directions

In this paper we have presented and discussed the design of an interface for a library for building data analysis and data discovery tools. We believe that the key components of design for a toolkit that will be useful for analysts are that the interface is easy to use but still generally applicable to a wide variety of science domains. Also, the interface must support post-processing data access methods, the most important of which are features. The toolkit itself should minimize low-level processing and provide a variety of building blocks for creating new data analyses.

For future work, we plan to expand the set of building blocks and to build more specialized analyses for analysts. Most of our recent analyses have been for mechanics simulations but there are many interesting analysis to be done for flow and chemical systems. We are also working on a GUI for the library to do all prebuilt analyses and allow some chaining of building blocks into simple analyses. Other interesting avenues of work include integration of visualization with the library and performance issues.

Acknowledgments

Sandia is a multiprogram laboratory operated by Sandia Corporation, a Lockheed Martin Company, for the United States Department of Energy's National Nuclear Security Administration under contract DE-AC04-94AL8500. This research was supported by the DOE Advanced Strategic Computing Initiative (ASCI) Data and Visualization Sciences (DVS) Data Discovery Program.

References

1. W.S. Koegler, W.P. Kegelmeyer. One Users's Report on Sandia Data Objects: Evaluation of the DOL and PMO for Use in Feature Characterization. SAND2003-8591, Sandia National Laboratories, 2003.
2. M.C. Miller, J.F. Reus, R.P. Matzke, W.J. Arrighi, L.A. Schoof, R.T. Hitt, P.K. Espen. Enabling Interoperation of High Performance, Scientific Computing Applications: Modeling Scientific Data With the Sets & Fields (SAF) Modeling System. International Conference on Computational Science (ICCS-2001), San Francisco. March, 2001.
3. W. Schroeder, K. Martin, B. Lorensen. The Visualization Toolkit, 2nd Edition. Prentice Hall PTR, 1998. See also http://www.kitware.com.
4. TeraScale, LLC. The Parallel Mesh Object Annotated Reference Manual: Version 1.0. LLC Report TSC02-01, June 24, 2002. See also http://www.terscale.net.
5. University of Mannheim, University of Tennesse, NERSC/LBL. TOP500 Supercomputer Sites, http://www.top500.org.

A Knowledge-Based Model for Analyzing GSM Network Performance

Pasi Lehtimäki and Kimmo Raivio

Helsinki University of Technology,
Laboratory of Computer and Information Science,
P.O. Box 5400, FIN-02015 HUT, Finland

Abstract. In this paper, a method to analyze GSM network performance on the basis of massive data records and application domain knowledge is presented. The available measurements are divided into variable sets describing the performance of the different subsystems of the GSM network. Simple mathematical models for the subsystems are proposed. The model parameters are estimated from the available data record using quadratic programming. The parameter estimates are used to find the input-output variable pairs involved in the most severe performance degradations. Finally, the resulting variable pairs are visualized as a tree-shaped cause-effect chain in order to allow user friendly analysis of the network performance.

1 Introduction

The radio resource management in current mobile communication networks concentrates on maximizing the number of users for which the services can be provided with required quality, while using only limited amount of resources [7]. Once the network is designed and implemented, the goal is to find a network configuration parameters that use the existing resources as efficiently as possible from the user point of view. In practice, this means that a reasonable tradeoff between the coverage and capacity of the network must be found. Good coverage allows users to initiate services at any location with acceptable service quality, while high capacity allows many network subscribers to use services simultaneously. However, improving the coverage tends to diminish the capacity and vice versa. A good tradeoff between coverage and capacity is obtained when the number of service denials (blocking) and abnormal service interruptions (dropping) are at the minimum, i.e the performance of the network is well optimized.

In this paper, the performance of a GSM network is analyzed based on massive data records and application domain knowledge. Next, the GSM network infrastructure is shortly outlined. In Section 3, a hierarchical model for describing the network performance is proposed. Then, the usage of the proposed model as a part of an analysis process is presented. In Section 5, results of the experiments are presented.

A.F. Famili et al. (Eds.): IDA 2005, LNCS 3646, pp. 204–215, 2005.

2   The GSM Network

A GSM network consists of high number of sites, each usually having three base station transceivers (BTS) positioned to cover separate sectors around the site (see Figure 1). Each BTS has one or more transceiver/receiver pairs (TRX), each allocated on a single physical radio frequency. Base station controller (BSC) manages the operation of several BTSs connected to it through the Abis interface. A single mobile services switching center (MSC) is connected to several BSCs through the A interface.

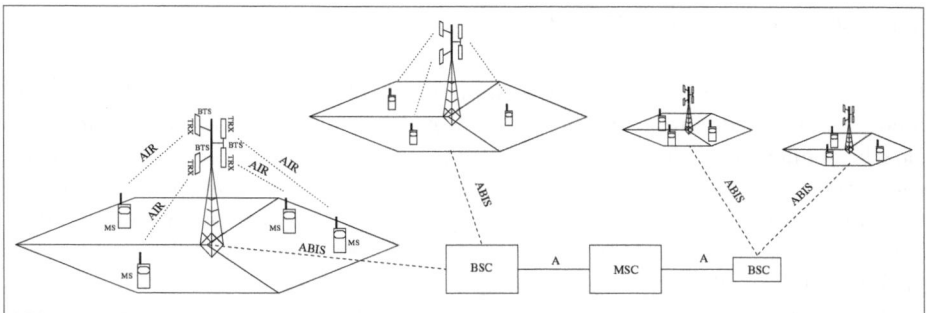

Fig. 1. GSM network architecture

The performance of the mobile network is measured based on thousands of counters, describing the numbers of the most important events over a measurement period (typically one hour). In order to allow more efficient performance monitoring, a set of high-level key performance indicators (KPIs) are derived from the counter data. Typically, the KPIs describe the success/failure rates of the most important events such as service blocking, service dropping and handovers.

Such indicators are traditionally used in resource management [2] and they are well suited for performance monitoring [3,5,6], but there are several drawbacks when they are used in fault diagnosis [4]. For example, most KPIs compute the sum of attempts failed due to different causes and divide the result by the number of all attempts (failed and succeeded). Therefore, the information about the actual causes of the failures is lost. Also, the most widely used performance indicators describe the operation of the network at the BTS level. As a result, the performance degradations originating from interaction between several BTSs become very difficult to observe. In many cases, however, the operation of the close-by BTSs are highly dependent on each other. Examples of operation in which several BTSs interact are handovers between close-by BTSs and interference between BTSs having TRXs on the same physical frequency.

In this work, we aim to avoid the above mentioned KPI-related problems by using a novel performance analysis approach based on counter data. Due to the significant increase in number of variables, a knowledge-based model is used to divide the analysis process into a set of small system identification problems in order to keep the overall analysis process tractable.

3   A Model for GSM Network Performance

The basic idea in the developed model is to measure performance in terms of the number of failed operations in the network. Examples of such operations are an attempt to allocate a signaling channel, an attempt to allocate a traffic channel for a call, or to perform a handover between two neighboring BTSs. In order to obtain good performance, the number of failing operations within the network must be minimized.

3.1   Model Structure Selection

In order to make the analysis process more tractable, the set of available measurements were divided into counter groups. Within a counter group, the counters are clearly connected while the counters from different groups are independent on each other. Therefore, the modeling problem is more easily solved by identifying a subsystem model for each counter group separately. This phase of the model construction process consisted of extensive literature research and repeated data visualization, until a realistic grouping for the counters were obtained.

In Figure 2, the memberships of the variables in different counter groups (subsystems) are shown. The subsystems tend to form a hierarchical structure (see Figure 3), i.e the outputs of a subsystem describing some low-level phenomena can be an input to a higher-level subsystem. Next, the principles used to divide the data generating system into subsystems is described.

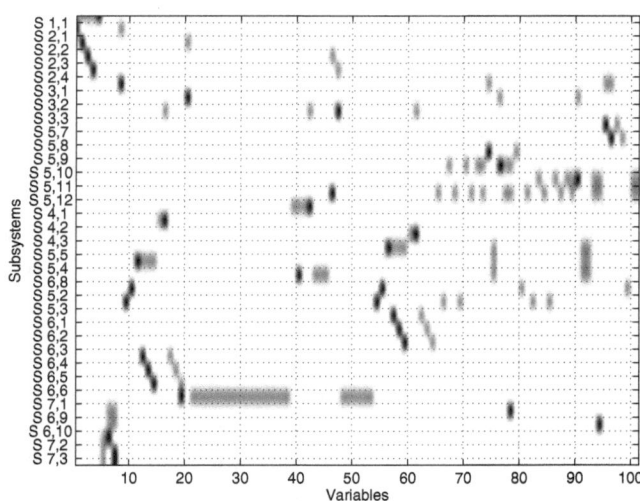

Fig. 2. The data set contains 101 variables (x-axis) and 33 subsystems (y-axis). This plot shows the set of input variables (gray) and output variables (black) that belong to each of the subsystems.

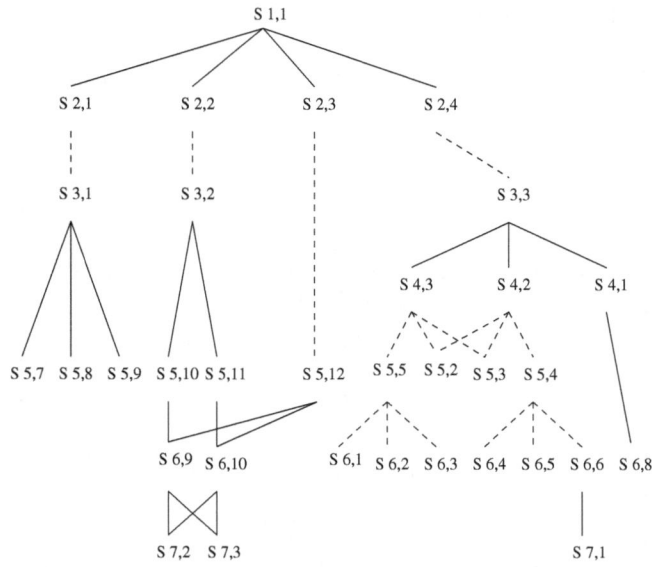

Fig. 3. The subsystem hierarchy. The solid lines indicate that the input of the upper level system contains outputs of the lower level system from the same BTS only. The dashed line indicates, that the input of the upper level system contains input signals from lower level systems of the other BTSs also.

User Perceived Quality. The main purpose of the analysis is to locate bottlenecks in the network performance that have direct impact on the user perceived quality. The system $S_{1,1}$ describes the number of user perceived quality problems by summing the problems from four different categories, each focusing on different part of the transaction. This model is of type I (see Table 1), in which the number of user perceived quality problems in the whole network is the output variable $y(t)$ and the set of input variables $x_i(t)$ consist of number of blocked channel requests, the number of call setup failures, the number of calls dropped during transaction, and the number of failed handovers, each computed over the whole analyzed network. The contribution of each problem type into the overall user perceived quality is described by the parameter a_i, describing the percentage of the failures of type x_i to the total number of failures y.

Blocking. The number of blocked channel requests (rejects) measure the networks ability to satisfy the demand generated by the network users. System $S_{2,1}$ with a model of type I describes how many blocked requests $y(t)$ in the network originate from BTS i of the network (the variable $x_i(t)$ is the number of blocked requests in BTS i at time t). The contribution of BTS i to the number of all blocked requests in the analyzed network is described by the parameter a_i.

The system $S_{3,1}$ is described by a model of type I, in which the proportions of stand-alone dedicated control channel (SDCCH) rejects $x_1(t)$, full rate traffic channel (FR-TCH) rejects $x_2(t)$ and half rate traffic channel (HR-TCH) rejects $x_3(t)$ to the total number of rejected channel requests $y(t)$ is computed. This

Table 1. The different types of models for the subsystems

Type	Systems	Model	Parameter estimation
I	$S_{1,1}$, $S_{2,1}$, $S_{2,2}$, $S_{2,3}$, $S_{2,4}$, $S_{3,1}$, $S_{3,2}$, $S_{3,3}$	$y(t) = \sum_i x_i(t)$	$a_i = \sum_t x_i(t)/\sum_t y(t)$
II	$S_{5,7}$, $S_{5,8}$,	$y(t) = ax(t) + b$ $= [\,a\ b\,]\begin{bmatrix} x(t) \\ 1 \end{bmatrix}$ $= \tilde{\mathbf{a}}^T \tilde{\mathbf{x}}(t)$	$\min_{\tilde{\mathbf{a}}} \frac{1}{2}\tilde{\mathbf{a}}^T \tilde{\mathbf{X}}^T \tilde{\mathbf{X}}\tilde{\mathbf{a}} - \mathbf{y}^T \tilde{\mathbf{X}}\tilde{\mathbf{a}}$ $\tilde{a}_1 \in [0,1]$ $\tilde{a}_2 \geq 0$
III	$S_{5,9}$ $S_{5,10}$, $S_{5,11}$, $S_{5,12}$, $S_{6,9}$, $S_{6,10}$, $S_{7,1}$	$y(t) = \mathbf{a}^T \mathbf{x}(t) + b$ $= [\,a\ b\,]\begin{bmatrix} \mathbf{x}(t) \\ 1 \end{bmatrix}$ $= \tilde{\mathbf{a}}^T \tilde{\mathbf{x}}(t),$	$\min_{\tilde{\mathbf{a}}} \frac{1}{2}\tilde{\mathbf{a}}^T \tilde{\mathbf{X}}^T \tilde{\mathbf{X}}\tilde{\mathbf{a}} - \mathbf{y}^T \tilde{\mathbf{X}}\tilde{\mathbf{a}}$ $\sum_{i \neq N} \tilde{a}_i = 1,$ $\tilde{a}_{i \neq N} \in [0,1]$ $\tilde{a}_N \geq 0$
IV	$S_{7,2}$, $S_{7,3}$	$y(t) = \mathbf{a}^T \mathbf{x}(t)$	$\min_{\mathbf{a}} \frac{1}{2}\mathbf{a}^T \mathbf{X}^T \mathbf{X}\mathbf{a} - \mathbf{y}^T \mathbf{X}\mathbf{a}$ $a_i \geq 0$
V	$S_{4,1}$, $S_{4,2}$, $S_{4,3}$, $S_{5,2}$ $S_{5,3}$, $S_{6,1}$, $S_{6,2}$, $S_{6,3}$ $S_{6,4}$, $S_{6,5}$, $S_{6,6}$	$\mathbf{y}(t) = \mathbf{x}(t)\mathbf{A},$	$\min_{\mathbf{A}} \sum_i \frac{1}{2}\mathbf{A}_i^T \mathbf{X}^T \mathbf{X}\mathbf{A}_i - \mathbf{y}_i^T \mathbf{X}\mathbf{A}_i$ $\sum_i a_{ij} = 1,$ $a_{ij} \in [0,1]$
VI	$S_{5,4}$, $S_{5,5}$, $S_{6,8}$	$y(t) = a_1(c_1(t) + c_2(t) + c_3(t))x_1(t)$ $+a_2(c_1(t) + c_2(t) + c_3(t))x_2(t)$ $+a_3(c_1(t) + c_2(t) + c_3(t))x_3(t),$	$\min_{\mathbf{a}} \frac{1}{2}\mathbf{a}^T \mathbf{X}^T \mathbf{X}\mathbf{a} - \mathbf{y}^T \mathbf{X}\mathbf{a}$ $\sum_i a_i = 1,$ $a_i \in [0,1]$

model is estimated for each BTS separately. That is, the parameter a_i describes the contribution of the channel type i to the total number of blocked channel requests.

Finally, the systems $S_{5,7}$, $S_{5,8}$ and $S_{5,9}$ describe the above mentioned channel type rejects due to congestion vs. other possible reasons for blocking. These subsystems are described by a model of type II, in which the output variable $y(t)$ describes the number of blocked requests and the input variable $x(t) = C(t)R_{tot}(t)$ describes the proportion of channel requests assumed to have occurred during congestion ($C(t)$ denotes the percentage of time in congestion in time period t and $R_{tot}(t)$ is the total number of channel requests). These models include a bias b since it is not expected that all request rejects are due to congestion, but also other causes may exist (but measurements are not available). The minimization of the mean square prediction error $\frac{1}{T}\sum_t e^2(t) = \frac{1}{T}\sum_t(y(t) - \hat{y}(t))^2$ with the corresponding constraints (see Table 1) leads to a standard quadratic programming problem. For more information about algorithms to solve such problems, see [1].

Call Setup Failures. It is possible, that the user request is not served due to problems in the resource allocation phase (call setup) of the transaction. As in the case of service blocking, a model describing the contributions of each BTS to the total number of call setup failures in the network is defined (model $S_{2,2}$). Basically, the call setup phase includes allocation of a signaling channel in which the negotiation for the actual traffic channel is performed. Model $S_{3,2}$ of type II divides the call setup failures in a single BTS into SDCCH and TCH setup failures separately.

Both the SDCCH and TCH signaling may fail due to problems in different network elements (or interfaces between them) being involved in the signaling procedure. The number of SDCCH and TCH signaling failures due to problems in different network elements are described by the models $S_{5,10}$ and $S_{5,11}$, respectively. These models are of type III and include a bias since it is possible, that call setup failures are caused by other reasons for which measurements are not available. Also, an equality constraint is introduced since it is necessary to require that a single failure in a network element causes the failure of the call setup phase of exactly one transaction.

The most common reasons for failures during the call setup phase or actual service are the inadequate radio signal propagation conditions (problems in radio channel in the air interface). The failures in radio channel are usually due to bad signal quality, i.e the transmitted data includes too many bit errors. The models $S_{6,9}$ and $S_{6,10}$ of type III describe the number of SDCCH and TCH radio channel failures due to bad signal quality vs. other reasons.

The radio signal quality is mostly affected by two components. Firstly, the propagation environment causes attenuation to the transmitted radio signal due to path loss, shadow fading and multipath fading. Secondly, the radio signal may be attenuated by the other radio signals originating from other BTSs having a TRX on the same physical frequency (interference). The purpose of the models $S_{7,2}$ and $S_{7,3}$ of type IV is to compute how many bit errors in uplink (from MS to BTS) and downlink (from BTS to MS) traffic are due to difficult propagation conditions in the BTS's coverage area and how many are likely the result of interference.

Call Dropping. When the call setup phase is successfully completed, the actual service (usually speech in GSM networks) is started. However, the service may be abnormally interrupted (dropped) due to several reasons. The purpose of the model $S_{2,3}$ is to describe how many calls are dropped in BTS i w.r.t the number of dropped calls in the whole analyzed network. The model is of type I and the contributions of each BTSs to the total number of dropped calls is described by the parameter a_i (similarly to the blocking and call setup failures).

The call may be dropped due to internal failures in the network elements or interfaces between network elements. The purpose of the model $S_{5,12}$ is to describe the contributions of the possible causes (very similar to the causes of call setup failures). This model is of type III, having a bias since a call may be dropped due to reasons for which measurements are not available. As in the case of call setup failures, most of the dropped calls are expected to result from radio channel (air interface) problems. Therefore, the same models explaining the number of call setup failures due to bad signal quality also describe the number of dropped calls due to radio channel problems.

Handover Failures. As in the previous cases, also the (outgoing) handover failures (HO) are divided into network level variable and BTS level variables. The model $S_{2,4}$ is used to compute the value for parameters a_i describing the percentage of handover failures originating from BTS i. The model $S_{3,3}$ divides the handover failures according to the handover type (within BTS HO, BSC controlled HO and MSC controlled HO), and a model of type I is used to obtain the contributions of the different handover types into the total number of handover failures.

Both the BSC and MSC controlled outgoing handovers may fail due to problems in the source (serving) BTS or the target BTS. The serving BTS problems can be various BSS problems (very rare in practice) and the target BTS problems may be due to lack of resources, BSS level problems (rare) or problems with the connection (radio link) to the target BTS. Models $S_{4,2}$ and $S_{4,3}$ explicitly describe the dependencies between these failures in different BTSs, i.e the cause for failed outgoing handover may be in lack of resources or connection in any of the BTSs around the same operation area. Model $S_{4,1}$ describes the handover failures within one BTS due to BSS problems or lack of resources. These three models are of type V in which both the equality constraint and box constraints for the parameters are used.

Model $S_{5,2}$ describes the causes for the target BTS radio channel failures and the model $S_{5,3}$ describes the reasons for BSS problems in the target BTS that caused the failed outgoing handover from the serving BTS. Models $S_{5,4}$ and $S_{5,5}$ describe the causes for the failing BSC and MSC controlled outgoing handovers due to lack of resources in the target BTS, respectively. Similarly, model $S_{6,8}$ describes the causes of the lack of resources in within BTS HO attempts. The purpose of these three latter models is to analyze the number of HO failures per HO type (SDCCH-SDCCH, SDCCH-TCH, TCH-TCH) due to SDCCH, HR-TCH and FR-TCH congestion vs. other unmeasured causes. These models are of type VI, and contain signals $c_1(t)$, $c_2(t)$ and $c_3(t)$ that describe the percentages of SDCCH, FR-TCH and HR-TCH congestion w.r.t the length of the measurement period (one hour in our case) and $x_i(t)$ denotes the number of HO attempts.

Since both the BSC and MSC controlled handovers may be of one of the three above mentioned types, there are six different kinds of handovers involving distinct target and source BTS. Models at the level six all describe the percentage of incoming handovers originating from different source BTSs, one model per each HO type. These models can be used to find out which close-by BTSs are generating the major portion of the handover load to the target BTS during handovers. Finally, the model $S_{7,1}$ describes the causes for BSC controlled TCH-TCH handovers (most typical type of handover).

4   Model Based Analysis Process

4.1   Preprocessing

In order to estimate all subsystem models, the data must be carefully preprocessed. In this work, the preprocessing phase includes outlier removing, data segmentation and constant variable pruning. The number of outliers (points clearly differing from other measurements) is quite high in this type of application. Such samples are generated during network reconfigurations or hardware breakdowns, typically lasting only few hours. Such time points should be removed since they do not help in finding the major bottlenecks in network performance. In TRX level model construction, it is also necessary to segment the data into subsegments (time periods) during which the number of TRXs on a certain physical frequency do not change.

After outlier detection and data segmentation, the model parameters can be estimated from the data. However, in some network elements rarely suffering from any types of failures, some or most of the signals in the model are nearly constant. In such a case, the data is not rich enough for estimating a model. Instead, the (nearly) constant input signals are pruned before the model is estimated. In case of a constant signal being an output variable, the model is not estimated at all.

4.2 Visualization of the Dependencies

After the data is cleaned, the models can be estimated using standard quadratic programming techniques. After the parameters of the subsystem models have been estimated, an item to a dependency list is generated per each input-output variable pair. Each item in the dependency list include the strength of the dependency between the input and output variable (the value of the parameter a), a measure of model accuracy (root mean square prediction error (RMSE) of the model), and a measure of models importance in overall network performance analysis (the average number of failures stored in the output variable of the input-output variable pair).

After all the models have been estimated and the properties of each input-output variable pairs are stored in the dependency list, a tree-shaped graph is constructed in order to analyze the cause-effect chains generating the major performance degradations of the network. Since the number of theoretically possible dependencies is extremely large, only the most important dependencies are included to the dependency tree.

Three criteria are used to prune uninteresting dependencies from the tree. Firstly, the model accuracy from which the dependency originates must be at a reasonable level. Otherwise, the analysis might be mislead by very inaccurate models having large values for parameter a (which is forced in several models due to the equality constraints for the parameter vector). Secondly, the output variable of the dependency must be interesting enough (i.e relatively large number of failures must be observed in the output variable). Finally, only the dependencies that belong to the cause-effect chains contributing most to the overall network performance degradations are included into the dependency tree. For each subsystem, different minimum and maximum values for strength of dependency, model accuracy and model interestingness are defined.

5 Experiments

The analyzed GSM network data contained 120 BTSs, in which 101 most important variables (counters) were measured during a two-month time period.

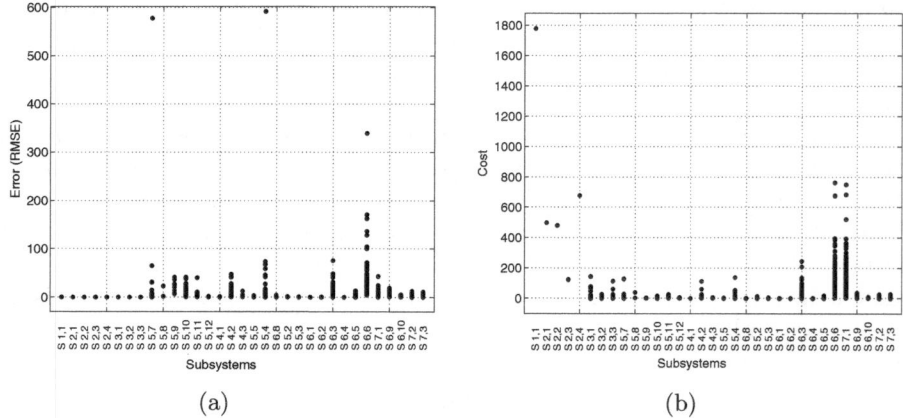

Fig. 4. (a) The prediction errors (RMSE) of the subsystems. (b) The interestingness (cost) of the subsystems.

Since the variables were divided into 33 subsystems consisting of 5 network level systems, 24 BTS level systems, 4 TRX level systems and 143 frequency related segments with non-changing frequency plan, we have $5 + (120 \times 24) + (143 \times 4)$ models with a corresponding parameter vector **a** estimated using quadratic programming.

Figure 4(a) shows the root mean square errors (RMSE) of the models. Note, that the RMSE of the models can be interpreted as the number of user perceived quality problems that could not be explained by the model. Models in which the RMSE is below ~ 50 failures can be regarded as accurate enough in order to make useful inferences about the data. Clearly, there are lots of models that are accurate enough, but also many models are not accurate enough to allow any justified conclusions to be made. Also, three models seem to be very inaccurate.

Figure 4(b) shows the means of the output variables of the models. These values measure the number of user perceived failures of the subsystems. Therefore, it can be regarded as a measure of interestingness or importance of the subsystem in the performance analysis.

Figures 5(a)-(d) show the pruned dependency trees in four separate cases. In Figure 5(a), the cause-effect chains of the most significant blocking problems are shown. Clearly, there are 4 BTSs (6,11,18,85) that suffer from lack of resources. BTSs (6,11,18) suffer from lack of half rate traffic channels and BTS 85 suffers from lack of full rate traffic channels. Only in BTS 6, the causes for blocking can be said to result regularly from congestion.

In Figure 5(b), the results of the corresponding analysis for the call setup failures are shown. Here, four BTSs (17,52,66,74) seem to suffer from call setup failures regularly. In all these four BTSs, the failures tend originate during SD-CCH signaling and fail due to radio link problems. In BTSs 17 and 74 the radio link failures can be said to result from bad downlink signal quality and in BTSs 52 and 66 they are due to bad uplink signal quality.

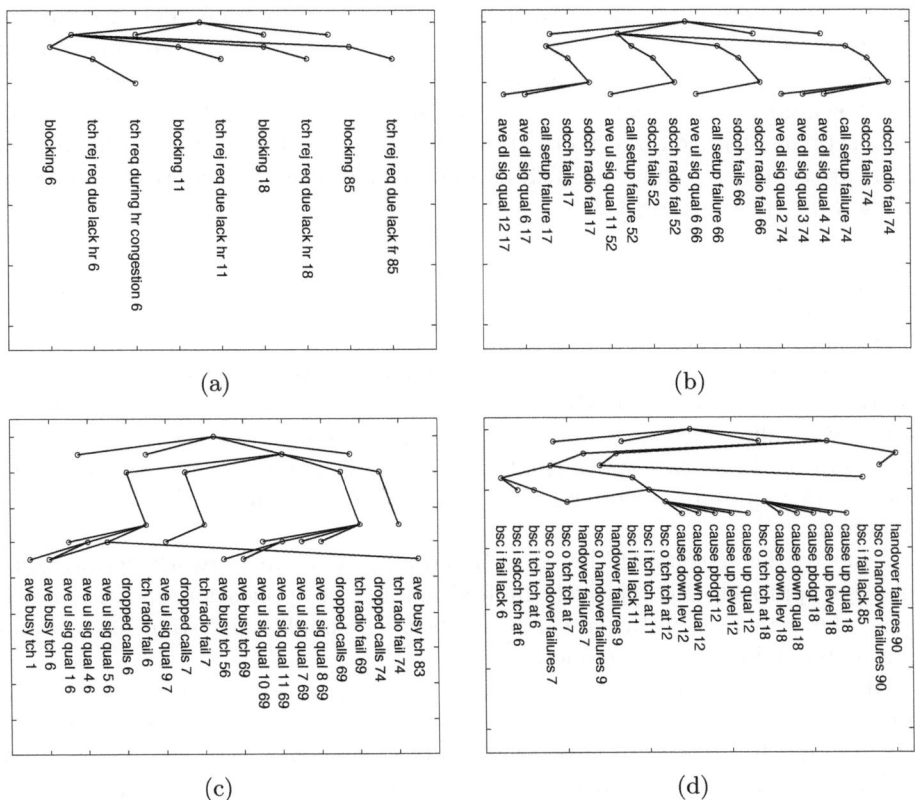

Fig. 5. The analysis of the four main components of the user perceived quality. (a) Cause-effect chains for blocking of services, (b) call setup failures, (c) call dropping and (d) handover failures.

Figure 5(c) shows the corresponding results for call dropping problems. Again, the cause-effect chains for describing the reasons for dropped calls in four BTSs (6,7,69,74) are shown. The reasons for call dropping seem to be radio link failures. In BTSs 6, 7 and 69 the radio failures are likely to result from bad signal quality in uplink. In two TRXs of BTS 6 and in one TRX of BTS 69 the number of bit errors seem to correlate with the amount of traffic in both the own BTS as well as in interfering TRXs on the same radio frequency.

Finally, in Figure 5(d) the analysis of the handover problem sources are shown. The results for three BTSs (7,9,90) having the worst handover performance are shown, indicating that the problems are in BSC controlled outgoing handovers. In BTSs 7 and 9, the problems seem to be in lack of resources in the target BTSs (6,11,85). In target BTS 6 suffering from lack of resources, there seems to be high amount of incoming TCH-TCH handover attempts from BTS 7. This same BTS tend to cause problems also for target BTS 11 suffering from lack of resources during handover attempts. For target BTS 11, two additional

BTSs (12,18) are found that can be said to generate high number of handover attempts. The handover attempts of these two BTSs are due to very similar reasons: the quality of the uplink and downlink radio connection and the uplink and downlink signal strength are not reasonable in these two BTSs, and some of the users are switched into more appropriate BTSs. Also, significant number of users are switched to another BTS in order to minimize the energy consumption of the MSs (power budget).

6 Conclusions

In this paper, a knowledge-based model for analyzing the performance of the GSM network is presented. The presented model is based on application domain knowledge, allowing a logical division of the system into a set of subsystems with appropriate input and output variables. Also, the type of the model per subsystem required a priori knowledge about the semantics of the subsystem variables.

Due to the novelty of the counter data based GSM network performance analysis approach and relatively limited amount of application domain knowledge, the emphasis of our research has been in simple subsystem models and parameter estimation techniques.

In the experiments, a data record from an operational GSM network was used to estimate the parameters of the subsystems. The estimated parameters were interpreted to describe the strength of dependency between input-output variable pairs. After parameter estimation, the most important input-output variable pairs were analyzed further by constructing a hierarchical dependency tree. The dependency tree was constructed for four major problem types in order to analyze the cause-effect chains generating the user perceived quality problems. The provided information can be used to enhance the current radio resource usage in the network.

Acknowledgment

Nokia Foundation is gratefully acknowledged for their financial support.

References

1. S. Mokhtar Bazaraa, D. Hanif Sherali, and C. M. Shetty. *Nonlinear Programming: theory and algorithms*. John Wiley and Sons, Inc., 1993.
2. Sofoklis A. Kyriazakos and George T. Karetsos. *Practical Radio Resource Management in Wireless Systems*. Artech House, Inc., 2004.
3. Jaana Laiho, Kimmo Raivio, Pasi Lehtimäki, Kimmo Hätönen, and Olli Simula. Advanced analysis methods for 3G cellular networks. *IEEE Transactions on Wireless Communications*, 4(3):930–942, May 2005.

4. Pasi Lehtimäki and Kimmo Raivio. A SOM based approach for visualization of GSM network performance data. In *Proceedings of the 18th Internation Conference on Industrial and Engineering Applications of Artificial Intelligence and Expert Systems (IEA/AIE) (to appear)*, 2005.
5. Pasi Lehtimäki, Kimmo Raivio, and Olli Simula. Mobile radio access network monitoring using the self-organizing map. In *Proceedings of the 10th European Symposium on Artificial Neural Networks (ESANN)*, pages 231–236, Bruges, Belgium, April 2002.
6. Pasi Lehtimäki, Kimmo Raivio, and Olli Simula. Self-organizing operator maps in complex system analysis. In *Proceedings of the Joint International Conference on Artificial Neural Networks and International Conference on Neural Information Processing (ICANN/ICONIP)*, pages 622–629, Istanbul, Turkey, June 2003.
7. Jens Zander. *Radio Resource Management for Wireless Networks*. Artech House, Inc., 2001.

Sentiment Classification Using Information Extraction Technique

Jian Liu, Jianxin Yao, and Gengfeng Wu

Department of Computer Science,
Shanghai University, 149 Yanchang Road,
Shanghai, PR China 200072
{liujian, jianxin_yao, gfwu}@mail.shu.edu.cn

Abstract. This paper explores the sentiment classification with Information Extraction (IE) approach. The IE approach here is required to detect the sentiment expressions on specific subject (person, product, company and so on) and then to evaluate the sentiment strength and/or the validation of them. Our method can be illustrated logically as: (1) From a given text, extract the sentiment expressions on the specific subjects and attach certain sentiment tag and weight to each of them; (2) Calculate the sentiment indicator for each sentiment genre by accumulating the weights of all the expression with the corresponding tag; (3) Given the indicators on different sentiment genres, use a classifier to predict the sentiment label of the given text. To extract expression robustly when encounter some complex linguistic phenomena (such as ellipsis, anaphora), a new parsing idea named *super parsing* is proposed. It enables some non-adjacent linguistic constituents to be merged to deduce a new one. As an incremental implementation of super parsing, a system named *Approximate Text Analysis* (ATA) is described in this paper. As for the classification task, two different classifiers are used: simple linear classifier (called SLC here) and SVM. The experiments show the reasonable performance of our approach.

1 Introduction

Today, with the rapid expansion of Internet and e-commerce, increasing documents, in the form of news, report, BBS post and so on, appear on the web. As part of the effort to better organize this information for users, researchers have been actively investigating the problem of automatic text categorization.

Traditional work has focused on *topical* categorization, attempting to sort documents according to their subject matter (e.g., sports vs. politics). However, another important area of text categorization, i.e. Sentiment Classification, has become a hotspot. Attributing to the rapid growth of online forum, discussion groups and review sites, an increasing number of posted articles are written by people to express their *sentiments* or overall opinions towards the subject matters – for example, whether supporting a president candidate. Labeling these articles with their sentiments would provide an overall image of public opinions on certain issues. Sentiment classification would be helpful in recommender systems (e.g. Terveen et al.[1], Tatemura[2]). There are also potential applications to message filtering (e.g. Spertus[3]). One might be able to use the sentiment information to recognize or discard some offensive utterance.[4]

A.F. Famili et al. (Eds.): IDA 2005, LNCS 3646, pp. 216–227, 2005.

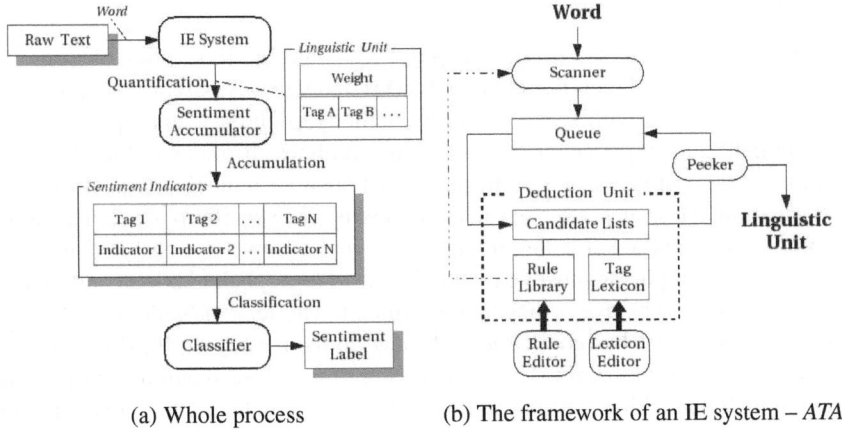

(a) Whole process   (b) The framework of an IE system – *ATA*

Fig. 1. Sentiment Classification with IE approach

This paper explores to employ Information Extraction (IE) approach in Sentiment Classification. The IE method is required to detect the sentiment expressions on specific subject and to evaluate the sentiment strength and/or the validation of each expression. As shown in Figure 1(a), the whole process can be illustrated logically as the following steps:

1. *Quantification*: from a given text, extract the sentiment expressions on the specific subjects and attach certain sentiment tag and weight to each expression;
2. *Accumulation*: calculate the sentiment indicator for each sentiment genre by accumulating the weights of all the expression with the corresponding tag;
3. *Classification*: given the indicators on different sentiment genres, use a classifier to predict the sentiment label of the given text.

The rest of this paper is organized as following: Section 2 reviews the previous work related to our research, such as Sentiment Classification and Information Extraction. Section 3 introduces the idea of super parsing, which enables some non-adjacent linguistic constituents to be merged to deduce a new one. In section 4, an incremental Super-Parsing-based IE approach, named *Approximate Text Analysis* (ATA) (see Figure 1(b)) is revealed in detail. It enforces the Quantification task in Figure 1(a). Section 5 briefly introduces the component for sentiment accumulation. And section 6 illustrates the classifiers used in this work. The experiments on web articles are illustrated and remarked in section 7. Finally, some conclusions are given in section 8.

2   Previous Work

This section gives a brief survey about previous work on Sentiment Classification and Information Extraction.

2.1 Sentiment Classification

Hearst [5] and Sack [6] use models inspired by cognitive linguistics on sentiment-based classification of the entire documents. Das and Chen [7] use a manually crafted lexicon in conjunction with several scoring methods to classify stock postings on an investor bulletin. Tong [8] generates sentiment timelines. It tracks online discussions about movies and displays a plot of the number of positive and negative sentiment messages over time. Messages are classified by looking for specific phrases that indicate the author's sentiment towards the movie (e.g. 'great acting','wonderful visuals', 'uneven editing'). Each phrase must be manually added to a special lexicon and manually tagged as indicating positive or negative sentiment. The lexicon is domain dependent (e.g. movies) and must be rebuilt for each new domain. Pang et al.[4] examine several supervised machine learning methods for sentiment classification of movie views and conclude that machine learning techniques outperform the method that is based on human-tagged features, although none of existing methods could handle the sentiment classification with a reasonable accuracy.

Some researchers use semantic orientation (SO) to evaluate the sentiment strength of a text or a word, and apply it in Sentiment Classification. Turney calculates the SO of each word with the mutual information of each word between document phrases and the words 'excellent' and 'poor', where the mutual information is computed using statistics gathered by a search engine. The SO of the whole text is obtained by summing up the SO of all the words[9]. The semantic orientation methods suppose that all the sentiment words are semantically related with one subject or one subject class in the text. Such simplified hypothesis may lead to a wrong conclusion when one subject is blamed and another is eulogized in same text. So this work tries to employ IE method to find the relation between sentiment words and topical subjects.

2.2 Information Extraction

The goal of Information Extraction (IE) is to transform text into a structured format and thereby reduce the information in a document to a tabular structure. Unseen texts are taken as input to produce fixed-format, unambiguous data as output. Specified information can then be extracted from different documents with a heterogeneous representation and be summarized and presented in a uniform way. IE systems do not attempt to understand the whole text in the input documents, but they analyze those portions of each document that contain relevant information. Relevance is determined by predefined domain guidelines which specify what types of information the system is expected to find.[10]

Recently, many extraction systems are developed [11], such as AutoSlog, LIEP, PALKA, WHISK, RAPIER, SRV, etc. The extracted information is domain-specific events varied from the house renting, job hunting to the terrorist activities. Most extracting systems use templates to detect the relevant events. The templates are designed to detect some event description in a fixed form, so some indirect expressions caused by ellipsis or anaphora are easily neglected. This work uses super parsing to search the potential relations among existing linguistic elements, so the indirect relations (e.g. inter-sentential semantic relations) can be recognized by our approach.

On the other hand, some researchers have explored to extract evaluation-related expressions. Such expressions can be sentimental review or neutral description on some subjects. Yi and Nasukawa manually construct the tag patterns for the sentiment expression extraction on specific subject. Their method extracts the expression in the form of ternary tuples and binary tuples[12,13]. Kobayashi and Inui [14] explore the sentiment expression extraction at a finer granularity in which the expressions about the specific features and products can be identified with some co-occurrence patterns. Hu and Liu [15,16] use Semantic Orientation to pick up evaluative sentences on various features of pre-determined product, and classify these sentences on the polarity, feature to generate sentimental summary. Different from the above research, this work does not only detect the sentiment expressions, but also evaluates the validation and/or strength of expressions with weights. Then the quantified sentiment strength will be generated for prediction of sentiment label in the later process.

3   Super Parsing

Due to some linguistic phenomena, such as anaphora and ellipsis, the syntactically extracting approaches can hardly discover some inter-sentential semantic relations. Ignoring these relations will always lose some valuable expressions, and eventually leads to a partial evaluation on author's attitude.

Taking the following sentence as an example,

> The new *government* consists of many elites. It has <u>achieved</u> <u>greatly</u> in the recent years.

where the word 'government' is a specific subject which the readers are interested in, and each underlined word is a sentiment word. From the view of human reader, the words 'achieve' and 'greatly' semantically depend on 'government'. But many extracting apparatus, which analyze the text with syntactical knowledge, easily neglect these relations. Some extraction methods such as [12] use moving window to seek the relevant expressions. They can detect inter-sentential linguistic relations in some extent, but how to adjust the length of window is still challenging for the extraction performance.

Motivated by the problem, this work proposes *super parsing* to search the potential association among the linguistic constituents.

3.1   Formalism

The super parsing takes a loose policy in deduction, so it enables non-adjacent constituents to be merged to create a new one. If the operator $x \preceq y$ denotes 'x matches y', a loose deduction can be described as following:

Definition 1. *Given a rule r*

$$X_1 \cdots X_n \to C$$

*and constituents $y_i = \langle x_i, [l_i, r_i] \rangle$ ($i = 1, \cdots, n$), if (1) $x_i \preceq X_i$ (2) $l_i < r_i$ (3) $r_i \leqslant l_{i+1}$, then a constituent $z = \langle c, [l_1, r_n] \rangle$ is generated by **loose deduction**, where $c \preceq C$. Such generation is denoted as*

$$\langle y_1, \cdots, y_n \rangle \stackrel{r}{\Rightarrow} z$$

S-PARSE(S, R)
1   **while** $F(S, R) = false$
2   **do**
3   **for** $r \in R$
4   **do**
5           $S \leftarrow S \cup G(S, r)$

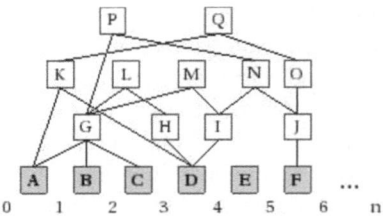

(a) Pseudo code of Super Parsing

(b) Super Parsing can be viewed as constructing DAG of constituents

Fig. 2. Super parsing

In the definition, each of $X_1 \cdots X_n$ is a **daughter** of rule r, while C is the **mother**. A constituent can be a word, a phrase, a grammar structure or a semantic concept. The assignment $y = \langle x, [l, r] \rangle$ means the constituent y takes the text region from l to r and its linguistic information is stored as x. The storage form of x can be varied from a single symbol to a complex data structure. This work uses 'linguistic unit' to store the text region and the said linguistic information (details in section 4.2).

Definition 2. $G(S, r) = \{\exists y_1, \cdots, y_n \in S; \nexists z \in S | \langle y_1, \cdots, y_n \rangle \overset{r}{\Rightarrow} z \bullet z\}$

Given a constituent set S, the **Generating Set** $G(S, r)$ is the constituents that can be generated by rule r via loose deduction.

Definition 3. $F(S, R) \iff (\forall r \in R | G(S, r) = \emptyset)$

A boolean function $F(S, R)$ is constructed to decide whether the parsing process should be ceased. It is true if and only if no more new constituents can be generated by rule set R on constituent set S.

Figure 2(a) describes the whole process of *Super Parsing*. Given the initial constituent set S and the rule set R, a super parser iteratively casts different rules on existing constituents to generate new ones and add them back into S, until no more new creation. The procedure can be viewed as the construction of a Directed Acyclic Graph (DAG), whose nodes are constituents and arcs are deducing relations. Figure 2(b) gives a straightforward view. The gray boxes are the initial constituents (for example, all the words in a given text), while the white boxes are the deduced constituents during parsing. And the numeric subscripts are the text location. As shown in the figure, the constituent G is loosely deduced by merging A, B and C and it covers the text region $\langle 0, 3 \rangle$. However, according to Definition 1, the constituent K is possible to be deduced from non-adjacent constituents A and D and its text region of K is $\langle 0, 4 \rangle$.

Different from the known parsing ideas, Super Parsing enables merging non-adjacent constituents for the creation of new one. Moreover, a whole text, not only a sentence, is processed in one parsing. So it is possible and reasonable to find some inter-sentential relations among constituents.

3.2 Combinational Explosion

The Super Parsing visits each rule circularly in the given rule set. And for a specific rule, it searches the existing constituents to find the qualified ones to match the different daughters of the rule. If no restrictions are set when seeking, it definitely leads to a combinational explosion. So, it is vital to limit the searching scale in Super Parsing.

To overcome the obstacle, a threshold-based solution is given. First, the variable 'weight' is attached to each constituent to evaluate its validation. If a constituent y is produced by constituents $x_1, \cdots, x_n$ with a certain rule, the weight of y is less when $x_1, \cdots, x_n$ are more sparsely distributed and/or they are less weighted. In this work, the weight of initial constituent is 1. While the weight of a deduced constituent is calculated as:

$$w(y) = W(x_1, \cdots, x_n) = \prod_{i=1}^{n} w(x_i) \bullet \prod_{i=1}^{n-1} log_2(Dist(x_i, x_{i+1}) + 2) \qquad (1)$$

where $w(x)$ denotes the weight of x, and the function $Dist(x, y)$ means the distance between constituents x and y. If x and y take the region $\langle l_x, r_x \rangle$ and $\langle l_y, r_y \rangle$ respectively, the distance between x and y can be calculated by

$$Dist(x, y) = min(l_y - r_x, l_x - r_y) \qquad (2)$$

Then, with the weight on each constituent, a threshold can be set to eliminate the newly-born constituent that has poor reliability. Meanwhile, another threshold can be set to eliminate a combination member when it is distant from its neighbors in a combination. Hence, the scale of constituent combination can be restricted effectively.

4 Approximate Text Analysis (ATA)

In this work, the IE approach is required to extract the sentimental expressions on specific subjects and to evaluate the strength of different sentiments. We adopt an incremental Super-Parsing-based IE approach, named Approximate Text Analysis (ATA), to obtain the goal. For the readers who are interested in ATA, a more general embodiment can be found in [17].

Figure 1(b) shows the process of sentiment review extraction with ATA. The approach receives a raw text as input, and outputs the linguistic units which represent the sentiment reviews. A raw text can be pure text, or tagged document such as HTML and XML. While each outputted unit has at least one sentiment tag to denote the sentiment genre of the corresponding review.

The **Scanner** is a component to provide initial linguistic units for ATA. When the Queue is empty, the scanner will be informed to offer a unit. It scans the raw text by moving the reading pointer forward until finding a word. Then it creates a new unit for the word, sends it into the Queue and waits for the next request. For instance, when a word 'congress' is found between location 1 and 2, the component creates a unit $\langle \langle 1, 2 \rangle, \textbf{WRD}, \emptyset, \textbf{'congress'} \rangle$. The detail about linguistic unit is in section 4.2.

The **Queue** provides a temporary storage for linguistic units which are waiting to be accessed by Deducing Unit. Its source includes the initial units from Scanner and the deduced units from Deduction Unit.

The **Peeker** monitors any units sent from Deduction Unit into the queue. If the Peeker finds a linguistic unit containing at least one sentiment tag ('POS' or 'NEG' in this work), it means the unit is a sentiment expression and will be outputted.

The rest part of this section will reveal the details of tag, Linguistic Unit, Deduction Rule and Deducing Unit.

4.1   Tag Lexicon

To make the rule matching more powerful and flexible, ATA uses a hierarchical tag system called **Tag Lexicon**. This work designs it as a tree structure. However, it can be also implemented as a simple tag set, a directed acyclic graph (DAG), and so on.

ATA uses tags to describe the different linguistic meaning of a constituent. In this work, a tag 'SBJ' denotes that the attached constituent is a subject, probably a product, a service or something else. A tag 'COM' denotes that the attached constituent is a commendation to something, while a tag 'DER' is derogatory. More specifically, tags 'COM_S' and 'COM_O' is commendatory to subject constituent and object constituent respectively. And the tags 'DER_S' and 'DER_O' can be understood in a similar way. There are two sentiment tags 'POS' and 'NEG' in our work, representing positive and negative evaluation respectively.

4.2   Linguistic Unit

This work uses the term **Linguistic Unit** to denote the data structure corresponding to a constituent. A Linguistic Unit contains three components: (1) text region, (2) key feature, (3) additional feature set. **Text Region** is the text area occupied by the given constituent. **Key Feature** stores the tag denoting the essence of a linguistic unit. For example, a unit attached with a 'WRD' as key feature, means that it is a word. **Additional Feature Set** is a tag set storing the secondary descriptive information for the unit.

In the following sentence,

$$_0 \text{ The } _1 \text{ congress } _2 , _3 \text{ I } _4 \text{ think } _5 , _6 \text{ is } _7 \text{ wise } _8$$

the *'congress is wise'* is a constituent about expression. Its text region can be described in different forms. For example, a set of continuous location $\{\langle 1, 2 \rangle, \langle 6, 8 \rangle\}$, or a bit-pattern $\langle 0100011 \rangle$ where a '1' indicates that constituent occupies this position [18]. More roughly, it can be depicted as a pair $\langle 1, 8 \rangle$. For convenience, this work uses the location pair to represent the region. And the Key Feature of the unit is set as 'EXP' to denote a sentiment expression. The Additional Feature Set can contain several tags to describe a constituent in different meanings. To mark the constituent as a positive expression, we can add tag 'POS' into the Additional Feature Set.

4.3   Deducing Rules

The **Deduction Rules** are used in the Deduction Unit to generate new linguistic units with the given unit combination. They contribute to the detection of the words, phrases, grammar structures and semantic relations relevant to the sentiment expression.

1. $\langle \mathbf{WRD} : congress \rangle \{\} \Rightarrow \langle \mathbf{ENT} \rangle \{\langle SBJ \rangle\}$
2. $\langle \mathbf{WRD} : wise \rangle \{\} \Rightarrow \langle \mathbf{ADJ} \rangle \{\langle COM \rangle\}$
3. $\langle \mathbf{WRD} : defeat \rangle \{\} \Rightarrow \langle \mathbf{VER} \rangle \{\langle COM\_S \rangle \langle DER\_O \rangle\}$
4. $\langle \mathbf{WRD} : attack \rangle \{\} \Rightarrow \langle \mathbf{VER} \rangle \{\langle DER\_S \rangle\}$
5. $\langle \mathbf{ADJ} \rangle \{\langle DER \rangle\} + \langle \mathbf{ENT} \rangle \{\langle SBJ+ \rangle\} \Rightarrow \langle \mathbf{ENT} \rangle \{\langle NEG \rangle\}$
6. $\langle \mathbf{ADJ} \rangle \{\langle COM \rangle\} + \langle \mathbf{ENT} \rangle \{\langle SBJ+ \rangle\} \Rightarrow \langle \mathbf{ENT} \rangle \{\langle POS \rangle\}$
7. $\langle \mathbf{ENT} \rangle \{\langle SBJ \rangle\} + \langle \mathbf{ADJ} \rangle \{\langle DER \rangle\} \Rightarrow \langle \mathbf{EXP} \rangle \{\langle NEG \rangle\}$
8. $\langle \mathbf{ENT} \rangle \{\langle SBJ \rangle\} + \langle \mathbf{ADJ} \rangle \{\langle COM \rangle\} \Rightarrow \langle \mathbf{EXP} \rangle \{\langle POS \rangle\}$
9. $\langle \mathbf{ENT} \rangle \{\langle SBJ \rangle\} + \langle \mathbf{VER} \rangle \{\langle COM\_S \rangle\} \Rightarrow \langle \mathbf{VER} \rangle \{\langle POS \rangle\}$
10. $\langle \mathbf{ENT} \rangle \{\langle SBJ \rangle\} + \langle \mathbf{VER} \rangle \{\langle DER\_S \rangle\} \Rightarrow \langle \mathbf{VER} \rangle \{\langle NEG \rangle\}$
11. $\langle \mathbf{VER} \rangle \{\langle COM\_O \rangle\} + \langle \mathbf{ENT} \rangle \{\langle SBJ \rangle\} \Rightarrow \langle \mathbf{EXP} \rangle \{\langle POS \rangle\}$
12. $\langle \mathbf{VER} \rangle \{\langle DER\_O \rangle\} + \langle \mathbf{ENT} \rangle \{\langle SBJ \rangle\} \Rightarrow \langle \mathbf{EXP} \rangle \{\langle NEG \rangle\}$

Fig. 3. Some toy rules for ATA

Figure 3 shows some sample rules for ATA. For example, the rule 1 means that the word 'congress' is a subject for analyzing. The word 'wise' is a positive evaluation, written as rule 2. The word 'defeat' always implies the desirable affect to the *subject* constituent while the undesirable affect to the *object* constituent, so the tags 'COM_S' and 'DER_O' are used to denote respectively in rule 3. Similarly, the word 'attack' expresses negative feeling to the *subject* constituent, so the tag 'DER_S' is marked, as defined in rule 4.

As shown in rule 8, a sentence like 'the congress is wise' can be recognized as a positive expression on the specific subject 'congress'. Because of the super parsing, the word 'is' is ignored, but it affects little to the recognition of the expression. In the case of 'the congress attacks the new policy', a human reader can feel the author's negative sentiment to the 'congress', it can be interpreted by the rule 10.

In some cases, a postfix '+' is attached to a specific tag. It means that both the unit matched and the new unit generated by the rule should contain the tag. For example, the constituent matches the second daughter of rule 5 should have the tag 'SBJ', and the new unit generated by this rule should also have the tag 'SBJ'. However, if a tag is presented without any postfix, it should be contained only by the matched unit, not the new unit. And all the other additional features of matched units not mentioned in the rule will be add to the new unit.

4.4 Deduction Unit

The **Deduction Unit** is the core of ATA to enforce the super parsing. It maintains candidate list for each daughter of each rule (see Figure 4). When a new unit is obtained from Queue, the Deduction Unit traverses all the daughters of all the Deduction Rules, and appends the unit into the candidate list of each *matched* daughter.

For a specific n-daughter rule r, if the new unit u matched the last daughter D_n, a deduction process will be activated. The Deduction Unit searches the valid unit combination among the candidate lists of r. For each combination, the i-th member is chosen from list $CL_{r,i}$ and the last member should constantly be u, the tail of $CL_{r,n}$. Following

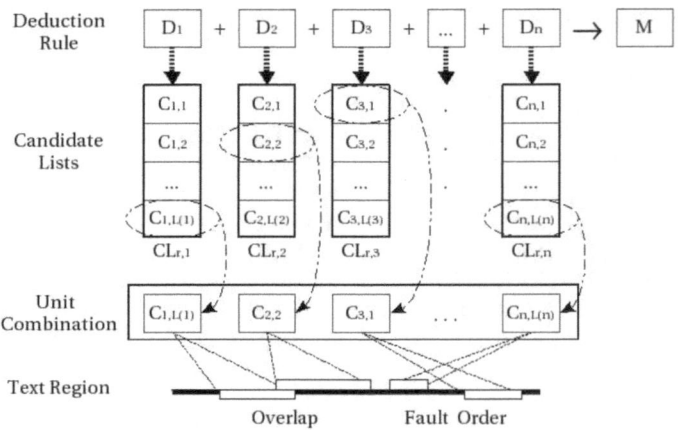

Fig. 4. Candidate Lists of a given rule r

the definition of loose deduction, the units in combination should satisfy the constraints on text region: daughter order and no overlap. As shown in Figure 4, $C_{3,1}$ for D_3 and $C_{n,L(n)}$ for D_n are in a fault order, because the daughter ID implies the precedence of text region. However, the text regions of $C_{1,L(1)}$ and $C_{2,2}$ are overlapped. Once a unit combination is found, Deduction Unit will create a new unit according to the mother of rule r and output it via Peeker back into Queue.

5   Sentiment Accumulator

The **Sentiment Accumulator** accepts the review-related linguistic units from the IE system, and accordingly update the different sentiment indicators, whose value are initially 0. As shown in Figure 1(a), a unit to the accumulator has weight and one or more sentiment tags. The accumulator adds the weight up to the sentiment indicator of each tag. Take a unit as example. If it is weighted as 0.8 and has sentiment tag 'POS', the accumulator will add the value 0.8 into the sentiment indicator of 'POS'. However, if a same-weighted unit has two sentiment tags 'POS' and 'NEG', the value 0.8 will be add into the indicators of both 'POS' and 'NEG'.

The accumulator processes each inputted unit until the whole text has passed through the IE system. Then the eventual value of each sentiment indicator is the strength evaluation of different sentiment genre on the text.

6   Classifier

As illustrated in Figure 1(a), the classifier accepts the sentiment indicators as input from Sentiment Accumulator, and outputs a class label as the overall author attitude in the given text. In this work, two classifiers are used: a simple linear classifier (SLC for short) and Support Vector Machine (SVM).

Corpus	Accuracy	Sentiment Quantification	
		VSM + SO	ATA
Religion	Positive	0.6120	0.8398
	Negative	0.5196	0.9920
Politics	Positive	0.7071	0.8571
	Negative	0.3458	0.8235

(a) Classified with SLC

Corpus	Accuracy	Sentiment Quantification	
		VSM + SO	ATA
Religion	Positive	0.8085	0.8609
	Negative	0.9486	0.9888
Politics	Positive	0.8786	0.9598
	Negative	0.8966	0.9863

(b) Classified with SVM

Fig. 5. Experimental results

The classifier SLC is defined as following:

$$SLC(P, N; c) = \begin{cases} positive & N \leqslant P \vee N \leqslant c \\ negative & \text{otherwise} \end{cases} \tag{3}$$

It accepts indicators on two tags 'POS' and 'NEG', and returns a label 'positive' or 'negative'. The threshold c is a parameter learned by training process, where it can discriminate positive and negative instances to the maximum extent.

Support vector machines (SVMs) [19] are a set of related supervised learning methods, applicable to both classification and regression. Basically, a SVM algorithm creates a maximum-margin hyper plane, to which the distance from the closest examples (the margin) is maximized. The SVM software package used in this work is *SVMTorch* [1]. It is developed by Collobert[20,21].

7   Experiments

This section uses some on-line documents to evaluate the performance of ATA approach. The experimental documents cover two domains: politics and religion. All the documents are collected from www.google.com and groups-beta.google.com. The latter is a news-group search engine. For each domain, two subject classes are formed by manually selected keywords for subjects such as political organizations, politicians, religious concepts or religious leaders. These words are used to retrieve web pages from the search engines. Each web page is reviewed by human reader to ensure an overall attitude ('positive' for supporting Class A and 'negative' for supporting Class B). The corpus *politics* contains 826 articles, 672 of them are positive and 154 are negative. The corpus *religion* contains 2856 articles, 620 are positive and 2236 are negative.

Turney's semantic orientation (SO) approach [9] is implemented as a contrast approach against ATA. Since a text in this work may involve plural subject classes and it cannot be processed well by Turney's method, a topical classification approach, i.e. Vector Space Model (VSM) [22], is used to determine the major subject class of the text. Both the major subject class ID and the sentiment orientation will be sent to a classifier to predict the sentiment label of the text. Two classifiers are used in the experiments: SLC and SVM. The classification takes 5-flod cross validation for the corpus politics, while 10-flod for the corpus religion.

---

[1] http://www.idiap.ch/machine_learning.php?content=Torch/en_SVMTorch.txt

As for the sentiment words, some English sentiment words are collected from General Inquirer (GI) [2] at first. And then, from WordNet (or HowNet for Chinese text), the synonyms of known sentiment words are extracted. Since their Part-Of-Speech tags and sentiment labels are given, it is easy to provide keywords for VSM+SO and to generate rule for ATA.

Figure 5(a) shows the accuracy of two sentiment analyzing approaches when taking different classifiers. The accuracy of ATA is above 80%, while the VSM+SO get poor performance. However, when the classifier is changed to SVM (see Figure 5(b)), the performances of both VSM+SO and ATA are promoted. The accuracy of former approach is up to 80%, but is still inferior to our approach.

8   Conclusions

This paper explores the sentiment classification with Information Extraction approach. The IE approach should detect the expressions with different attitudes toward specific subjects and evaluate the validation of each expression with weight. We propose the *super parsing*, which enables some non-adjacent linguistic constituents to be merged to deduce a new one. As an incremental implementation of super parsing, a system named *Approximate Text Analysis* (ATA) is revealed. The experiments show the reasonable performance of our approach against the Semantic Orientation approach.

Acknowledgments

The authors deeply thank Xuanjin Huang, Jin Min, and Xiaochun Wu in Fudan University for their valuable corpora and the implementation of semantic orientation approach in the experiments. Zhongchao Fei in Shanghai University and the anonymous reviewers of IDA2005 are also greatly thanked for their informative advice. Meanwhile, the gratefulness is given to Science and Technology Commission of Shanghai Municipal Government for their financial support (Project NO: 035115028).

References

1. Terveen, L., Hill, W., Amento, B., McDonald, D., Creter, J.: Phoaks: a system for sharing recommendations. Communications of ACM **40** (1997) 59–62
2. Tatemura, J.: Virtual reviewers for collaborative exploration of movie reviews. In: Proceedings of the 5th international conference on Intelligent user interfaces. (2000) 272–275
3. Spertus, E.: Smokey: Automatic recognition of hostile messages. In: Proceedings of Innovative Application of Artificial Intelligence (IAAI). (1997) 1058–1065
4. Pang, B., Lee, L., Vaithyanathan, S.: Thumbs up? Sentiment classification using machine learning techniques. In: Proceedings of the 2002 Conference on Empirical Methods in Natural Language Processing (EMNLP). (2002) 79–86
5. Hearst, M.A.: Direction-based text interpretation as an information access refinement. In Jacobs, P.S., ed.: Text-Based Intelligent Systems: Current Research and Practice in Information Extraction and Retrieval. Erlbaum, Hillsdale (1992) 257–274

---

[2] http://www.wjh.harvard.edu/~inquirer

6. Sack, W.: On the computation of point of view. In: Proc. of AAAI-94, Seattle, WA (1995) 1488

7. Das, S., Chen, M.: Yahoo! for amazon:extracting market sentiment from stock message boards. In: Asia Pacic Finance Association Annual Conference (APFA). (2001)

8. Tong, R.M.: An operational system for detecting and tracking opinions in on-line discussion. In: SIGIR Workshop on Operational Text Classification. (2001)

9. Turney, P.D.: Thumbs up or thumbs down? semantic orientation applied to unsupervised classification of reviews. In: Proceedings of the 40th Annual Meeting of the Association for Computational Linguistics (ACL). (2002) 417–424

10. Eikvil, L.: Information extraction from world wide web - a survey. Technical Report 945, Norweigan Computing Center (1999)

11. Muslea, I.: Extraction patterns for information extraction tasks: A survey. In: The AAAI Workshop on Machine Learning for Information Extraction. (1999)

12. Nasukawa, T., Yi, J.: Sentiment analysis: Capturing favorability using natural language processing. In: The Second International Conferences on Knowledge Capture (K-CAP 2003), Sanibel Island, FL, USA. (2003) pp 70 – 77

13. Yi, J., Nasukawa, T., Bunescu, R., Niblack, W.: Sentiment analyzer: Extracting sentiments about a given topic using natural language processing techniques. In: The Third IEEE International Conference on Data Mining. (2003)

14. Kobayashi, N., Inui, K., Matsumoto, Y., Tateishi, K., Fukushima, T.: Collecting evaluative expressions for opinion extraction. In: Proceedings of the 1st International Joint Conference on Natural Language Processing,Hainan,China. (2004) pp. 584–589

15. Hu, M., Liu, B.: Mining and summarizing customer reviews. In: Proceedings of the ACM SIGKDD International Conference on Knowledge Discovery and Data Mining(KDD-2004), Seattle, Washington, USA. (2004)

16. Hu, M., Liu, B.: Mining opinion features in customer reviews. In: Proceedings of Nineteeth National Conference on Artificial Intellgience (AAAI-2004), San Jose, USA. (2004)

17. Liu, J., Wu, G.: Apparatus and method for approximate text analysis (2005) Application NO: 200510023589.8, Chinese Patent.

18. Johnson, M.: Parsing with discontinuous constituents. In: Proceedings of the 23rd conference on Association for Computational Linguistics,Chicago, Illinois. (1985) 127 – 132

19. Boser, B.E., Guyon, I.M., Vapnik, V.N.: A training algorithm for optimal margin classifiers. In: Proceedings of the 5th annual ACM workshop on Computational Learning Theory, ACM Press (1992) 144–152

20. Collobert, R., Bengio, S.: SVMTorch: Support vector machines for large-scale regression problems. Journal of Machine Learning Research 1 (2001) 143–160

21. Collobert, R., Bengio, S.: SVMTorch: A support vector machine for large-scale regression and classification problems. Journal of Machine Learning Research 1 (2001) 143–160

22. Salton, G., Wong, A., Yang, C.S.: A vector space model for automatic indexing. Communications of the ACM 18 (1975) 613–620

Extending the SOM Algorithm to Visualize Word Relationships

Manuel Martín-Merino[1] and Alberto Muñoz[2]

[1] Universidad Pontificia de Salamanca,
C/Compañía 5, 37002, Salamanca, Spain
mmerino@ieee.org
[2] Universidad Carlos III de Madrid,
C/Madrid, 126, 28903 Getafe, Spain
albmun@est-econ.uc3m.es

Abstract. Self Organizing Maps (SOM) are useful tools to discover the underlying structure of high dimensional data. However the algorithms proposed in the literature rely on the use of symmetric measures such as the Euclidean. Therefore when asymmetry arises they fail to reflect accurately the object proximities and the resulting maps become often meaningless. This is a serious drawback for several applications such as text mining in which the object relations are strongly asymmetric.

In this paper, we propose two variants of the original SOM algorithm that are able to deal successfully with asymmetric relations. The algorithms are tested using real document collections, and the performance is reported using appropriate measures. The asymmetric algorithms improve significantly the maps generated by their symmetric counterpart.

1 Introduction

Self Organizing Maps (SOM) [10] have been widely used to visualize multidimensional object relationships. In particular, they have been successfully applied to discover semantic relations between words or documents in textual databases [11,16]. However, as far as we know, all the variants presented in the literature rely on the use of symmetric dissimilarities (usually the Euclidean distance). Therefore, they are not able to handle asymmetric relations in an appropriate fashion.

Let (δ_{ij}) be the dissimilarity matrix between the objects. Asymmetry arises when $(\delta_{ij} \neq \delta_{ji})$. The Multidimensional Scaling (MDS) community has proposed several models to deal with asymmetric dissimilarities (see for instance [6,22,19,9]). They usually adjust independently the symmetric and skew symmetric component of the similarity matrix. However, as we have mentioned in [14,18] when the asymmetry is strong the symmetric component gives frequently smaller values than expected. This fact degrades the quality of the term relations suggested by any mapping algorithm.

Term relations in textual databases are strongly asymmetric. Consider for instance a broad term such as 'mathematics' and a specific term like 'statistics'.

A.F. Famili et al. (Eds.): IDA 2005, LNCS 3646, pp. 228–238, 2005.

It is obvious that 'mathematics' subsumes the semantic meaning of 'statistics', but the reverse relation is much weaker. In this case, a symmetric similarity would indicate that 'mathematics' is hardly related to 'statistics' [18] and the terms would be located too far in the map. So symmetric dissimilarities such as the Euclidean distance should be modified to reflect accurately the object proximities when relations are asymmetric. However notice that the asymmetry is not the only factor by which textual distances become meaningless [1] but it stands as a problem that deserves more attention.

In this paper we first study the influence of the asymmetry over the quality of the textual maps and present new dissimilarities that are less sensitive to this problem. Next new versions of the SOM algorithm [10] are proposed that incorporate the new dissimilarities keeping the simplicity of the symmetric counterpart. The derivation of the algorithms from an energy function provides a strong theoretical foundation for the models. Finally, the algorithms proposed are tested on the interesting problem of word relations visualization.

This paper is organized as follows. In section 2 we study the problem of asymmetry. Section 3 proposes new versions of the SOM algorithm that are able to deal with asymmetric relations. In section 4 the new algorithms are tested using two real textual collections. Finally section 5 gets conclusions and outlines future research trends.

2   Asymmetry

Symmetric measures have been widely used in the context of information retrieval [20,5]. In this section we study the impact that the asymmetry has on a number of commonly used symmetric similarity measures. Next a coefficient of asymmetry is defined that allow us to model the information conveyed by the asymmetry. Finally the meaning and relevance of asymmetry in the field of textual data analysis is discussed.

Consider a set of n objects and let $D = (\delta_{ij})$ be the dissimilarity matrix made up of object proximities. If a similarity matrix (s_{ij}) is given instead it can be transformed easily into a dissimilarity using any of the transformations provided in [6] ($\delta_{ij} = 1 - s_{ij}$). Asymmetry arises when $\delta_{ij} \neq \delta_{ji}$. In this case the dissimilarity matrix can be decomposed into a symmetric and skew-symmetric component ($D = S + A$) [22] where $s_{ij} = (\delta_{ij} + \delta_{ji})/2$ and $a_{ij} = (\delta_{ij} - \delta_{ji})/2$. The first term represents the object proximities and the second one the deviation from symmetry (it equals 0 if D is symmetric).

When asymmetry arises symmetric similarities usually considered in the literature produce often too small values and fail to reflect the object proximities [18,15]. To get a deeper understanding of this problem we are going to study a text mining example.

Consider a collection of abstracts from scientific journals where, the broad term 'mathematics' appears for instance in 500 documents while the more specific term 'regularization' appears only in a subset of 10 documents. Obviously, the relation between 'mathematics' and 'regularization' is highly asymmetric in the sense that 'mathematics' subsumes the semantic meaning of 'regularization' while the reverse relation is much weaker. Let x_m and x_r be the binary vector space representation [2] of both terms.

Consider a similarity such as the cosine that has been widely used in the information retrieval literature [20,5]. This measure is equivalent to the Euclidean distance (commonly used by SOM algorithms) if the objects are normalized previously by the L_2 norm. Besides that, this similarity is strongly correlated with several popular measures inside the field such as *Dice*, *Kulczynski* or *Jaccard* ($\rho = 0.99$ for the textual collection considered in the experimental section). Therefore, the cosine similarity represents somewhat the behavior of a broad range of symmetric similarities over textual data. Now, if the cosine similarity is computed for the example considered above we get:

$$\cos(x_m, x_r) = \frac{\sum_k x_{mk} x_{rk}}{\|x_m\| \|x_r\|} = \frac{10}{\sqrt{500}\sqrt{10}} = 0.14\,, \tag{1}$$

that would suggest that 'mathematics' is hardly related to 'regularization', which is not true. Notice that other distances such as the χ^2 [13] seem to be more robust to this problem but the empirical results [5] suggest that the performance is only slightly better. Therefore they suffer from the same drawback.

The previous example suggests that the symmetric similarities become meaningless (too small) when the asymmetry grows large. Moreover, this bias toward small values tends to reduce the variance of the similarity histogram. In particular, the cosine similarity has a standard deviation as low as 0.03 for the datasets considered in this paper (see figure 2). Consequently, usual similarities become almost constants over textual data and thus any algorithm based on distances will be highly distorted [4].

Next, we are going to study the relation between the asymmetry and the L_1 norm. This will allow us to derive an asymmetry coefficient suitable to model the skew-symmetric component of a given similarity measure.

Consider the fuzzy logic similarity measure [12] defined as

$$s_{ij} = |\, x_i \wedge x_j \,| \,/\, |\, x_i \,|\,, \tag{2}$$

where $\wedge$ denotes the standard fuzzy intersection and $|\,|$ the L_1 norm. Obviously this similarity is asymmetric ($s_{ij} \neq s_{ji}$) and the skew-symmetric component can be written as follows:

$$a_{ij} = \frac{1}{2}(s_{ij} - s_{ji}) \propto |\, x_j \,| - |\, x_i \,|\,. \tag{3}$$

This equation suggests that the asymmetry is a property associated to individual objects and may be modeled by the following coefficient of asymmetry: $\omega_i = \frac{|x_i|}{\max_k |x_k|}$. In the context of text mining this coefficient will become large for broad terms.

According to equation (3), a_{ij} will become large for relations between broad terms (large L_1 norm) and specific terms (small L_1 norm). Therefore, the asymmetry will be an important factor in many applications such as text mining in

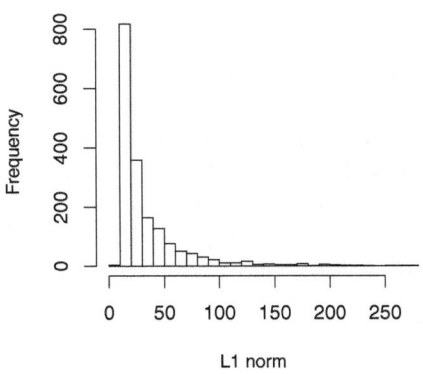

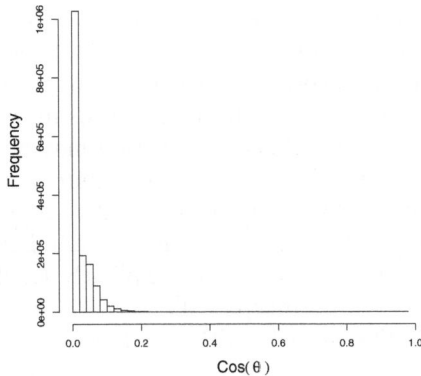

Fig. 1. Term L_1 norm histogram for a textual database

Fig. 2. Cosine similarity histogram for a textual database

which the object L_1 norm obeys a Zipf's law [2] (see figure 1). In this case the L_1 norm histogram is very skew and a_{ij} will become large quite often.

Finally it has been pointed out in [1] that for sparse databases such as textual datasets, the relations between specific (low norm) terms should be established through their associations with broader terms. Therefore if this kind of asymmetric relations are underestimated, the position of specific terms in the map will become meaningless. To avoid this problem the proximities corresponding to asymmetric relations should be compensated proportionally to the degree of asymmetry as we will see in the next section.

3   Asymmetric Variants of Self Organizing Maps

In this section we first introduce shortly the SOM algorithm proposed originally by [10] and interpreted as a principal curve by [17]. Next we propose two asymmetric variants of the SOM algorithm that take advantage of the asymmetry coefficients to reflect accurately the object proximities. The new models keep the simplicity of original SOM algorithm.

The SOM [10] is a nonlinear visualization technique for high dimensional data. Input vectors are represented by neurons arranged according to a regular grid (usually 1D-2D) in such a way that similar vectors in input space become spatially close in the grid.

The experimental results obtained by the SOM are equivalent to that obtained by optimizing the following energy function [7]:

$$E(\mathcal{W}) = \sum_{r} \sum_{x_{\mu} \in V_r} \sum_{s} h_{rs} D(\mathbf{x}_{\mu}, \boldsymbol{w}_s), \qquad (4)$$

where D denotes the square Euclidean distance and V_r is the Voronoi region corresponding to prototype $\boldsymbol{w}_r$. h_{rs} is a neighborhood function (for instance a

Gaussian kernel) that transforms nonlinearly the neuron distances (see [10] for other possible choices). The SOM energy function (4) is minimized when objects that are close together in input space (according to the Euclidean distance) are mapped to neighboring neurons in the grid.

The SOM energy function may be optimized by an iterative algorithm made up of two steps [7]. First a quantization algorithm is run that represents each pattern by the nearest neighbor prototype. Next, the prototypes are organized along the grid of neurons by minimizing the error function (4). The optimization problem can be solved explicitly resulting in a simple iterative adaptation rule for each prototype [10].

The kernel width is adapted in each iteration using the rule proposed by [17] ($\sigma(t) = \sigma_i(\sigma_f/\sigma_i)^{t/N_{iter}}$), where $\sigma_i \approx M/2$ is usually considered in the literature [10] and σ_f is a parameter that determines the degree of smoothing of the principal curve generated by SOM [17].

Next, two asymmetric variants of the original SOM are proposed. The new models improve the position of the more specific terms (low L_1 norm) incorporating similarities that reflect accurately their (asymmetric) relations with broader terms (large L_1 norm). To this aim, a new asymmetric similarity based on the Euclidean distance is first defined. Next an energy function which incorporates the asymmetric similarity is introduced. Finally the error function is optimized keeping the simplicity of the original algorithm.

Let $d(\boldsymbol{x}_i, \boldsymbol{x}_j) = \|\boldsymbol{x}_i - \boldsymbol{x}_j\|^2$ be the square Euclidean distance usually considered in the Self Organizing Maps. This dissimilarity can be easily transformed into a similarity [22,6] using for instance the following transformation:

$$s_{ij} = K - \|\boldsymbol{x}_i - \boldsymbol{x}_j\|^2, \tag{5}$$

where the constant K is an upper bound for the square Euclidean distances. Now an asymmetric index is defined as follows:

$$s_{ij} = (K - \|\boldsymbol{x}_i - \boldsymbol{x}_j\|^2)\omega_i, \tag{6}$$

where ω_i denotes the asymmetry coefficient introduced in section 2. The object proximities induced by the symmetric component of s_{ij} have now the following form:

$$s_{ij}^{(s)} = (K - \|\boldsymbol{x}_i - \boldsymbol{x}_j\|^2)\frac{\omega_i + \omega_j}{2} \tag{7}$$

If the object relations are highly asymmetric then $\omega_i \gg \omega_j$ or conversely. In this case the similarity (7) compensates the value of the Euclidean proximity proportionally to the degree of asymmetry given by $|\omega_i - \omega_j| \approx \max(\omega_i, \omega_j)$.

Notice that a number of asymmetric measures may be defined just substituting the Euclidean distance in (6) by other possible choices such as for instance the χ^2. However, this would increase significantly the complexity of the resulting optimization problem losing the simplicity of the original SOM. Therefore, the measure proposed in (6) achieves a balance between accuracy and simplicity of the optimization problem.

Substituting the similarity (7) into equation (4) the error function for the asymmetric SOM can be written as

$$E(\mathcal{W}) = \sum_r \sum_{x_\mu \in V_r} \sum_s h_{rs} \omega_\mu (K - \|\boldsymbol{x}_\mu - \boldsymbol{w}_s\|^2). \tag{8}$$

As we have mentioned earlier, this asymmetric error decomposes into a symmetric component that represents the object proximities and a skew-symmetric component that equals 0 in this case. From equation (8) it can be easily seen that when the object relations are asymmetric (ω_μ large) the similarities are compensated proportionally to the degree of asymmetry. Therefore the corresponding distances along the grid of neurons will shrink reflecting more accurately the object proximities.

The error function (8) can be optimized in two steps as in the symmetric case. First a quantization algorithm is run that generates the SOM prototypes $\boldsymbol{w}_s$. Next the function error is maximized with respect to the weights $\boldsymbol{w}_s$. This yields a simple adaptation rule for the network prototypes:

$$\boldsymbol{w}_s = \frac{\sum_{r=1}^M \sum_{x_\mu \in V_r} \omega_\mu h_{rs} \boldsymbol{x}_\mu}{\sum_{r=1}^M \sum_{x_\mu \in V_r} \omega_\mu h_{rs}} \tag{9}$$

where h_{rs} is a Gaussian kernel of parameter $\sigma(t)$ which is adapted using the same rules considered for the symmetric version. Notice that the simplicity of the original SOM algorithm is maintained.

The SOM algorithm proposed earlier improves the relative position of the objects in the map when their relations are asymmetric. However, the similarity (7) reduces also the Euclidean distance between objects of similar and large L_1 norm (broad terms). This behavior may eventually increase the overlapping among the main topics of the database which is an undesirable effect. To avoid this problem we introduce an alternative similarity defined as:

$$s_{ij} = (K - \|\boldsymbol{x}_i - \boldsymbol{x}_j\|^2)[1 + (\omega_i - \omega_j)^2], \tag{10}$$

where ω_i, ω_j denote the asymmetry coefficients defined in section 2. This similarity becomes larger than the Euclidean proximity measure just when ($\omega_i \neq \omega_j$). In this case, the similarity is compensated proportionally to the degree of asymmetry $|\omega_i - \omega_j|$. Substituting this similarity into equation (4) we get the error function to be optimized:

$$E(\mathcal{W}) = \sum_r \sum_{x_\mu \in V_r} \sum_s h_{rs} (K - \|\boldsymbol{x}_\mu - \boldsymbol{w}_s\|^2)[1 + (|\boldsymbol{x}_\mu| - |\boldsymbol{w}_s|)^2]. \tag{11}$$

The optimization of this error is quite complex, because the parameter $|\boldsymbol{w}_s|$ depends on the prototype coordinates. To overcome this problem, the original space is enlarged with a new coordinate, the L_1 norm. In this space, the error function (11) can be easily derived with respect to the last coordinate. Using this trick and solving the set of linear equations $\frac{\partial E(\mathcal{W})}{\partial |\boldsymbol{w}_s|} = 0$, we get the following updating rule for the last prototype coordinate:

$$| \boldsymbol{w}_s | = \frac{\sum_r \sum_{x_\mu \in V_r} h_{rs} \alpha'_{\mu s} \, | \, \boldsymbol{x}_\mu \, |}{\sum_r \sum_{x_\mu \in V_r} h_{rs} \alpha'_{\mu s}}, \tag{12}$$

where $\alpha'_{\mu s} = (K - \|\boldsymbol{x}_\mu - \boldsymbol{w}_s\|^2)$ and h_{rs} is a Gaussian kernel defined as usual. The updating rule for $\boldsymbol{w}_s$ can be derived in the same way and is given by the following expression:

$$\boldsymbol{\omega}_s = \frac{\sum_r \sum_{x_\mu \in V_r} h_{rs} \boldsymbol{x}_\mu [1 + (| \, \boldsymbol{x}_\mu \, | - | \, \boldsymbol{w}_s \, |)^2]}{\sum_r \sum_{x_\mu \in V_r} h_{rs}[1 + (| \, \boldsymbol{x}_\mu \, | - | \, \boldsymbol{w}_s \, |)^2]} \tag{13}$$

We finish this section with a brief comment about the related work.

As far as we know, no asymmetric version of the SOM has been proposed earlier in the literature. However, the multidimensional scaling (MDS) community [6,19,22,9] has proposed several models to deal with asymmetric measures in the context of psychometric or sociometric data. Those algorithms optimize a quadratic error measure of the form $\sum_{ij} (\delta_{ij} - d_{ij})^2$, where δ_{ij} and d_{ij} denote the asymmetric dissimilarities in input and output spaces. However, it has been pointed out in the literature [22,15] that the optimization of this error function is equivalent to build two maps that approximate independently the symmetric and skew symmetric components of the dissimilarity matrix (δ_{ij}). Therefore the map that visualizes the object proximities is exclusively derived from the symmetric component of δ_{ij} and is degraded by asymmetry as well. Thus, the contribution of the work presented here is to improve the map that visualizes the object proximities taking advantage of the information conveyed by the asymmetry.

4   Experimental Results

In this section we apply the proposed algorithms to the construction of word maps that visualize term semantic relations. First we describe briefly the textual collections used in the experiments.

The first collection, is made up of 2000 *scientific abstracts* retrieved from three commercial databases 'LISA', 'INSPEC' and 'Sociological Abstracts'. For each database a thesaurus created by human experts is available. Therefore, the thesaurus induces a classification of terms according to their semantic meaning. This will allow us to exhaustively check the term associations created by the map.

The second collection is made up of 6702 abstracts corresponding to the journals of the ACM digital library. The collection was retrieved by means of a robot developed by our research team. In this case, no thesaurus is available for the collection and therefore the evaluation must rely on unsupervised measures. This is a real and interesting problem not previously considered in the literature.

Assessing the performance of algorithms that generate word maps is not an easy task. In this paper the maps are evaluated from different viewpoints through several objective functions. This methodology guaranty the objectivity and validity of the experimental results.

The first measure considered is the Spearman rank correlation coefficient [3] (Sp.). This coefficient checks if the neighbor's ordering in input space is preserved in the map. A complementary measure is the Sp. coefficient taking into account only the 10% of the first nearest neighbors. Notice that the first nearest neighbors of specific terms are frequently broad terms [18,14]. Therefore, this index provides more specific information about the preservation of dissimilarities corresponding to asymmetric relations.

The second group of measures quantifies the agreement between the semantic word classes induced by the map and the thesaurus. Therefore, once the objects have been mapped, they are grouped into topics with a clustering algorithm (for instance PAM [8]). Next the partition induced by the map is evaluated through the following objective measures:

The F measure [2] has been widely used by the Information Retrieval community and evaluates if words from the same class according to the thesaurus are clustered together. The entropy measure [18] evaluates the uncertainty for the classification of words from the same cluster. Small values suggest little overlapping among different topics in the map and are preferred. Finally the Mutual Information [21] is a nonlinear correlation measure between the word classification induced by the thesaurus and the word classification given by the clustering algorithm. This measure gives more weight to specific words and therefore provides valuable information about changes in the position of specific terms.

Table 1 shows the experimental results for the two problems considered: The abstracts of scientific journals and the ACM digital library. The symmetric SOM algorithm (row 1) has been taken as reference because it has been widely applied in text mining (see for instance the WEBSOM [11]). Term vectors have been codified using the vector space model [2] and normalized by the L_2 norm. The primary conclusions are the following:

The first asymmetric version of SOM proposed in section 3 (row 2) outperforms the symmetric counterpart. In particular the Mutual Information (I) is improved a 17% which suggests that the position of specific terms in the map is significantly better in the asymmetric model. This fact helps to avoid that

Table 1. Empirical evaluation of the asymmetric SOM algorithms for a collection of scientific abstracts and the journals of the ACM digital library

	Scientific abstracts					ACM corpus	
	Sp.	Sp. 10%	F	E	I	Sp.	Sp. 10%
[1] *Symmetric SOM*	0.43	0.64	0.70	0.38	0.23	0.43	0.74
[2] Asymmetric SOM	0.57	0.76	0.78	0.35	0.27	0.51	0.76
Improvement in %	33	16	11	8	17	19	3
[3] Asymmetric SOM (L_1 norm difference)	0.37	0.78	0.74	0.31	0.22	0.48	0.79
Improvement in %	-14	22	6	18	-4	12	7

Parameters: Nneur = 88, niter = 30; $\sigma_i^1 = \sigma_i^2 = 30$, $\sigma_i^3 = 33$; $\sigma_f^1 = \sigma_f^2 = 3$, $\sigma_f^3 = 2$.

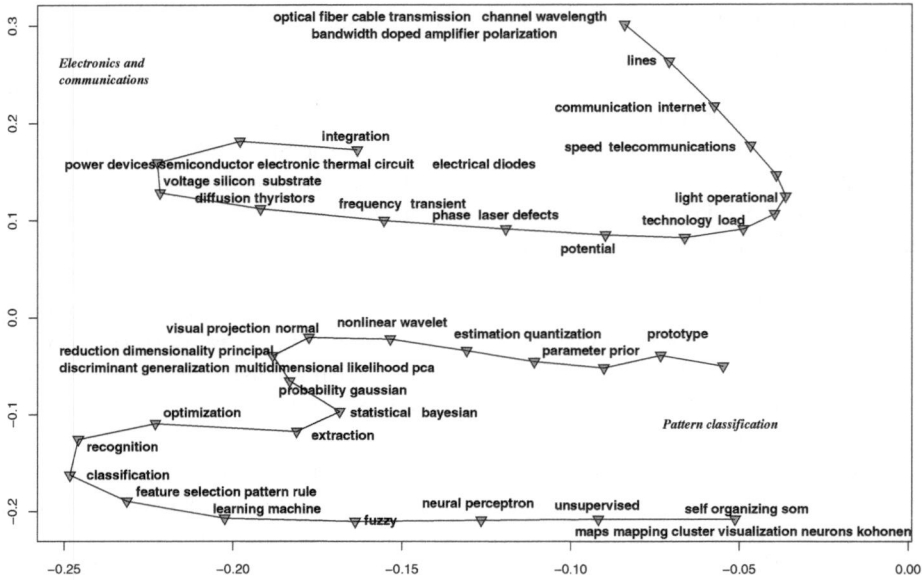

Fig. 3. Word map generated by the asymmetric SOM for a collection of scientific abstracts

the more specific terms (low norm) concentrate in some specific area of the map regardless of their semantic meaning (see [4] for a detailed analysis of this problem). Consequently the overlapping among terms belonging to different topics in the map is reduced ($\Delta E = 8\%$). Finally the overall word map quality (F) is a 10% better than in the symmetric version.

The unsupervised measures (Sp.) and (Sp. 10) show that the organization of the network is even better than for the classic algorithm. Therefore it is easier to preserve the asymmetric similarity probably because the histogram is smoother.

The second asymmetric version of SOM (row 3) improves also the map generated by the symmetric counterpart. Notice that as it was suggested in section 3 the overlapping in the map is reduced more than in the previous algorithm ($\Delta E = 18\%$). However, the (Sp.) and I measures suggest that the network organization is more problematic particularly for terms of medium L_1 norm. Finally we point out that the overall word map quality (F) is improved a 6%.

The experimental results for the journals of the ACM digital library collection corroborate the superiority of the asymmetric algorithms proposed in this paper.

Finally figure 3 shows a visual map generated by our asymmetric SOM for the first textual collection considered in this paper. This figure illustrates the performance of the asymmetric SOM algorithm and allow us to check a number of asymmetric associations induced by the map.

For the sake of clarity only a small subset of terms that belong to two different topics have been drawn. The SOM prototypes have been projected using the Sammon algorithm [10] and those one corresponding to the neighboring neurons have been joined together by continuous trace. Terms with L_1 norm > 30 and ≤ 30 are visualized in different colors.

The figure 3 shows that the terms spread along the map regardless of their frequency (L_1 norm). The term associations induced are consistent with the thesaurus even for words of disparate degree of generality (L_1 norm). Notice that, as we have mentioned in section 2, the symmetric algorithms fail to reflect frequently this kind of asymmetric relations. Finally figure 3 suggests that the ordering of the network prototypes is good for the datasets and parameters considered in this paper.

5   Conclusions and Future Research Trends

In this paper we have proposed two asymmetric variants of the SOM algorithm that improve the visualization of the object proximities when relations are asymmetric. The algorithms have been tested in the challenging problem of word relation visualization using real problems such as the ACM digital library. The algorithms have been exhaustively evaluated through several objective functions.

The experimental results show that the asymmetric algorithms improve significantly the map generated by a SOM algorithm that relies solely on the use of a symmetric distance. In particular, our asymmetric models achieve a remarkable improvement of the position of specific terms in the map. Besides the new models keep the simplicity of the original SOM algorithm.

Future research will focus on the development of new asymmetric techniques for classification purposes.

References

1. C. C. Aggarwal and P. S. Yu. Redefining clustering for high-dimensional applications, IEEE Transactions on Knowledge and Data Engineering, 14(2):210-225, March/April 2002.
2. R. Baeza-Yates and B. Ribeiro-Neto. Modern information retrieval. Addison Wesley, Wokingham, UK, 1999.
3. J. C. Bezdek and N. R. Pal. An index of topological preservation for feature extraction, Pattern Recognition, 28(3):381-391, 1995.
4. A. Buja, B. Logan, F. Reeds and R. Shepp. Inequalities and positive default functions arising from a problem in multidimensional scaling, Annals of Statistics, 22, 406-438, 1994.
5. Y. M. Chung and J. Y. Lee. A corpus-based approach to comparative evaluation of statistical term association measures, Journal of the American Society for Information Science and Technology, 52, 4, 283-296, 2001.
6. T. F. Cox and M. A. A. Cox. Multidimensional scaling. Chapman & Hall/CRC, 2nd edition, USA, 2001.
7. T. Heskes. Self-organizing maps, vector quantization, and mixture modeling, IEEE Transactions on Neural Networks, 12, 6, 1299-1305, 2001.
8. L. Kaufman and P. J. Rousseeuw. Finding groups in data. An introduction to cluster analysis. John Wiley & Sons. New York. 1990.

9. H. A. L. Kiers and Y. Takane. A generalization of GIPSCAL for the analysis of nonsymmetric data. Journal of Classification, 11:79-99, 1994.
10. T. Kohonen. Self-organizing maps. Springer Verlag, Berlin, Second edition, 1995.
11. T. Kohonen, S. Kaski, K. Lagus, J. Salojarvi, J. Honkela, V. Paatero and A. Saarela. Organization of a massive document collection, IEEE Transactions on Neural Networks, 11(3):574-585, 2000.
12. B. Kosko. Neural networks and fuzzy systems: A dynamical approach to machine intelligence. Prentice Hall, Englewood Cliffs, New Jersey, 1991.
13. L. Lebart, A. Morineau and J. F. Warwick. Multivariate descriptive statistical analysis. John Wiley, New York, 1984.
14. M. Martin-Merino and A. Muñoz. Self organizing map and Sammon mapping for asymmetric proximities, ICANN, LNCS 2130, 429-435, Springer Verlag, 2001.
15. M. Martin-Merino and A. Muñoz.Visualizing asymmetric proximities with SOM and MDS models, Neurocomputing, 63, 171-192, 2005.
16. D. Merkl. Text classification with self-organizing maps. Some lessons learned. Neurocomputing, 21:61-77, 1998.
17. F. Mulier and V. Cherkassky. Self-organization as an iterative kernel smoothing process. Neural Computation, 7, 1165-1177, 1995.
18. A. Muñoz. Compound key word generation from document databases using a hierarchical clustering ART model. Journal of Intelligent Data Analysis, 1(1), 25-48, 1997.
19. A. Okada. Asymmetric multidimensional scaling of two-mode three-way proximities, Journal of Classification, 14, 195-224, 1997.
20. M. Rorvig. Images of similarity: A visual exploration of optimal similarity metrics and scaling properties of TREC topic-document sets, Journal of the American Society for Information Science, 50, 8, 639-651, 1999.
21. Y. Yang and J. O. Pedersen. A comparative study on feature selection in text categorization, Proc. of the 14th International Conference on Machine Learning, Nashville, Tennessee, USA, July, 412-420, 1997.
22. B. Zielman and W.J. Heiser. Models for asymmetric proximities, British Journal of Mathematical and Statistical Psychology, 49:127-146, 1996.

Towards Automatic and Optimal Filtering Levels for Feature Selection in Text Categorization*

E. Montañés, E.F. Combarro, I. Díaz, and J. Ranilla

Artificial Intelligence Center, University of Oviedo, Spain
ir@aic.uniovi.es

Abstract. Text Categorization (TC) is an important issue within Information Retrieval (IR). Feature Selection (FS) becomes a crucial task, because of the presence of irrelevant features causing a loss in the performance. FS is usually performed selecting the features with highest score according to certain measures. However, the disadvantage of these approaches is that they need to determine in advance the number of features that are selected, commonly defined by the percentage of words removed, which is called Filtering Level (FL). In view of that, it is usual to carry out a set of experiments manually taking several FLs representing all possible ones. This process does not guarantee that any of the FLs chosen are the optimal ones, even not an approximation. This paper deals with overcoming this difficulty proposing a method that automatically determines optimal FLs by means of solving a univariate maximization problem.

1 Introduction

Text Categorization (TC) [1] consists of assigning the documents of a corpus to a set of prefixed categories. Since the documents in TC are normally represented by a great number of features and most of them could be irrelevant [2], a previous Feature Selection (FS) usually improves the performance of the classifiers, also reducing the computational time and the storage requirements.

A common way of tackling FS in TC consists in scoring the features using a certain measure, ordering them according to this measure and keeping or removing a predefined number or percentage of them [3,4]. Let consider what is called Filtering Level (FL), that is, the percentage of features removed. Then, a swapping of FLs are commonly taken to approximate all possible ones. This could lead to run the risk of not choosing either the optimal ones nor an approximation of them.

This paper proposes an approach to automatically determine optimal FLs by means of solving a univariate maximization problem. Several non-linear optimization methods are adapted for this purpose.

* This work has been supported in part under MEC and FEDER grant TIN2004-05920.

A.F. Famili et al. (Eds.): IDA 2005, LNCS 3646, pp. 239–248, 2005.

The rest of the paper is organized as follows: In Section 2 some related work is briefly exposed. Section 3 deals with the main stages of FS in TC using scoring measures. The description of the method is exhaustively detailed in Section 4. The description of the corpus and the experiments are presented in Section 5. Finally, in Section 6 some conclusions and ideas for further research are described.

2   Related Work

FS is one of the approaches commonly adopted in TC. FS could be performed by filtering or by wrappering. In the former, a feature subset is selected independently of the classifier. In the latter, a feature subset is selected using an evaluation function based on the classifier. A widely adopted approach in TC is the filtering one based on selecting the features with higher score granted by a certain measure [3,5,4]. The reason of choosing filtering approaches for TC is because wrapper ones can result in a rather time consuming process.

However, only in [6] it is automatically selected an adequate FL based on the typical deviation of the values reached by the measure. By contrast, most of the works choose a set of FLs representing all possible ones to perform experiments [3,5,4]. This paper proposes an method based on an univariate maximization to obtain optimal FLs, allowing the method to be automatic.

3   Using Scoring Measures for Feature Selection

The process of TC involves the stages described in this section.

3.1   Document Representation and Previous Feature Reduction

The most common way of representing the documents is the so called *bag of words* [2]. In this paper, we adopted the absolute frequency (tf) of the word, used in [7,4], to weight the importance of a word in a document.

Previous to FS, other kinds of feature reduction are typically carried out. They consist of removing *stop words* (words without meaning) and of performing *stemming* (reducing each word to its root or *stem*). We have conducted this last task according to Porter [8]. Also, following [1], we have only considered local sets of features, that is, words occurring in documents of the category (local approach), rather than a global set.

3.2   Feature Selection Using Scoring Measures

This section deals with the measures adopted for FS in TC.

  i) **Information Retrieval Measures.** In the IR field, several measures [2] have been used to determine the relevance of a word. The absolute frequency

(tf), the document frequency (df) or the combination of them $tfidf$, which considers the dispersion of the word, are measures of this kind.

ii) **Information Theory Measures.** One of main properties of IT measures is that they consider the distribution of the words over the categories. One of the most widely adopted [5] is the *information gain (IG)*, which takes into account either the presence of the word in a category or its absence. A similar measure is the *expected cross entropy for text (CET)* [3], which only takes into account the presence of the word in a category.

iii) **ML Measures.** In [5], some measures of this kind are proposed for FS in TC. They quantify the importance of a word w in a category c by means of evaluating the quality of the rule $w \to c$, assuming that it has been induced by a Machine Learning (ML) algorithm. Some of these measures are based on the percentage of successes and failures of its application, for instance the Laplace measure (L), which adds a little modification to the percentage of success and the *difference (D)*. Besides, other measures also take into account the number of documents of the category in which the word occurs and the distribution of the documents over the categories, for instance the *impurity level (IL)*. Some variants of these measures which also take into account the absence of the term in the rest of the categories were also adopted [5], leading respectively to L_{ir}, D_{ir} and IL_{ir} measures.

3.3   Classification and Evaluation of the Performance

The classification in TC is usually performed adopting the *one-against-the-rest* [9] approach. It involves converting the original problem into a set of binary problems, each one determining whether a document belongs to a certain category or not.

Here, the classification is performed using SVM, since they have shown to perform fast and well in TC [10]. They satisfactorily deal with many features and with sparse examples. They are binary classifiers which find out threshold functions to separate the documents of a certain category from the rest. We adopt a linear threshold since most TC problems are linearly separable [7].

The effectiveness of the classification is usually quantified with the F_1 measure [1], defined by

$$F_1 = \frac{1}{0.5\frac{1}{P} + 0.5\frac{1}{R}}$$

P quantifies the percentage of documents that are correctly classified as belonging to the category while R quantifies the percentage of documents of the category that are correctly classified. Two different approaches are typically adopted [1] to evaluate the global performance. The first one is *macroaverage* (averaging giving the same importance to each category) and the second one is *microaverage* (averaging proportionally to the size of each category). It is said that *macroaverage* gives information about the smaller categories, mean *microaverage* does about the biggest ones.

4   Finding Optimal Filtering Levels

This section deals with the raising and the solution of the problem.

4.1   Raising the Problem

The problem of finding an optimal FL o^* for a scoring measure $m_w(w)$ could be converted into a univariate maximization problem with F_1 as target function. Formally,

$$Find \quad o^* \in [o_1, o_2) \quad such \quad that \quad F_1(o^*) \geq F_1(o) \quad \forall o \in [o_1, o_2)$$

Notice that $o_1 = 0$, since keeping all the words could improve the classification. Analogously, $o_2 = 100$ with open interval, since it makes no sense to consider 100 as FL because this means not to keep any word.

The main disadvantage of this approximation is that choosing an arbitrary point $o \in [o_1, o_2)$ could cause that a word w_1 is taken, mean another one w_2 such that $m_w(w_2) = m_w(w_1)$ is not. A solution to solve this handicap could be to define a function g in the interval $[m_1, m_2]$ such that:

  i) $m_1 = \min_{w \in V} m_w(w)$ with V the set of features.
 ii) $m_2 = \max_{w \in V} m_w(w)$ with V the set of features.
iii) $g(m)$ is the FL that considers the words w such that $m_w(w) \leq m$

Hence, the original problem is converted into this new one:

$$Find \quad m^* \in [m_1, m_2] \quad such \quad that \quad F_1 \circ g(m^*) \geq F_1 \circ g(m) \quad \forall m \in [m_1, m_2]$$

Notice that g is not an injective function, since there could exist different values of the interval $[m_1, m_2]$ that grant the same FL. But, in fact, the domain of g is $m_w(V/ \sim)$, that is, the image of m_w defined over the quotient set $V/ \sim$ induced by the following equivalence relation:

$$w_1 \sim w_2 \quad if \quad and \quad only \quad if \quad m_w(w_1) = m_w(w_2)$$

Indeed, $\sim$ is an equivalence relation, since it is obvious that it satisfies reflexivity, symmetry and transitivity properties.

Therefore, defining g over $m_w(V/ \sim)$, g becomes injective. Finally, making effective these transformations yields to the following univariate maximization problem:

$$Find \quad m^* \in m_w(V/ \sim) \, such \, that \quad F_1 \circ g(m^*) \geq F_1 \circ g(m) \quad \forall m \in m_w(V/ \sim)$$

4.2   Approaches to Solve the Problem

Let consider the following univariate maximization problem:

$$Find \quad u^* \in [a, b] \quad such \quad that \quad f(u^*) \geq f(u) \quad \forall u \in [a, b]$$

There exist several methods to tackle univariate maximization [11]. All of them start in an interval $[a, b]$, called uncertainty interval, containing a local maximum u^*. These methods could be classified into:

i) **Methods based on nonlinear equations resolution.** They obtain the roots of the first derivative of f, which are candidates for being local maximum and minimum points of f, like Secant and Newton-Raphson methods [11]. As seen, they require that f is derivable.

ii) **Comparison methods.** The comparison methods, like Fibonacci, Golden Section and Uniform Search methods, evaluate the objective function at points over the interval and iteratively obtain smaller intervals of uncertainty. They use the property that evaluating four points in the uncertainty interval, there always exists at least a subinterval that does no contain the optimal, and hence it could be removed, obtaining a smaller interval [11]. In general, these methods do not demand that f is derivable or continuous, otherwise they assume that f is unimodal[1]. If f is multimodal, they only guarantee that a local maximum is found, but there could exist other local maximums whose value of f are greater.

iii) **Polynomial interpolation methods.** These methods iteratively approximate the target function in an interval by a quadratic or cubic polynomial that has the same function and/or gradient values at a number of points over the interval. The maximum of the polynomial is then used to predict the maximum of the target function [11]. Quadratic interpolation requires that f is continuous, mean cubic interpolation requires that f is derivable.

iv) **Hybrid methods.** These methods combine the best features of the comparison methods and the polynomial interpolation, like Brent's and the Modified Brent's methods [11].

4.3   Solving the Problem

Unfortunately, not all of the approaches cited above could be applied to our problem, since the target function that take up us ($F_1 \circ g$) has not enough regularity. The assumption of being continuous could not be even made, since it is a step function. Hence, only some comparison methods could be applied, like Fibonacci, Uniform and Golden Search methods. Also they are adequate since they have first order rate of convergence.

i) **Uniform Search.** This method consists in evaluating a predefined number of points uniformly distributed in each interval $[a_k, b_k]$ selecting that with higher value of the evaluation function. The interval of the next iteration is $[a_{k+1}, b_{k+1}] = [\lambda - \delta, \lambda + \delta]$ with $\delta = \frac{b-a}{n+1}$ where $[a, b]$ is the uncertainty initial interval, n is the predefined number of points to evaluate in each interval and λ is the point in the interval $[a_k, b_k]$ with higher value of the evaluation function among those in which the function is evaluated.

ii) **Fibonacci and Golden Search.** These methods will require two function evaluations per iteration (the two internal points $c_k < d_k$ in the interval

---

[1] f is an unimodal function in $[a, b]$ if it admits an unique maximum u^* in $[a, b]$ and if $\forall u_1, u_2 \in [a, b]$ such that $u_1 < u_2$ it satisfies that $u_1 \leq u^* \leq u_2$, $f(u_1) < f(u^*)$ and $f(u^*) > f(u_2)$.

$[a_k, b_k]$ of the k-iteration). This can be halved if one of the points is used again as an internal point for the following iteration. It is also desirable that the two possible subintervals $[a_k, d_k]$ and $[c_k, b_k]$ at each iteration be equal in size, since the theoretical interval reduction properties of the method are independent of the objective function and the rate of convergence can be analyzed much more conveniently. An interesting property is that the relationship $I_k = I_{k+1} + I_{k+2}$ holds between the interval size at successive iterations. Particular methods will differ in the way in which the last two intervals are selected:

ii.1) **Fibonacci.** After $m - 2$ iterations, the length I_{m-1} is obtained. This method imposes that $I_{m-1} = 2I_m$, that is, that $c_{m-1} = d_{m-1}$, which is the most that could be imposed. Hence, the general term of the sequence of interval lengths that has been built is $I_k = F_{m-k+1}I_m$ where F_{m-k+1} is obtained from the Fibonacci sequence, whose general term F_n satisfies the following difference equation

$$F_n = F_{n-1} + F_{n-2}$$

$$F_0 = F_1 = 1$$

Then, the explicit expression of F_n is

$$F_n = \frac{1}{\sqrt{5}} \left[\left(\frac{1 + \sqrt{5}}{2} \right)^{n+1} - \left(\frac{1 - \sqrt{5}}{2} \right)^{n+1} \right]$$

The reduction rate of the interval is variable in each iteration and it equals

$$\frac{I_{k+1}}{I_k} = \frac{F_{n-k}}{F_{n-k+1}}$$

ii.2) **Golden Search.** Imposing that the reduction rate of the interval be constant, the Golden Search method arises.

$$\frac{I_{k+1}}{I_k} = \tau$$

Dividing the following equation by I_{k+1}

$$I_k = I_{k+1} + I_{k+2}$$

it is obtained that τ satisfies $\tau^2 + \tau - 1 = 0$ whose positive root is $\frac{\sqrt{5}-1}{2} \approx 0,618$ and it is called the golden number.

The stop criterium for Fibonacci method could be either the number of iterations or a threshold for the final uncertainty interval, mean for Uniform and Golden Search methods it could only be the second one. This paper proposes to automatically define for all them a threshold as the minimum difference between the score of two words granted by the measure in question. In this way, the target of automatically defining an optimal FL is maintained.

Additionally, just the Uniform Search method need to defined the number of evaluation points in each iteration, which makes it a non strictly automatic method. We made the common election of 3 evaluations per iteration.

Another method, proposed by us and called Absolute method, consists in arbitrarily choosing a sequence of FLs, obtaining a sample of all possible FLs, and selecting that with higher value of the target function.

5   Experiments

The Absolute, Fibonacci, Uniform and Golden Search methods are compared among them. Also, they are compared with *reference values* which are obtained by selecting the maximum and the average of F_1 in test. The *Maximum Reference* is the one to which we can aspire at most and the *Average Reference* is a good reference to compare with. The FLs chosen for the Absolute method are the ones reported in [5], which are 20%, 40%, 60%, 80%, 85%, 90%, 95%, 98%. The methods are applied to the best measures obtained for each corpus in [5] (tf, IG, L_{ir}, D_{ir} and IL_{ir} for Reuters and $tfidf$, CET, L, D_{ir} and IL for Ohsumed). The evaluation function is chosen to be F_1 of a *Cross Validation* (CV) with 2 folds and 5 repetitions over SVM in the training phase. Then, the FL obtained is applied to the test phase.

5.1   The Corpus

This section describes the corpora used in the experiments.

i) **Reuters-21578 Collection.** The Reuters-21578 corpus is a set of economic news published by Reuters in 1987[2]. They are distributed over 135 categories. Each document belongs to one or more of them. The distribution of the documents is quite unbalanced and the words in the corpus are little scattered. The split into train and test documents chosen is that of Apté [9]. Removing some documents without body or topics, 7063 train and 2742 test documents assigned to 90 categories are obtained.

ii) **Ohsumed Collection.** Ohsumed is a MEDLINE subset of references from 270 medical journals over 1987-1991[3]. They are classified into the 15 fixed categories of MeSH [4]: A, B, C ... Each category is in turn split into subcategories. We have taken the first 20000 documents of 1991 with abstract, labelling the first 10000 documents as training and the rest as test ones. We split them into the 23 subcategories of category C of MeSH. The distribution of documents over the categories is much more balanced and the words are quite more scattered.

---

[2] It is publicly available at http://www.research.attp.com/lewis/reuters21578.html
[3] It can be found at http://trec.nist.gov/data/t9-filtering
[4] Available at www.nlm.nih.gov/mesh/2002/index.html

Table 1. *Macroaverage* and *Microaverage* of F_1 for Reuters-21578 and Ohsumed

Reuters

	Macroaverage				Microaverage			
	Absolute	Fibonacci	Golden	Uniform	Absolute	Fibonacci	Golden	Uniform
tf	47.34%	37.73%	36.53%	47.80%	86.12%	83.12%	82.92%	85.59%
IG	46.99%	47.19%	47.59%	48.07%	85.96%	83.70%	83.83%	85.51%
L_{ir}	50.35%	50.36%	50.50%	50.97%	86.16%	85.66%	85.87%	86.02%
D_{ir}	46.43%	49.85%	50.34%	50.10%	85.16%	85.67%	85.95%	85.75%
IL_{ir}	40.20%	18.20%	18.20%	16.49%	84.80%	32.66%	32.66%	74.53%

Ohsumed

	Macroaverage				Microaverage			
	Absolute	Fibonacci	Golden	Uniform	Absolute	Fibonacci	Golden	Uniform
$tfidf$	40.68%	16.64%	16.61%	41.98%	50.00%	27.32%	27.29%	50.69%
CET	41.22%	20.44%	20.62%	38.77%	50.38%	32.24%	32.32%	49.41%
L	46.18%	46.06%	45.39%	51.49%	51.54%	51.25%	50.73%	57.25%
D_{ir}	52.08%	51.27%	51.78%	51.76%	57.56%	57.36%	57.49%	57.47%
IL	41.17%	17.42%	17.42%	19.34%	51.10%	32.42%	32.42%	33.70%

Table 2. Maximum and Average Reference

Reuters					**Ohsumed**			
	Maximum		Average			Maximun		Average	
	Macrova.	Microav.	Macroav.	Microav.		Macroav.	Microav.	Macroav.	Microav.
tf	46.13%	85.28%	44.56%	84.74%	$tfidf$	42.43%	51.42%	41.31%	50.51%
IG	49.19%	85.53%	47.58%	85.14%	CET	48.91%	55.69%	46.99%	54.23%
L_{ir}	49.01%	85.26%	47.41%	84.81%	L	51.32%	56.92%	47.59%	53.55%
D_{ir}	46.71%	84.87%	43.60%	80.76%	D_{ir}	52.37%	58.04%	50.19%	56.61%
IL_{ir}	49.01%	85.07%	47.51%	84.66%	IL	45.76%	52.21%	49.79%	56.11%

5.2 Results

Table 1 shows the *macroaverage* and *microaverage* of F_1. Table 2 shows the *Maximum Reference* and the *Average Reference* and Table 3 does the results of a one-tail paired t-test over F_1 at a significance level of 95% among all methods.

With regard to Reuters-21578, the results reveal that for tf, IG and L_{ir} the Uniform Search method performs better in the *macroaverage*, mean the Absolute method does in the *microaverage*. Golden Search improves the rest of the methods for D_{ir} and Absolute method does for IL_{ir}. Also, *Maximum Reference* and *Average Reference* are improved by most of the measures and methods. The hypothesis contrasts show that, in general, there not exist appreciable differences among the methods for IG, L_{ir} and D_{ir}, that the Absolute method is significantly better than the rest for IL_{ir} and that both Absolute and Uniform Search methods do for tf. By contrast, for Ohsumed, Uniform Search improves the rest of the methods for $tfidf$ and L, being the results almost the same for all the methods for D_{ir} and being the Absolute method better for CET and IL, either for the *macroaverage* and the *microaverage*. Here, only L and Uniform method

Table 3. *t-test* of F_1 between Absolute, Fibonacci, Golden and Uniform Search methods ("+" means that the method in the left is significantly better than the method in the right , "=" means that there are not appreciable differences between both methods and "-" means that the method in the right is significantly better that method in the left)

	Reuters					Ohsumed				
	tf	IG	Lir	Dir	ILir	tfidf	CET	L	Dir	IL
Uni-Abs	=	=	=	+	-	=	=	=	=	-
Fib-Abs	-	=	=	=	-	-	-	=	=	-
Gol-Abs	-	=	=	=	-	-	-	=	=	-
Fib-Uni	-	=	=	=	=	-	-	-	=	=
Gol-Uni	-	=	=	=	=	-	-	-	=	=
Gol-Fib	=	=	=	+	=	=	=	=	=	=

improve both references, although D_{ir} for all methods and $tfidf$ for Uniform Search method do with regard to the *Average Reference*. The t-test reveals that, in general, the Absolute and Uniform methods are significantly better than the rest for $tfidf$ and CET, only the Absolute method does for IL and only the Uniform Search method does for L, being all the methods statistically similar for D_{ir}.

The differences in the behaviour of the *macroaverage* and *microaverage* for Reuters-21578 is due to its unbalanced distribution of the documents over the categories, being for Ohsumed quite similar. Also, the little scattered distribution of words the best the performance is, since there exists a higher probability of selecting specific words and hence an adequate FL.

Those methods assume that the target function is unimodal, and we are not able to assure it. Hence, the risk of finding a local maximum is latent. This could be reduced locating first a smaller uncertainty interval and then applying the methods over it. But the methods available for this purpose require the values of the gradient or in the case of the comparison methods, they require the manually setting of some parameters.

Among the methods proposed, only Fibonacci and Golden Search methods are strictly automatic methods, since Absolute method needs to fix the FLs and Uniform Search method requires to define the number of evaluations in each iteration.

6 Conclusions and Future Work

This paper proposes an automatic method based on an univariate maximization to obtain optimal FLs when FS is performed in TC using scoring measures.

Among the methods available, only comparison methods could be applied, since our target function is not even continuous. Four methods were selected, called Absolute, Fibonacci, Golden and Uniform Search methods.

The results reveal that, in general, Absolute and Uniform Search methods are better. For Reuters-21578 the results reached, for most measures and methods, the *Maximum Reference* and the *Average Reference* values, mean Ohsumed only does it for some.

Among the methods, only Fibonacci and Golden Search ones are strictly automatic methods.

As future work, we plan to study some alternatives to automatically avoid the local maximums. We are also interested in applied other maximization methods like genetic algorithms or simulated annealing.

References

1. Sebastiani, F.: Machine learning in automated text categorisation. ACM Computing Survey **34** (2002)
2. Salton, G., McGill, M.J.: An introduction to modern information retrieval. McGraw-Hill (1983)
3. Mladenic, D., Grobelnik, M.: Feature selection for unbalanced class distribution and naive bayes. In: Proc. 16th International Conference on Machine Learning ICML-99, Bled, SL (1999) 258–267
4. Yang, Y., Pedersen, J.O.: A comparative study on feature selection in text categorisation. In: Proc. 14th International Conference on Machine Learning ICML-97. (1997) 412–420
5. Montañés, E., Díaz, I., Ranilla, J., Combarro, E., Fernández, J.: Scoring and selecting terms for text categorization. IEEE Intelligent Systems (to appear)
6. Gabrilovich, E., Markovitch, S.: Text categorization with many redundant features: using aggressive feature selection to make svms competitive with c4.5. In: ICML '04: Twenty-first international conference on Machine learning, New York, NY, USA, ACM Press (2004)
7. Joachims, T.: Text categorization with support vector machines: learning with many relevant features. In Nédellec, C., Rouveirol, C., eds.: Proc. 10th European Conference on Machine Learning ECML-98. Number 1398, Chemnitz, DE, Springer-Verlag (1998) 137–142
8. Porter, M.F.: An algorithm for suffix stripping. Program (Automated Library and Information Systems) **14** (1980) 130–137
9. Apte, C., Damerau, F., Weiss, S.: Automated learning of decision rules for text categorization. Information Systems **12** (1994) 233–251
10. Yang, Y., Liu, X.: A re-examination of text categorization methods. In Hearst, M.A., Gey, F., Tong, R., eds.: Proc. 22nd ACM International Conference on Research and Development in Information Retrieval SIGIR-99, Berkeley, US, ACM Press, New York, US (1999) 42–49
11. Scales, M.: Introduction to Non-Linear Optimization. McMillan (1986)

Block Clustering of Contingency Table and Mixture Model

Mohamed Nadif[1] and Gérard Govaert[2]

[1] LITA EA3097, Université de Metz, Ile du Saulcy, 57045 Metz, France
mohamed.nadif@univ-metz.fr
[2] HEUDIASYC, UMR CNRS 6599, Université de Technologie de Compiègne,
BP 20529, 60205 Compiègne Cedex, France
gerard.govaert@utc.fr

Abstract. Block clustering or simultaneous clustering has become an important challenge in data mining context. It has practical importance in a wide of variety of applications such as text, web-log and market basket data analysis. Typically, the data that arises in these applications is arranged as a two-way contingency or co-occurrence table. In this paper, we embed the block clustering problem in the mixture approach. We propose a Poisson block mixture model and adopting the classification maximum likelihood principle we perform a new algorithm. Simplicity, fast convergence and scalability are the major advantages of the proposed approach.

1 Introduction

Cluster analysis is an important tool in a variety of scientific areas such as pattern recognition, information retrieval, microarray, data mining, and so forth. Although many clustering procedures such as hierarchical clustering, k-means or self-organizing maps, aim to construct an optimal partition of objects or, sometimes, of variables, there are other methods, called block clustering methods, which consider simultaneously the two sets and organize the data into homogeneous blocks. If $\mathbf{x}$ denotes a data matrix defined by $\mathbf{x} = \{(x_i^j); i \in I \text{ and } j \in J\}$, where I is a set of objects (rows, observations, cases) and J is a set of variables (columns, attributes), the basic idea of these methods consists in making permutations of objects and variables in order to draw a correspondence structure on $I \times J$.

These last years, block clustering (also called biclustering) has become an important challenge in data mining context. In the text mining field, Dhillon [3] has proposed a spectral block clustering method by exploiting the duality between rows (documents) and columns (words). In the analysis of microarray data where data are often presented as matrices of expression levels of genes under different conditions, block clustering of genes and conditions has permitted to overcome the problem of the choice of similarity on the two sets found in conventional clustering methods [2]. Also, these kinds of methods have practical importance in a wide variety of applications such as text, web-log and market

A.F. Famili et al. (Eds.): IDA 2005, LNCS 3646, pp. 249–259, 2005.
© Springer-Verlag Berlin Heidelberg 2005

basket data analysis. Typically, the data that arises in these applications is arranged as a two-way contingency or co-occurrence table.

In this paper, we will focus on these kinds of data. In exploiting the clear duality between rows and columns, we will study the block clustering problem in embedding it in the mixture approach. We will propose a *block mixture model* which takes into account the block clustering situation and perform an innovative co-clustering algorithm. This one is based on the alternated application of Classification EM [1] on intermediate data matrices. To propose this algorithm, we set this problem in the classification maximum likelihood (CML) approach [9]. Results on simulated data are given, confirming that this algorithm works well in practice.

This paper is organized as follows. Section 2 begins with a description of the *Croki2* algorithm proposed by Govaert [5] to partitioning simultaneously the rows and columns of a contingency table. As we are interested in the modeling of our problem, we review briefly the block mixture model in Section 3. In Section 4, we propose a *Poisson block mixture model* adapted to our situation and a new algorithm based on the Classification EM algorithm. And to achieve our aim we study the behavior of our algorithm and compare it with the *Croki2* algorithm in Section 5. Finally, the last section summarizes the main points of this paper.

Notation: we now define the notation that is used consistently throughout this paper. The two-way contingency table will be denoted $\mathbf{x}$; it is a $r \times s$ data matrix defined by $\mathbf{x} = \{(x_{ij}); i \in I, j \in J\}$, where I is a categorical variable with r categories and J a categorical variable with s categories. We shall denote the row and columns total of $\mathbf{x}$ by $x_{i.} = \sum_j x_{ij}$ and $x_{.j} = \sum_i x_{ij}$ and the overall total simply by $n = \sum_{ij} x_i^j$. We will also use the frequency table $\{(f_{ij} = x_{ij}/n); i \in I, j \in J\}$, the marginal frequencies $f_{i.} = \sum_j f_{ij}$ and $f_{.j} = \sum_i f_{ij}$, the row profiles $f_J^i = (f_{i1}/f_{i.}, \ldots, f_{is}/f_{i.})$ and the average row profile $f_J = (f_{.1}, \ldots, f_{.s})$. We represent a partition of I into g clusters by $\mathbf{z} = (\mathbf{z}_1, \ldots, \mathbf{z}_r)$ where $\mathbf{z}_i$, which indicates the component of the row i, is represented by $\mathbf{z}_i = (z_{i1}, \ldots, z_{ig})$ with $z_{ik} = 1$ if row i is in cluster k and 0 otherwise. Then, the kth cluster corresponds to the set of rows i such that $z_{ik} = 1$. We will use similar notation for a partition $\mathbf{w}$ into m clusters of the set J. In the following, to simplify the notation, the sums and the products relating to rows, columns or clusters will be subscripted respectively by letters i, j or k without indicating the limits of variation, which will be thus implicit. Thus, for example, the sum $\sum_i$ stands for $\sum_{i=1}^r$ or $\sum_{i,j,k,\ell}$ stands for $\sum_{i=1}^r \sum_{j=1}^s \sum_{k=1}^g \sum_{\ell=1}^m$.

2   The *Croki2* Algorithm

2.1   Aim of *Croki2*

To measure the information brought by a table of contingency, one seeks to evaluate the links existing between the two sets I and J. There are several measures of association and the most employed is the chi-square χ^2. This criterion, used for example in the correspondence analysis, is defined as follows

$$\chi^2(I, J) = \sum_{i,j} \frac{(x_{ij} - \frac{x_{i.}x_{.j}}{n})^2}{\frac{x_{i.}x_{.j}}{n}} = n \sum_{i,j} \frac{(f_{ij} - f_{i.}f_{.j})^2}{f_{i.}f_{.j}}.$$

This measure usually provides statistical evidence of a significant association, or dependence, between rows and columns of the table. This quantity represents the deviation between the theoretical frequencies $f_{i.}f_{.j}$, that we would have if I and J were independent, and the observed frequencies f_{ij}. If I and J are independent, the χ^2 will be zero and if there is a strong relationship between I and J, the χ^2 will be high. So, a significant chi-square indicates a departure from row or column homogeneity and can be used as a measure of heterogeneity. Then, the chi-square can be used to evaluate the quality of a partitions $\mathbf{z}$ of I and $\mathbf{w}$ of J: for this, we will associate to these partitions $\mathbf{z}$ and $\mathbf{w}$ the chi-square $\chi^2(\mathbf{z}, \mathbf{w})$ of the contingency table with g rows and m columns obtained from the initial table in making the sum of rows and columns of each cluster. It is straightforward that we have

$$\chi^2(I, J) \geq \chi^2(\mathbf{z}, \mathbf{w}) \tag{1}$$

which shows that the proposed regrouping necessarily leads to a loss of information. The objective of classification is to find the partitions $\mathbf{z}$ and $\mathbf{w}$ which minimize this loss, i.e. which maximizes $\chi^2(\mathbf{z}, \mathbf{w})$. Let us notice that when for each cluster the row profiles and the column profiles are equal, the inequality (1) becomes $\chi^2(\mathbf{z}, \mathbf{w}) = \chi^2(I, J)$ and in this particular case there is no loss of information. In addition, the problem we define has a sense only when the number of clusters is fixed. In the opposite case, the optimal partition is just the partition where each element of I form a cluster.

To maximize $\chi^2(\mathbf{z}, \mathbf{w})$, Govaert [5] has proposed the *Croki2* algorithm. The author has shown that the maximization of $\chi^2(\mathbf{z}, \mathbf{w})$ can be carried out by the alternated maximization of $\chi^2(\mathbf{z}, J)$ and $\chi^2(\mathbf{w}, J)$ which guarantees the convergence. The *Mndki2* algorithm that we describe hereafter can perform these maximizations.

2.2   The Mndki2 Algorithm

The *Mndki2* algorithm is based on the same geometrical representation of a contingency table as that which is used by correspondence analysis. This representation is justified for several reasons, in particular for the similar roles reserved for each of the two dimensions of the analyzed table and the property of distributional equivalence allowing for a great stability of the results when agglomerating elements with similar profiles. In this representation, each row i is associated to a point vector $\mathbb{R}^s$ defined by the profile f_J^i weighted by the marginal frequency $f_{i.}$. The distances between profiles is not defined by the usual Euclidean metric but rather by the weighted Euclidean metric, called the *chi-squared metric*, defined by the diagonal matrix $\text{diag}(\frac{1}{f_{.1}}, \ldots, \frac{1}{f_{.s}})$.

If $\mathbf{z}$ is a partition of the rows, we can define the frequencies $f_{kj} = \sum_{i/z_{ik}=1} f_{ij}$ and the average row profile of the kth cluster $f_J^k = (f_{k1}/f_{k.}, \ldots, f_{ks}/f_{k.})$ where

$f_{k.} = \sum_j f_{kj}$. With this representation, we can show after some calculus that the total of squared distances T, the between cluster sums of squares $B(\mathbf{z})$ and the within cluster sums of squares $W(\mathbf{z})$ can be written

$$T = \sum_i f_{i.} d^2(f_J^i, f_J) = \frac{1}{n}\chi^2(I, J),$$

$$B(\mathbf{z}) = \sum_k f_{k.} d^2(f_J^k, f_J) = \frac{1}{n}\chi^2(\mathbf{z}, J) \text{ and } W(\mathbf{z}) = \sum_k \sum_{i|z_{ik}=1} f_{i.} d^2(f_J^i, f_J^k).$$

Then, the traditional relation between the total of squared distances, the within cluster sums of squares and the between cluster sums of squares $T = W(\mathbf{z}) + B(\mathbf{z})$ leads to the following relation

$$\chi^2(I, J) = n.W(\mathbf{z}) + \chi^2(\mathbf{z}, J).$$

Thus, $nW(\mathbf{z})$ represents the information lost in regrouping the elements according the partition $\mathbf{z}$, and $\chi^2(\mathbf{z}, J)$ corresponds to the preserved information. Consequently, since the quantity $\chi^2(I, J)$ does not depend on the partition $\mathbf{z}$, the research of the partition maximizing the criterion $\chi^2(\mathbf{z}, J)$ is equivalent to the research of the partition minimizing criterion $W(\mathbf{z})$. To minimize this criterion, it is then possible to apply k-means to the set of profiles with the χ^2 metric. One thus obtains an iterative algorithm, called *Mndki2*, maximizing locally $\chi^2(\mathbf{z}, J)$.

2.3   Description of *Croki2*

Finally the different steps of the *Croki2* algorithm are the following.

1. Start from an initial position $(\mathbf{z}^{(0)}, \mathbf{w}^{(0)})$.
2. Computation of $(\mathbf{z}^{(c+1)}, \mathbf{w}^{(c+1)})$ starting from $(\mathbf{z}^c, \mathbf{w}^c)$:
   (a) Computation of $\mathbf{z}^{(c+1)}$. From $\mathbf{z}^{(c)}$, we use *Mndki2* on the contingency table $(I, \mathbf{w})$ obtained by making the column sums of each cluster of $\mathbf{w}$.
   (b) Computation of $\mathbf{w}^{(c+1)}$. From $\mathbf{w}^{(c)}$, we use *Mndki2* on the contingency table $(\mathbf{z}, J)$ by making the row sums of each cluster of $\mathbf{z}$.
3. Iterate the steps 2 until the convergence.

3   Block Mixture Model

The mixture model is undoubtedly one of the greatest contributions to clustering [8]. It offers a great flexibility and solutions to the problem of the number of clusters. Its associated estimators of posterior probabilities allow one to obtain a fuzzy or hard clustering by using the maximum a posterior principle.

For the classical mixture model, we have shown [6] that the probability density function of a mixture sample $\mathbf{x}$ defined by $f(\mathbf{x}; \boldsymbol{\theta}) = \prod_i \sum_k \pi_k \varphi(\mathbf{x}_i; \boldsymbol{\alpha}_k)$ where the π_k's are the mixing proportions, the $\varphi(\mathbf{x}_i; \boldsymbol{\alpha}_k)$ are the densities of each component k, and $\boldsymbol{\theta}$ is defined by $(\pi_1, \ldots, \pi_g, \boldsymbol{\alpha}_1, \ldots, \boldsymbol{\alpha}_g)$, can be written as

$$f(\mathbf{x}; \boldsymbol{\theta}) = \sum_{\mathbf{z} \in \mathcal{Z}} p(\mathbf{z}; \boldsymbol{\theta}) f(\mathbf{x}|\mathbf{z}; \boldsymbol{\theta}), \tag{2}$$

where $\mathcal{Z}$ denotes the set of all possible partitions of I in g clusters, $p(\mathbf{z}; \boldsymbol{\theta}) = \prod_i \pi_{z_i}$ and $f(\mathbf{x}|\mathbf{z}; \boldsymbol{\theta}) = \prod_i \varphi(\mathbf{x}_i; \boldsymbol{\alpha}_{z_i})$. With this formulation, the data matrix $\mathbf{x}$ is assumed to be a sample of size 1 from a random (r, s) matrix.

To study the block clustering problem, we have extended the formulation (2) to propose a block mixture model defined by the following probability density function $f(\mathbf{x}; \boldsymbol{\theta}) = \sum_{\mathbf{u} \in U} p(\mathbf{u}; \boldsymbol{\theta}) f(\mathbf{x}|\mathbf{u}; \boldsymbol{\theta})$ where U denotes the set of all possible partitions of $I \times J$ and $\boldsymbol{\theta}$ is the parameter of this mixture model. In restricting this model to a set of partitions of $I \times J$ defined by a product of partitions of I and J, which will be supposed to be independent, we obtain the following decomposition

$$f(\mathbf{x}; \boldsymbol{\theta}) = \sum_{(\mathbf{z},\mathbf{w}) \in \mathcal{Z} \times \mathcal{W}} p(\mathbf{z}; \boldsymbol{\theta}) p(\mathbf{w}; \boldsymbol{\theta}) f(\mathbf{x}|\mathbf{z}, \mathbf{w}; \boldsymbol{\theta})$$

where $\mathcal{Z}$ and $\mathcal{W}$ denote the sets of all possible partitions $\mathbf{z}$ of I and $\mathbf{w}$ of J.

Now, extending the latent class principle of local independence to our block model, the x_i^j will be supposed to be independent once $\mathbf{z}_i$ and $\mathbf{w}_j$ are fixed; then, we have $f(\mathbf{x}|\mathbf{z}, \mathbf{w}; \boldsymbol{\theta}) = \prod_{i,j} \varphi(x_{ij}; \boldsymbol{\alpha}_{z_i w_j})$ where $\varphi(x, \boldsymbol{\alpha}_{k\ell})$ is a probability density function defined on the real set $\mathbb{R}$. Denoting $\boldsymbol{\theta} = (\boldsymbol{\pi}, \boldsymbol{\rho}, \boldsymbol{\alpha}_{11}, \ldots, \boldsymbol{\alpha}_{gm})$ where $\boldsymbol{\pi} = (\pi_1, \ldots, \pi_g)$ and $\boldsymbol{\rho} = (\rho_1, \ldots, \rho_m)$ are the vectors of probabilities π_k and ρ_ℓ that a row and a column belong to the kth component and to the ℓth component respectively, we obtain a block mixture model with the following probability density function

$$f(\mathbf{x}; \boldsymbol{\theta}) = \sum_{(\mathbf{z},\mathbf{w}) \in \mathcal{Z} \times \mathcal{W}} \prod_i \pi_{z_i} \prod_j \rho_{w_j} \prod_i \prod_j \varphi(x_{ij}; \boldsymbol{\alpha}_{z_i w_j}).$$

To tackle the simultaneous partitioning problem, we will use the CML approach, which aims to maximize the classification log-likelihood called complete data log-likelihood associated to the block mixture model. With our model, the complete data are $(\mathbf{z}, \mathbf{w}, \mathbf{x})$ and the classification log-likelihood is given by

$$L(\boldsymbol{\theta}; \mathbf{x}, \mathbf{z}, \mathbf{w}) = \log(p(\mathbf{z}; \boldsymbol{\theta}) p(\mathbf{w}; \boldsymbol{\theta}) f(\mathbf{x}|\mathbf{z}, \mathbf{w}; \boldsymbol{\theta})).$$

4   Block Mixture Model for Contingency Table

4.1   The Model

Counts in the $r \times s$ cells of a contingency table are typically modelled as random variables. In our situation, we assume that for each block $k\ell$ the values x_{ij} are distributed according the Poisson distribution $\mathcal{P}(\alpha_i \beta_j \delta_{k\ell})$ and the probability mass function is

$$\frac{e^{-\alpha_i \beta_j \delta_{k\ell}} (\alpha_i \beta_j \delta_{k\ell})^{x_{ij}}}{x_{ij}!}.$$

The Poisson parameter is split into α_i, β_j the effects of the row i and the column j and $\delta_{k\ell}$ the effect of the block $k\ell$. Because the aim is to maximize the complete data log-likelihood not only depending on $\boldsymbol{\theta}$ but on $\mathbf{z}, \mathbf{w}$, an adapted re-parametrization of the Poisson distribution becomes necessary. To this end, we impose some constraints and we assume that

$$\sum_\ell \beta_\ell \delta_{k\ell} = 1 \quad \text{and} \quad \sum_k \alpha_k \delta_{k\ell} = 1 \quad \text{with} \quad \alpha_k = \sum_{i,k} z_{ik} \alpha_i, \beta_\ell = \sum_{j,\ell} w_{j\ell} \beta_j. \tag{3}$$

The classification log-likelihood $L(\boldsymbol{\theta}; \mathbf{x}, \mathbf{z}, \mathbf{w})$ takes the following form

$$\sum_{i,k} z_{ik} \log \pi_k + \sum_{j,\ell} w_{j\ell} \log \rho_\ell$$

$$+ \sum_{k,\ell} (x_{k\ell} \log \delta_{k\ell} - \alpha_k \beta_\ell \delta_{k\ell}) + \sum_i x_{i.} \log \alpha_i + \sum_j x_{.j} \log \beta_j + cste. \tag{4}$$

From (4), it is straightforward that for $(\mathbf{z}, \mathbf{w})$ fixed, a solution for the maximization of $L(\boldsymbol{\theta}; \mathbf{x}, \mathbf{z}, \mathbf{w})$ is given by

$$\alpha_k = x_{k.} \quad \beta_\ell = x_{.\ell} \quad \text{and} \quad \delta_{k\ell} = \frac{x_{k\ell}}{x_{k.} x_{.\ell}} \quad \text{with} \quad x_{k\ell} = \sum_{i,j,k,\ell} z_{ik} w_{j\ell} x_{ij}, \tag{5}$$

and therefore $\alpha_i = x_{i.}$ and $\beta_j = x_{.j}$ which do not depend on the blocks. Finally, the maximization of $L(\boldsymbol{\theta}; \mathbf{x}, \mathbf{z}, \mathbf{w})$ reduces to minimizing the following criterion

$$L_c(\mathbf{z}, \mathbf{w}, \boldsymbol{\theta}) = \sum_{i,k} z_{ik} \log \pi_k + \sum_{j,\ell} w_{j\ell} \log \rho_\ell + \sum_{k,\ell} x_{k\ell} \log \delta_{k\ell}$$

$$= \sum_{i,k} z_{ik} \log \pi_k + \sum_{j,\ell} w_{j\ell} \log \rho_\ell + \sum_{i,j,k,\ell} z_{ik} w_{j\ell} x_{ij} \log \delta_{k\ell}$$

where $\boldsymbol{\theta} = (\boldsymbol{\pi}, \boldsymbol{\rho}, \delta_{11}, \dots, \delta_{gm})$ with $\sum_\ell x_{.\ell} \delta_{k\ell} = 1$ and $\sum_k x_{k.} \delta_{k\ell} = 1$.

4.2   Cemcroki2 Algorithm

To maximize $L_c(\mathbf{z}, \mathbf{w}, \boldsymbol{\theta})$, we propose to maximize alternatively the classification log-likelihood with $\mathbf{w}$ and $\boldsymbol{\rho}$ fixed and then with $\mathbf{z}$ and $\boldsymbol{\pi}$ fixed. By noting $u_{i\ell} = \sum_j w_{j\ell} x_{ij}$, the classification log-likelihood can be written as

$$L_c(\mathbf{z}, \mathbf{w}, \boldsymbol{\theta}) = \sum_{i,k} z_{ik} \log \pi_k + \sum_{j,\ell} w_{j\ell} \log \rho_\ell + \sum_{i,k} z_{ik} \sum_\ell u_{i\ell} \log \delta_{k\ell}.$$

Let $\gamma_{k\ell}$ denote $u_{.\ell} \delta_{k\ell}$, since $u_{.\ell} = x_{.\ell}$ we have $\sum_\ell \gamma_{k\ell} = 1$ and $L_c(\mathbf{z}, \mathbf{w}, \boldsymbol{\theta})$ breaks into two terms

$$L_c(\mathbf{z}, \mathbf{w}, \boldsymbol{\theta}) = L_c(\mathbf{z}, \boldsymbol{\theta}/\mathbf{w}) + g(\mathbf{x}, \mathbf{w}, \boldsymbol{\rho})$$

where the first one corresponds to classification data log-likelihood of classical mixture model

$$L_c(\mathbf{z}, \boldsymbol{\theta}/\mathbf{w}) = \sum_{i,k} z_{ik} \log(\pi_k \Phi(\mathbf{u}_i, \boldsymbol{\gamma}_k))$$

where $\Phi(\mathbf{u}_i, \boldsymbol{\gamma}_k)$ is the multinomial for $u_{i1}, \ldots, u_{im}$ with the probabilities $\gamma_{k1}, \ldots,$ γ_{km} and the second one which does not depend on $\mathbf{z}$

$$g(\mathbf{x}, \mathbf{w}, \boldsymbol{\rho}) = \sum_{j,\ell} w_{j\ell} \log \rho_\ell - \sum_\ell u_{.\ell} \log u_{.\ell}.$$

Hence, the conditional classification log-likelihood $L_c(\mathbf{z}, \boldsymbol{\theta}/\mathbf{w})$ corresponds to the complete log-likelihood associated to multinomial mixture applied on the samples $\mathbf{u}_1, \ldots, \mathbf{u}_r$ where $\mathbf{u}_i = (u_{i1}, \ldots, u_{im})$. Maximizing $L_c(\mathbf{z}, \mathbf{w}, \theta)$ for $\mathbf{w}$ fixed is equivalent to maximize the conditional classification log-likelihood $L_c(\mathbf{z}, \theta/\mathbf{w})$, which can be done by the CEM algorithm applied to the multinomial mixture model. The different steps of CEM are

- E-step: compute the posterior probabilities $t_{ik}^{(c)}$;
- C-step: the kth cluster of $\mathbf{z}^{(c+1)}$ is defined with

$$z_{ik}^{(c+1)} = 1 \text{ if } k = \mathrm{argmax}_{k=1,\ldots,g} \ t_{ik}^{(c)} \text{ and } z_{ik}^{(c+1)} = 0 \text{ otherwise}$$

- M-step: by standard calculations, one arrives at the following re-estimations parameters.

$$\pi_k^{(c+1)} = \frac{\# z_k^{(c+1)}}{r} \qquad \text{and} \qquad \gamma_{k\ell}^{(c+1)} = \frac{u_{k\ell}}{x_{k.}} \Longrightarrow \delta_{k\ell}^{(c+1)} = \frac{u_{k\ell}}{x_{k.}x_{.\ell}} = \frac{x_{k\ell}}{x_{k.}x_{.\ell}}$$

where $\#$ denotes the cardinality.

In the same way, taking the sufficient statistic $v_{kj} = \sum_{i,k} z_{ik} x_{ij}$, we can easily show that $L_c(\mathbf{z}, \mathbf{w}, \boldsymbol{\theta}) = L_c(\mathbf{w}, \boldsymbol{\theta}/\mathbf{z}) + g(\mathbf{x}, \mathbf{z}, \boldsymbol{\pi})$ and therefore develop the differents steps of the CEM algorithm applied on $\mathbf{v}^j = (v^{j1}, \ldots, v^{jg})$. Finally, we can describe easily the different steps of the algorithm called *Cemcroki2*.

1. Start from an initial position $(\mathbf{z}^{(0)}, \mathbf{w}^{(0)}, \boldsymbol{\theta}^{(0)})$.
2. Computation of $(\mathbf{z}^{(c+1)}, \mathbf{w}^{(c+1)}, \boldsymbol{\theta}^{(c+1)})$ starting from $(\mathbf{z}^{(c)}, \mathbf{w}^{(c)}, \boldsymbol{\theta}^{(c)})$:
   (a) Computation of $\mathbf{z}^{(c+1)}, \boldsymbol{\pi}^{(c+1)}, \delta^{(c+\frac{1}{2})}$ using the CEM algorithm on the data $(\mathbf{u}_1, \ldots, \mathbf{u}_r)$ starting from $\mathbf{z}^{(c)}, \boldsymbol{\pi}^{(c)}, \delta^{(c)}$.
   (b) Computation of $\mathbf{w}^{(c+1)}, \boldsymbol{\rho}^{(c+1)}, \delta^{(c+1)}$ using the CEM algorithm on the data $(\mathbf{v}^1, \ldots, \mathbf{v}^s)$ starting from $\mathbf{w}^{(c)}, \boldsymbol{\rho}^{(c)}, \delta^{(c+\frac{1}{2})}$.
3. Iterate the steps 2 until the convergence.

If we substitute $\gamma_{k\ell}$ by their M-step re-estimation formula, the criterion takes the following form

$$L_c(\mathbf{z}, \mathbf{w}, \boldsymbol{\theta}) = \sum_k \# z_k \log \pi_k + \sum_\ell \# w_\ell \log \rho_\ell + \sum_{k,\ell} x_{k\ell} \log \frac{x_{k\ell}}{x_{k.}x_{.\ell}}.$$

Having found the estimate of the parameters and noting $f_{k\ell} = \frac{x_{k\ell}}{x_{..}}$, the criterion is expressed as

$$L_c(\mathbf{z}, \mathbf{w}, \boldsymbol{\theta}) = \sum_k \# z_k \log \pi_k + \sum_\ell \# w_\ell \log \rho_\ell + x_{..} \sum_{k,\ell} f_{k\ell} \log \frac{f_{k\ell}}{f_{k.}f_{.\ell}} + cste \quad (6)$$

Note that the term $\sum_{k,\ell} f_{k\ell} \log \frac{f_{k\ell}}{f_{k.} f_{.\ell}}$ is the mutual information $I(\mathbf{z}, \mathbf{w})$ quantifying the information shared between $\mathbf{z}$ and $\mathbf{w}$. It is easy to show it from the definition in terms of entropies $I(\mathbf{z}, \mathbf{z}) = H(\mathbf{z}) + H(\mathbf{w}) - H(\mathbf{z}, \mathbf{w})$ where $H(.)$ is the entropy. Furthermore, using the approximation $2x \log x \approx x^2 - 1$, the expression of $L(\boldsymbol{\theta}; \mathbf{x}, \mathbf{z}, \mathbf{w})$ can be approximated by

$$L_c(\mathbf{z}, \mathbf{w}, \theta) = \sum_k \#z_k \log \pi_k + \sum_\ell \#w_\ell \log \rho_\ell + \frac{x_{..}}{2} \chi^2(\mathbf{z}, \mathbf{w}) + cste. \qquad (7)$$

Then, from (6) and (7), when the proportions are fixed the maximization of $L(\boldsymbol{\theta}; \mathbf{x}, \mathbf{z}, \mathbf{w})$ is equivalent to the maximization of the mutual information $I(\mathbf{z}, \mathbf{w})$ and approximately equivalent to the maximization of the chi-square criterion $\chi^2(\mathbf{z}, \mathbf{w})$: the use of the both criteria $\chi^2(\mathbf{z}, \mathbf{w})$ and $I(\mathbf{z}, \mathbf{w})$ assumes implicitly that the data arise from a mixture of Poisson distributions.

5   Numerical Experiments

5.1   Synthetic Data

To illustrate the behavior of our algorithms *Croki2* and *Cemcroki2*, we studied their performances on simulated data. We selected thirty kind of data arising from 3×2-component Poisson block mixture in considering two situations : equal proportions ($p_1 = p_2 = p_3$ and $q_1 = q_2$), and not equal ($p_1 = 0.70, p_2 = 0.20, p_3 = 0.10$). These data are obtained by varying the following parameters: the degree of overlap depending on $\boldsymbol{\theta} = (\boldsymbol{\pi}, \boldsymbol{\rho}, \delta)$ and the size. The *overlap* can be measured by the Bayes error corresponding to our model. Its computation being theoretically difficult we used Monte Carlo simulations and evaluated this error by comparing the simulated partitions and those we obtained by applying a C-step. Six *overlap* have been considered and are approximatively equal to $5\%, 11\%, 16\%, 20\%, 27\%, 34\%$. Concerning the size, we took $r \times s = (30 \times 20), (50 \times 20), (100 \times 20), (500 \times 20)$ and (1000×20).

For each of these 30 data structures, we generated 30 samples and for each sample, we ran *Cemcroki2* and *Croki2* 30 times starting from random situations and selected the best solution for each method. In order to summarize the behavior of these algorithms, we used the proportion of misclassified points "error rate" occurring for each sample.

The results obtained are displayed in Tables 1,2. For each data set and each algorithm, we summarize the 30 trials with the means and standard deviations of error rates obtained by comparing the partitions obtained by the both methods and the simulated partitions.

From these first experiments, the main points arising are the following. When the proportions are equal (see Table 1), the both algorithms are equivalent. In contrary when the proportions are dramatically different (see Table2), in all situations *Cemcroki2* outperforms clearly *Croki2* which does not hold account of the proportions.

Table 1. *Cemcroki2* vs. *Croki2* for 30 kinds of data when $p_1 = p_2 = p_3$ and $q_1 = q_2$: means and standard deviations of error rates

Size		Overlap					
		1	2	3	4	5	6
30	Cemcroki2	.070(.049)	.131(.079)	.221(.094)	.269(.124)	.349(.152)	.413(.167)
	Croki2	.071(.052)	.133(.079)	.222(.092)	.269(.129)	.348(.136)	.430(.161)
50	Cemcroki2	.063(.029)	.112(.050)	.170(.063)	.223(.078)	.261(.105)	.334(.167)
	Croki2	.064(.029)	.112(.050)	.166(.052)	.222(.073)	.246(.096)	.328(.158)
100	Cemcroki2	.058(.019)	.109(.026)	.153(.045)	.184(.059)	.210(.062)	.253(.089)
	Croki2	.059(.019)	.108(.025)	.156(.044)	.193(.099)	.206(.056)	.261(.094)
500	Cemcroki2	.055(.011)	.094(.013)	.143(.020)	.169(.040)	.195(.023)	.247(.115)
	Croki2	.055(.012)	.094(.013)	.150(.060)	.165(.020)	.204(.062)	.244(.107)
1000	Cemcroki2	.054(.008)	.094(.013)	.138(.013)	.174(.046)	.188(.044)	.267(.151)
	Croki2	.054(.008)	.094(.013)	.138(.013)	.169(.016)	.183(.014)	.281(.167)

Table 2. *Cemcroki2* vs. *Croki2* for 30 kinds of data when $\mathbf{p} = (.70, .20, .10)$ and $q_1 = q_2$: means and standard deviations of error rates

Size		Overlap					
		1	2	3	4	5	6
30	Cemcroki2	.075(.074)	.167(.148)	.407(.199)	.429(.233)	.474(.146)	.511(.158)
	Croki2	.145(.077)	.245(.145)	.432(.208)	.463(.189)	.519(.147)	.570(.193)
50	Cemcroki2	.054(.041)	.112(.077)	.318(.194)	.353(.177)	.376(.150)	.521(.137)
	Croki2	.141(.067)	.203(.068)	.333(.168)	.406(.159)	.453(.170)	.541(.158)
100	Cemcroki2	.044(.023)	.109(.081)	.200(.128)	.266(.144)	.373(.171)	.415(.172)
	Croki2	.144(.061)	.192(.043)	.258(.120)	.305(.074)	.422(.140)	.520(.163)
500	Cemcroki2	.045(.010)	.099(.092)	.173(.129)	.196(.101)	.273(.129)	.287(.110)
	Croki2	.131(.028)	.194(.080)	.245(.106)	.407(.150)	.513(.144)	.507(.115)
1000	Cemcroki2	.043(.008)	.072(.010)	.177(.150)	.215(.124)	.299(.119)	.301(.122)
	Croki2	.133(.016)	.180(.020)	.266(.122)	.405(.160)	.490(.137)	.622(.164)

5.2   Real Data

To illustrate the Cemcroki2 algorithm on real data, we choose the SMART collection from Cornell (ftp.cs.cornell.edu/pub/smart). The SMART collection consists of Medline, a set of 1033 abstracts from medical journals, CISI, a set of 1460 abstracts from information retrieval papers and CRANFIELD sub-collection, a

Table 3. Cemcroki2 vs BSGP and IT algorithms

	$Med.$	Cis.	Cra.	$Med.$	Cis.	Cra.	$Med.$	Cis.	Cra.
z_1	**1008**	23	2	**965**	0	0	**977**	22	34
z_2	2	**1453**	6	65	**1458**	0	1	**1444**	16
z_3	4	12	**1383**	3	2	**1390**	0	15	**1384**

set of 1400 abstracts from aerodynamic systems. After removing stop words and numeric characters, Dhillon et al. [3] selected the top 2000 words by mutual information as part of their pre-processing. The authors refer to this data as Classic3. Note that for this example, Dhillon [3] and Dhillon et al. [4] have proposed two block clustering algorithms. The first one (BSGP) deals to cluster documents and words by using biparte spectral graph partioning and the second one (IT) is based on the theory of information. In our experiment, since we know the number of document clusters, we can give that as input of Cemcroki2, BSGP and IT, $g = 3$ and we have taken $m = 3$. Table 3 shows the three confusion matrices matrices obtained on the Classic3 data using these algorithms. It appears clearly that Cemcroki2 outperforms BSGP and IT. The number of documents misclassified are 49 for Cemcroki2, 70 for BSGP and 64 for IT.

6   Conclusion

Most of methods of statistical analysis are concerned with understanding relationships among variables. With categorical variables, these relationships are usually studied from data that has been summarized by a contingency table, giving the frequencies of observations cross-classified by two variables. To classify the rows and the columns simultaneously of this contingency table, we can use *Croki2* which can be employed jointly with the correspondence analysis.

In this paper, using a Poisson block mixture model, we have proposed the *Cemcroki2* algorithm which can be viewed as an extension of *Croki2*. In this setting, the probabilistic interpretation of *Croki2* constitutes an interesting support to consider various situations and avoids the development of ad hoc methods: for example, it allows one to take into account situations in which the clusters are ill-separated or situations in which the proportions of clusters are different by applying *Cemcroki2* whereas the χ^2 and the mutual information criteria assume equal proportions implicitly. From our experiments, the new algorithm appears clearly better than *Croki2* in real situations when the proportions are not necessary equal. In addition it has several advantages such as the simplicity, the fast convergence and the scalability. Now, it would be interesting to consider the block clustering problem under the ML approach and develop an adapted version block EM [7].

References

1. Celeux, G., Govaert, G.: A Classification EM Algorithm for Clustering and two Stochastic Versions. Computational Statistics and Data Analysis, **14** (1992) 315–332
2. Cheng, Y., Church, G.: Biclustering of expression data. In: Proceedings of the Eighth International Conference on Intelligent Systems for Molecular Biology (ISMB) (2000) 93–103
3. Dhillon, I.: Co-clustering documents and words using bipartite spectral graph partitioning. In: ACM SIGKDD International Conference, San Francisco, USA (2001) 269–274
4. Dhillon, I., Mallela, S., Modha, D.S.: Information-Theoretic Co-clustering. In: ACM SIGKDD International Conference on Knowledge Discovery and Data Mining (KDD) (2003) 89–98
5. Govaert, G.: Classification de tableaux binaires. In Data analysis and informatics 3, North-Holland, Amsterdam (1984) 223–236
6. Govaert, G., Nadif, M.: Clustering with block mixture models. Pattern Recognition **36** (2003) 463–473
7. Govaert, G., Nadif, M.: An EM algorithm for the block mixture model. IEEE Transactions on Pattern Analysis and Machine Intelligence **27** (2005) 643–647
8. McLachlan, G.J., Peel, D.: Finite Mixture Models. Wiley, New York (2000)
9. Symons, M.J.: Clustering criteria and multivariate normal mixtures. Biometrics **37** (1981) 35–43

Adaptive Classifier Combination for Visual Information Processing Using Data Context-Awareness

Mi Young Nam and Phill Kyu Rhee

Dept. of Computer Science & Engineering , Inha University,
253, Yong-Hyun Dong, Nam-Gu,
Incheon, South Korea
rera@im.inha.ac.kr, pkrhee@inha.ac.kr

Abstract. This paper addresses a novel method of classifier combination for efficient object recognition using data context-awareness called "Adaptable Classifier Combination (ACC)". The proposed method tries to distinguish the context category of input image data and decides the classifier combination structure accordingly by Genetic algorithm. It stores its experiences in terms of the data context category and the evolved artificial chromosome so that the evolutionary knowledge can be used later. The proposed method has been evaluated in the area of face recognition. Most previous face recognition schemes define their system structures at the design phases, and the structures are not adaptive during operation. Such approaches usually show vulnerability under varying illumination environment. Data context-awareness, modeling and identification of input data as data context categories, is carried out using SOM(Self Organized Map). The face data context are described based on the image attributes of light direction and brightness. The proposed scheme can adapt itself to an input data in real-time by identifying the data context category and previously derived chromosome. The superiority of the proposed system is shown using four data sets: Inha, FERET and Yale DB.

1 Introduction

Much research has been devoted on this problem. However, most object recognition methods today can only operate successfully only under strongly constrained images captured in controlled environments. In this paper, we discuss about evolvable classifier combination that can behave in a robust manner under such variations of input image data. It employs the concept of context-awareness and Genetic algorithm, and determines a most effective structure of classifier combination for an input data.

The context-awareness consists of context modeling and identification. Context modeling can is be performed by an unsupervised learning method such as SOM. Context identification can be implemented by a normal classification method such as NN, k-nn, etc. Classifier structure is encoded in terms of artificial chromosome, and Genetic algorithm is used to explore a most effective classifier combination structure for each identified data context category. The knowledge of an individual context category and its associated chromosomes of effective classifiers is stored in the context knowledge base in order to preventing repetitive search.

A.F. Famili et al. (Eds.): IDA 2005, LNCS 3646, pp. 260–271, 2005.

Classifier combination scheme is expected to produce a superior performance to a single classifier in terms of accuracy and reliability [1]. Even though some research reported that classifier combination methods might not be a primary factor for improving accuracy [2], they might be at least a secondary factor for that [3]. The classifier combination approaches can be found in the literature of classifier fusion [4,5,6], classifier ensemble [7], etc. Several researchers have studied theoretically the area of classifier combination, however, they treated only some special cases [3, 8]. Kuncheva studied the limited type of classifier combination where only the aggregation of individual classifier outputs [1] is discussed.

In classifier fusion, individual classifiers are activated in parallel, and group decision is used to combine the output of the classifiers. In classifier selection, the selection of a proper classifier that is most likely to produce an accurate output for a given environment (sample data) is attempted. Most classifier fusion approaches assume that all classifiers employ the same feature space. Some classifier selection approaches need prior knowledge which classifier is specialized in which region of the feature space. Contrary to the previous approaches, the proposed method is relatively general in a sense that it can combine classifiers with the strategy of static classifier selection, dynamic classifier selection, classifier fusion, hybrid, etc.

We will deal with image objects the spacial boundaries of which can be well estimated in prior, called "spacially well-defined object classes" without loss of generality. Face images are in the class of well-defined image objects, the spacial boundaries of which can be well estimated in prior. Recently, face recognition becomes a popular task in visual information processing research. It is one of the most promising application areas of computer vision. Face recognition technology has been motivated from the application areas of physical access control, face image surveillance, visual communication for human computer interaction, and humanized robot vision. Many face recognition methods are proposed such as PCA [9], FLD [10, 11], ICA(Independent Component Analysis) [12], and Gabor based approaches [13]. Even though many algorithms and techniques are invented, face recognition still remains a difficult problem yet, and existing technologies are not sufficiently reliable, especially under diversity of input image quality. Recently, several researchers have tried to attack on this problem [14, 15]. Liu and Wechsler [15] have introduced EP(Evolutionary Persuit) for face image encoding, and have shown its successful application. However, EP needs too large search space, i.e. time-consuming, to be employed in real world applications. The illumination cone approach [14] has proposed a generative model that can be used to render face images under novel illumination conditions. The proposed method has been tested using four data sets and their virtual data sets: Inha, FERET and Yale database where face images are exposed to different lighting condition (see Fig. 1). We achieve encouraging experimental results showing that the performance of the proposed method is superior to those of most popular methods.

The major contributions of this paper are: 1) it achieves highly robust and real time classifier scheme under varying lighting condition by providing the capability of adaption/evolution and 2) it solves the time-consuming problem of the multiple classifier based on a conventional GA by introducing the context-aware multiple classifier method. The paper is organized as follows. In the section 2, we present the proposed architecture for context-aware evolutionary computation and the overview of the proposed face recognition scheme. In the section 3, we discuss about the illumination modeling and illumination identification using Kohonen's Self Organization Map. In the section 4, we present the adaptive classifier combination for face recognition. Finally, we give the experimental results and the concluding remarks in the section 5 and 6, respectively.

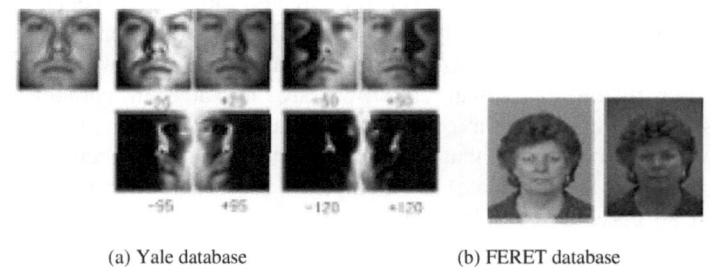

(a) Yale database   (b) FERET database

Fig. 1. Various face database

2   Adaptive Classifier Combination Scheme

The proposed ACC (Adaptable Classifier Combination) scheme consists of the context identification module (CIM), the evolution control module (ECM), the Action module (AM), the evolutionary module (EM), and the context knowledge base (CKB) (see Fig.2).

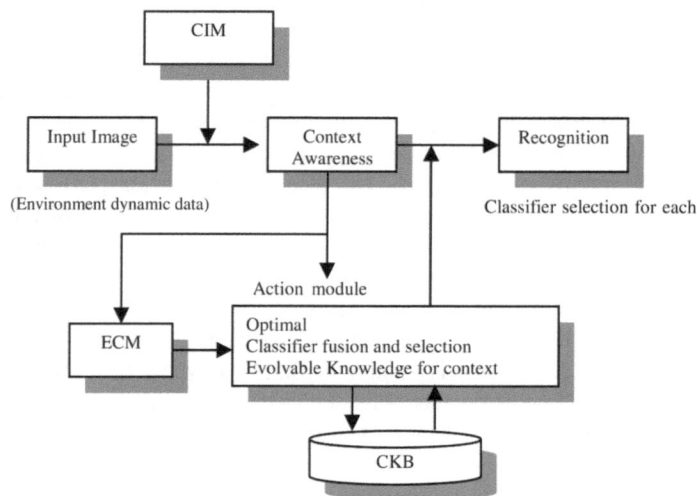

Fig. 2. The block diagram of the proposed ECC scheme

The CIM identifies a current context using context input data. Context can be various configurations, computing resource availability, dynamic task requirement, application condition, environmental condition, etc.. Context describes a trigger of the scheme action using the previously accumulated knowledge of context-action relation in the CKB. The CKB over a period of time and-or the variation of a set of context informations of the system over a period of time. Context data is defined as any observable and relevant attributes, and its interaction with other entities and/or surrounding environment at an instance of time.

The AM consists of one or more action primitives. The action primitives can be heterogeneous, homogeneous, or hybrid operational entities. For example, the action primitives of a pattern classifier are divided into preprocessing, feature representation, class decision, post processing primitives. The ECM searches for a best combining structure of action primitives for an identified context. Initially, the scheme accumulates the knowledge in the CKB that guarantees optimal performance for individual identified context. The CKB stores the expressions of identifiable contexts and their matched actions that will be performed by the AM. The matched action can be decided by either experimental trial-and-error or some automating procedures. In the operation time, the context expression is determined from the derived context representation, where the derived context is decided from the context data. The ECM searches the matched action in the CKB, and the AM performs the action.

3   Context Identification Modeling (CIM)

The CIM identifies context or context category from context input data in order to determine the structure of the AM. The CIM is implemented by the hybrid of Kohnen's self-organizing map (SOM) [7] and decides the category of current context. The ECM searches for the best chromosome from the CKB. The CIM is trained to distinguish input image quality in term of data context such as brightness and light direction. SOM is selected to be the most promising algorithm for constructing the model of face images under changing lighting condition. SOM can be used to create an intuitive model of the important concepts contained in information [21, 22].

Continuous-valued vectors of face image features which are presented sequentially without specifying the desired output. After a sufficient number of input vectors have been presented, network connection weights specify clusters, the point density function of which tends to approximate the probability density function of the input vectors. In addition, the connection weights will be organized such that topologically close nodes are sensitive to inputs that are similar. SOM is used to model image-based visual thesaurus identifying changing lighting condition. An example of training data for the SOM is shown Fig.3.

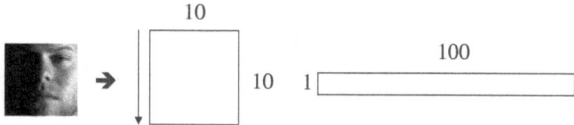

Fig. 3. Face data vectorization is 1x100 dimensions

Next figure shows images of three clusters various illuminant face dataset(Yale database), we define 3 step illuminant environment.

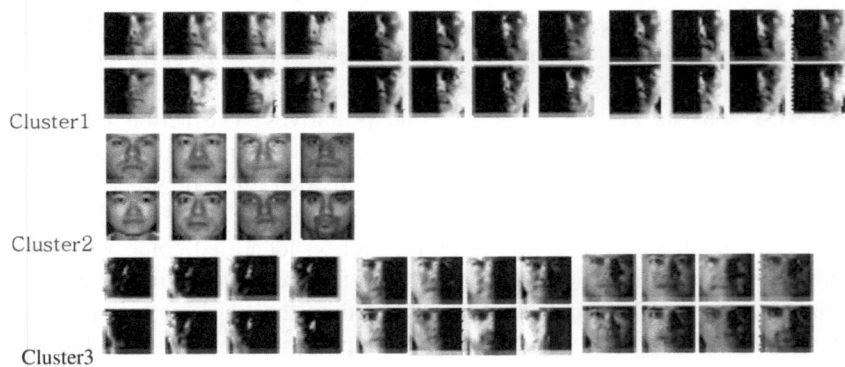

Cluster1

Cluster2

Cluster3

Fig. 4. Discriminant result for illumination conditions in Yale databse

4   The Adaptive Classifier Combination for Face Recognition

The proposed ECC method has been tested in the area of object recognition. We deal with image objects the spacial boundaries of which can be well estimated in prior, called spacially well-defined object classes. Face images are in the class of well-defined image objects, the spacial boundaries of which can be well estimated in prior.

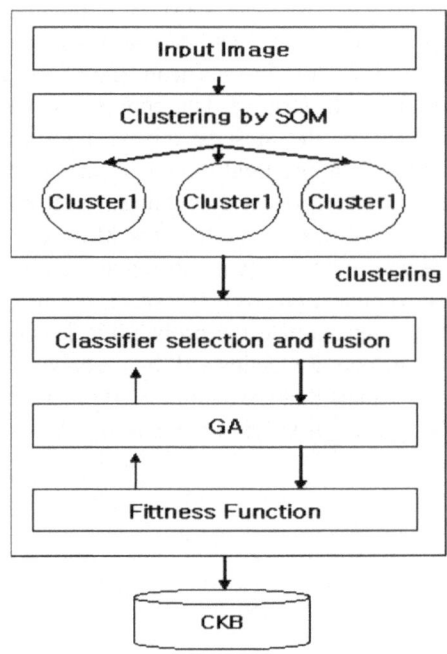

Fig. 5. An example of situation-aware classifier fusion

4.1 ACC Implementation for Face Recognition

In general, it is almost impossible or very difficult to decide an optimal classifier or classifier structure at the design step considering all possible factors of operational time variations. We employ the strategy that the classifier structure, is allowed to evolve or adapt itself dynamically during operation in accordance with changing quality of input image data, i.e.data context. Changes in image data can include lighting direction, brightness, contrast, and spectral composition. The architecture of face recognition using the ECC is given in Fig. 5.

4.2 Chromosome Encoding and Fitness Function

The GA is employed to search among the different combinations of feature representations (if any) and combining structure of classifiers. The optimality of the chromosome is defined by classification accuracy and generalization capability. Fig. 6 shows a possible encoding of chromosome description.

CF1	CF2	...	CFn	CLS1	CLS2	...	CLS flag	CLF flag

Fig. 6. A possible chromosome description of the proposed scheme

As the GA searches the genospace, the GA makes its choices via genetic operators as a function of probability distribution driven by fitness function. The genetic operators used here are selection, crossover, and mutation [7]. The GA needs a salient fitness function to evaluate current population and chooses offspring for the next generation. Evolution or adaption will be guided by a fitness function defined in terms of the system accuracy and the class scattering criterion.

The evolutionary module derives the classifier being balanced between successful recognition and generalization capability. The fitness function adopted here is defined as follows:

$$\eta(V) = \lambda_1 \eta_s(V) + \lambda_2 \eta_g(V) \tag{1}$$

where $\eta_k^{(V)}$ is the term for the system correctness, i.e., successful recognition rate and $\eta_k^{(V)}$ is the term for class generalization. λ_1 and λ_2 are positive parameters that indicate the weight of each term, respectively.

Let $\omega_1, \omega_2, .., \omega_k$ be the classes and $N_1, N_2, ..., N_k$ be the number of images in each class, respectively. Let $M_1, M_2, .., M_k$ be the means of corresponding classes, and $M_{\omega k}$ be the total mean in the Gabor feature space. Then, M_i can be calculated,

$$M_i = \frac{1}{N_i} \sum_{j=1}^{N_i} S_j^{(i)} i = 1,2,3,.....,L \tag{2}$$

Where $S_j^{(i)} = j = 1,2, ..., N$, denotes the sample data in class wi, and

$$M_{avg} = \frac{1}{n}\sum_{i=1}^{k} N_i M_i \qquad (3)$$

4.3   The Face Recognition Scheme Using ACC

The recognition system learns an optimal structure of multi-classifier and Gabor representation by restructuring its structure and parameters. Preprocessing is performed for providing nice quality images as much as possible using conventional image filtering techniques. The image filters employed here is the lighting compensation, histogram equalization, opening operation, boost-filtering. We use 5 classifiers, Eigenface, Gabor3, Gabor13, Gabor 28, and Gabor30, for the AM. The details of classifiers are given in the followings.

Classifier Gabor3, Gabor13, Gabor 28, Gabor30
The Gabor wavelet transform guided by an evolutionary approach has been employed to adapt the system for variations in illumination. The proposed approach employs Gabor feature vector, which is generated from the Gabor wavelet transform. The kernels of the Gabor wavelets show biological relevance to 2-D receptive field profiles of mammalian cortical cells. The receptive fields of the neurons in the primary visual cortex of mammals are oriented and have characteristic frequencies. Gabor wavelet is known to be efficient in reducing redundancy and noise in images.Face gabor vector is generated as shown Fig.8. We adopt 4 Gabor based classifiers: Gabor3, Gabor13, Gabor28, Gabor30 and weighted Gabor32. They are different only in the number of feature points.

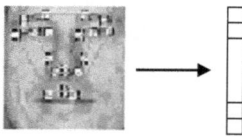

Fig. 7. An example of feature points for face recognition

Classifier Eigenface Based Face Classifier
The eigenface is constructed registration images of FERET, Yale, our Lab database. first step : We made in covariance matrix of registration data. Next figure is shown eigenface of each person. The eigenface is belong to global recognition. registration data computed covariance matrix.

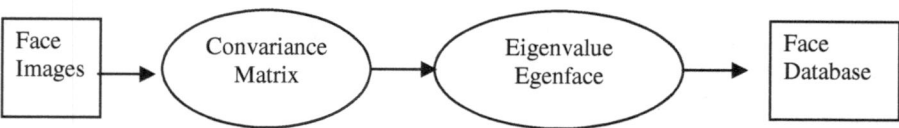

Fig. 8. Face recognition architecture using PCA

The details of the face recognition scheme constructing the CKB is given in the following:

Step 1. Cluster input images into the illumination categories using the SOM in the CAM (Data context analysis).

Step 2. Start to search for an optimization of classifier structure for each illumination category until a criterion is met, where the criterion is the fitness does not improve anymore or the predefined maximum trial limitation is encountered as follows.

    1) Generate initial population of classifier structures.

    2) Evaluate the fitness function of the scheme using the newly derived population of the classifier structures. If the criterion is met, Go to Step 3.

    3) Search for the population of the classifier structures that maximize the fitness function and keep those as the best chromosomes.

    4) Applying GA's genetic operators to generate new population from the current classifier structures. Go to Step 2.2).

Step 3. Update the CKB (Context Knowledge Base) for the identified illumination category and the derived classifier structure.

The recognition task is carried out using the knowledge of the CKB evolved from the evolutionary mode as follows:

Step 1. Identify the illumination situation in the CAM.

Step 2. Search for the chromosome from the CKB representing the optimal classifier structure corresponding to the identified illumination category.

Step 3. Perform the task of recognition using the restructured feature vector.

Step 4. If the system performance is measured to fall down below the predefined criterion, the system activates the evolution mode, and/or evolves the system periodically or when it is needed.

5 Experimental Results

The feasibility of the proposed method has been tested in the area of face recognition using Inha, FERET[16], Yale [17]. Experiments have been carried out to compare the performance of the proposed evolvable classifier combination, that the best among individual classifiers. We used 1000 images of 100 persons from our lab data set, 330 images of 33 persons excluding 99 images of wearing sunglasses from AR face data set, 60 images of 15 persons from Yale Face DB [17], and 2418 images of 1209 persons from FERET data set. The above data sets are merged for training and testing the CAM (see session 3). The data context of the merged data is analyzed by the SOM. Fig. 10 shows the examples of six data context clusters.

    The data clustering process:

1. Design the individual classifiers D1..DL using the labeled data set Z

2. Disregarding the class labels, cluster Z into C clusters, using, e.g., the SOM clustering procedure. Find the cluster centroids $V_1, ..., V_k$ as the arithmatic means of the points in the respective clusters.

Face Images

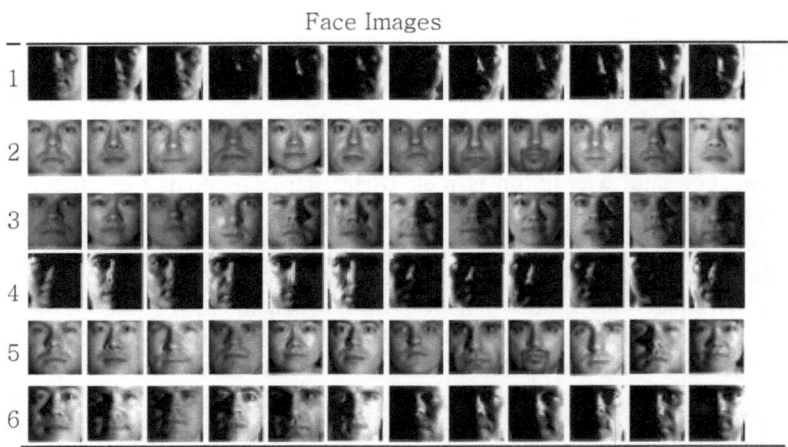

Fig. 9. The examples of face image clustered into six categories using SOM

The first experiment was performed using the data set accumulated by our lab InhaDB. The data set has 1000 face images from 100 people. We used 5 images for registration of each person. The remaining 500 images are used as the test images. For the Yale data set, we used 15 registration face images and 45 test images. The FERET gallery images of 1196 people were used for registration and 1196 probe_fafb_expression images were used for test. Table 1 shows the recognition performance of individual classifiers for each cluster. One can note that single classifier cannot be the winner of all clusters. For example, Gabpr13 and Gabor30 show the highest performance in cluster0. PCA shows the highest performance in Cluster2. The Fig. 11. is shown the recognition rate of individual classifiers for individual clusters. Thus, the proposed adaptive method should be useful under uneven illumination environments.

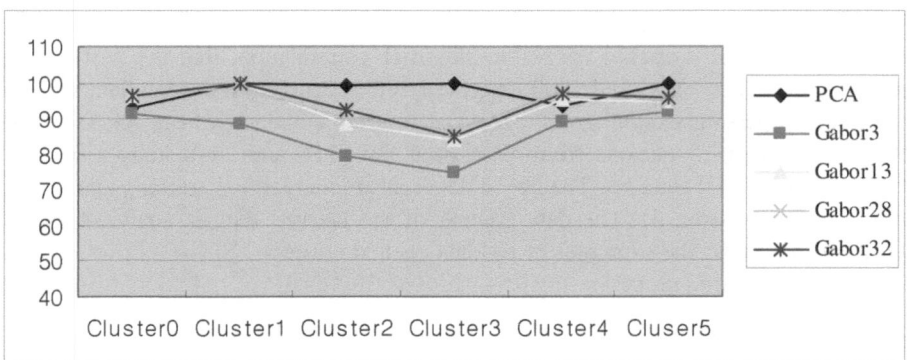

Fig. 10. The recognition rate of individual classifiers for individual clusters

Table 1. The context-based face recognition using classifier fusion by GA in six cluster

Database	CF (Proposed Method)	Majority voting	PCA	Ganor3	Gabor13	Gabor28	Gabor32
FERET(1196)	95%	91%	90%	25%	50.25%	90.25%	94%
Yale(10)	98.5%	92.30%	85%	10%	30%	85.25%	96.5%
Our Lab(100)	99%	94%	92%	30%	65%	93%	97%

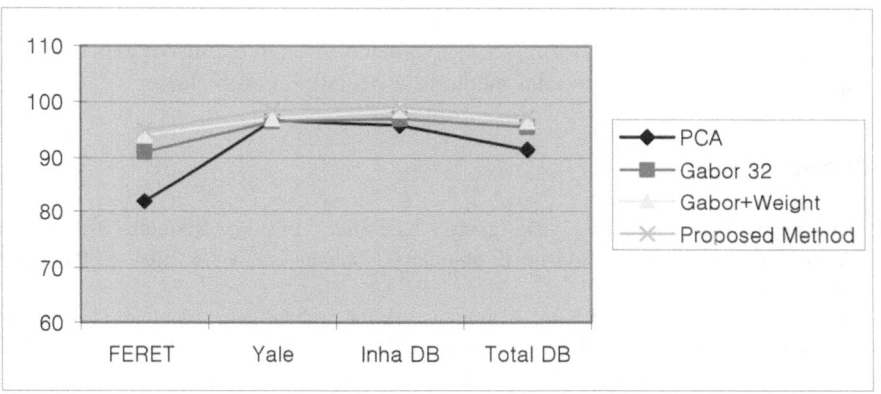

Fig. 11. The recognition rate for the each database

Fig 11 shows a recognition rate of proposed method and comparison with other methods. It is 99% for our Lab DB, 98.5% for Yale dataset and 95 % for FERET dataset.

Table 2. Comparative Testing Performance: FERET database

Method	Rank1 correct acceptance
Eigenface[18]	83.4%
Eigenface by Bayesian[19]	94.8%
Evolutionary Pursuit[20]	92.14%
Proposed method	95%

Table 2 shows the comparative test performance by Eigenface using Bayesian theory, linear discriminant, elastic graph matching, and evolutionary pursuit. The recognition rate of Eigenface is 83.4 % and Evolutionary Pursuit is 92.14 %. In own experimental results, the proposed method shows recognition rate of over 95 % for FERET dataset, which exceeds the performance of the other popular methods. The context-based classifier fusion method performs better than single classifier method because the image feature is different according context information. From Table 2, it becomes apparent that the proposed method shows good recognition performance.

6 Conclusion

In this paper, adaptive classifier combination (ACC), a novel method of classifier combination using data context-awareness is proposed and applied to object recognition problem. The research generated clustering and classifier fusion by GA. The proposed method tries to distinguish its input data context and evolves the classifier combination structure accordingly by Genetic algorithm. This included the use of the clustering by SOM for data context-awareness modeling and identification of input data as data context categories. The proposed scheme can optimize itself to a given data in real-time by using the identified data context and previously derived chromosome in varying illumination. We tested using three datasets: Inha DB, FERET DB, Yale DB. Its performance is evaluated through extensive experiments to be superior to those of most popular methods, especially in each cluster.

References

1. Kuncheva L., Jain LC,: Designing Classifier Fusion Systems by Genetic Algorithms. IEEE Tranaction on Evoltionalry Computation, vol.4, no.4, SEPTEMBER (2000), pp 327-335
2. H.-J.Kang, K.Kim, and J.H.Kim: A frame work for probabilistic combination of multiple classifier satanab stract level. Eng.Applicat.Artif.Intell., vol.10, no.4, (1997) pp.379–385
3. L. Kuncheva: Switching Between Selection and Fusion in Combining Classifiers. AnExperiment, IEEE Transaction on Systems, Man and Cybernetics—PARTB, vol..32, no.2, APRIL (2002) pp146-156
4. P.D.Gader, M.A.Mohamed , and J.M.Keller: Fusion of handwritten wordclassifiers. Pattern Recognit. Lett., vol.17, (1996) pp.577–584
5. J.M.Keller, P.Gader, H.Tahani, J.-H.Chiang, and M.Mohamed: Advances in fuzzy integration for pattern recognition. Fuzzy SetsSyst., vol.65, pp.273–283, 1994
6. H.-J.Kang, K.Kim, and J.H.Kim: A frame work for probabilistic combination of multiple classifier satanabstract level. Eng.Appl.Artif.In-tell., vol.10, no.4, pp.379–385, 1997
7. D. Goldberg: Genetic Algorithm in Search. Optimization, and Machine Learning, Addison-Wesley, (1989)
8. B.V.Dasarathy and B.V.Sheela: A composite classifier system design: Concept sand methodology. Proc.IEEE, vol.67, (1978) pp.708–713
9. M. Turk and A. Pentland: Eigenfaces for recognition. J. Cong. Neurosci. vol. 13, no. 1, (1991) pp.71-86
10. D. Swets and J. Weng: Using discriminant eigenfeatures for image retrieval. IEE Trans. PAMI, 18(8), (1996) pp. 831-836
11. C. Liu and H. Wechsler: Robust coding Schemes for indexing and retrieval from large database. IEEE Trans. Image Processing," vol. 9, no. 1, 2000, pp. 132-137
12. G. Donato, M. Bartlett, J. Hager, P. Ekman, andSejnowski, "Classifying facial actions," IEE Trans. PAMI, 21(10), (1999) pp. 974-989
13. M. Potzsch, N. Kruger, and C. Von der Malsburg: Improving Object recognition by Transforming Gabor Filter reponses. Network: Computation in Neural Systems, vol.7, no.2, pp. 341-347

14. A. S. Georghiades, P. N. Belhumeur, and D. J. Kriegman: From Few to Many: Illumination COn·e Models for face recognition under Variable Lighting and Pose. IEEE Trans. on PAMI, vol. 23 no. 6, June (2001) pp. 643-660
15. C. Liu and H. Wechsler: Evolutionary Pursuit and Its Application to Face recognition. IEEE Trans. on PAMI, vol. 22, no. 6, (2000) pp. 570-582
16. P.Phillips: The FERET database and evaluation procedure for face recognition algorithms. Image and Vision Computing, vol.16, no.5, (1999) pp.295-306
17. http://cvc.yale.edu/projects/yalefaces/yalefaces.html
18. M. Turk and A. Pentland: Eigenfaces for recognition. J. Cong. Neurosci. vol. 13, no. 1, (1991) pp.71-86
19. B.Moghaddam, C.Nastar, and A,Pentland : A Bayesian similarity Measure for direct Image Matching. Proc. of Int. Conf. on Pattern Recogntion, (1996)
20. Chengjun Liu and Harry Wechsler: Evolutionary Pursuit and Its Application to Face Recognition. IEEE Trans. on Pattern Analysis and Machin Intelligent, Vol.22, No.6, (2000) pp.570∽582
21. T. Kohonen, J. Hymminen, J. Kangras, and J. Laaksonan. SOM PAK: The Self-Organizing Map program package. Technical Report A31, Helsinki University of Technology, (1996)
22. Markus Koskela. Interactive Image Retrieval Using Self-Organizing Maps. PhD thesis, Helsinki University of Technology, November (2003)

Self-poised Ensemble Learning[*]

Ricardo Ñanculef[1], Carlos Valle[1], Héctor Allende[1], and Claudio Moraga[2]

[1] Universidad Federico Santa María,
Departamento de Informática, CP 110-V Valparaíso, Chile
{jnancu, cvalle, hallende}@inf.utfsm.cl
[2] Dortmund Universitaet, 44221 Dortmund, Deutschland
claudio.moraga@udo.edu

Abstract. This paper proposes a new approach to train ensembles of learning machines in a regression context. At each iteration a new learner is added to compensate the error made by the previous learner in the prediction of its training patterns. The algorithm operates directly over values to be predicted by the next machine to retain the ensemble in the target hypothesis and to ensure diversity. We expose a theoretical explanation which clarifies what the method is doing algorithmically and allows to show its stochastic convergence. Finally, experimental results are presented to compare the performance of this algorithm with boosting and bagging in two well-known data sets.

1 Introduction

As long as problems modern data analysis has to tackle on become harder, machine learning tools reveal more important abilities for the process of extracting useful information from data [7].

Intuitively, learning systems are such that they can modify their actual behavior using information about their past behavior and their performance in the environment to achieve a given goal [9]. Mathematically speaking, the supervised learning problem can be put in the following terms: given a sample of the form $T = \{(x_1, y_1) \dots (x_n, y_n)\}$, $x_i \in X$, $y_i \in Y$ obtained sampling independently a probability measure P over $X \times Y$, we are asked to recover a function $f_0(x)$, such that it minimizes the risk functional

$$R(f) = E_{X \times Y}\left[Q(f(x), y)\right] = \int_{X \times Y} Q(f(x), y) dP(x, y) \tag{1}$$

where Q is a problem specific loss function, integrable in the measure P. The space X is usually named "the input space" and Y "the output space" of the problem, when the function f_0 -to be recovered- is thought as a mapping from X to Y, underlying the particular sample pairs (x_i, y_i). Since in realistic scenarios, the measure P is not known, the risk functional cannot be computed and

---

[*] This work was supported in part by Research Grant Fondecyt (Chile) 1040365 and 7040051, and in part by Research Grant DGIP-UTFSM (Chile). Partial support was also received from Research Grant BMBF (Germany) CHL 03-Z13.

A.F. Famili et al. (Eds.): IDA 2005, LNCS 3646, pp. 272–282, 2005.
© Springer-Verlag Berlin Heidelberg 2005

neither its minimum f_0. So, it is necessary to develop some induction criterion to be minimized in order to get an approximation to this, which is typically the empirical risk

$$R(f) = \hat{E}_{X \times Y}\left[Q(f(x), y)\right] = \sum_i Q(f(x_i), y_i) \tag{2}$$

Machine learning deals with the problem of proposing and analyzing such inductive criteria. One of the most successful approaches introduced recently in this field, relies on the idea of using a set of simple learners to solve a problem instead of using a complex single one. The term used to describe this set of learning machines is "an ensemble" or "a committee machine".

The practice and some theoretical results [5] [6] have revealed that diversity is a desirable property in the set of learners to get real advantages of combining predictors. Ensemble learning algorithms can then be analyzed describing the measure of diversity they are - implicitly or explicitly- using and the way in which they are looking for the maximization of this quantity. Boosting and bagging, for example, introduce diversity perturbing the distribution of examples each machine uses to learn. In [14] an algorithm to introduce diversity in neural networks ensembles for regression was described. In this approach we encourage decorrelation between networks adding a penalty term to the loss function. [17] and [5] are examples where negative correlation between learners of the ensemble is looked for.

In the present paper, an algorithm to generate ensembles for regression is examined. It is shown that this algorithm introduces a strictly non-positive correlation between the bias of learner at time t and the average bias of previous learners. This approach allows to introduce diversity without damaging the performance of the machine on the training set. Theory shows that this approach stochastically converges, while experimental results show that this works well in practice. The structure of this paper is as follows: in the next section we present the ensemble model for learning from examples discussing the notion of diversity in a regression context. We also introduce for comparing purposes, boosting and bagging. Following with this idea, we present in the third section the proposed algorithm and discuss the theoretical aspects of the same. Finally we show a set of experiments on two difficult benchmarks, to test the final algorithm and compare the results with boosting and bagging. The fifth section is devoted to some concluding remarks and future work.

2   Ensemble Learning and Diversity

To solve the problem of learning from examples -formalized in the latter section- one needs to choose an hypothesis space H for searching the desired f_0 or an approximation to this. Statistical learning theory establishes that the particular structure of this space is fundamental for guaranteeing the well-behavior of the inductive criterion selected for the learning machine.

The strategy chosen in ensemble learning is to set a base space H and get a hypothesis from the convex hull $co(H)$ of this space, defined as

$$co(H) = \left\{ \sum_{i=1}^{n} a_i h_i(x) : \sum_{i=1}^{n} a_i = 1, h_i \in H, n \in \mathbb{N} \right\} \quad (3)$$

When modular versions of ensemble learning are wanted it is enough to modify the latter definition to allow a_i's to be functions in some space $\mathcal{A}$. We are interested, however, in non-modular versions of ensemble learning.

The problem in designing ensemble learning algorithms is how to navigate through this space to look for the desired function. In other words, What set of functions must one choose and combine in the ensemble?. Studies tend to show that diversity in errors is a key characteristic to get better generalization performance, although the way they define this concept is highly heterogeneous.

In [1] the following property for an ensemble $f_{ens} = \sum_i w_i f_i$ and the quadratic loss function was proved, $\forall y$

$$(f_{ens} - y)^2 = \sum_i w_i (f_i - y)^2 - \sum_i w_i (f_i - f_{ens})^2 \quad (4)$$

The last result is very clarifying about the benefits of error decorrelation between learners in an ensemble. It states that the quadratic loss of the ensemble is the weighted average of individual errors minus the weighted average of individual deviations with respect to the ensemble. The latter term, also called "ambiguity" guarantees that the ensemble error is always less than the average error of members. Although, in general there may always be a learner better than the ensemble in a particular point y, we cannot know a priori which is, and then the ensemble is a reasonable gamble. If we define as measure of "diversity" the second term of the decomposition we have a concrete explanation of the phrase "diversity is good for ensembles", at least in the case of regression with quadratic loss.

The intuitive search for diversity can be seen clearly in the classical algorithms for ensemble learning: boosting and bagging [8]. In Bagging [3], one perturbs data distribution each machine uses to learn by resampling uniformly with replacement the original empirical distribution of the training set. This procedure is particulary effective to generate diversity when the predictors to be joined are unstable. In boosting [15] [16], machines are trained sequentially passing all the training patterns through the last machine and noting which ones are most in error. For these patterns, their sampling probabilities are adjusted so that they are more likely to be picked as members of the training set of the next machine. Empirical results in classification [13] and regression [4] tend to show that boosting, although more sensitive to outliers, is better than bagging. Some researchers however, have informed problems with overfitting in boosting for regression.

3   The Proposed Algorithm

3.1   Motivation

Let us consider an ensemble that is a uniformly weighted convex combination of t hypotheses, that is

$$\bar{f}_t = (1/t) \sum_{i=1}^{t} f_i(x) \tag{5}$$

If we are learning with the quadratic loss function, the bias-variance decomposition states that the generalization error of the estimator can be decomposed in two terms: bias and variance. The bias can be characterized as a measure of how close, on average over different training sets T, the estimator is to the target. The variance is a measure of how stable the estimator is with respect to the random training set T. Formally, for any fixed realization y of the output random variable Y we have[1]

$$
\begin{aligned}
E\left[(f-y)^2\right] &= E\left[((f-E[f]) + (E[f]-y))^2\right] \\
&= E\left[(f-E[f])^2\right] + 2E\left[(f-E[f])(E[f]-y)\right] + E\left[(E[f]-y)^2\right] \\
&= E\left[(f-E[f])^2\right] + (E[f]-y)^2 \\
&= var(f) + bias(f)^2
\end{aligned}
$$

For our ensemble (5) we can break down the variance term and get the *Bias-Variance-Covariance decomposition* described in [10]. This is

$$E\left[(\bar{f}_t - y)^2\right] = bias^2(\bar{f}_t) + \tfrac{1}{t}var(\bar{f}_t) + \tfrac{t-1}{t}covar(\bar{f}_t) \tag{6}$$

where ensemble bias, variance and covariance are defined as

$$bias(\bar{f}_t) = \frac{1}{t} \sum_{i=1}^{t} E[f_i - y] \tag{7}$$

$$var(\bar{f}_t) = \frac{1}{t} \sum_{i=1}^{t} E\left[(f_i - E[f_i])^2\right] \tag{8}$$

$$covar(\bar{f}_t) = \frac{1}{t(t-1)} \sum_{i=1}^{t} \sum_{j \neq i} E\left[(f_i - E[f_i])(f_j - E[f_j])\right] \tag{9}$$

The connection of (6) with the ambiguity decomposition can be obtained after a bit of algebra making $w_i = 1/t$ in (4)

$$E\left[\sum_i \frac{1}{t}(f_i - \bar{f}_t)^2\right] = \frac{1}{t} \sum_{i=1}^{t} E[f_i - \bar{f}_t]^2 - \left(1 - \frac{1}{t}\right)\left(var(\bar{f}_t) + covar(\bar{f}_t)\right) \tag{10}$$

---

[1] Expectations are taken with respect to T.

$$E\left[\sum_i \frac{1}{t}(f_i - y)^2\right] = \frac{1}{t}\sum_{i=1}^{t} E[f_i - \bar{f}_t]^2 + bias(\bar{f}_t)^2 + var(\bar{f}_t) \qquad (11)$$

So, it seems clear that a negative covariance term is desirable to get diversity in the sense introduced in section 2.

Now, using the symmetry of the covariance function we can write the last term as

$$\frac{2}{t(t-1)}\sum_{i=2}^{t}\sum_{j<i} E\left[(f_i - E[f_i])(f_j - E[f_j])\right] \qquad (12)$$

Let be $e_i = f_i - E(f_i)$ and $\bar{e}_i = \sum_{j=1}^{i-1} e_j$ Then we can rearrange (9) to get

$$\frac{2}{t(t-1)}E\left[\sum_{i=2}^{t} e_i \sum_{j<i} e_j\right] = \frac{2}{t(t-1)}\sum_{i=2}^{t} E\left[e_i \bar{e}_i\right] \qquad (13)$$

If we think the process of building an ensemble as a sequential process indexed by the time $i = 1, 2, \ldots$, the term

$$E\left[e_i \bar{e}_i\right] \qquad (14)$$

is the expected correlation between the bias of the actual machine i and the cumulated bias of the previous machines. Since (9) contributes positively to the generalization error of the ensemble, it seems natural to encourage a decorrelation between these biases, forcing that the new machine makes errors in the opposite direction to the ensemble error. Based on this idea, it is possible to build an ensemble with a non-positive term (14) and (9), assuring diversity of their members. A detailed exposition of this idea is the matter of the next section.

3.2 Formal Setting

Now, we will formalize the mathematical structure of the proposed algorithm. Let be $(\Omega, \mathcal{F}, \mathcal{P})$ a probability space and $X : \Omega \to R_1$, $Y : \Omega \to R_2$, $W_i : \Omega \to R_2$, $i = 1, \ldots$ random variables, where $R_1 = \mathbb{R}^d$, $R_2 = \mathbb{R}^q$, $d, q \in \mathbb{N}$, equipped with the standard Borel σ-field; W_i, W_j mutually uncorrelated for $i \neq j$. (X, Y) will represent the random sample from which a function is estimated to minimize the risk functional (1); and W_i a random noise whose meaning will be soon clarified. The process of sequentially learning an ensemble can be viewed as the selection, from some hypothesis space H, of a set of random variables $f_1, f_2, \ldots$, $f_i : R_1 \to R_2$ to be aggregated in some way that we select to be the uniform convex combination. Each function f_i is obtained from a particular realization of a random variable (X_i, Y_i) which in general is a transformation of the original (X, Y). For example, in bagging, (X_i, Y_i) are obtained resampling (X, Y). In a regression problem, the functions f_i are asked to satisfy

$$Y_i = f_i(X_i) + W_i \qquad (15)$$

where $f_i \in H$ minimizes

$$E_{X_i \times Y_i} \{Q(Y_i - g_i(X_i))\} = \int_{X_i \times Y_i} Q(Y_i - g_i(X_i)) \, dP(X_i, Y_i) \qquad (16)$$

such that W_i models the error of the approximation given by f_i over (X_i, Y_i).

Our final purpose is to minimize at time t the ensemble error. If $|g|$ is the number of machines in the ensemble g it is stated as

$$
\begin{aligned}
\bar{f}^t &= \underset{g \in co(H), |g| = t}{\arg\min} \ E_{X \times Y} \{Q(Y - g(X))\} \\
&= \underset{f_i \in H}{\arg\min} E_{X \times Y} \left\{ Q\left(Y - \tfrac{1}{t} \sum_{i=1}^{t} f_i(X)\right) \right\}
\end{aligned}
\qquad (17)
$$

In our algorithm we choose to start by building a function $f_{t=1}$ using (X,Y), that is we make $X_1 = X, Y_1 = Y$ and we wonder what (X_2, Y_2) must be, to get the minimum of (17) at time $t = 2$ given that f_1 was already chosen and knowing that f_2 satisfies (16). More generally, what (X_{t+1}, Y_{t+1}) must be after time t such that the ensemble at time $t + 1$ pursues to be the optimal ensemble. If we choose Q to be the quadratic loss and derive the ensemble error at a given point (x, y) with respect to the free variable after time t we get

$$\frac{\delta}{\delta f_{t+1}} (y - \bar{f}_t(x))^2 = 2 \left(y - \frac{1}{t+1} \sum_{i=1}^{t+1} f_i(x) \right) \frac{-1}{t+1} \qquad (18)$$

If f_{t+1}^* is optimal it has to satisfy

$$y = \frac{1}{t+1} \sum_{i}^{t+1} f_i^*(x) \implies f_{t+1}^*(x) = y + ty - t\bar{f}_t \qquad (19)$$

If we name $y - \bar{f}_t = \bar{\epsilon}_t$ we have

$$f_{t+1}^*(x) = y + t\bar{\epsilon}_t \qquad (20)$$

that is, except by modelling noise W_{t+1}, the machine in time $t + 1$ has to approximate Y perturbed as in (20). If we define

$$
\begin{aligned}
\epsilon_t &= f_t^*(x) - \bar{f}_t, t > 1 \\
\epsilon_1 &= y - f_1
\end{aligned}
\qquad (21)
$$

that is, ϵ_t is the estimation error in time t, we obtain applying some steps of algebra that

$$t\bar{\epsilon}_t = \epsilon_t \qquad (22)$$

In summary, the ensemble learning algorithm works with the following recursive structure of data sets:

$$
\begin{aligned}
Y_1 &= Y, X_1 = X \\
Y_{t+1} &= Y + W_t, X_t = X
\end{aligned}
\qquad (23)
$$

where random variables X_i, Y_i, f_i, W_i satisfy (15). Equivalently, the stochastic process $(X_t, Y_t, W_t)_t$ can be made explicitly Markovian with respect to the natural filtration writing Y_{t+1} as $Y_t - W_{t-1} + W_t$. We now make the following remarks with respect to the learning process built using this scheme.

Proposition 1. *The ensemble process* $(\bar{f}_t)_t$, $t = 1, 2, \ldots$ *defined by*

$$\bar{f}_t = (1/t) \sum_{i=1}^{t} f_i \tag{24}$$

where f_i *satisfies (15) and* X_i, Y_i *are defined by (23) satisfies*

$$|\bar{f}_t - Y| = \frac{|W_t|}{t} \tag{25}$$

Then, if the errors W_i *are almost surely bounded by* $C \in R_2$ *or by a non-increasing process* $(C_i)_i$, $(\bar{f}_t)_t$ *converges in probability to* Y. *Moreover, the process of ensemble errors* $(Y - \bar{f}_t)^2$ *is almost surely* $O(1/t^2)$.

Proposition 2. *Consider an ensemble* $(\bar{f}_t)_t$, $t = 2, \ldots$ *built as in proposition (1). If* $\forall t\ E(f_t) = Y$, *term (14) is equal to* $-E(W_{t-1}^2)$ *and so the covariance in (9) is strictly non-positive.*

The proof of both propositions is omitted for space limitations, but can be obtained in [12].

3.3   The Algorithm

We now expose the final algorithm for ensemble regression with self-poised learning. As we have remarked previously we are interested in training a uniformly weighted ensemble (5).

Algorithm 1 Self-Poised Learning

1: Let be M the required number of learners and $z^1 = \{(x_1, y_1), \ldots, (x_n, y_n)\}$ a training set.
2: Generate an initial predictor f_1 training a learner with z^1 and the quadratic loss.
3: **for** $t = 1$ to $M - 1$
4:    Set the ensemble at time t to be $\bar{f}_t = 1/t \sum_{i=1}^{t} f_i(x)$
5:    Compute the difference between the prediction and the target at time t, for each point $j = 1, \ldots, n$ of the training set z^t as:

$$\epsilon_j^t = y_j^t - f_t(x_j) \tag{26}$$

6:    Generate a new sample $z^{t+1} = \{(x_1, y_1^{t+1}), \ldots, (x_n, y_n^{t+1})\}$ by modifying the targets to be predicted in time $t + 1$ as

$$y_j^{t+1} = y_j + \epsilon_j^t \tag{27}$$

7:    Generate a new predictor f_{t+1} training a learner with z^{t+1} and the quadratic loss.
8: **end for**

If we examine algorithm 1 we will note that the key step is the generation of a new data set to train the next machine added to the ensemble (step number 7). This is based in the perturbation of the original data set with an innovation ϵ_j^t which corrects the error of the ensemble at time $t - 1$. It should be noticed however, that in some cases the perturbation introduced generates a data too much complicated for being approximated by the following machine, such that the error of the ensemble becomes worse when adding the former to this. To alleviate this problem it is possible to consider a smoothing strategy which moderates the innovation in some way, for example we can replace this by $\gamma(t)\epsilon_j^t$ where $\gamma(t)$ is a polynomial in t. The linear function $\gamma(t) = t/M$, where M is the final number of machines in the ensemble, seems to be a good selection in practice, although a formal study is required based on the smoothness of the final data and the capacity of the trained learner. Another option to alleviate the effect of a "bad" machine is to consider a rejection criterion like the used by boosting. A possibility is to reject a machine when the sample probability of improving the performance of the ensemble is lower than $1/2$.

As we report in the following section, the presented algorithm exhibits a competitive performance compared with classical algorithms in complicated data sets, yet without a smoothing strategy and a rejection criterion. It was also noted that self-poised learning shows to be very stable. It is important to remark that our algorithm does not optimize the aggregation procedure like boosting.

4   Experimental Results

Our experimental analysis will be made with respect to two well-known data sets, *Boston* and *Building2* which are reported as problematic by the presence of outliers in the first case and by the time-series structure in the second. A detailed description of both data sets can be obtained from the *UCI Machine Learning Repository* [2]. The proposed algorithm is compared with bagging and boosting as defined in [3] and [4] respectively.

Each reported experiment, that is, the algorithm and parameters combination, was repeated a minimum of 50 times, randomly reordering data each time, to compute performance statistics with the quadratic loss. As learning machines we select neural networks with one hidden layer trained using standard gradient descent with learning parameter $\alpha = 0.2$. We follow [11] to select a training set equivalent to 75% of the total number of examples in the data and 25% for testing, when validation was omitted. In the opposite case, the sizes was 50% for learning, 25% for testing and 25% for validation. Validation was implemented as proposed in [11], that is, when in two successive sample points the validation error became worse, the learning was stopped. For training purposes, the data was scaled to the unitary hypercube centered in the origin. We get a better performance of the gradient descent algorithm with this transformation. Reported errors were however computed in the original scale. Table 1 shows the results for the proposed approach with different number of epochs to show the effect of

Table 1. Comparing the proposed approach with a smoothing strategy (S) or without one, and with different number of epochs (E) over data Boston (D1) and Building2 (D2). We computed over 50 trials: the average training mse (TR.Mean) and testing mse (TS.Mean), the standard deviation of testing mse (TS.Std) and the minimum testing mse (TS.Min). The ensemble was trained with 5 hidden neurons and 10 machines.

S	E	TR.Mean		TS.Mean		TS.Min		TS.Std	
		D1	D2	D1	D2	D1	D2	D1	D2
no	100	9,8157	0,002930	16,8259	0,002998	7,3392	0,002827	5,2733	7,444E-5
no	200	7,7295	0,002701	16,4226	0,002874	8,5854	0,002843	5,7670	2,620E-5
no	500	5,8755	0,002538	15,6913	0,002730	7,9909	0,002679	6,8561	3,754E-5
no	val	10,9654	0,002986	17,6639	0,003180	7,8914	0,002974	6,0502	7,861E-5
yes	100	9,5444	0,003126	14,8049	0,003203	8,1670	0,003116	4,3724	4,420E-5
yes	200	7,6429	0,002968	16,3222	0,003145	8,6203	0,002953	5,7258	6,897E-5
yes	500	6,1399	0,002380	14,4320	0,002450	8,0227	0,002410	4,1232	2,890E-5
yes	val	8,6377	0,002934	16,6239	0,003002	7,3950	0,002827	5,7187	8,388E-5

smoothing. Table 2 compare the proposed algorithm with boosting and bagging, using different structural parameters. For comparing, we select the strategy with the best generalization rate (TS.Mean/TR.Mean).

5   Future Work

In this paper an algorithm to build an ensemble for regression was proposed. The method is based on the idea of adding an artificial innovation to the map to be predicted by each machine such that it compensate the error incurred by the previous one. It was shown that this approach ensures diversity, furthermore stochastic convergence was also proved. Future work has to consider a more sound study of how to bound these innovations in order to avoid possible unstabilities. The convergence of the method is guaranteed only if the local errors incurred by each individual learner are bounded by the same constant or are non-increasing in time. If, on the other hand, the capacity of each learner is kept fix but the map to be estimated is increasedly complex or irregular, the ensemble could become worse with time because each machine must compensate the deviations of the previous one with respect to its data. A possibility is to analytically characterize the appropriate smoothing functions to compute the innovations. In this work only a linear function was tested with moderate results. An alternative approach could incorporate the smoothing function in the learning process itself, estimating the amount of innovation to be applied in the next learner, similar for example to the lateral feedbacks in cascade learning for neural networks.

It is also required to study a rejection criteria to reduce the influence of bad machines or an optimization of the aggregation procedure to take in account the quality of each machine, like boosting for regression.

Another issue to analyze is the generalization behavior of the method and possible overfitting phenomena although the generalization rates exhibited by

Table 2. Performance results for ensemble learning algorithms with different number of neurons (N) and ensemble sizes (M) over data Boston (D1) and Building2 (D2). We computed over 50 trials the statistics in table 1 and the computation time in seconds incurred by each algorithm (Time).

Statistic	M	N	Self-Poised		Bagging		Boosting	
			D1	D2	D1	D2	D1	D2
TR.Mean	5	5	9,5324	0,002504	11,0582	0,002657	10,6459	0,002743
	5	10	8,8845	0,002327	10,8725	0,002336	9,8158	0,002394
	10	5	9,5444	0,002968	11,5588	0,002562	8,9136	0,002828
	10	10	7,9763	0,002395	9,7304	0,002264	7,2538	0,002480
	20	5	9,8657	0,002423	10,5024	0,002529	8,4366	0,002735
	20	10	8,7424	0,002596	9,8524	0,002235	6,7876	0,002360
TS.Mean	5	5	17,1740	0,002588	17,5992	0,002762	18,0869	0,002918
	5	10	15,7029	0,002593	17,2186	0,002539	17,0666	0,002698
	10	5	14,8049	0,003145	16,0329	0,002687	16,9364	0,002970
	10	10	15,6729	0,002817	15,9964	0,002467	16,6778	0,002751
	20	5	17,2894	0,002767	16,2457	0,002629	16,0660	0,002952
	20	10	15,9519	0,002654	14,2401	0,002431	14,7379	0,002690
TS.Min	5	5	9,7550	0,002516	9,5704	0,002473	9,0966	0,002673
	5	10	7,1433	0,002544	8,1380	0,002539	8,7070	0,002494
	10	5	8,1670	0,002953	10,0320	0,002318	10,1059	0,002683
	10	10	6,9258	0,002759	6,3937	0,002254	9,8874	0,002476
	20	5	9,2342	0,002676	9,4113	0,002354	8,6129	0,002606
	20	10	8,1676	0,002583	6,4513	0,002152	8,3938	0,002528
TS.Std	5	5	5,0867	4,211E-5	5,3182	1,586E-4	6,0985	1,162E-4
	5	10	4,9682	2,516E-5	6,0785	2,539E-4	4,4294	1,112E-4
	10	5	4,3724	6,897E-5	4,1598	1,381E-4	5,0401	1,693E-4
	10	10	5,1772	3,585E-5	5,6799	1,111E-4	4,4665	1,303E-4
	20	5	6,4968	4,244E-5	4,2255	1,284E-4	3,7182	1,006E-4
	20	10	5,4697	3,377E-5	4,3362	1,365E-4	3,7456	0,851E-4
Time	5	5	28,7810	715,6410	27,6090	704,8280	32,9180	701,1560
	5	10	30,2570	741,7350	28,2190	727,2190	39,5270	733,1880
	10	5	58,3940	1454,1250	54,4800	1409,2660	65,4850	1408,8750
	10	10	61,2650	1506,2650	56,4140	1461,5160	73,3480	1466,2500
	20	5	120,0470	2990,5470	110,2970	2822,9840	161,3320	2823,5780
	20	10	127,8050	3103,3750	119,7420	2934,7500	164,7190	2935,4850

the algorithm in the reported experiments is comparable with those exhibited by boosting and bagging.

References

1. J. Vedelsby A. Krogh, *Neural network ensembles, cross-validation and active learning*, Neural Information Processing Systems **7** (1995), 231–238.
2. C.L. Blake and C.J. Merz, *UCI repository of machine learning databases*, 1998.
3. L. Breiman, *Bagging predictors*, Machine Learning **24** (1996), no. 2, 123–140.

4. H. Drucker, *Improving regressors using boosting techniques*, Fourteenth International Conference on Machine Learning, 1997, pp. 107–115.
5. R. Harris G. Brown, J. Wyatt and X. Yao, *Diversity creation methods: A survey and categorisation*, Information Fusion Journal (Special issue on Diversity in Multiple Classifier Systems) **6** (2004), no. 1, 5–20.
6. C. Whitaker L. Kuncheva, *Measures of diversity in classifier ensembles*, Machine Learning **51** (2003), 181–207.
7. D. Hand M. Berthold (ed.), *Intelligent data analysis*, 2 ed., Springer-Verlag, 2003.
8. R. Meir and G. Rätsch, *An introduction to boosting and leveraging*, Advanced lectures on machine learning, Springer-Verlag New York, 2003, pp. 118–183.
9. T. Mitchell (ed.), *Machine learning*, 1 ed., Mc Graw-Hill, 1997.
10. R. Nakano N. Ueda, *Generalization error of ensemble estimators*, Proceedings of International Conference on Neural Networks, 1996, pp. 90–95.
11. L. Prechelt, *Proben1 - a set of benchmarks and benchmarking rules for neural training algorithms*, Tech. Report 21/94, Fakultat fur Informatik, Universitat Karlsruhe, D-76128 Karlsruhe, Germany, 1994.
12. C. Valle R. Ñanculef, *Self-poised ensemble learning*, Tech. Report 2005/01, Departamento de Informática, Universidad Federico Santa María, CP 110-V, Valparaíso, Chile, 2005.
13. D. Opitz R. Maclin, *An empirical evaluation of bagging and boosting*, AAAI/IAAI, 1997, pp. 546–551.
14. B. Rosen, *Ensemble learning used decorrelated neural networks*, Connection Science (Special Issue on Combining Artificial Neural Networks: Ensemble Approaches) **8** (1999), no. 3-4, 373–384.
15. R. Schapire, *The stregth of weak learnability*, Machine Learning **5** (1990), 197–227.
16. R. Schapire Y. Freud, *A decision-theoretic generalization of on-line learning and application to boosting*, Journal of Computer and System Sciences **55** (1997), no. 1, 119–137.
17. X. Yao Y.Lui, *Ensemble learning via negative correlation*, Neural Networks **12** (1999), no. 10, 1399–1404.

Discriminative Remote Homology Detection Using Maximal Unique Sequence Matches

Hasan Oğul[1] and Ü. Erkan Mumcuoğlu[2]

[1] Department of Computer Engineering, Baskent University, 06530, Ankara, Turkey
hogul@baskent.edu.tr
[2] Information Systems and Health Informatics, Informatics Institute,
Middle East Technical University, 06531, Ankara, Turkey
mumcuoglu@ii.metu.edu.tr

Abstract. We define a new pairwise sequence comparison scheme for distantly related proteins and report its performance on remote homology detection task. The new scheme compares two protein sequences by using the *maximal unique matches* (MUM) between them. Once identified, the length of all non-overlapping MUMs is used to define the similarity between two sequences. To detect the homology of a protein to a protein family, we utilize the feature vectors containing all pairwise similarity scores between the test protein and the proteins in the training set. Support vector machines are employed for the binary classification in the same way that the recent works have done. The new method is shown to be more accurate than the recent methods including SVM-Fisher and SVM-BLAST, and competitive with SVM-Pairwise. In terms of computational efficiency, the new method performs much better than SVM-Pairwise.

1 Introduction

Automated categorization of proteins into their structural or functional classes is an important challenge for computational biology. Most current protein classification methods rely on computational solution for homology modeling via sequence similarity. The main assumption used here is that the primary sequence of proteins determines the structure and the structure determines the functional properties. Using this assumption, a new protein sequence is searched in large databases containing previously annotated proteins. If similar proteins are found, the new protein is assigned to same structural or functional classes with the associated ones. There are two main problems with this assumption. First, the target protein may be entirely new and its structure is different from all of the proteins available in the databases. Second, in spite of the weak similarity between two protein sequences they may still have evolutionary relationships. The first problem is a bottleneck of computational biology and there is no method that works well at the moment. The second problem is known as *remote homology detection* problem, and various methods have been proposed in recent years. In spite of several successful attempts, they are either computationally inefficient or insufficient to work for all cases.

Since the introduction of dynamic programming based sequence alignment algorithm [12], many methods have been proposed for the comparison of protein

A.F. Famili et al. (Eds.): IDA 2005, LNCS 3646, pp. 283–292, 2005.

sequences. While the dynamic programming approach finds the optimal alignment between the sequences, it suffers from a long computation time for relatively long sequences. To speed up the alignment, some heuristic methods such as BLAST [1], have been defined to find a near-optimal alignment in a reasonable time. Although these methods are very successful in the search of homolog proteins, they do not perform well for the detection of remote homologies since the alignment scores fall into a twilight zone when the sequence similarity is below 40% [11]. The later methods have incorporated the family information to detect the more distant homologies and achieved approximately three times as accurate results as simple pairwise comparison methods [10]. These methods are based on the similarity statistics derived upon more than one homolog examples, that is, all statistical information is generated from a set of sequences that are known or posited to be evolutionary related to another. These probabilistic methods are often called as generative because they induce a probability distribution over the protein family and try to generate the unknown protein as a new member of the family from this stochastic model. Further improvements have also achieved by iteratively collecting homolog proteins from a large database and incorporating the resulting statistics into a central model [2,7]. The main problem with generative approaches is the fact that they produce so much false positives, that is, they report a number of homologs though they are not homolog.

The recent works on remote homology detection have begun to use a discriminative framework to make separation between homolog (positive) and non-homolog (negative) classes. In contrast to generative methods, the discriminative methods focus on learning the combination of features that discriminate between the classes. These methods try to establish a model that differentiates between positive and negative examples. In other words, non-homologs are also taken into account.

In discriminative homology detection methods, there are two main phases: training and testing. The training phase constructs a machine learning classifier for the specified family, and the testing phase uses this classifier to decide whether the test protein is belonging to this family or not. In general, a machine learning classifier is constructed for each family in the database and the protein is checked if it is belonging to any of those known protein classes. Both phases require the extraction of some informative features from the protein sequence and the representation of these features in a suitable way. Fig. 1 gives an overview of the discriminative homology detection approach.

The current methods using the discriminative approach differ in the feature extraction methods, the feature representation forms and the type of the machine learning classifiers they have used. Among k-Nearest Neighbor Method, Neural Networks and Support Vector Machine, the last one has been reported as outperforming to the others for homology detection purposes [8].

Discriminative methods are more successful than generative methods in terms of separation accuracy between true positives and false positives. However, the training phase requires so much time with conventional workstations, which makes them inappropriate to use in practice. Thus, more efficient methods are required while preserving the classification accuracy.

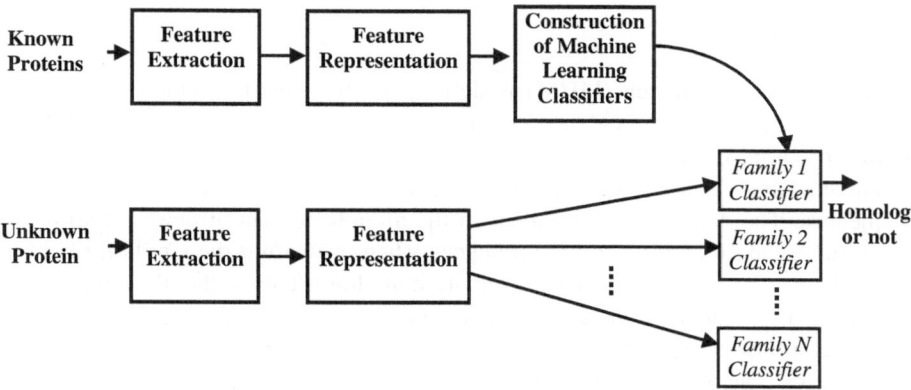

Fig. 1. Discriminative homology detection model

The first discriminative approach (SVM-Fisher) represents each protein by a vector of Fisher scores extracted from a profile Hidden Markov model constructed for a protein family and utilizes Support Vector Machines to classify the protein with those feature vectors [6]. A recent and more successful work, called SVM-Pairwise [8], combines the sequence similarity with the Support Vector Machines to discriminate between positive and negative examples. In SVM-Pairwise, both the training and test sets include positive and negative examples. Each protein, P_x, in the data set is vectorized by $\varphi(P_x)$ with the following equation:

$$\varphi(P_x) = S(P_x, P_1),\ \ S(P_x, P_2),...,S(P_x, P_n) \tag{1}$$

where P_i is i^{th} protein sequence in the training set, n is the total number of proteins in the training set, including both positives and negatives, and $S(P_x, P_i)$ is the alignment score between any protein sequences P_x and P_i. This method has been tested for dynamic programming based alignment scores [12] and BLAST scores [1]. Note that the latter one is referred as SVM-BLAST in the following sections. SVM-Pairwise approach is among the best methods in terms of accuracy, but it suffers from computational inefficiency since the alignment takes too much time for long sequences. Another drawback of this approach is that the alignment may force some residues to match even if they are evolutionary not related.

In this work, we try a more conservative approach to compare the protein sequences. Instead of using alignments, we define a new similarity scoring scheme based on more conserved sequence patterns, called maximal unique matches. This scheme does not require the alignment of the sequences but it quickly finds the matches between them. By combining the similarity scores obtained using the maximal unique matches with a binary classifier, Support Vector Machine, we perform protein family classification tests on a subset of SCOP families [9] and compare our results with those given by recent methods; SVM-Fisher, SVM-BLAST and SVM-Pairwise. The new method is better than SVM-Fisher and SVM-BLAST and competitive with SVM-Pairwise. In terms of computational efficiency, the new method performs much better than SVM-Pairwise.

2   Methods

A maximal unique match (MUM) is defined as the substring which appears only once in both sequences and not contained in a larger such substring. With this definition, a maximal unique match can be considered as one of the core part in an alignment and yields an important evidence for the homology between the sequences. The definition is stricter than the alignment since it does not allow any substitution and repetition in the sequences when evaluating the similarity between them. However, since we deal with remote homolog proteins, the allowance of any mutation may lead to by-chance matches between the sequences. Therefore, this strict definition is expected to show more evidently the local relationships between the proteins. The definition of MUM has been originally introduced by Delcher et al. [3] to accelerate the alignment of long DNA strands. In this study, we adopt their definition for the protein sequences to represent the conservative relationships between them.

Once identified, the length of all non-overlapping maximal unique matches can be used to compare two protein sequences; $S(P_x,P_i)$ score in Eq.2 is replaced by the total length of all MUMs, $M(P_x,P_i)$, in protein vectorization step;

$$\varphi(P_x) = M(P_x,P_1),\ M(P_x,P_2),...,\ M(P_x,P_n) \tag{2}$$

The simple sum of all MUM lengths leads to a serious problem when two or more MUMs overlap since the overlapping residues would be counted more than once. To overcome this problem, we modify the definition of $M(P_x,Pi)$ as the number of the residues contained in a maximal unique match between P_x and P_i.

2.1   Finding MUMs

To find the maximal unique matches, we used a special data structure called suffix tree. A suffix tree is a compacted tree that stores all suffixes of a given text string. (An example suffix tree is shown in Fig. 2.). It is a powerful and versatile data structure which finds application in many string processing algorithms [5].

Let A be string of n characters, $A=s_1s_2...s_n$, from an ordered alphabet Σ except s_n. Let $\$$ be a special character, matching no character in Σ, and s_n be $\$$. The suffix tree T of A is a tree with n leaves such that;

- Each path from the root to a leaf of T represents a different suffix of A.
- Each edge of T represents a non-empty string of A.
- Each non-leaf node of T, except the root, must have at least two children.
- Substrings represented by two sibling edges must begin with different characters.

The definition of a suffix tree can be easily extended to represent the suffixes of a set $\{A_1,A_2,...,A_n\}$ of strings. This kind of suffix tree is called as a generalized suffix tree.

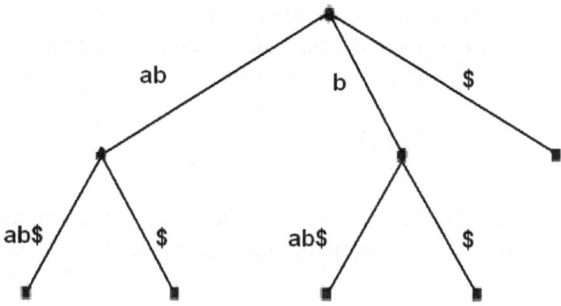

Fig. 2. Suffix tree of "abab$".

To find maximal unique matches between any two sequences, first, a generalized suffix tree is constructed for the sequences. This is simply done by concatenating two sequences with a dummy character (not contained in the alphabet) between them and constructing a suffix tree for the newly created sequence. More formally, the generalized suffix tree GT of $\{A,B\}$, where A and B are the sequences being compared, is a suffix tree Tg of the sequence "$a_1a_2...a_n\#b_1b_2...b_n\$$".

In our representation, a maximal unique match is a *maximal pair* in the concatenated sequence one of which appears before the dummy character and the other appears after that. The algorithm to find maximal pairs is given by Gusfield [5]. We used a variation of this algorithm considering the fact that each of the pair should appear in different sequences. To satisfy this property, the following fact is used; in Tg, any internal node v which has exactly two childs both of which are leaves and represent the suffixes from different sequences is a *matching node,* and the path from root to a *matching node* gives a unique match.

The leaves of the suffix tree are numbered according to the position of the suffix which they represent. The node number of the leaf which represents the suffix "$\#b_1b_2...b_n$" is *separating-point*. Thus, we can find easily which sequence is represented by a leaf just looking whether its node number is smaller or grater than *separating-point*. Therefore, a bottom to up traversal of tree to find matching nodes gives us the unique matches in linear time. Since the construction of a suffix tree can also be completed in linear time [13], the algorithm for finding maximal unique matches would have a linear time complexity. The maximality of unique matches can be determined simply by mismatches at their left and right ends.

2.2 The Classification

To discriminate between positive and negative examples Support Vector Machines are used. SVMs are binary classifiers that work based on the structural risk minimization principle. An SVM non-linearly maps its n-dimensional input space into a high dimensional feature space. In this high dimensional feature space a linear classifier is constructed. To train the Support Vector Machines, SVM-Gist software, available at www.cs.columbia.edu/compbio/svm, is used. In SVM-Gist software, a kernel function acts as the similarity score between pairs of input vectors. The base kernel is normalized in order to make that each vector has a length of 1 in the feature space, that is,

$$K(X,Y) = \frac{X.Y}{\sqrt{(X.X)(Y.Y)}} \tag{3}$$

where X and Y are the input vectors, $K(.,.)$ is the kernel function, and "." denotes the dot product. This kernel is then transformed into a radial basis kernel $K'(X,Y)$, as follows:

$$K'(X,Y) = e^{-\frac{K(X,X)-2K(X,Y)+K(Y,Y)}{2\sigma^2}} + 1 \tag{4}$$

where the width σ is the median Euclidean distance from any positive training example to the nearest negative example. Since the separating hyperplane of SVM is required to pass from the origin, the constant 1 is added to the kernel so that the data goes away from the origin. An asymmetric soft margin is implemented by adding to the diagonal of the kernel matrix a value $0.02*\rho$, where ρ is the fraction of training set proteins that have the same label with the current protein, as done in the previous SVM classification methods (SVM-Pairwise, SVM-BLAST, SVM-Fisher). The SVM output is a list of discriminant score values corresponding to each protein in the test set.

3 Results

There are two issues to be considered in remote homology detection task: accuracy and efficiency. The recent remote homology detection methods have been tested to see their ability to classify proteins into families in a subset of SCOP family database [9]. The accuracies of the methods are evaluated using *sensitivity* and *specificity* measures. Sensitivity is defined as the ability to detect true positives (homolog proteins which are also reported as homolog). On the other hand, specificity is described as the ability to reject false positives. As all classification tasks do, the homology detection methods have to deal with the trade-off between specificity and sensitivity.

For the cases in which the positive and negative examples are not evenly distributed, the best way to evaluate the trade-off between the specificity and sensitivity is to use a Receiver Operating Characteristics (ROC) curve [4]. A ROC score may be defined as the area under the ROC curve, where the ROC curve is plotted as the number of true positives as a function of false positives for varying classification thresholds. A score of 1 indicates that the positives are perfectly separated from negatives whereas the score of 0 yields that no positives are reported.

All the tests are performed on a subset of the SCOP1.53 database including no protein pair with a pairwise similarity higher than an E-value of 10^{-25}. The training and test sets are separated as done in the previous works [8] resulting with 54 families to test. The ROC scores obtained for the SCOP families, using four different methods including our method (SVM-MUM), are given in the Table 1.

Table 1. ROC scores for 54 SCOP families

family	SVM-Pairwise	SVM-Fisher	SVM-BLAST	SVM-MUM
1.27.1.1	**0,971**	0,511	0,890	0,950
1.27.1.2	**0,918**	0,629	0,779	0,889
1.36.1.2	0,935	0,845	0,870	**0,955**
1.36.1.5	**0,976**	0,641	0,708	0,913
1.4.1.1	0,968	0,708	0,878	**0,980**
1.4.1.2	0,814	0,795	0,810	**0,834**
1.4.1.3	0,944	0,635	**0,999**	0,970
1.41.1.2	0,999	0,956	**1,000**	0,954
1.41.1.5	**0,998**	0,935	0,996	0,927
1.45.1.2	**0,971**	0,547	0,729	0,921
2.1.1.1	**0,978**	0,840	0,949	0,883
2.1.1.2	**0,994**	0,756	0,972	0,970
2.1.1.3	**0,985**	0,844	0,907	0,966
2.1.1.4	**0,974**	0,876	0,947	0,886
2.1.1.5	**0,832**	0,647	0,790	0,799
2.28.1.1	**0,815**	0,490	0,389	0,559
2.28.1.3	**0,829**	0,596	0,412	0,543
2.38.4.1	0,697	0,501	0,702	**0,780**
2.38.4.3	0,707	0,419	**0,764**	0,681
2.38.4.5	**0,877**	0,539	0,668	0,786
2.44.1.2	0,146	0,533	**0,925**	0,403
2.5.1.1	**0,925**	0,680	0,899	0,840
2.5.1.3	**0,896**	0,669	0,826	0,782
2.52.1.2	0,643	0,472	0,641	**0,793**
2.56.1.2	0,844	0,612	**0,878**	0,839
2.9.1.2	0,874	0,485	0,543	**0,887**
2.9.1.3	0,970	0,655	0,909	**0,989**
2.9.1.4	0,918	0,431	0,645	**0,926**
3.1.8.1	0,963	0,323	0,406	**0,990**
3.1.8.3	0,931	0,445	0,345	**0,986**
3.2.1.2	0,838	0,371	**0,842**	0,806
3.2.1.3	**0,898**	0,611	0,746	0,807
3.2.1.4	0,964	0,847	**0,969**	0,850
3.2.1.5	**0,932**	0,597	0,854	0,879
3.2.1.6	**0,912**	0,624	0,776	0,822
3.2.1.7	0,909	0,536	0,812	**0,922**
3.3.1.2	**0,937**	0,733	0,847	0,836
3.3.1.5	**0,917**	0,448	0,709	0,828
3.32.1.1	**0,946**	0,777	0,866	0,826
3.32.1.11	0,880	0,899	0,888	**0,947**
3.32.1.13	0,836	0,727	0,646	**0,901**
3.32.1.8	**0,901**	0,759	0,776	0,781
3.42.1.1	0,886	0,687	**0,923**	0,795
3.42.1.5	**0,811**	0,586	0,580	0,665
3.42.1.8	0,760	0,425	**0,930**	0,710
7.3.10.1	0,986	0,898	**0,997**	0,978
7.3.5.2	**0,996**	0,850	0,992	0,919
7.3.6.1	0,998	0,985	**0,999**	0,945
7.3.6.2	0,994	0,955	**0,997**	0,969
7.3.6.4	0,992	0,864	**1,000**	0,993
7.39.1.2	**0,928**	0,713	0,877	0,898
7.39.1.3	**0,990**	0,820	0,985	0,922
7.41.5.1	0,791	0,798	**0,916**	0,505
7.41.5.2	0,943	0,979	**0,999**	0,605
Average	0,893	0,676	0,817	0,846
Std. Dev.	0,133	0,174	0,171	0,133

According to the table, our method achieves the best accuracy in 13 of the 54 families among the four methods compared. Both average ROC scores and pairwise comparison plots (Fig. 3) show that the new method performs better than SVM-BLAST and SVM-Fisher and it is comparable with SVM-Pairwise. It should also be noted that the standard deviation on the ROC scores is the same as SVM-Pairwise. Since it has been already shown that the discriminative methods outperform the generative methods [6], we do not need to include them in the results.

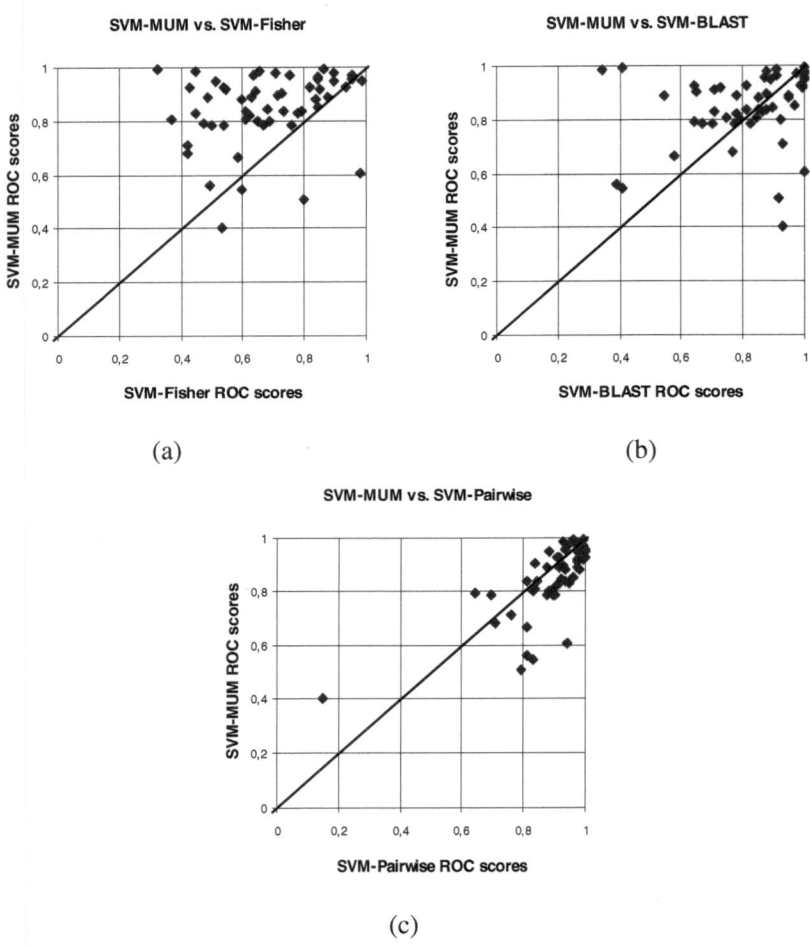

Fig. 3. Pairwise comparison of SVM-MUM with (a) SVM-Fisher (b) SVM-BLAST and (c) SVM-Pairwise

Computational efficiency is another important aspect in the evaluation of methods. In this respect, SVM-MUM method is much more efficient than SVM-Pairwise. Considering that all other parts of the methods except the extraction of similarity scores are the same, the only factor to be compared is the calculation time for pairwise similarities. In SVM-Pairwise, dynamic programming algorithm for two sequences with the lengths of n and m takes $O(nm)$ time. On the other hand, all maximal unique matches can be identified in $O(n+m)$ time, as described in Methods section. According to our experiments, calculation of all pairwise MUM scores for a protein is completed in 2 minutes on average, whereas the SVM prediction takes only a few seconds with a workstation having 2.0GHz CPU and 1GB RAM. Therefore, the stage for similarity calculations is the crucial part of the prediction system. Meanwhile, a straightforward ANSI C implementation for dynamic programming can complete the vectorization of a protein in between 7 and 9 minutes, i.e. approximately four times larger than MUM scoring. The same difference can also be observed in training stage. While SVM-MUM completes all training stage in 4 days, it takes up to 20 days with SVM-Pairwise. This clearly shows the superiority of our method to SVM-Pairwise in terms of computational efficiency.

4 Conclusion

We define a new scheme based on the maximal unique sequence matches for the pairwise comparison of remote homolog proteins. As a result of protein family classification tests on a subset of SCOP database, it is observed that the maximal unique matches are very simple and efficient way of detecting remote homologies. When it is compared with the similar methods which utilize the dynamic programming based alignment and BLAST comparison, we have found that the new scheme is more accurate than BLAST and as successful as dynamic programming based alignment. In terms of computational efficiency, we have shown that the new method is much better than the dynamic programming algorithm.

References

1. Altschul S., Gish W., Miller W., Myers E. W., Lipman D.: A basic local alignment search tool. Journal of Molecular Biology, (1990) 251, 403-10.
2. Altschul S., Madden T., Schaffer A., Zhang J., Zhang Z., Miller W., Lipman D.: Gapped BLAST and PSI-BLAST: a new generation of protein database search programs. Nucleic Acids Research, (1997) 25, 3389-3402.
3. Delcher A., Kasif S., Fleishmann R., Peterson J., White O., Salzberg S.: Alignment of whole genomes. Nucleic Acids Research, (1999) 27, 2369-2376.
4. Gribskov, M., and Robinson, N.L.: Use of receiver operating characteristic (ROC) analysis to evaluate sequence matching. Computers and Chemistry, (1996) 20(1), 25–33.
5. Gusfield D.: Algorithms on Strings, Trees, and Sequences: Computer science and Computational Biology. Cambridge University Press, New York (1997).
6. Jaakola T., Diekhans M., Haussler D.: A discriminative framework for detecting remote protein homologies. Journal of Computational Biology, (2000) 7, 95-114.
7. Karplus K., Barrett C., Hughey R.: Hidden Markov Models for detecting remote protein homologies. Bioinformatics, (1998) 14, 846-856.
8. Liao L., Noble W. S.: Combining pairwise sequence similarity and support vector machines for detecting remote protein evolutionary and structural relationships. Journal of Computational Biology, (2002) 10, 857-868.

9. Murzin A. G., Brenner S. E., Hubbard T., Chothia C.: SCOP: A structural classification of proteins database for the investigation of sequences and structures, Journal of Molecular Biology, (1995) 247, 536-40.

10. Park J., Karplus K., Barrett C., Hughey R., Haussler D., Hubbard T., Chothia C.: Sequence comparisons using multiple sequences detect tree times as many remote homologues as pairwise methods. Journal of Molecular Biology, (1998) 284, 1201-1210.

11. Rost B.: Twilight zone of protein sequence alignments. Protein engineering, (1999) 12, 85-94.

12. Smith T. F., Waterman M. S.: Identification of common molecular subsequences. Journal of Molecular Biology, (1981) 147,195-97.

13. Ukkonen E.: On-line construction of suffix-trees. Algorithmica, (1995) 14, 249-60.

From Local Pattern Mining to Relevant Bi-cluster Characterization

Ruggero G. Pensa and Jean-François Boulicaut

INSA Lyon, LIRIS CNRS,
UMR 5205 F-69621, Villeurbanne cedex, France
{Ruggero.Pensa, Jean-Francois.Boulicaut}@insa-lyon.fr

Abstract. Clustering or bi-clustering techniques have been proved quite useful in many application domains. A weakness of these techniques remains the poor support for grouping characterization. We consider eventually large Boolean data sets which record properties of objects and we assume that a bi-partition is available. We introduce a generic cluster characterization technique which is based on collections of bi-sets (i.e., sets of objects associated to sets of properties) which satisfy some user-defined constraints, and a measure of the accuracy of a given bi-set as a bi-cluster characterization pattern. The method is illustrated on both formal concepts (i.e., "maximal rectangles of true values") and the new type of δ-bi-sets (i.e., "rectangles of true values with a bounded number of exceptions per column"). The added-value is illustrated on benchmark data and two real data sets which are intrinsically noisy: a medical data about meningitis and Plasmodium falciparum gene expression data.

1 Introduction

Clustering has been proved extremely useful for exploratory data analysis. Its main goal is to identify a partition of objects and/or properties such that an objective function which specifies its quality is optimized (e.g., maximizing intra-cluster similarity and inter-cluster dissimilarity). Looking for optimal solutions is intractable such that heuristic local search optimizations are performed [1]. Many efficient algorithms can provide good partitions but suffer from the lack of explicit cluster characterization. For example, considering gene expression data analysis, clustering is used to look for sets of co-expressed genes and/or sets of biological situations or experiments which seem to trigger this co-expression (see, e.g., [2]). In this context, an explicit characterization would be a symbolic statement which "explains" why genes and/or situations are within the same groups. Once such characterizations are available, it supports the understanding of gene regulation mechanisms. Our running example **r** (see Table 1) concerns a toy Boolean data set. For instance, it encodes gene expression properties (e.g., over-expression) in various biological situations and, genes denoted by p_1, p_3, p_4 are considered over-expressed in situation o_1.

The crucial need for characterization has motivated the research on conceptual clustering [3]. Among others, it has been studied in the context of co-clustering or bi-clustering [4,5,6,7], including for the special case of categorial

A.F. Famili et al. (Eds.): IDA 2005, LNCS 3646, pp. 293–304, 2005.

Table 1. A Boolean context **r**

	p_1	p_2	p_3	p_4	p_5
o_1	1	0	1	1	0
o_2	0	1	0	0	1
o_3	1	0	1	1	0
o_4	0	0	1	1	0
o_5	1	1	0	0	1
o_6	0	1	0	0	1
o_7	0	0	0	0	1

or Boolean data. The goal is to identify bi-clusters or bi-partitions in the data, i.e., a mapping between a partition of situations (more generally objects) and a partition of gene expression properties (more generally, Boolean properties of objects). For instance, an algorithm like COCLUSTER [6] can compute the interesting bi-partition $\{\{\{o_1, o_3, o_4\}, \{p_1, p_3, p_4\}\}, \{\{o_2, o_5, o_6, o_7\}, \{p_2, p_5\}\}\}$ from **r**. The first bi-cluster indicates that the characterization of objects from $\{o_1, o_3, o_4\}$ is that they almost always share properties from $\{p_1, p_3, p_4\}$. Also, properties in $\{p_2, p_5\}$ are characteristics for objects in $\{o_2, o_5, o_6, o_7\}$. Unfortunately, this first step towards characterization is not sufficient to support the needed interactivity with the end-users who have to interpret the resulting (bi-)partitions. Our thesis is that it is useful to look for bi-sets, i.e., sets of objects associated to sets of properties, that exhibit strong and characteristic relations between bi-cluster elements. For instance, once a bi-partition of a Boolean gene expression data set has been found, one can be interested in studying all the interactions between genes involved in a "cancer" bi-cluster, and these interactions might imply genes which are involved in "non cancerous" processes as well.

Given a bi-partition on a Boolean data set, our goal is to provide characterizing patterns for each bi-cluster and our contribution is twofold. First, we introduce an original and generic cluster characterization technique which is based on constraint-based bi-set mining, i.e., mining bi-sets whose set components satisfy some constraints, and a measure of the accuracy of a given extracted bi-set as a characterization pattern for a given bi-cluster (see Section 2). We also discuss the opportunity to shift from the characterization by bi-sets towards a characterization based on association rules. The method is then illustrated on two kinds of bi-sets, the well-known formal concepts (i.e., associated closed sets [8] or, intuitively, "maximal rectangle of true values") and a new class, the so-called δ-bi-sets. This later pattern type is new and it is based on a previous work about approximate condensed representations for frequent patterns [9]. Intuitively, a δ-bi-set is a "rectangle of true values with a bounded number of exceptions per column" (see Section 3). We illustrate the added-value of our characterizing method not only on a benchmark data set but also on two real-life data sets. The obtained characterizations are consistent with the available knowledge (see Section 4).

2   Bi-cluster Characterization Using Bi-sets

Let us consider a set of objects $\mathcal{O} = \{o_1, \ldots, o_m\}$ and a set of Boolean properties $\mathcal{P} = \{p_1, \ldots, p_n\}$. The Boolean context to be mined is $\mathbf{r} \subseteq \mathcal{O} \times \mathcal{P}$, where $r_{ij} = 1$ if the property p_j is true for object o_i. Formally, a bi-set is an element of $2^{\mathcal{O}} \times 2^{\mathcal{P}}$. We assume that a bi-clustering algorithm, e.g., [6], provides a mapping between k clusters of objects (say $\{C_1^o \ldots C_k^o\}$) and k clusters of properties (say $\{C_1^p \ldots C_k^p\}$). A first characterization comes from this mapping.

Our goal is to support each bi-cluster interpretation by collections of bi-sets which are locally pointing out interesting associations between groups of objects and groups of properties. Therefore, we assume that a collection of N bi-sets $\mathcal{C} = c_1, \ldots, c_N$ has been extracted from the data. First, we associate each of them to one the k bi-clusters to obtain a collection of k groups of bi-sets $\{C_1, \ldots, C_k\}$, where $C_i \subseteq \mathcal{C}$. Each bi-set $\in C_i$ characterizes the bi-cluster (C_i^o, C_i^p) with some degree of accuracy.

Let us first define the signature in $\mathbf{r}$ of each bi-cluster (C^o, C^p) denoted $\mu(C^o, C^p) = (\tau, \gamma)$ where $\tau = \{o_i \in C^o\}$ and $\gamma = \{p_i \in C^p\}$. We can now define a similarity measure between a bi-set $c = (T, G)$ and a bi-cluster signature:

$$sim(c, \mu(C^o, C^p)) = \frac{|(T, G) \cap (\tau, \gamma)|}{|(T, G) \cup (\tau, \gamma)|} = \frac{|T \cap \tau| \cdot |G \cap \gamma|}{|T| \cdot |G| + |\tau| \cdot |\gamma| - |T \cap \tau| \cdot |G \cap \gamma|}$$

Intuitively, bi-sets (T, G) and (τ, γ) denote rectangles in the matrix (modulo permutations over the lines and the columns) and we measure the area of the intersection of the two rectangles normalized by the area of their union.

Each bi-set c which is a candidate characterization pattern can now be assigned to the bi-cluster (C^o, C^p) for which $sim(c, \mu(C^o, C^p))$ is maximal. Doing so, we get k groups of potentially characterizing bi-sets. Finally, we can use an accuracy measure to select the most relevant ones. For that purpose, we propose to measure the exception ratios for the two set components of the bi-sets.

Given a bi-set (T, G) and a bi-cluster (C^o, C^p), it can be computed as follows:

$$\epsilon_o = \frac{|\{o_i \in T| \; o_i \notin C^o\}|}{|T|}, \quad \epsilon_p = \frac{|\{p_i \in G| \; p_i \notin C^p\}|}{|G|}$$

It is then possible to consider thresholds to select only the bi-sets that have little exception ratios, i.e., $\epsilon_o < \varepsilon_o$ and $\epsilon_p < \varepsilon_p$ where $\varepsilon_o, \varepsilon_p \in [0, 1]$. There are several possible interpretations for these measures. If we are interested in characterizing a cluster of objects (resp. properties), we can look for all the sets of properties (resp. objects) for which the ϵ_o (resp. ϵ_p) values of the related bi-sets are less than a threshold ε_o (resp. ε_p). Alternatively, we can consider the whole bi-cluster and characterize it with all the bi-sets for which the two exception ratios ϵ_o and ϵ_p are less than two threshold ε_o and ε_p.

3   Looking for Candidate Characterizing Bi-sets

We now discuss the type of bi-sets which will be post-processed for bi-cluster characterization. It is clear that bi-clusters are, by construction, interesting characterizing bi-sets but they only support a global interpretation. We are interested in strong associations between sets of objects and sets of properties that can locally explain the global behavior. Clearly, formal concepts can be used [8].

Definition 1 (formal concept). *If $T \subseteq \mathcal{O}$ and $G \subseteq \mathcal{P}$, assume $\phi(T, \mathbf{r}) = \{g \in \mathcal{P} \mid \forall t \in T, (t, g) \in \mathbf{r}\}$ and $\psi(G, \mathbf{r}) = \{t \in \mathcal{O} \mid \forall g \in G, (t, g) \in \mathbf{r}\}$. A bi-set (T, G) is a formal concept in $\mathbf{r}$ when $T = \psi(G, \mathbf{r})$ and $G = \phi(T, \mathbf{r})$. By construction, G and T are closed sets, i.e., $G = \phi \circ \psi(G, \mathbf{r})$ and $T = \psi \circ \phi(T, \mathbf{r})$. Intuitively, (T, G) is a maximal rectangle of true values.*

$(\{o_1, o_3\}, \{p_1, p_3, p_4\})$, $(\{o_1, o_3, o_4\}, \{p_3, p_4\})$, and $(\{o_5, o_6\}, \{p_2, p_5\})$ are examples of formal concepts among the 8 ones which hold in $\mathbf{r}$ (see Table 1). Efficient algorithms have been developed to extract complete collections of formal concepts which satisfy also user-defined constraints, e.g., [10,11]. A fundamental problem with formal concepts is that the Galois connection (ϕ, ψ) is, in some sense, a too strong one: we have to capture every maximal set of objects and its maximal set of associated properties. As a result, the number of formal concepts even in small matrices can be huge. A solution is to look for "dense" rectangles in the matrix, i.e., bi-sets with mainly true values but also a bounded (and small) number of false values or exceptions. Well-defined collections of dense bi-sets can be obtained by merging formal concepts [12], i.e., a post-processing over collections of formal concepts. This turns to be intractable when the number of formal concepts is too large. We propose a new type of bi-set which can be efficiently extracted, including in noisy data in which it is common to have several millions of formal concepts.

3.1 Mining δ-Bi-sets

We want to compute efficiently smaller collections of bi-sets which still capture strong associations. We recall some definitions about the association rule mining task [13] since it is used for both the definition of the δ-bi-set pattern type and for bi-cluster characterization.

Definition 2 (association rule, frequency, confidence). *An association rule R in $\mathbf{r}$ is an expression of the form $X \Rightarrow Y$, where $X, Y \subseteq \mathcal{P}$, $Y \neq \emptyset$ and $X \cap Y = \emptyset$. Its absolute frequency is $|\psi(X \cup Y, \mathbf{r})|$ and its confidence is $|\psi(X \cup Y, \mathbf{r})| / |\psi(X, \mathbf{r})|$.*

In an association rule $X \Rightarrow Y$ with high confidence, the properties in Y are almost always true for an object when the properties in X are true. Intuitively, $X \cup Y$ associated to $\psi(X, \mathbf{r})$ is then a dense bi-set: it contains a few false values. We now consider our technique for computing association rules with high confidence, the so-called δ-strong rules [14,9].

Definition 3 (δ-strong rule). *Given an integer δ, a δ-strong rule in $\mathbf{r}$ is an association rule $X \Rightarrow Y$ $(X, Y \subset \mathcal{P})$ s.t. $|\psi(X, \mathbf{r})| - |\psi(X \cup Y, \mathbf{r})| \leq \delta$, i.e., the rule is violated in no more than δ objects.*

Interesting collections of δ-strong rules with minimal left-hand side can be computed efficiently from the so-called δ-free-sets [14,9,15] and their δ-closures.

Definition 4 (δ-free set, δ-closure). *Let δ be an integer and $X \subset \mathcal{P}$, X is a δ-free-set in $\mathbf{r}$ iff there is no δ-strong rule which holds between two of its own and proper subsets. The δ-closure of X in $\mathbf{r}$, $h_\delta(X, \mathbf{r})$, is the maximal superset Y of X s.t. $\forall p \in Y \setminus X$, $|\psi(X \cup \{p\})| \geq |\psi(X, \mathbf{r})| - \delta$. In other terms, the frequency of the δ-closure of X in $\mathbf{r}$ is almost the same than the frequency of X when $\delta << |\mathcal{O}|$. Moreover, $\forall p \in h_\delta(X) \setminus X$, $X \Rightarrow p$ is a δ-strong rule.*

For example, in Table 1, the 1-free itemsets are $\{p_1\}$, $\{p_2\}$, $\{p_3\}$, $\{p_4\}$, $\{p_5\}$, $\{p_1, p_2\}$, and $\{p_1, p_5\}$. An example of 1-closure for $\{p_1\}$ is $\{p_3, p_4\}$. The association rules $\{p_1\} \Rightarrow \{p_3\}$ and $\{p_1\} \Rightarrow \{p_4\}$ have only one exception.

δ-freeness is an anti-monotonic property such that it is possible to compute δ-free sets (eventually combined with a minimal frequency constraint) in very large data sets. Notice that $h_0 \equiv \phi \circ \psi$, i.e., the classical closure operator. Looking for a 0-free-set, say X, and its 0-closure, say Y, provides the closed set $X \cup Y$ and thus the formal concept $(\psi(X \cup Y, \mathbf{r}), X \cup Y)$.

Definition 5 (δ-bi-set). *A δ-bi-set (T, G) in $\mathbf{r}$ is built on each δ-free-set $X \subset \mathcal{P}$ with $T = \psi(X, \mathbf{r})$ and $G = h_\delta(X, \mathbf{r})$.*

In Table 1, the 1-bi-sets derived from the 1-free-sets $\{p_3\}$ and $\{p_5\}$ are $(\{o_1, o_3, o_4\}, \{p_1, p_3, p_4\})$ and $(\{o_2, o_5, o_6, o_7\}, \{p_2, p_5\})$. When $\delta << |T|$, δ-bi-sets are dense bi-sets with a small number of exceptions per column. In order to experiment, we implemented a straightforward extension of AcMiner [9] which provides the supporting set for each extracted δ-free-set.

3.2 Concepts vs. δ-Bi-sets

To study the relevancy of δ-bi-sets w.r.t. formal concepts, we have considered the addition of noise to a synthetical data set. Hereafter, $\mathbf{r}$ denotes a reference data set from which we generate noisy data sets by adding a given quantity of uniform random noise. Then, we compare the collection of formal concepts which are "built-in" within $\mathbf{r}$ with various collections of formal concepts and δ-bi-sets extracted from the noised matrices. To measure the relevancy of each extracted collection w.r.t the reference one, we look for subsets of the reference collection in each of them. Since both set components of each formal concept can be changed when adding noise, we identify those having the largest area in common with the reference ones, and we compute the σ measure which takes into account the common area:

$$\sigma(\mathcal{C}_r, \mathcal{C}_a) = \frac{1}{N_r} \sum_{i=1}^{N_r} max_j \left(\frac{|(T_i, G_i)_r \cap (T_j, G_j)_a|}{|(T_i, G_i)_r \cup (T_j, G_j)_a|} \right)$$

where $\mathcal{C}_r$ is the collection of formal concepts in reference $\mathbf{r}$, $N_r = |\mathcal{C}_r|$, $\mathcal{C}_a$ is a noised collection of bi-sets, $(T_i, G_i)_r \in \mathcal{C}_r$ and $(T_j, G_j)_a \in \mathcal{C}_a$. When $\sigma(\mathcal{C}_r, \mathcal{C}_a) = 1$, all the bi-sets $\in \mathcal{C}_r$ have identical instances in $\mathcal{C}_a$.

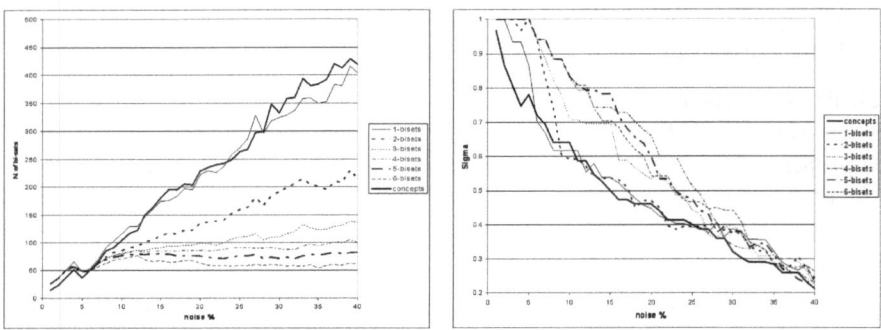

Fig. 1. Size of different collections of bi-sets (left) and related values of σ (right) depending on noise level

In the experiment, **r** has 30 objects and 15 properties and it contains 3 formal concepts of the same size which are pair-wise disjoints. In other terms, the formal concepts are $(\{o_1, \ldots, o_{10}\}, \{p_1, \ldots, p_5\})$, $(\{o_{11}, \ldots, o_{20}\}, \{p_6, \ldots, p_{10}\})$, and $(\{o_{21}, \ldots, o_{30}\}, \{p_{11}, \ldots, p_{15}\})$. We generated 40 different data sets by adding to **r** increasing quantities of noise (from 1% to 40% of the matrix). Then, for each data set, we have extracted a collection of formal concepts and different collections of δ-bi-sets with increasing values of δ (from 1 to 6). Finally, we looked for the occurrence of the 3 formal concepts in each of these extracted collections by using our σ measure. Results are in Fig. 1.

The σ measure decreases when the noise level increases. Interestingly, its values for δ-bi-set collections are always greater or similar to the values for the collection of formal concepts. The collections of δ-bi-sets contain always less patterns than the collection of formal concepts (for a noise level greater than 7%). For $\delta = 2$, the size is halved. For greater values of δ, noise does not influence the size of the collections of δ-bi-sets. This experiment confirms that δ-bi-sets are more robust to noise than formal concepts. Furthermore, it enables to reduce significantly the size of the extracted collections and this is important to support the interpretation process.

3.3 Using Association Rules

Association rules can be derived from extracted bi-sets and used for bi-cluster characterization. For characterization but also classification, heuristics have been studied which select relevant association rules based on their frequency and confidence values [16,17,18]. In our case, we propose to use exception ratios on the extracted bi-sets to provide characterization rules. They have the form $X \Rightarrow v$ where X is a set of properties (resp. objects if the transposed matrix is used) and v is a variable denoting a cluster of objects (resp. properties). When considering formal concepts, deriving characterization rules from them is straightforward.

Property 1. Given a bi-cluster (C^o, C^p), if (T, G) is a formal concept, then $G \Rightarrow C^o$ (resp. $T \Rightarrow C^p$) is a rule with frequency equal to $|T| \cdot (1 - \epsilon_o)$ (resp. $|G| \cdot (1 - \epsilon_p)$) and confidence equal to $1 - \epsilon_o$ (resp. $1 - \epsilon_p$).

When we use δ-bi-sets instead of formal concepts, Property 1 does not hold because $|\psi(G, \mathbf{r})| < |T|$. However, if we are interested in characterizing a cluster of objects, we can use the following property:

Property 2. Given a cluster C^o, if (T, G) is a δ-bi-set, and $X \subseteq G$ is a δ-free-set then $X \Rightarrow C^o$ is a rule with frequency equal to $|T| \cdot (1 - \epsilon_o)$ and confidence equal to $1 - \epsilon_o$.

Such rules are interesting in practice because X is often a rather small set such that its interpretation is easier. However, this approach can not be applied to data sets with large numbers of properties (e.g., for gene expression data sets where thousands of properties are common). In such cases, we propose to use the ϵ_o and ϵ_p measures.

3.4 Examples of Characterizing Queries

So far, we have a methodology for characterizing (bi-)clusters by using different kinds of bi-sets or association rules which can be derived from them. Proposed accuracy measures can be used for a direct selection of characterizing patterns by means of queries:

- Select all the bi-sets which characterize bi-cluster (C^o, C^p) with a maximum exception ratio of ε for both objects and properties;
- Select all the rules with minimal body characterizing bi-cluster (C^o, C^p) with a minimal frequency f, a minimal confidence c, and a maximal exception ratio ε for the set of properties;
- Select all the rules with minimal body characterizing bi-cluster (C^o, C^p) with a minimal frequency f, a minimal confidence c, and a minimal exception ratio ε for the set of properties.

The last example is interesting since it returns bi-sets (or rules) that are exceptions, i.e., they concern objects belonging to bi-cluster (C^o, C^p) that are characterized by some properties from other bi-clusters.

4   Experimental Validation

First, we applied our characterization method to the well-known benchmark voting-records [19]. It contains 435 objects and 48 Boolean attributes (removing class variables). We used COCLUSTER [6] to get 2 bi-clusters:

bi-cluster	$|\tau|$	rep.	dem.	$|\gamma|$
bi-cluster1	193	153	40	16
bi-cluster2	242	15	227	32
total	435	168	267	48

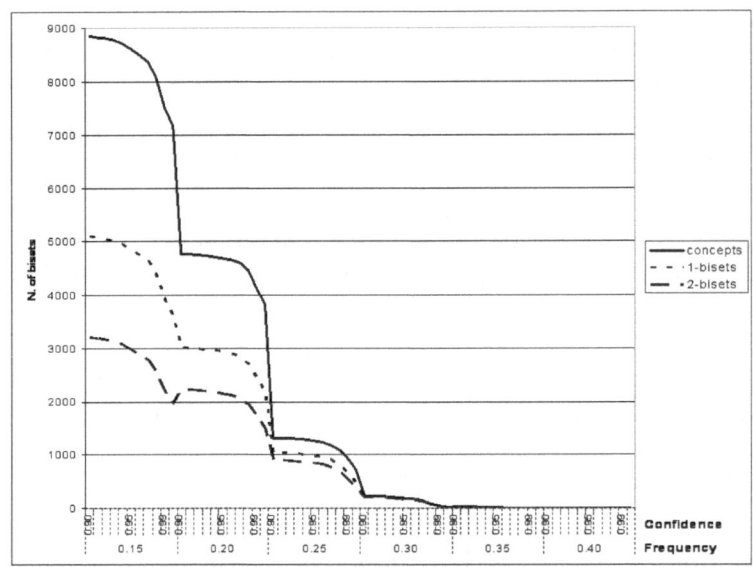

Fig. 2. Characterizing patterns for bi-cluster1 in voting-records w.r.t. different values of minimal frequency and confidence

To characterize each bi-cluster, we used D-MINER [11] to extract all formal concepts, and our slight extension of ACMINER to extract two collections δ-bi-sets (δ=1,2). We obtained 227 031 formal concepts, 130 313 1-bi-sets and 66 908 2-bi-sets. The collections have been post-processed by looking for rules with increasing values of the relative minimal frequency (15% up to 40%) and confidence (90% up to 100%). Results for the first bi-cluster are in Fig. 2. Results for the second one look similar. The number of characterizing rules decreases when we increase the frequency and confidence thresholds. When we use δ-bi-sets, we have to process significantly smaller collections. Two examples of characterizing rules which are consistent with the domain knowledge associated to voting-records are now given. The first one (resp. the second) has a 42% relative frequency (resp. 31%) and both have a 100% confidence, i.e., we have $\epsilon_o = 0$.

```
el-salvador-aid = yes ∧ anti-satellite-test-ban = yes
            ∧ aid-to-nicaraguan-contras = yes ⇒ bi-cluster2
handicapped-infants = no ∧ physician-fee-freeze = yes
            ∧ el-salvador-aid = yes ⇒ bi-cluster1
```

Then, we applied the method to the real world medical data set meningitis already used in [18]. It has been gathered from children hospitalized for acute meningitis. The pre-processed Boolean data set is composed of 329 examples described by 60 Boolean attributes encoding clinical signs (hemodynamic troubles, consciousness troubles, . . .), cytochemical analysis of the cerebrospinal fluid (C.S.F proteins, C.S.F glucose, . . .), and blood analysis (sedimentation rate,

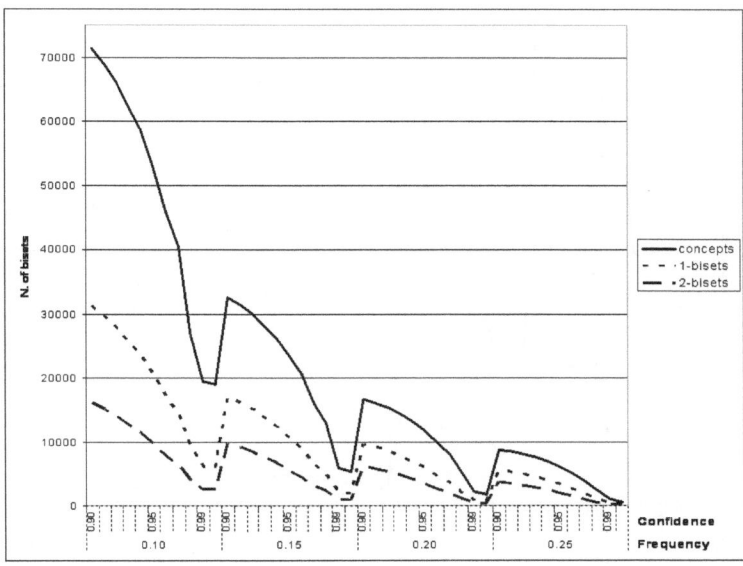

Fig. 3. Characterizing patterns for the bi-cluster2 in meningitis w.r.t. different values of minimal frequency and confidence

white blood cell count, ...). In meningitis, the majority of the cases are known to be viral infections whereas about one quarter are are known to be caused by bacteria. Furthermore, medical knowledge is available which can be used to assess characterization relevancy. Using COCLUSTER, we got two bi-clusters:

bi-cluster	$|\tau|$	bact.	vir.	$|\gamma|$
bi-cluster1	100	81	19	21
bi-cluster2	229	3	226	39
total	329	84	245	60

The first bi-cluster contains a majority of bacterial cases while the second one contains almost only viral cases. We selected characterization rules based on a collection of formal concepts and 2 collections of δ-bi-sets (δ=1,2). We obtained the results in Fig. 3. Here again, using δ-bi-sets leads to smaller collections of candidate characterization patterns. The number of characterization rules for the first bi-cluster is always very low and it does not significantly change when using δ-bi-sets instead of formal concepts. If we select the rules with a minimal body, a 10% frequency threshold, a 98% confidence threshold, and for which the property exception ratio ϵ_p is zero, we obtain only 9 rules which are consistent with the medical knowledge (see [18] for details). Examples of rules are:

```
presence of bacteria in C.S.F. analysis = yes ⇒ bi-cluster1
polynuclear percent > 80 ∧ C.S.F. proteins > 0.8 ⇒ bi-cluster1
C.S.F. proteins > 0.8 ∧ C.S.F. glucose < 1.5 ⇒ bi-cluster1
```

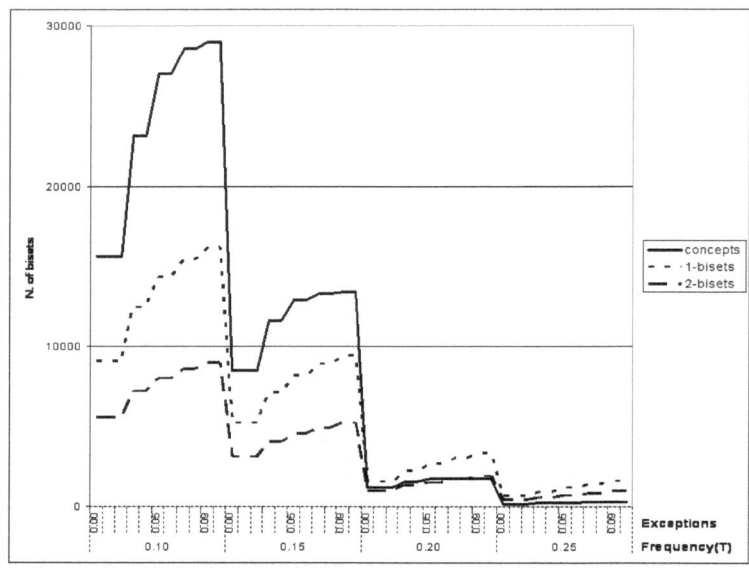

Fig. 4. Characterizing bi-sets for bi-cluster1 in plasmodium w.r.t. different values of minimal size and maximal exception ratio

Finally, our last experiment concerns the analysis of plasmodium, a public gene expression data set concerning Plamodium falciparum (i.e., a causative agent of human malaria) described in [20]. It records the expression profile of 3 719 genes in 46 biological samples. Each sample corresponds to a time point of the developmental cycle. It is divided into 3 phases: the ring, the trophozoite and the schizont stages. The numerical expression data have been preprocessed by using one of the property encoding methods described in [21]. We used Co-CLUSTER to get the following bi-clusters.

bi-cluster	$|\tau|$	ring	troph	schiz.	$|\gamma|$
bi-cluster1	20	15	5	0	558
bi-cluster2	16	0	5	11	1699
bi-cluster3	10	6	0	4	1462
total	46	21	10	15	3719

We extracted collections of bi-sets to characterize clusters of samples by means of sets of genes. In this case however, the number of properties (columns) was too large to be processed and we extracted the collections of δ-bi-sets on the transposed matrix. Obviously, the frequency and confidence measure do not make sense any more because they are computed on sets of samples and we are looking for sets of genes. Therefore, we have used the size of the bi-sets $|T|$ and $|G|$, and their exception ratios ϵ_o and ϵ_p. Results for a minimal size from 10% up to 25% of $\mathcal{O}$ and for maximal values of ϵ_o from 0% up to 10% are in Fig. 4.

Considering bi-cluster1, we analyzed the characterizing 2-bi-sets whose the minimal size for their sets of objects was 25% of $\mathcal{O}$ and for a maximal exception ratio $\varepsilon_o = 0$. Among the 442 bi-sets characterizing bi-cluster1, only 4 of them concern genes that belong to the same bi-cluster. In each of them, we found at least one gene belonging to the cytoplasmic translation machinery group which is known to be active in the ring stage (see [20] for details), i.e., the majority developmental stage within bi-cluster1.

5   Conclusion

We presented a new (bi-)cluster characterization method based on extracted local patterns, more precisely formal concepts and δ-bi-sets. One motivation is that it is now possible to use quite efficient constraint-based mining techniques for various local patterns and it makes sense to consider their multiple uses. While a bi-partition provides a global and generally expected characterization, selected collections of characterizing bi-sets point out local association which might lead to unexpected but relevant information. We strongly believe in the complementarity between global pattern and local pattern mining techniques when considering the whole knowledge discovery process. Our perspective is now to consider the somehow convergent techniques developed for (conceptual) clustering, subgroup discovery [22], summarization by association rules in order to support real-life knowledge discovery processes in functional genomics.

Acknowledgements. The authors wish to thank P. Francois and B. Crémilleux who provided the data set meningitis. They also thank C. Rigotti, J. Besson and C. Robardet for exciting discussions. This research is partially funded by ACI MD 46 (CNRS STIC 2004-2007) BINGO (Bases de Données Inductives pour la Génomique).

References

1. Jain, A., Dubes, R.: Algorithms for clustering data. Prentice Hall, Englewood cliffs, New Jersey (1988)
2. Eisen, M., Spellman, P., Brown, P., Botstein, D.: Cluster analysis and display of genome-wide expression patterns. PNAS **95** (1998) 14863–14868
3. Fisher, D.H.: Knowledge acquisition via incremental conceptual clustering. Machine Learning **2** (1987) 139–172
4. Cheng, Y., Church, G.M.: Biclustering of expression data. In: Proceedings ISMB 2000, San Diego, USA, AAAI Press (2000) 93–103
5. Robardet, C., Feschet, F.: Efficient local search in conceptual clustering. In: Proceedings DS'01. Volume 2226 of LNCS., Springer-Verlag (2001) 323–335
6. Dhillon, I.S., Mallela, S., Modha, D.S.: Information-theoretic co-clustering. In: Proceedings ACM SIGKDD 2003, Washington, USA, ACM Press (2003) 89–98
7. Madeira, S.C., Oliveira, A.L.: Biclustering algorithms for biological data analysis: A survey. IEEE/ACM Trans. Comput. Biol. Bioinf. **1** (2004) 24–45
8. Wille, R.: Restructuring lattice theory: an approach based on hierarchies of concepts. In Rival, I., ed.: Ordered sets. Reidel (1982) 445–470

9. Boulicaut, J.F., Bykowski, A., Rigotti, C.: Free-sets: a condensed representation of boolean data for the approximation of frequency queries. Data Mining and Knowledge Discovery journal **7** (2003) 5–22
10. Stumme, G., Taouil, R., Bastide, Y., Pasqier, N., Lakhal, L.: Computing iceberg concept lattices with TITANIC. Data & Knowledge Engineering **42** (2002) 189–222
11. Besson, J., Robardet, C., Boulicaut, J.F.: Constraint-based mining of formal concepts in transactional data. In: Proceedings PaKDD'04. Volume 3056 of LNAI., Sydney (Australia), Springer-Verlag (2004) 615–624
12. Besson, J., Robardet, C., Boulicaut, J.F.: Mining formal concepts with a bounded number of exceptions from transactional data. In: Proceedings KDID'04. Volume 3377 of LNCS., Springer-Verlag (2004) 33–45
13. Agrawal, R., Imielinski, T., Swami, A.: Mining association rules between sets of items in large databases. In: Proceedings of ACM SIGMOD'93, Washington, D.C., USA, ACM Press (1993) 207–216
14. Boulicaut, J.F., Bykowski, A., Rigotti, C.: Approximation of frequency queries by mean of free-sets. In: Proceedings PKDD'00. Volume 1910 of LNAI., Lyon, F, Springer-Verlag (2000) 75–85
15. Crémilleux, B., Boulicaut, J.F.: Simplest rules characterizing classes generated by delta-free sets. In: Proceedings ES 2002, Cambridge, UK (2002) 33–46
16. Liu, B., Hsu, W., Ma, Y.: Integrating classification and association rule mining. In: Proceedings KDD'98, New York, NY (1998) 80–86
17. Li, W., Han, J., Pei, J.: CMAR: Accurate and efficient classification based on multiple class-association rules. In: Proceedings ICDM'01, San Jose, CA (2001) 369–376
18. Robardet, C., Crémilleux, B., Boulicaut, J.F.: Characterization of unsupervized clusters by means of the simplest association rules: an application for child's meningitis. In: Proceedings IDAMAP'02 co-located with ECAI'02, Lyon, F (2002) 61–66
19. Blake, C., Merz, C.: UCI repository of machine learning databases (1998)
20. Bozdech, Z., Llinás, M., Pulliam, B.L., Wong, E., Zhu, J., DeRisi, J.: The transcriptome of the intraerythrocytic developmental cycle of plasmodium falciparum. PLoS Biology **1** (2003) 1–16
21. Pensa, R.G., Leschi, C., Besson, J., Boulicaut, J.F.: Assessment of discretization techniques for relevant pattern discovery from gene expression data. In: Proceedings ACM BIOKDD'04, Seattle, USA (2004) 24–30
22. Gamberger, D., Lavrac, N.: Expert-guided subgroup discovery: Methodology and application. JAIR **17** (2002) 501–527

Machine-Learning with Cellular Automata

Petra Povalej, Peter Kokol, Tatjana Welzer Družovec, and Bruno Stiglic

Faculty of Electrical Engineering and Computer Science,
Smetanova ulica 17, 2000 Maribor, Slovenia
{Petra.Povalej, Kokol, Welzer, Stiglic}@uni-mb.si

Abstract. As the possibility of combining different classifiers into Multiple Classifier System (MCS) becomes an important direction in machine-learning, difficulties arise in choosing the appropriate classifiers to combine and choosing the way for combining their decisions. Therefore in this paper we present a novel approach – Classificational Cellular Automata (CCA). The basic idea of CCA is to combine different classifiers induced on the basis of various machine-learning methods into MCS in a non-predefined way. After several iterations of applying adequate transaction rules only a set of the most appropriate classifiers for solving a specific problem is preserved.

We empirically showed that the superior results compared to AdaBoost ID3 are a direct consequence of self-organization abilities of CCA. The presented results also pointed out important advantages of CCA, such as: problem independency, robustness to noise and no need for user input.

1 Introduction

In recent decades researchers have developed several machine-learning approaches and many of them were successfully moved from research laboratories into practice. The number of applications in different fields is still rising proving the growing necessity for the development and use of machine-learning tools. However, the majority of the most successful applications of machine-learning are made specifically to solve a fixed problem and thus the result from experienced usage of expert knowledge. The most likely reason for that is a vast and still growing availability of machine-learning models, their complexity and the lack of methods to compare them. For example: several data-mining tools already implement different machine-learning methods but the choice of the appropriate method or combination methods to solve a specific problem is still left to the user. Therefore the user must possess a lot of knowledge and experience to choose a method which would produce the best results for a specific problem. On the other hand, the user also has to possess knowledge for using the methods available and to interpret the obtained results. Obviously, the practical application of such necessary technology is rather demanding and expensive for experienced and for inexperienced users alike.

For the purpose of diminishing the problem described above, we present our novel idea of Classificational Cellular Automata (CCA). CCA takes advantage of the benefits of Multiple Classifier Systems (MCS) and the self-organizing abilities of cellular automata with the aim to improve its performance.

A.F. Famili et al. (Eds.): IDA 2005, LNCS 3646, pp. 305–315, 2005.

In CCA, multiple machine-learning approaches (such as: decision rules, neural networks, decision trees, Support Vector Machines (SVM), etc.) are implemented to obtain a diversity of induced classifiers combined in MCS.

This paper is organized as follows. In the next section different methods for combining classifiers into MSC are discussed. The basics of cellular automata (CA) are shortly presented in section 3. The new Classificational Cellular Automata approach is presented in section 4. Empirical evaluation of different CCA methods and comparison to Quinlan's ID3 and AdaBoost ID3 using data from UCI repository is presented in section 5. The paper concludes with some final remarks and directions for future work.

2 Multiple Classifier Systems

In recent years there has been a growing interest in the area of combining classifiers into MCS (also known as ensembles or committees). An important characteristic of MCS is that using the classification capabilities of multiple classifiers (experts), where each classifier may make different and perhaps complementary errors, tends to yield an improved performance over single experts. Therefore many researchers have focused on developing various approaches for combining classifiers by selection and/or fusion. However, the diversity of combined classifiers is emphatically a key factor for the success of the combination approach.

Some MSC approaches actively try to perturb some aspects of the training set, such as training samples, attributes or classes, in order to force classifier diversity. One of the most popular perturbation approaches are Bootstrap Aggregation – Bagging and Boosting. Bagging, first introduced by Breiman [1] in 1996, works by manipulating the training samples and forming replicate training sets. The final classification is based on a majority vote.

Boosting was introduced by Freund and Schapire in 1996 [2]. Boosting combines classifiers with weighted voting and is more complex since the distribution of training samples in the training set is adaptively changed according to the performance of sequentially constructed classifiers.

3 Cellular Automata

The concept of cellular automata was firstly proposed in early 1960's by J. Von Neumann [3] and Stan Ulam. From those early years to the recent Wolfram's book "A New Kind of Science" [4], the CA have attracted researchers from all science branches – physical and social. The reasons for the popularity of CA are their simplicity and the enormous potential for modeling complex systems.

CA can be viewed as a simple model of a spatially extended decentralized system made up of a number of individual components (cells) [5]. Each individual cell is in a specific state which changes through the time depending on the state of neighborhood cells and according to the transaction rules. In spite of their simplicity, when iterated several times, the dynamics of CA are potentially very rich, and range from attracting stable configurations to spatio-temporal chaotic features and pseudo-random generation abilities. Those abilities enable a diversity that can possibly overcome local optima when solving engineering problems. Moreover, from the computational viewpoint, they are universal, and as powerful as Turing machines and, thus, classical Von Neumann architectures. These structural and dynamical features make them very powerful: fast CA-based algorithms are developed to solve engineering problems in cryptography and microelectronics for instance, and theoretical CA-based models are built in ecology, biology, physics and image-processing. On the other hand, these powerful features make CA difficult to analyze. Almost all long-term behavioral properties of dynamical systems, and cellular automata in particular, are unpredictable. However, in this paper the aim is not to analyze the process of CA rather to use it for superior classification tasks.

3.1 Cellular Automata as a MSC Model

In general MCS approaches can be divided into two groups: (1) MCS approaches that combine different independent classifiers (such as: Bayesian Voting, Majority Voting, etc.) and (2) MCS approaches which construct a set of classifiers on the basis of one base classifier with perturbation of training set (such as: Bagging, Boosting, Windowing, etc.). A detailed empirical study is presented in [6].

When studying both groups we came across some drawbacks. The most essential deficiency in group 1) is their restriction of a predefined way of combining classifiers induced on the basis of different predefined methods of machine-learning. On the contrary, the MCS approaches that are based on improving one base classifier group (2), use only one method for constructing all classifiers in MCS. Therefore the problem of choosing the appropriate method for solving a specific task arises.

Therefore our idea was to combine different classifiers induced on the basis of various methods of machine-learning into MCS in a non-predefined way. After several iterations of applying adequate transaction rules only the set of the most appropriate classifiers will be preserved. Consequently the problem of choosing the right machine learning method or a combination of them would be solved automatically.

4 Classificational Cellular Automata

CCA is presented as a classifier. Generally, learning a classifier is based on samples from a learning set. Every learning sample is completely described by a set of attributes (sample properties) and class (decision).

CCA is initially defined as a 2D lattice of cells. Each cell can contain a classifier and according to it's classification of an input sample the cell can be in one of the k states, where k is the number of possible classes and the state "can not classify". The last state has an especially important role when such a classifier is used, which is not defined on a whole learning set, i. e. when using the *if-then* rule. Therefore, from the classification point of view in the learning process the outcome of each cell can be: (1) the same as the learning sample's class, (2) different from the learning sample's class or (3) "cannot be classified". However, a cell with an unknown classification for the current learning sample should be treated differently as a misclassification.

In addition to the cell's classification ability, the neighborhood plays a very important role in the self-organization ability of a CCA. Transaction rules depend on the specific neighborhood state to calculate a new cell's state and must be defined in such a way that enforces self-organization of CCA, which consequently should lead to generalization process.

In general we want to group classifiers that support similar hypotheses and consequently have similar classification on learning samples. Therefore even if a sample is wrongly classified, the neighborhood can support a cell by preventing the elimination of its classifier from the automata. With that transaction rule we encourage a creation of decision centers for a specific class and thus we can overcome the problem of noisy learning samples.

As for all MCS approaches it is clear that if all classifiers are identical or even similar, there can be no advantage in combining their decisions, therefore some difference among base classifiers is a necessary condition for improvement. The diversity of classifiers in CCA cells is ensured by using different machine-learning methods for classifier induction. However, there is only a limited number of machine-learning methods, which can be a problem for a large CCA. But most methods have some tuning parameters that affect classification and therefore, by changing those parameters, many different classifiers can be obtained. Another possibility for obtaining several different classifiers is by changing the expected probability distributions of the input samples, which may also result in different classifiers, even by using the same machine learning method with the same parameters.

4.1 Learning Algorithm

Once the diversity of induced classifiers is ensured by the methods discussed above, the classifiers are placed into a pool. For each classifier the basics statistical information such as confidence and support is preserved. In the process of filling CCA for each cell a classifier is randomly chosen from the pool. After filling a CCA the following learning algorithm is applied:

```
Input: learning set with N learning samples
Number of iterations: t=1,2,…T
For t=1,2…T:
    -choose a learning sample I
    -each cell in automaton classifies the learning sam-
     ple I
    -change cells energy according to the transaction
     rules
    -a cell with energy bellow zero does not survive
    -probabilistically fill the empty cells with classi-
     fiers from the pool
```

Beside its classifier information, each cell also contains statistical information about its successfulness in a form of cell's energy. Transaction rules can increase or decrease the energy level depending on the success of classification and on the state of the cell's neighbors. Each neighborhood cell influences the energy of the current cell dependent on its *score* (Eq. 1).

$$score = support \bullet conficence \bullet distance ; \tag{1}$$

where *distance* is an Euclidian distance from the neighborhood cell to the current cell.

The sum of *scores* of all neighborhood cells that equally classified the learning sample as the current cell (*eqScore*) is used in transaction rules to calculate the cells new energy (*e*). Similarly, the sum of *scores* of all the neighborhood cells which can not classify the learning sample (*noClassScore*) is calculated and used in the following transaction rules:

- If a cell has the same classification as the sample class:

  (a) if *noClassScore*>0 than increase the energy of the cell using the equation (Eq. 2).

  $$e = e + \left(100 - \frac{eqScore \bullet 100}{noClassScore}\right) \tag{2}$$

  (b) if all neighbourhood cells classified the learning sample (*noClassScore=0*) than increase cell's energy according to (Eq. 3).

  $$e = e + 400 \tag{3}$$

- If a cell classification differs from the learning sample class:

  (a) if *noClassScore*>0 than decrease energy of the cell using (Eq. 4).

  $$e = e - \left(100 - \frac{eqScore \bullet 100}{noClassScore}\right) \tag{4}$$

  (b) if *noClassScore=0* than decrease cell's energy using (Eq. 5).

  $$e = e - 100 \tag{5}$$

- If a cell cannot classify the learning sample then slightly decrease the energy state of the cell (Eq. 6).

  $$e = e - 10 \tag{6}$$

Through all iterations all cells use one point of energy (to live). If energy drops below zero the cell is terminated (blank cell). A new cell can be created dependent on learning algorithm parameters with its initial energy state and a classifier used from the pool of classifiers or from a newly generated classifier. Of course if a cell is too different from the neighborhood it will ultimately die out and the classifier will be returned to the pool.

The learning of a CCA is done incrementally by supplying samples from the learning set. Transaction rules are executed first on the whole CCA with a single sample and then continued with the next until the whole problem is learned by using all samples - that is a technique similar to that used in neural networks [7].

Transaction rules do not directly imply learning, but the iteration of those rules creates the self-organizing ability. However, this ability depends on classifiers used in CCAs cells, and its geometry. Stopping criteria can be determined by defining fixed number of iterations or by monitoring accuracy.

4.2 Inference Algorithm

An Inference algorithm differs from a learning algorithm, because it does not use self-organization.

```
Input: a sample for classification
Number of iterations: t=1,2,…V
For t=1,2…V
    -each cell in automaton classifies the sample
    -change cells energy according to the transac-
    tion rules
    -each cell with energy bellow zero does not sur-
    vive
Classify the sample according to the weighted vot-
ing of the survived cells
Output: class of the input sample
```

The simplest way to produce single classification would be to use the majority voting of cells in CCA. However some votes can be very weak from the transaction rule point of view. Therefore transaction rules which consider only the neighborhood majority vote as a sample class are used in order to eliminate all weak cells. After several iterations of the transaction rules only cells with strong structural support survive. The final class of an input sample is determined by a weighted voting where the energy state of each survived cell is considered as a weight.

5 Empirical Evaluations

The CCA model was evaluated on a collection of 9 randomly chosen datasets from the UCI Machine Learning Repository [8].

CCA was initialized with the following parameters, which experimentally had proved to produce best results:

- Size: 10x10 matrix (bounded into torus)
- Neighborhood radius $r=5$
- Initial energy of each cell $e=100$
- Ending criteria: number of iterations $T=V=1000$

Through all iterations all cells use one point of energy (to live). If the energy level of a cell drops below zero the cell is terminated.

In the experiments presented in this section we used the following evaluation criteria: accuracy (Eq. 7) and average class accuracy (Eq. 8).

$$accuracy = \frac{num\ of\ correctly\ classified\ objects}{num.\ of\ all\ objects} \tag{7}$$

$$average\ class\ accuracy = \frac{\sum_c accuracy_c}{num.\ of\ classes} \tag{8}$$

where $accuracy_c$ is defined in (Eq. 9).

$$accuracy_c = \frac{num\ of\ correctly\ classified\ objects\ in\ class\ c}{num.\ of\ all\ objects\ in\ class\ c} \tag{9}$$

5.1 Simple CCA

In our first experiment we followed the basic concept of CA – simplicity can produce complex behavior. Therefore we introduced simple rule classifiers in the following form:

$$\textbf{if } (attribute \leq value) \textbf{ then } decision ;$$

where *decision* is a class of a current sample.

A CCA which combines simple classifiers in the form of rules is therefore named Simple CCA. The rule classifiers are randomly created and saved in the pool. In the process of filling the CCA's cells classifiers are randomly selected from the pool. After filling the CCA the learning algorithm followed by the inference algorithm is applied.

The classification accuracy of Simple CCA on testing sets can be seen in Table 1 and Table 2. When compared to Quinlan's ID3, Simple CCA performed better on 5 databases and in one case it achieved equal results.

5.2 Complex CCA

The next experiment involved implementing more complex CCA where different classifiers induced on the basis of various machine-learning methods are included. The demand for classifier diversity forced an implementation of several different machine-learning methods.

At this point we implemented the following methods:

- Greedy decision tree induction methods based on different purity measures (for detailed information please see [9]): Information gain ratio (ID3), Gini, Chi-square, J-measure, sum of pairs, linear combination of purity measures and voting of purity measures.
- AdaBoost method for boosting decision trees induced on the basis of purity measures listed above
- Support Vector Machine (SVM)
- Neural Network

Al together 65 different classifiers are induced and placed in the pool of classifiers before initializing CCA. In the process of filling CCA each classifier is randomly chosen from the pool and placed into the selected cell.

The results on 9 databases are showed in Table 1 and Table 2 where they can be compared to other methods presented in the paper.

Closer comparison implies that in one third of databases the benefits from complex classifiers can be observed (see Graph 1). However, when an average accuracy through all databases is considered a small, but not significant benefit compared to Simple CCA is noticed. Since inducing Complex CCA is much more time and source consuming the legitimate question can be posed whether such a little difference is worth the additional computational time.

At this point, we demonstrate that simplicity is really a key concept in using cellular automata.

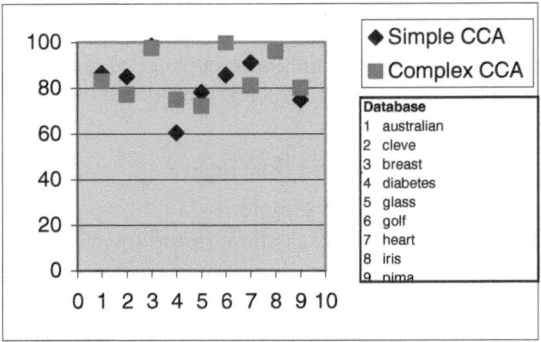

Fig. 1. Comparison between Simple and Complex CCA according to the testing accuracies

5.3 Comparison to Boosting

Since Boosting is one of the most successful methods for combining classifiers, a comparison to CCA approach was made. Therefore the AdaBoost algorithm [2] based on ID3 decision tree induction with 10 individual classifiers was used on all databases (see Table 1 and Table 2).

Table 1. Comparison of different methods in the terms of accuracy on the test set

Data	ID3	Ada Boost ID3	Simple CCA	Boosted ID3 CCA	Complex CCA
australian	77,39	84,78	86,52	83,91	83,04
cleve	92,70	96,14	85,14	76,24	77,23
breast	71,29	74,26	98,48	95,71	97,43
diabetes	68,75	66,41	60,55	73,44	75,00
glass	58,33	69,44	78,18	78,18	72,22
golf	100,00	100,00	85,71	100,00	100,00
heart	68,89	72,22	91,11	85,56	81,11
iris	96,00	92,00	96,00	94,00	96,00
pima	68,36	72,66	75,00	74,60	80,08
Average	**77,97**	**80,88**	**84,08**	**84,63**	**84,68**

Table 2. Comparison of different methods in the terms of average class accuracy on the test set

Data	ID3	Ada Boost ID3	Simple CCA	Boosted ID3 CCA	Complex CCA
australian	76,99	84,78	87,86	84,42	83,33
cleve	88,94	95,58	83,91	77,30	77,05
breast	72,81	72,58	97,85	96,95	97,75
diabetes	65,57	61,22	50,00	70,82	71,77
glass	43,67	60,93	76,97	76,97	56,93
golf	100,00	100,00	80,00	100,00	100,00
heart	68,83	72,42	90,54	86,08	80,82
iris	96,08	92,38	96,08	94,22	96,08
pima	64,05	67,23	68,96	71,18	76,49
Average	**75,22**	**78,57**	**81,35**	**84,22**	**82,25**

Table 1 presents the test accuracy of induced classifiers on 9 databases. When compared to AdaBoost ID3, Simple CCA was more accurate in 5 cases. However, Complex CCA was even more successful, since it outperformed AdaBoost ID3 in 6 cases and in one case it obtained the same result. When the average class accuracy is considered the results are similar (see Table 2).

On average through all 9 databases both Simple CCA and Complex CCA performed better than AdaBoost ID3. However, we wanted to prove that better performance is a direct consequence of CCA's self-organization abilities. Therefore in the next experiment we included only classifiers from the AdaBoost MCS into the pool and used them as a source of diverse classifiers for CCA (Boosted ID3 CCA). The results are also presented in the Table 1 and Table 2. A closer look shows that AdaBoost ID3 outperformed Boosted ID3 CCA only on the first two databases. Graph 1 shows a more direct comparison of both methods in terms of the average class accuracy. Since in both methods the same individual classifiers were used as a base, we can conclude, that improvement of classification accuracy results only from organization and transaction rules of cellular automata.

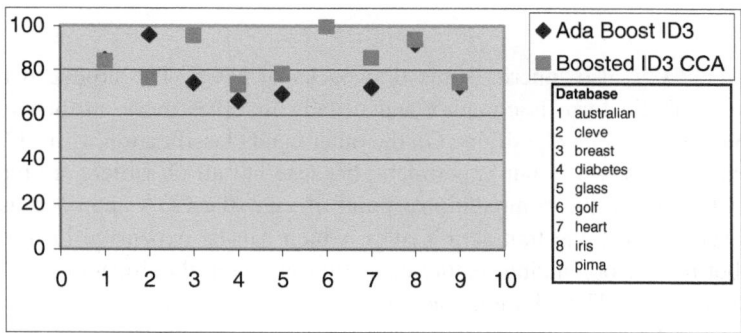

Fig. 2. Comparison between AdaBoost ID3 and Boosted ID3 CCA according to accuracy on test set

5   Discussion and Conclusion

In this paper we presented a new approach for combining diverse classifiers induced on the basis of various machine-learning methods into MCS using the model of cellular automata. The CCA approach was empirically evaluated on 9 randomly chosen datasets from UCI Repository.

The key observation from the experiments with the presented CCA approach is that the self-organization ability of CCA is very promising.

In Graph 3 the average classification accuracy on the test sets is presented for all CCA methods and additionally for ID3 and AdaBoost ID3. We can see that on average all CCA methods performed better compared to ID3 and AdaBoost ID3. Boosted ID3 CCA method performed best according to average class accuracy and it also has negligible difference between average accuracy and class accuracy. The direct comparison to AdaBoost ID3 showed that improvement of classification accuracy results only from organization and transaction rules of cellular automata.

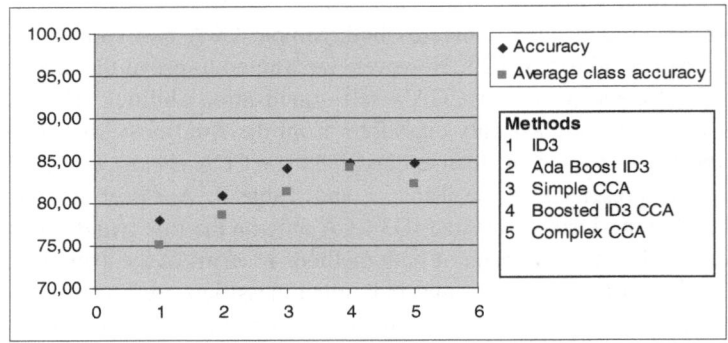

Fig. 3. Comparison of average accuracy and class accuracy through all 9 databases among all methods

However, CCA also inherits some drawbacks of MCS. The produced result although in symbolic form is complex and usually involves more attributes than the simpler but less accurate classifiers. On the other hand classification with CCA can be even cheaper, as shown in our experiment, because not all classifiers are required in the final CCA. From the computational point of view the CCA approach uses additional power to apply the transaction rules, which can be expensive in the learning process, but its self-organizing feature can result in better classification, and that can also mean less cost. The additional advantages of the resulting self-organizing structure of cells in CCA is problem independency, robustness to noise and no need for user input. A comparison to other approaches is presented in Table 3.

Important research directions in the future are to analyze the resulting self-organized structure, the impact of transaction rules on classification accuracy, the introduction of other social aspects for cell survival and enlarging the classifier diversity by implementing even more machine-learning methods.

Table 3. Advantages of CCA compared to other approaches

Method	automatically chosen machine-learning method	non-predefined way of combining classifiers	deterministic	global optimum	required user input
Single method approach	-		+	o	o
MSC with independent classifiers	O	-	+	+	o
MSC with basic classifier	-	-	+	+	o
CCA	+	+	-	+	+

References

1. Breiman, L.: Bagging Predictors. Machine Learning. Vol. 24. No. 2 (1996) 123-140 [http://citeseer.ist.psu.edu/breiman96bagging.html]
2. Freund, Y., Schapire, R.E.: Experiments with a new boosting algorithm. In: Proceedings Thirteenth International Conference on Machine Learning, Morgan Kaufman, San Francisco (1996) 148-156
3. Von Neumann, J.: Theory of Self-Reproducing Automata. A. W. Burks (ed.), Univ. of Illinois Press, Urbana and London (1966)
4. Wolfram, S.: A new kind of science. Wolfram Media (2002)
5. Ganguly, N., Sikdar, B. K., Deutsch, A., Canright, G., Chaudhuri, P. P.: A survey on cellular automata. Technical Report, Centre for High Performance Computing, Dresden University of Technology (2003) [http://www.cs.unibo.it/bison/pub.shtml]
6. Dietterich, T.G.: Ensemble Methods in Machine Learning. In: Kittler, J., Roli, F. (Ed.) First International Workshop on Multiple Classifier Systems, Lecture Notes in Computer Science. Springer Verlag, New York (2000) 1-15
7. Towel, G., Shavlik, J.: The Extraction of Refined Rules From Knowledge Based Neural Networks. Machine Learning (1993) 131, 71-101
8. Blake, C. L., Merz, C. J.: UCI Repository of machine learning databases, Irvine, CA: University of California, Department of Information and Computer Science.[http://www.ics.uci.edu/~mlearn/MLRepository.html].
9. Lenič, M., Povalej, P., Kokol, P.: Impact of purity measures on knowledge extraction in decision trees. In: Lin Tsau Y., Ohsuga Setsuo (Ed.). Third IEEE International conference on data mining, Foundations and new directions in data mining : Workshop notes. [S.l.]: IEEE Computer society (2003) 106-111

MDS$_{polar}$: A New Approach for Dimension Reduction to Visualize High Dimensional Data

Frank Rehm[1], Frank Klawonn[2], and Rudolf Kruse[3]

[1] German Aerospace Center, Braunschweig, Germany
[2] University of Applied Sciences, Braunschweig/Wolfenbuettel, Germany
[3] Otto-von-Guericke-University of Magdeburg, Germany

Abstract. Many applications in science and business such as signal analysis or costumer segmentation deal with large amounts of data which are usually high dimensional in the feature space. As a part of preprocessing and exploratory data analysis, visualization of the data helps to decide which kind of method probably leads to good results. Since the visual assessment of a feature space that has more than three dimensions is not possible, it becomes necessary to find an appropriate visualization scheme for such datasets. In this paper we present a new approach for dimension reduction to visualize high dimensional data. Our algorithm transforms high dimensional feature vectors into two-dimensional feature vectors under the constraints that the length of each vector is preserved and that the angles between vectors approximate the corresponding angles in the high dimensional space as good as possible, enabling us to come up with an efficient computing scheme.

1 Introduction

Many applications in science and business such as signal analysis or costumer segmentation deal with large amounts of data which are usually high dimensional in the feature space.

Before further analysis or processing of data is carried out with more sophisticated data mining techniques, data preprocessing and exploratory data analysis is an important step. As a part of this process, visualization of the data helps to decide which kind of method probably leads to good results. Since the visual assessment of a feature space that has more than three dimensions is not possible, it becomes necessary to find an appropriate visualization scheme for such datasets.

The general data visualization problem we consider here is to map high dimensional data to a two-dimensional plane – usually a computer screen – trying to preserve as many properties or as much information of the high dimensional data as possible. In other words, we have to face a dimension reduction problem. A very simple approach is to look at scatter plots obtained from projections to two selected features. However, each scatter plot will only contain the information of the two chosen features and with a high number of features it is infeasible to inspect the scatter plots resulting from all possible combinations of two features.

A.F. Famili et al. (Eds.): IDA 2005, LNCS 3646, pp. 316–327, 2005.

Before finding a mapping of the high dimensional data to the two-dimensional plane, an error or quality measure must be defined in order to evaluate the suitability of such a mapping. Principal component analysis is one possible choice, producing an affine transform that preserves as much of the variance in the data as possible. However, instead of the variance, other criteria like the distance between the single data vectors might be of higher interest. Multidimensional scaling (see for instance [1,4]) is a technique that aims at preserving the distances between the data, when mapping them to lower dimensions. Although multidimensional scaling and related approaches yield promising and interesting results, they suffer from high computational needs concerning memory as well as computation time. In recent years some research has been done in this regard [2,6,7].

In this paper we present a new approach for dimension reduction to visualize high dimensional data. Instead of trying to preserve the distances between feature vectors directly, our algorithm transforms high dimensional feature vectors into two-dimensional feature vectors under the constraints that the length of each vector is preserved and that the angles between vectors approximate the corresponding angles in the high dimensional space as good as possible, enabling us to come up with an efficient computing scheme. After a brief review of the concept of multidimensional scaling and related approaches, we explain the theoretical background of our approach and discuss some illustrative examples in the end of the paper.

2   Multidimensional Scaling

Multidimensional scaling (MDS) is a method that estimates the coordinates of a set of objects $Y = \{y_1, \ldots, y_n\}$ in a feature space of specified (low) dimensionality that come from data $X = \{x_1, \ldots, x_n\} \subset \Re^p$ trying to preserve the distances between pairs of objects. Different ways of computing distances and various functions relating the distances to the actual data are commonly used. These distances are usually stored in a distance matrix

$$D^x = \left(d_{ij}^x\right), \quad d_{ij}^x = \|x_i - x_j\|, \quad i, j = 1, \ldots, n.$$

The estimation of the coordinates will be carried out under the constraint, that the error between the distance matrix D^x of the dataset and the distance matrix $D^y = \left(d_{ij}^y\right)$, $d_{ij}^y = \|y_i - y_j\|$, $i, j = 1, \ldots, n$ of the corresponding transformed dataset will be minimized.

Thus, different error measures to be minimized were proposed, i.e. the absolute error, the relative error or a combination of both. A commonly used error measure, the so called *Sammon's mapping*

$$E = \frac{1}{\sum\limits_{i=1}^{n} \sum\limits_{j=i+1}^{n} d_{ij}^x} \sum\limits_{i=1}^{n} \sum\limits_{j=i+1}^{n} \frac{\left(d_{ij}^y - d_{ij}^x\right)^2}{d_{ij}^x}$$

describes the absolute and the relative quadratic error. To determine the transformed dataset Y by means of minimizing error E a gradient descent method can be used. By means of this iterative method, the searched parameter y_k, will be updated during each step proportional to the gradient of the error function E. Calculating the gradient of the error function leads to

$$\frac{\partial E}{\partial y_k} = \frac{2}{\sum\limits_{i=1}^{n} \sum\limits_{j=i+1}^{n} d_{ij}^x} \sum_{j \neq k} \frac{d_{kj}^y - d_{kj}^x}{d_{kj}^x} \frac{y_k - y_j}{d_{kj}^y}.$$

After random initialization, for each projected feature vector y_k a gradient descent is carried out and the distances d_{ij}^y as well as the gradients $\frac{\partial d_{ij}^y}{\partial y_k}$ will be recalculated again. The algorithm terminates when E becomes smaller than a certain threshold.

The complexity of MDS is $O(c \cdot n^2)$, where c is the (unknown) number of iterations needed for convergence of the gradient descent scheme. Thus MDS is usually not applicable to larger datasets. Another problem of MDS is that it does not construct an explicit mapping from the high dimensional space to the lower dimensional space, but just tries to position the lower dimensional feature vectors in a suitable way. Therefore, when new data have to be considered, they cannot be mapped directly into the lower dimensional space, but the whole MDS procedure has to be repeated. NeuroScale [5] is a scheme that tries to construct an explicit mapping for MDS in the form of a neural network. However, it does not reduce the complexity of MDS. In [3] a more efficient, but still iterative approach was proposed making use of a step-by-step reduction by one dimension based on determining the best projection in each step.

In this paper, we propose a different algorithm, not needing any iterative scheme and whose complexity can be reduced to $O(n \cdot \log n)$.

3   Multidimensional Scaling with Polar Coordinates

Multidimensional scaling suffers from several problems. Besides the quadratic need of memory, MDS, as described above is solved by an iterative method, expensive with respect to computation time. Furthermore, a completely new solution must be calculated, if a new object is added to the dataset.

With MDS_{polar} we present a new approach to find a two-dimensional projection of a p-dimensional dataset X. MDS_{polar} tries to find a representation in polar coordinates $Y = \{(l_1, \varphi_1), \ldots, (l_n, \varphi_n)\}$, where the length l_k of the original vector x_k is preserved and only the angle φ_k has to be optimized. Thus, our solution is defined to be optimal if all angles between pairs of data objects in the projected dataset Y coincide as good as possible with the angles in the original feature space X.

A straight forward definition of an objective function to be minimized for this problem would be

$$E = \sum_{k=2}^{n} \sum_{i=1}^{k-1} (|\varphi_i - \varphi_k| - \psi_{ik})^2 \tag{1}$$

where φ_k is the angle of y_k, ψ_{ik} is the positive angle between x_i and x_k, $0 \leq \psi_{ik} \leq \pi$. E is minimal, if the difference of the angle of all pairs of vectors of dataset X and the corresponding two vectors in dataset Y are zero. The absolute value is chosen in equation (1) because the order of the minuends can have an influence on the sign of the resulting angle. The problem with this notation is that the functional E is not differentiable, exactly in those points we are interested in, namely, where the difference between angles φ_i and φ_k becomes zero. Another intuitive approach would be

$$E = \sum_{k=2}^{n} \sum_{i=1}^{k-1} ((\varphi_i - \varphi_k)^2 - \psi_{ik}^2)^2. \tag{2}$$

In this case the derivative can be determined easily, however, resulting in a system of nonlinear equations for which no analytical solution can be provided.

In order to overcome these difficulties, we propose an efficient method that enables us to compute an approximate solution for a minimum of the objective function (1) and related ones. In a first step we ignore the absolute value in (1) and consider

$$E = \sum_{k=2}^{n} \sum_{i=1}^{k-1} (\varphi_i - \varphi_k - \psi_{ik})^2 \tag{3}$$

instead. When we simply minimize (3), the results will not be acceptable. Although the angle between y_i and y_k might perfectly match the angle ψ_{ik}, $\varphi_i - \varphi_k$ can either be ψ_{ik} or $-\psi_{ik}$. Since we assume that $0 \leq \psi_{ik}$ holds, we always have $(|\varphi_i - \varphi_k| - \psi_{ik})^2 \leq (\varphi_i - \varphi_k - \psi_{ik})^2$. Therefore, finding a minimum of (3) means that this is an upper bound for the minimum of (1). Therefore, when we minimize (3) in order to actually minimize (1), we can take the freedom to choose whether we want the term $\varphi_i - \varphi_k$ or the term $\varphi_k - \varphi_i$ to appear in (3). Before we discuss techniques to minimize (3) with the freedom of reordering, we have to preprocess the data in order to fit them best to our approach.

3.1 Data Preprocessing

The following figure illustrates an important problem by means of a simple dataset. The table next to the graphics contains the values of the angles between all three feature vectors.

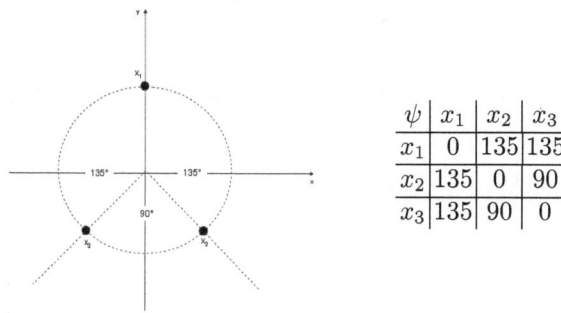

ψ	x_1	x_2	x_3
x_1	0	135	135
x_2	135	0	90
x_3	135	90	0

Even though, this feature space has only two dimensions and therefore an exact reproduction of the dataset should be possible, this cannot be achieved without additional preprocessing. Since we only want to preserve the angles between data vectors, it is obvious that any solution will be invariant with respect to rotation of the dataset. Thus, assuming without loss of generality $\varphi_1 = 0$ enforcing $\varphi_2 = 135$, then according to our objective function (1) $\varphi_3 = 180$ leads to the optimal solution, which is obviously not what we are looking for. This problem is caused by the fact that ψ_{ik} is defined as a positive angle which satisfies $\psi_{ik} \leq 180°$. This problem can be solved easily by translating all feature vectors into the first quadrant. More generally, for a higher dimensional dataset we apply a translation that makes all components of data vectors non-negative. For this we only have to determine for each component the largest negative value occurring in the dataset and using this as a positive value of the corresponding component of the translation vector. Note that, when the dataset is normalized, i.e. all components are between 0 and 1, no further preprocessing is required.

Thus, doing this kind of preprocessing, we actually do not preserve the original data properties but those after the transformation. Of course, rotation and translation is not changing any inter-data properties. The translation vector has to be stored so that for incremental adding of new objects the transformation can be performed accordingly. For most of the new objects the transformation will be as requested. It may occur that for new objects which have one or more extreme components the translation will not be sufficient to eliminate the negative components. In such a case, which is rather rare if the previous data is representative, the mapping of the respective object is still working, but not that exact sometimes.

3.2   Approximation of MDS$_{polar}$

When we are free to choose between $\varphi_i - \varphi_k$ and $\varphi_k - \varphi_i$ in (3), we take the following into account

$$(\varphi_k - \varphi_i - \psi_{ik})^2 = (-(\varphi_k - \varphi_i - \psi_{ik}))^2 = (\varphi_i - \varphi_k + \psi_{ik})^2.$$

Therefore, instead of exchanging the order of φ_i and φ_k, we can choose the sign of ψ_{ik}, leading to

$$E = \sum_{k=2}^{n} \sum_{i=1}^{k-1} (\varphi_i - \varphi_k - a_{ik}\psi_{ik})^2 \qquad (4)$$

with $a_{ik} = \{-1, 1\}$. In order to solve this modified optimization problem of equation (4) we take the partial derivatives of E, yielding

$$\frac{\partial E}{\partial \varphi_k} = -2 \sum_{i=1}^{k-1} (\varphi_i - \varphi_k - a_{ik}\psi_{ik}). \qquad (5)$$

Thus, on the one hand, neglecting that we still have to choose a_{ik}, our solution is described by a system of linear equations which means its solution can be calculated directly without the need of any iteration procedure. On the other hand, as described above, we have to handle the problem of determining the sign of the ψ_{ik} in the form of the a_{ik}-values.

To fulfil the necessary condition for a minimum we set equation (5) equal to zero and solve for the φ_k-values, which leads to

$$\varphi_k = \frac{\sum_{i=1}^{k-1} \varphi_i - \sum_{i=1}^{k-1} a_{ik}\psi_{ik}}{k-1}. \tag{6}$$

Different optimization strategies are conceivable. Of course, an important condition is the computational complexity of the respective approximation algorithm. In this paper we present a number of different strategies, starting with a greedy algorithm which is quadratic with the number of data objects in time, but is linear in space. Later on, we propose an algorithm that can even reduce the complexity to $O(n \cdot \log n)$.

3.3   A Greedy Algorithm for the Approximation of MDS$_{polar}$

As mentioned above, this solution describes a system of linear equations. Since the desired transformation is rotation invariant φ_1 can be set to any value, i.e. $\varphi_1 = 0$. By means of a greedy algorithm we choose $a_{ik} \in \{-1, 1\}$ such that for the resulting φ_k the error E of the objective function (4) is minimal. For φ_2 the exact solution can always be found, since a_{12} is the only parameter to optimize. For the remaining φ_k the greedy algorithm sets a_{ik} in turn either -1 or 1, verifying the validity of the result, setting a_{ik} the better value immediately and continuing with the next a_{ik} until all $k-1$ values for a_{ik} are set.

Algorithm 1 Greedy MDS$_{polar}$

$X = \{x_1, x_2, \ldots, x_n\}$
Let $\Psi_{n \times n}$ be a matrix with the pairwise angles ψ_{ij} between all (x_i, x_j)
$\varphi_1 = 0$
for $k = 2$ to n **do**
   $a_{ik} = 1$ for all $i = 1 \ldots k-1$
   for $i = 1$ to $k-1$ **do**
      $\varphi_k = \frac{\sum_{j=1}^{k-1} \varphi_j - \sum_{j=1}^{k-1} a_{jk}\psi_{jk}}{k-1}$ $e_k = \sum_{j=1}^{k-1}(\varphi_j - \varphi_k - a_{jk}\psi_{jk})^2$
      $t = \varphi_k$
      $a_{ik} = -1$
      $\varphi_k = \frac{\sum_{j=1}^{k-1} \varphi_j - \sum_{j=1}^{k-1} a_{jk}\psi_{jk}}{k-1}$ $f_k = \sum_{j=1}^{k-1}(\varphi_j - \varphi_k - a_{jk}\psi_{jk})^2$
      if $e_k < f_k$ **then**
         $a_{ik} = 1$
         $\varphi_k = t$
      end if
   end for
end for

Algorithm 1 describes in a simplified way the greedy method. When implementing the method, it can be optimized in that way, that the first φ_k in the *for*-loop has not always to be recalculated if in step $i - 1$ the parameter a_{ik} has not been changed to -1. In such cases φ_k holds the value from the previous step.

As mentioned above, φ_1 can be set to any value and φ_2 can always be chosen in such a way that the angle ψ_{12} is preserved exactly. For the remaining angles φ_k no guaranty can be given that the greedy algorithm finds the optimal solution. Incremental adding of feature vectors can be achieved by simply extending the outer *for*-loop for another iteration for each new object. The angle φ_k will be computed analogously as for previous feature vectors.

3.4   Relative MDS$_{polar}$

As for conventional MDS, also for MDS$_{polar}$ different approaches regarding the objective function are feasible. The solution described above minimizes the absolute difference of pairwise angles of the original dataset and the transformed dataset. Large angles, which cause in tendency a large E may effect the solution in that way, that the transformation will represent vectors with small angles to others less correctly. Considering the relative error leads to

$$E = \sum_{k=2}^{n} \sum_{i=1}^{k-1} \left(\frac{\varphi_i - \varphi_k - a_{ik}\psi_{ik}}{\psi_{ik}} \right)^2 \tag{7}$$

$$\frac{\partial E}{\partial \varphi_k} = -2 \sum_{i=1}^{k-1} \left(\frac{\varphi_i - \varphi_k - a_{ik}\psi_{ik}}{\psi_{ik}} \right) \frac{1}{\psi_{ik}}. \tag{8}$$

The greedy algorithm (1) can be applied only modifying the calculation specification for φ_k

$$\varphi_k = \frac{\sum_{i=1}^{k-1} \frac{\varphi_i}{\psi_{ik}^2} - \sum_{i=1}^{k-1} a_{ik} \frac{1}{\psi_{ik}}}{\sum_{i=1}^{k-1} \frac{1}{\psi_{ik}^2}}. \tag{9}$$

Because of the different objective functions the validity of solutions with the absolute MDS$_{polar}$ and the relative MDS$_{polar}$ can not be compared by means of E.

4   Weighted MDS$_{polar}$

In certain cases the objective when transforming data is to preserve relations of feature vectors of the original feature space in the target feature space. Thus, feature vectors that form a cluster should be represented as exact as possible in the target feature space, too. The transformation of feature vectors with a large distance to the respective feature vector can have a lower accuracy. An approach to achieve this goal is the introduction of weights w_{ik} to our objective function

$$E = \sum_{k=2}^{n} \sum_{i=1}^{k-1} w_{ik}(\varphi_i - \varphi_k - a_{ik}\psi_{ik})^2. \tag{10}$$

Determine the derivative leads to

$$\frac{\partial E}{\partial \varphi_k} = -2 \sum_{i=1}^{k-1} w_{ik}(\varphi_i - \varphi_k - \psi_{ik}) \tag{11}$$

and solving for φ_k

$$\varphi_k = \frac{\sum_{i=1}^{k-1} w_{ik}(\varphi_i - a_{ik}\psi_{ik})}{\sum_{i=1}^{k-1} w_{ik}}. \tag{12}$$

Note that this is a generalization of relative MDS$_{polar}$. For relative MDS$_{polar}$, we simply choose the weights as $w_{ik} = 1/\psi_{ik}^2$.

Since our transformation preserves the length of each data vector, it is guaranteed that vectors with a large difference in length will not be mapped to close points in the plane, even though their angle might not be matched at all. Therefore, we propose to use a small or even zero-weight for pairs of data vectors that differ significantly in their length. The weight could be defined as a function of the difference between the length values l_i and l_j of two data vectors:

$$w_{ik} = w(l_i, l_k) = w(z). \tag{13}$$

We can use the absolute difference for z, i.e.

$$z = z_a = |l_i - l_k|.$$

This might be useful if certain information about the structure of the data is known in advance. The argument z_r for relative weighting functions

$$z = z_r = \min\left\{\frac{l_i}{l_k}, \frac{l_k}{l_i}\right\}$$

might be useful if a certain threshold can be determined, beyond which difference in the relative distance between two feature vectors, the angle between them need not have any effect on the calculation of the respecting φ_k. To decrease the computational complexity, weights should be chosen in such a way, that for feature vectors with a certain (large) distance the respecting weights become zero. The following function describes a simple weighting function, which is the second function shown in Figure 1:

$$w(z_r) = \begin{cases} \sqrt{\left(\frac{z_r - \vartheta}{1 - \vartheta}\right)}, & \text{if } z_r \geq \vartheta \\ 0, & \text{otherwise} \end{cases} \tag{14}$$
$$\text{where } \vartheta \in [0,1].$$

With the threshold ϑ one can control indirectly the fraction of the data, that will be used to determine the respective angle φ_k. Thus, small values for ϑ lead to lots of weights w$\neq$ 0 which comes along with high computational complexity. Values

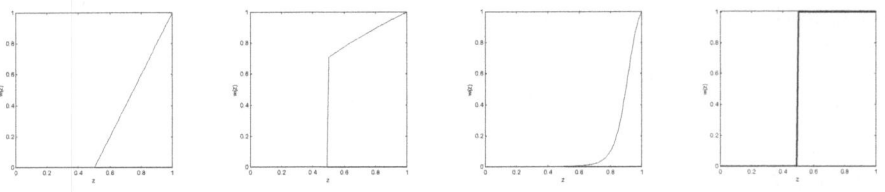

Fig. 1. Different Weighting Functions

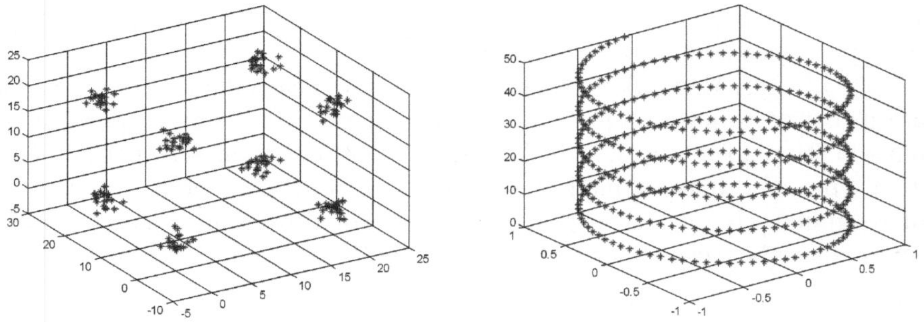

(a) Cube Dataset

(b) Coil Dataset

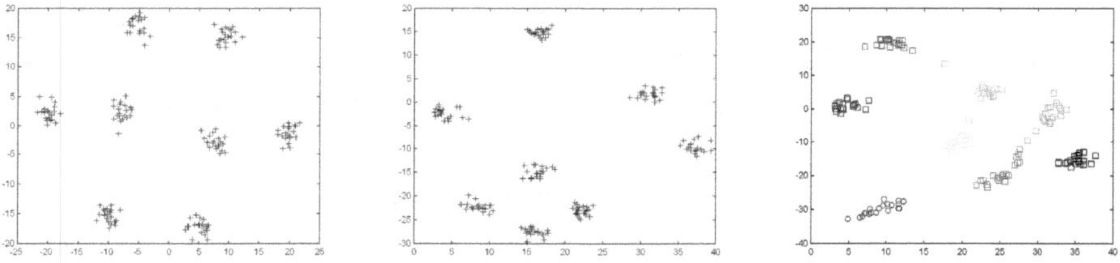

(c) Sammon Mapping

(d) relative MDS$_{polar}$

(e) weighted MDS$_{polar}$

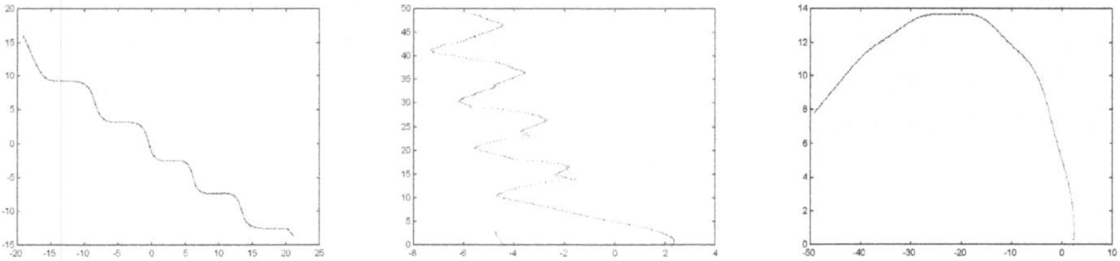

(f) Sammon Mapping

(g) relative MDS$_{polar}$

(h) weighted MDS$_{polar}$

Fig. 2. Different Transformations with MDS$_{polar}$

near 1 for ϑ lead to a quickly decreasing weighting function and to low computational complexity, respectively. Any other function can be used as weighting function. For reasons of an easy implementation and low computational complexity a decreasing function which leads to a more or less large fraction of zero weights should be used.

For an efficient implementation it is useful to sort the feature vectors by means of their length. Note that this can be achieved with $O(n \cdot \log n)$ time complexity. When determining the weights for the calculation of φ_k it is sufficient to consider the feature vectors starting from index k. Weights will be calculated stepwise. With every step the weights become smaller until a weight becomes zero. Since the weighting function is decreasing, a further iteration would lead to zero, too. Thus, the calculation of weights stops at this point. In cases where clusters with a large amount of data are expected in a dataset, it might be rather useful to limit the maximum number of iterations for the calculation of the weights than setting a larger threshold. In this case, the projected vectors will be forced to a proper position already by a significantly large fraction of other feature vectors in the dataset. It might also be useful to reduce ϑ locally, when only few vectors satisfy the condition in equation (14).

With a limitation of the number of weights $w > 0$ and a moderate ϑ at the same time, it can be achieved that the number of weights considered for the calculation of φ_k does not differ too much for different φ_k and limited computation time can be guarantied. Instead of considering the angles of all feature vectors with the greedy algorithm (1) it might be useful to consider only few feature vectors and calculating the exact solution of the sign problem. Using a weighting function enables the user of MDS$_{polar}$ to set a certain bin size which indicates the number of feature vectors that will be considered when calculating the desired φ_k. By means of this one can reduce the computation time and reinvest it in finding the exact solution of the sign problem. Thus, the upper bound for the complexity of our algorithm is due to sorting the data which is $O(n \cdot \log n)$. Solving the sign problem for a given maximum bin size b with the greedy strategy and using a certain number c of iterations, this accounts with $O(n \cdot b \cdot c)$ plus the costs for sorting to the entire algorithm.

Evaluation of the transformation can be done by determining the average deviation from the original angles. In general this can be obtained by dividing E by the number of terms summed up. For the error function (4) one has to divide by $\frac{n^2+n}{2}$. With this measured value one can compare different mappings even if they vary in the number of objects.

5 Results

Figure 2 shows some results of MDS$_{polar}$ in comparison with the Sammon Mapping. In favour of an easy verification of the results we applied MDS$_{polar}$ to some 3-dimensional datasets. The validity of the solution can be evaluated by visual inspection. The cube dataset (a) is about a synthetic dataset, where data points scatter around the corners of a 3-dimensional cube. Thus, the cube dataset contains eight well separated clusters. The dataset in (b) is comparable to a serpentine. As the figures (d) and (g) show, the transformations of MDS$_{polar}$ are similar to these of conventional MDS. Whereas MDS needs some thousand iterations until convergence, MDS$_{polar}$ finds an explicit solution after solving the system of equations. The transformations in figure (e) and (h) result from weighted MDS$_{polar}$ with weighting functions where at most twelve weights got values greater than zero. Thus, the transformation is based only on a relatively small number of angle comparisons. Therefore, locally these transformations are very accurate, but generally the loss of information is sometimes higher.

Since the value of φ_k is calculated from all preceding $\varphi_1 \dots \varphi_{k-1}$ according to equation (6) or equation (9) respectively, a solution with MDS$_{polar}$, either absolute or relative, depends to some degree on the order of the dataset. Our tests have shown that in such cases only few feature vectors lead to higher errors, while others will not. Thus, not the complete transformation will be wrong, but only some feature vectors. Initialization is also a matter of fact of conventional MDS.

6   Conclusion

We presented a new approach for dimension reduction. MDS$_{polar}$ bases on the reduction of the error of the pairwise angle between feature vectors comparing angles of the original feature space with the angles in the transformed feature space. With MDS$_{polar}$ it is possible to add new feature vectors to the dataset and find a transformation for this feature vector without re-calculating the whole transformation. Our solution is explicit, which leads here to short computation time. Furthermore, we presented a greedy algorithm to get an approximation of the exact solution.

With weighted MDS$_{polar}$ we have introduced a weighting function with the objective to differentiate ones feature vector's importance to the approximation of the respecting φ_k. Non-similar feature vectors contribute less to the accuracy of the result than similar feature vectors. If such weighting functions are designed in such a way that a (large) fraction of the angles φ_i gets zero weight, then an exact solution of the sign problem can be found within moderate computation time. Our tests have shown that good solutions can be already found with 10 non-zero weights. Our examples approve that this approach is promising. Developing appropriate approximation schemes will be subject of future work. Furthermore, we plan to modify this technique to learn a function that maps feature vectors to the 2-dimensional feature space. New objects could be mapped even simpler to the plane.

References

1. Borg, I., Groenen, P.: Modern Multidimensional Scaling : Theory and Applications. Springer, Berlin (1997).
2. Chalmers, M.: A Linear Iteration Time Layout Algorithm for Visualising High-Dimensional Data. Proceedings of IEEE Visualization 1996, San Francisco, CA (1996), 127–132.

3. Faloutsos, C., Lin, K.: Fastmap: A Fast Algorithm for Indexing, Data-Mining and Visualization of Traditional and Multimedia Datasets. In: Proceedings ACM SIG-MOD International Conference on Management of Data, San Jose, CA (1995), 163–174.
4. Kruskal, J.B., Wish, M.: Multidimensional Scaling. SAGE Publications, Beverly Hills (1978).
5. Lowe, D., Tipping, M.E.: Feed-Forward Neural Networks Topographic Mapping for Exploratory Data Analysis. Neural Computing and Applications, 4, (1996), 83–95.
6. Morrison, A., Ross, G., Chalmers, M.: Fast Multidimensional Scaling through Sampling, Springs and Interpolation. Information Visualization (2003) 2, 68–77.
7. Williams, M., Munzner, T.: Steerable, Progressive Multidimensional Scaling. 10th IEEE Symposium on Information Visualization, Austin, TX (2004), 57–64.

Miner Ants Colony: A New Approach to Solve a Mine Planning Problem*

María-Cristina Riff, Michael Moossen, and Xavier Bonnaire

Departamento de Informática, Universidad Técnica Federico,
Santa María, Valparaíso, Chile
{Maria-Cristina.Riff, mmoossen, Xavier.Bonnaire}@inf.utfsm.cl

Abstract. In this paper we introduce a simple ant-based algorithm for solving a copper mine planning problem. In the last 10 years this real-world problem has been tackled using linear integer programming and constraint programming. However, because it is a large scale problem, the model must be simplified by relaxing many constraints in order to obtain a near-optimal solution in a reasonable time. We now present an algorithm which takes into account most of the problem constraints and it is able to find better feasible solutions than the approach that has been used until now.

1 Introduction

Chile is the world's largest copper producer and the profit obtained by the copper extraction has an important role in the country economy. There are some approaches published in the literature related to mine problems, but they are usually applied to open pit mines, [9], [13]. Our particular problem is about an underground copper mine. In the last 10 years, the copper mine planning problem has been tackled using linear and mixed integer programming, and we have recently applied constraint programming techniques to it, [6], [17]. However, none of these techniques has been able to enterely solve our problem; thus it must be simplified by relaxing some geological and physical constraints. This problem belongs to large scale combinatorial optimization problems.

On the other hand, metaheuristics have solved complex problems succesfully like timetabling problems, [3], [11], scheduling [1], vehicle routing problems [16], travel salesman problems [4], constraint satisfaction problems [15], [5], [14], short-term electrical generation scheduling problems [12], and real-world applications [10], [7], [8], [2]. Our problem is similar to both, the scheduling problem and the travel salesman problem, but it has many other constraints that must be considered by the algorithm, in order to give better feasible solutions.

The purpose of this work goes in two directions: The first one is related to the problem; in this context our goal is not to obtain the optimal solution but a good one, which can be better than the solution found by the traditional approach.

---

* She was supported by the FONDEF Project: Complex Systems, and the other authors were supported by the Fondecyt Project 1040364.

A.F. Famili et al. (Eds.): IDA 2005, LNCS 3646, pp. 328–338, 2005.

The second one is related to the ant colony technique; our aim here is to show that it can be successfully applied to solve this real-world hard problem, using a simple version.

The paper is organized as follows: In the next section we define our real-world problem. In section three we present the linear integer programming model. The algorithm is introduced in section four. Section five presents the results obtained using random generated mine planning problems. Finally, in the last section we present the conclusions and the future issues that might come out of our work.

2 Problem Definition

For the purpose of resource modelling and mine planning, our mine has been divided into S sections. Each section is also subdivided into m blocks and each block is composed of 10 cells. The goal is to find the sequence of cells extraction that maximizes the profit. Our real-world problem is one section of an underground copper mine. The exploitation technique for this kind of mines requires the following two steps: To construct access tunnels, and to implement other facilities for extracting the cells of a block by using a bottom up procedure. The problem has many types of constraints, namely accessibility, geological, and capacity constraints.

Accessibility constraint: To have access to any cell within a block, its first cell must be previously extracted. We suppose that the access cost of a block is charged only once, when its first cell is extracted.

Geological constraint: The major set of constraints is called "subsidence constraints". Mine subsidence is the movement of the ground surface as a result of the collapse or failure of underground mine work, [18]. These constraints determine a physical relation among blocks.

The action of extracting a block implies the constraints: "the blocks that belong to its upper cone can not be exploited in a future time".

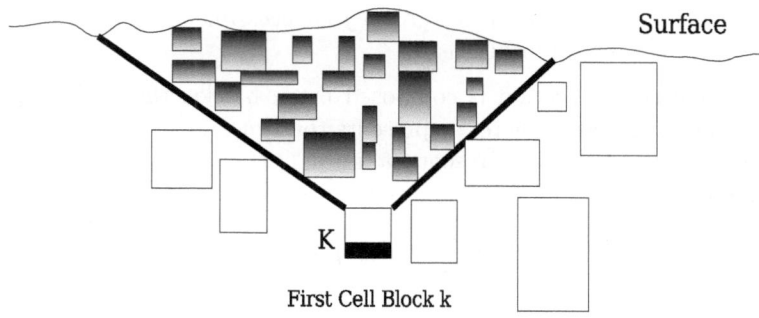

Fig. 1. Subsidence constraint

Figure 1 is a simplified 2D picture that shows the upper cone of block k. When the first cell of block k is extracted the blocks belonging to its upper cone become definitively *inaccesible blocks*, these blocks are painted in black. The white blocks could be exploited in a future time.

Capacity: The maximum number of cells extraction allowed is K_y cells by year.

The maximal profit value relates to both cell extraction and block access costs and to the cell copper concentration. The extraction of the first cell of block k implies the following actions:

1. To pay the access cost to block k
2. To avoid in the future the exploitation of the blocks belonging to the block k upper cone
3. To allow the extraction of the other cells of block k
4. To pay the extraction cost of each cell
5. To obtain the copper profit given by the cell copper concentration

We consider that the extraction of jth cell of a block, $j > 1$ implies only the points 4 and 5 listed above. The optimization is for a 20-year planning.

Remark 1. We use the Net Present Value as a way of comparing the value of money now with the value of money in the future. Thus, we apply a discount rate which refers to a percentage used to reflect the time value of money. Because of discounting the idea is to exploit particularly attractive blocks early but this makes the blocks in the cone inaccessible.

3   Problem Model

In this section we present the linear programming model. The idea is to find the sequence of cells extractions that maximizes the profit. It is evaluated computing the Net Present Value of the planning.

Variables:

$$z_{i,t} = \begin{cases} 1 \text{ if block } i \text{ is exploited at time } t \\ 0 \text{ otherwise} \end{cases} \tag{1}$$

$$h_{j,t} = \begin{cases} 1 \text{ if blocks group } j \text{ is accessible at time } t \\ 0 \text{ otherwise} \end{cases} \tag{2}$$

considering that a section is composed by m blocks, the horizon planning time equal to H, and where t represents the time.

Goal : Maximize NPV (Net Present Value)

Objective Function

$$NPV = \sum_{i,t} r^t U_i z_{i,t} - \sum_{j,t} r^t C_j h_{j,t} \tag{3}$$

where r is the discount rate, U_i is the benefit of the block i, C_j is the accessibility cost of the block j.

Constraints

 - Consistency

$$\sum_t z_{i,t} \leq 1, \forall i \tag{4}$$

The block i could be extracted at most once

$$\sum_t h_{j,t} \leq 1, \forall j \tag{5}$$

The access to the block is constructed only once
 - Accessibility

$$z_{i,t} \leq \sum_{s \leq t} h_{j,s}, \forall i \in j, \forall t \tag{6}$$

The access to the block must be done before its exploitation
 - Subsidence

$$z_{i,t} + z_{i',s} \leq 1, \forall i, \forall t, \forall i' \in I(i), \forall s \geq t \tag{7}$$

where $I(i)$ is the set of blocks belonging to the upper cone of block i. The blocks in the upper cone of block i cannot be exploited after the extraction of the block i.

 - Capacity

$$\sum_i \sum_t z_{i,t} \leq \sum_y K_y \tag{8}$$

There is a maximum number of blocks to be extracted.

Looking at the model we can observe that the problem has a lot of constraints. The model becomes very complex to be solved in its complete version. We tried to apply constraint programming techniques, [17], [6] in order to filter the domain of variables during the instantiations, but the problem is still hard to be solved with these techniques.

In the following sections we present our approach. This uses heuristics and is inspired by ant colonies. We have selected this metaheuristic for this work given the success of some reported real applications [2], [7], [8], using ants based approaches.

4   Miner Ants Colony

In the following sections we introduce the components of the algorithm based on Ant Colonies for Mine Planning.

4.1   Representation

In our approach the representation is a list where each element represents a block number. The same block number could appear more than once on the list. The number of a block appears as many times as the number of its cells have been extracted. For instance a list (3 6 7 4 7 5 5 ..) means that the first cell extracted is from block three, the third one is from block 7 and the fifth one is also from block 7. This representation is useful to manage the constraints that we named *consistency constraints*, equation 4 in the problem model section. With this representation we do not need to worry about the sequence of cells extraction inside a block, because it is directly deduced of the order of the list. Thus, by using this representation, all possible solutions satisfy the consistency constraints. Moreover, we can easily identify the scheduling of cell extractions of each block.

4.2   Evaluation Function

The hardest group of constraints is the subsidence constraints. The evaluation function has the two components of the equation 3 of the objective function described in the problem model:

- The profit obtained by cell extractions
- The cost of blocks accesibility

We need to point out that we also include in the evaluation function an opportunity cost, that takes into account the following issues: "when a block is exploited all the blocks belonging to its upper cone become inaccesible for ever". Thus, we consider a cost related to the physical impossibility of obtaining the copper concentration of these blocks in the future time. In each instantiation the algorithm is looking for the cell of a block k whose gain financially justifies the prohibition of extraction of the blocks in its k upper cone in the future. This cost is included as a penalization in the evaluation function. It is calculated by the addition of the profit expected from the blocks which belong to the upper cone of the blocks exploited before.

4.3   Algorithm

Miner Ants is an ant-based optimization approach. Generally speaking, each ant is looking for a complete block-cell extraction scheduling. Our method is inspired in the Ants-Solver approach proposed by Solnon in [15] to solve Constraint Satisfaction Problems. In her approach the ants look for a solution with a minimum number of constraints violations. After that, a local search procedure is applied to repair the solution found by the ants, in order to satisfy most of the constraints.

The ants in Miner Ants do not worry about the subsidence constraints. However, the solutions found by the ants must satisfy all the other constraints. That means that the ants solve the sub-problem respecting both the consistency and

Begin
For each block i **do**
    $MAA[i] = \text{profit}[i] - \text{cost extraction}[i] - \text{opportunity cost}[i]$
Identify the block l with the lower MAA
For each block i **do**
    $MAA[i] = MAA[i] - MAA[l]$
Identify the block b with the bigger MAA
For each block i **do**
    $MAA[i] = \frac{MAA[i]}{MAA[b]}$
End

Fig. 2. Maximum Adjusted Average

the capacity constraints. A local search algorithm is in charge of repairing the solutions found by the miner ants, in case that these solutions violate the subsidence constraints. This local search procedure is especially designed for solving this kind of constraint violation. In contrast to the Solnon algorithm the solution found by our approach must satisfy all the constraints. The ant-based algorithm uses a local function named Maximum Adjusted Average (MAA). It is shown in figure 2. The values of MAA belong to the interval $[0,1]$.

In the same way, the function MAA helps the algorithm to identify the largest profitable number of cells of a block to be extracted, taken into account the costs and the profit of the copper extraction of the cells still belonging to the block. This means that the profit must be greater than the addition of both the cost associated to the cells extraction and the opportunity cost. Given both a block and its MAA value the algorithm determines the number of cells to be extracted to obtain a net profit.

The algorithm is shown in figure 3. Each point is analyzed as follows:

(1) The blocks are ordered according to their MAA values.
(2) The Current feasible blocks queue is composed by all the feasible blocks which could be extracted. At the begining of the algorithm this queue is equal to the global blocks priority queue. It is modified at each step of the algorithm according to the last extraction carried out.
(3) L_c blocks with a MAA value within $[\gamma bestMAA, bestMAA]$ are selected by the algorithm, where $bestMAA$ is the MAA value of the best block in the list. These L_c blocks form the list of the best candidate blocks to be extracted. We have determined the best value of $\gamma = 0.75$ by tuning.
(4) As shown in Ant Colony System [4] to improve exploration, the block to be extracted is selected depending on q_0:
    If $q_0 \geq U[0,1]$ then the block with the best P_{ij}^k is selected. This value for the ant k which extracts the cell j inmediatly after cell i is determined by:

$$P_{ij}^k(t) = \begin{cases} \dfrac{\tau_{ij}(t)[\eta_{ij}]^\beta}{\sum_{h \in J_i^k(t)} \tau_{ih}(t)[\eta_{ih}]^\beta} & \text{if } j \in J_i^k \\ 0 & \text{else} \end{cases} \qquad (9)$$

Begin
Parameters setting
Global Pheromone Initialization to τ_0
Construct a global blocks priority queue using MAA   (1)
For each iteration **do**
    For each ant
    Repeat
        Update the Current feasible blocks queue   (2)
        Determine L_c, the length of the candidates blocks list   (3)
        Using q_0, select the next block k to extract from the L_c candidates blocks (4)
        Compute the number of cells of block k to be extracted using MAA   (5)
        Local pheromone update   (6)
    until a candidate a solution is completed
    Candidate solution subsidence reparation   (7)
    End For
    Global Pheromone update   (8)
    Elitist Ant Pheromone update   (9)
End For
End

Fig. 3. Miner Ants Algorithm

Where β is a parameter to control the pheromone intensity with the ant visibility. $\tau_{ij}(t)$ is the pheromone intensity, $\eta_{ij}(t)$ is the ant visibility (here MAA) and J_i^k the cells allowed to be extracted after the cell i. Otherwise it is randomly selected, with identical probability, from the candidate blocks list.

(5) MAA helps the algorithm to identify the most profitable number of cells of block k to be extracted, taken into account the costs (extraction and opportunity) and the copper concentration of the cells belonging to the block.

(6) The pheromone level is locally modified by $\tau_{ij}(t) = (1 - \rho)\tau_{ij}(t-1) + \rho\tau_0$, where ρ is the evaporation parameter.

(7) The reparation procedure will be explained in details in section 4.4.

(8) For each ant and for each block extracted:

Evaporation: $pheromone = pheromone \times (1 - \rho)$

Update Pheromone: $pheromone = pheromone + (Q \times \rho \times p)$
    where p is the rate between the evaluation function of the current ant and the evaluation function of the best ant, because we are maximizing the fitness value, and $pheromone$ is the pheromone (τ_{ij}) between two consecutive visited blocks (i, j) by an ant, and Q is a parameter.

(9) Similar to (8) but only considering the best found solution from the ants.

4.4   Local Search Subsidence Reparation

Generally speaking, the procedure changes the list position of a block *only if* it belongs to the subsidence cone of a block extracted before it. Figure 4 shows

Begin
For each block b_1:
   For each block b_2 after b_1 in the extraction sequence:
     if $(b_2 \in I(b_1))$:
      insert b_2 before b_1.
     end if
   end for
  end for
End

Fig. 4. Reparation Procedure

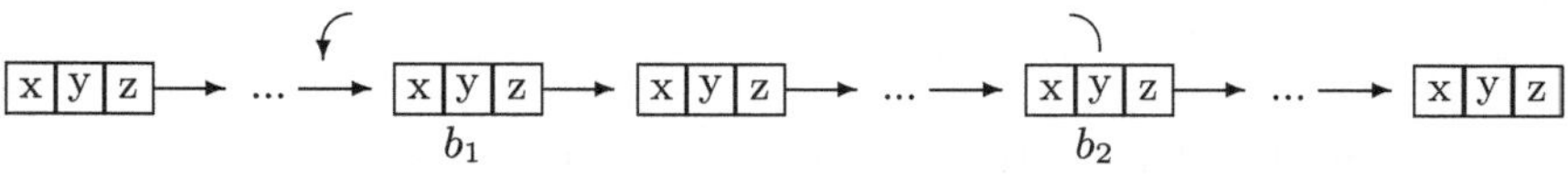

Fig. 5. Solution Repair

the algorithm, where $b_2 \in I(b_1)$ means that b_2 belongs to the subsidence cone of block b_1.

Figure 5 shows how the algorithm repairs this situation.

5   Tests

Because the information of our real mine is a confidential issue we are not able to report real results here. However, we have built a database of benchmarks[1], with 50 mines which have similar characteristics of a real one. The dimensions are in number of blocks in the three coordinates. The artificial mines have various kinds of copper concentration: at random, at the bottom, at the middle, at the borders, at the higher layers, both at bottom and in the middle. They have also different dimensions. Furthermore, some of these artificial mines are more complex than a real one. We report here the ten hardest mine configurations. The common parameters of these virtual mines are:

$H = 20$ Time in years.
$K_y = 5$ the maximum number of blocks to be extracted by year
$C_k = US\$8.0M$ Access cost of block k
$r = 10\%$ Annual discount rate.

In order to find the best parameters values for our ant algorithm we considered in the begining the most common values reported in the literature related to real-world applications and we have modified them by tuning. Tuning was a hard task. We modified a subset of the parameters in a systematic way using dichotomy. The parameters values of our ant algorithm are: $\alpha = 1, \beta = 5, \rho =$

---

[1] available in http://www.inf.utfsm.cl/~bonnaire/Mines-Benchmarks

Table 1. Tests with randomly generated mines

Mine	Characteristics	Dimensions	GLB	Miner Ants	PUB	Relaxed
M_2	Random	$18 \times 4 \times 10$	7400.00	8346.97	8593.71	11200.43
M_5	Random	$9 \times 8 \times 10$	7532.00	8498.24	8705.02	11230.42
M_{18}	Random	$18 \times 8 \times 5$	7714.65	8491.14	8645.87	11245.85
M_{22}	Bottom	$18 \times 8 \times 5$	7800.98	8583.90	8769.32	11199.63
M_{30}	Middle	$6 \times 6 \times 20$	7500.38	8258.26	8392.87	11582.02
M_{35}	Bottom	$6 \times 6 \times 20$	6010.30	10198.24	10302.73	11384.63
M_{40}	Borders	$4 \times 18 \times 10$	11408.05	11419.85	11429.50	11649.08
M_{42}	Middle	$8 \times 9 \times 10$	10606.80	10851.41	10890.21	11585.76
M_{45}	Higher layers	$9 \times 40 \times 2$	10150.30	11162.50	11163.03	11563.72
M_{50}	Bottom and Middle	$9 \times 8 \times 10$	9002.23	10090.08	10189.98	11229.53

$0.5, \rho_0 = 1E - 6, Q = 1, q_0 = 0.7, n = 5000, m = 20$. Where n is the number of iterations and m is the number of ants. Our tests were made on an Athlon XP 1.6 GHz computer with 256MB of RAM, running Linux RedHat 7.2 and the GNU G++ compiler and optimizer.
In order to evaluate the quality of the solutions found by the algorithm we define some metrics, one lower bound and one upper bound.

– Greedy Lower Bound (GLB): It is obtained using a greedy procedure to construct a feasible solution. This procedure considers that all blocks must be completely extracted, which means it does not work with the cell concept.
– Practical Upper Bound (PUB): We define a greedy procedure where the move uses the local heuristic MAA. It gives unfeasible solutions, but it is a realistic upper bound.

From table 1 we can observe that the traditional method working with a relaxed model found greater profit values than our algorithm. However, these values correspond to infeasible solutions and they must be repaired by an expert. In this context, Miner Ants gives more realistic solutions. For instance, if we consider the M_{22} mine we can observe that the basic greedy procedure gives a solution with a value of 7800.98. The Relaxed Model gives a better value equal to 11199.63. Considering the PUB value we know that the optimal value must be lower than 8769.32 when we take into account all the problem constraints. Finally, our algorithm find a solution with a $NPV = 8583.90$. Moreover, for the real mine the results obtained by Miner Ants are closer to the expert repaired solution. We have obtained solutions whose values are, in the worst cases, a 25% lower than the current solution given to the expert to be modified. Our real mine is similar to problem M_{50} with a copper concentration in both, at the bottom and in the middle. The problem with this kind of configurations is that we must be extremely careful about subsidence constraints, because if the algorithm begins extracting at the bottom levels, some blocks in the middle that can give a profit will be not accesible in the future. Thus, our algorithm is able to control factors such as: costs, opportunity costs and profits. In general, the algorithm is able to find solutions closer to the PUB in some seconds. It takes around 500 seconds for 1000 iterations.

6 Conclusion

We have obtained good results for a mine problem using an Ant colony based algorithm. In this work we have considered the mine divided into homogeneous blocks composed by the same number of cells. However, it is interesting to work with non-homogenous structures, because it is closer to reality. Obviously non homogenous blocks increase the complexity related to the subsidence constraints. Our research allows us to conclude that using an ant-based algorithm we are able to find better feasible solutions for different types of mine configurations. The algorithm takes into account all kinds of problem constraints. The tests showed that metaheuristic based techniques are very useful to solve complex real world problems in a reasonable computing time. Our future research intends to include some seismics conditions as hard constraints in order to tackle mines ground movements.

References

1. Burke E., Smith A., Hybrid Evolutionary Techniques for the Maintenance Scheduling Problem, IEEE Transactions on Power Systems, Vol. 15, N. 1, 122-128, 2000.
2. Casagrande N., Gambardella L.M., Rizzoli A.E., Solving the vehicle routing problem for heating oil distribution using Ant Colony Optimisation. ECCO XIV. Conference of the European Chapter on Combinatorial Optimisation, 2001.
3. Colorni A., Dorigo M., Maniezzo V., Metaheuristics for High-School Timetabling. Computational Optimization and Applications, 9, 3, 277-298, 1998.
4. Dorigo M., Gambardella L.M., Ant Colony System: A Cooperative Learning Approach to the Traveling Salesman Problem. IEEE Transactions on Evolutionary Computation, 1, 1, 53-66, 1997.
5. Eiben A.E., Van Hemert J.I, Marchiori E., Steenbeek A.G., Solving Binary Constraint Satisfaction Problems using Evolutionary Algorithms with an Adaptive Fitness Function. V PPSN, LNCS 1498, 196-205, 1998.
6. Freuder E., The Many Paths to Satisfaction. Constraint Processing, Ed. Manfred Meyer, LNCS 923, 103-119, 1995.
7. Gambardella L.M, Taillard E., Agazzi G., MACS-VRPTW: A Multiple Ant Colony System for Vehicle Routing Problems with Time Windows, In D. Corne, M. Dorigo and F. Glover, editors, New Ideas in Optimization. McGraw-Hill, 1999.
8. Gambardella L.M, Mastrolilli M., Rizzoli A.E., Zaffalon M., An integrated approach to the optimisation of an intermodal terminal based on efficient resource allocation and scheduling, Journal of Intelligent Manufacturing, 12 (5/6):521-534, October 2001.
9. Gunn E.A., Cunningham B., Forrester D., Dynamic programming for mine capacity planning, Proceedings of the 23nd APCOM Symposium, Montreal, Vol 1.,529-536. 1993.
10. Karanta I., Mikkola T., Bounsaythip C., Riff M-C., "Modeling Timber Collection for Wood Processing Industry. The case of ENSO", ERCIM Technical Report, TTE1-2-98, VTT Information Technology, Finland, Octobre 1998.

11. Newall J.P., Hybrid Methods for Automated Timetabling, PhD Thesis, Department of Computer Science, University of Nottingham, UK, 1999.
12. Maturana J, Riff M-C., An evolutionary algorithm to solve the Short-term Electrical Generation Scheduling Problem. In Proceedings of the Congress on Evolutionary Computation (CEC'2003), 1150-1156, 2003.
13. Ricciardi J, Chanda E., Optimising Life of Mine Production Schedules in Multiple Open Pit Mining Operations: A Study of Effects of Production Constraints on NPV, Mineral Resources Engineering, Vol. 10, No. 3, 301-314, 2001.
14. Riff M.-C., A network-based adaptive evolutionary algorithm for CSP, In the book "Metaheuristics: Advances and Trends in Local Search Paradigms for Optimisation", Kluwer Academic Publisher, Chapter 22, 325-339, 1998.
15. Solnon C., Ants Can Solve Constraint Satisfaction Problems, IEEE Transactions on Evolutionary Computation, Vol.6, 4, 347-357, 2002.
16. Taillard E., Heuristic Column Generation Method for the heterogenous VRP. Recherche-Operationnelle 33, 1-14, 1999.
17. Tsang E.P.K., Wang C.J., Davenport A., Voudouris C., Lau T.L., A family of stochastic methods for constraint satisfaction and optimization, The First International Conference on The Practical Application of Constraint Technologies and Logic Programming, London, pp. 359-383, 1999.
18. Waltham T., Waltham,A., Foundations of Engineering Geology. Publisher: Routledge mot E F & N Spon; 2nd edition, 2002.

Extending the GA-EDA Hybrid Algorithm to Study Diversification and Intensification in GAs and EDAs

V. Robles[1], J.M. Peña[1], M.S. Pérez[1], P. Herrero[2], and Ó. Cubo[1]

[1] Departamento de Arquitectura y Tecnología de Sistemas Informáticos,
Universidad Politécnica de Madrid, Madrid,Spain
{vrobles, jmpena, mperez, ocubo}@fi.upm.es
[2] Departamento de Lenguajes y Sistemas, Universidad Politécnica de Madrid, Madrid, Spain
pherrero@fi.upm.es

Abstract. Hybrid metaheuristics have received considerable interest in recent years. Since several years ago, a wide variety of hybrid approaches have been proposed in the literature including the new GA-EDA approach. We have design and implemented an extension to this GA-EDA approach, based on statistical significance tests. This approach had allowed us to make an study of the balance of diversification (exploration) and intensification (exploitation) in Genetic Algorithms and Estimation of Distribution Algorithms.

1 Introduction

Over the last years, interest in hybrid metaheuristics has risen considerably among researchers. The best results found for many practical or academic optimization problems are obtained by hybrid algorithms. Combination of algorithms such as descent local search [15], simulated annealing [10], tabu search [6] and evolutionary algorithms have provided very powerful search algorithms.

Two competing goals govern the design of a metaheuristic [19]: exploration and exploitation. Exploration is needed to ensure every part of the search space is searched thoroughly in order to provide a reliable estimate of the global optimum. Exploitation is important since the refinement of the current solution will often produce a better solution. Population-based heuristics (where genetic algorithms [9] and estimation of distribution algorithms [12] are found) are powerful in the exploration of the search space, and weak in the exploitation of the solutions found.

With the development of our new approach, GA-EDA, a hybrid algorithm based on genetic algorithms (GAs) and estimation of distribution algorithms (EDAs), we aim to improve the explorations power of both techniques.

This hybrid algorithm has been tested on combinatorial optimization problems (with *discrete* variables) as well as *real-valued* variable problems. Results of several experiments show that the combination of these algorithms is extremely promising and competitive.

This paper is organized in the following way: First, we will focus on different taxonomies of hybrid algorithms found in the literature; in section 3, the GA-EDA approach is reviewed with a complete performance study presented in section 4. Finally we close with our conclusions and further future work.

A.F. Famili et al. (Eds.): IDA 2005, LNCS 3646, pp. 339–350, 2005.

2   Taxonomy of Hybrid Algorithms

General taxonomies provides a mechanism to allow comparison of hybrid algorithms in a qualitative way and classifying new hybrid approaches. This section highlights some of the most important hybrid taxonomies.

[2] describes three different forms of hybridization:

- *Component Exchange Among Metaheuristics.*
  One of the most popular hybridization methods is the use of trajectory methods such as Local Search, Tabu Search, in population-based algorithms. These solutions combine the advantages of population based methods, which are better on diversification, and trajectory methods, which are better on intensification. For example [7] incorporates local search in a genetic framework.
- *Cooperative Search* [1,4,21].
  The second hybridization approach consists of a search performed with various algorithms that, typically, execute in parallel and exchange information about states, solutions, sub-problems or other characteristics.
- *Integrating Metaheuristics and Systematic Methods* .
  This approach has produced very effective algorithms. For instance [5] integrates metaheuristics and Constraint Programming.

A complementary taxonomy can be found in [19] which defines a hierarchical classification.

- *LRH (Low-level Relay Hybrid).*
  A given metaheuristic is embedded into a single-solution metaheuristic. For instance in [14] a LRH hybrid combines simulated annealing with local search.
- *LCH (Low-level Co-evolutionary Hybrid).*
  Algorithms consist in population based heuristics coupled with local search heuristics. The population based algorithms will try to optimize globally and the local search will try to optimize locally.
- *HRH (High-level Relay Hybrid).*
  The metaheuristics are executed in a sequence, one after another, each using the output of the previous as its input. In [13] annealing is used to improve the population obtained by a GA.
- *HCH (High-level Co-evolutionary Hybrid).*
  Several algorithm perform a search in parallel and cooperate in order to find the optimum. This approach is similar to the previous *cooperative search*. The use of parallel EDAs in a island model [18] is an of this.

The hybrid algorithm GA-EDA, can be classified as *cooperative search* in Blum and Roli's taxonomy. In Talbi's classification GA-EDA is heterogeneous; global because the algorithm search the whole state space, and general because both algorithms solve the same problem.

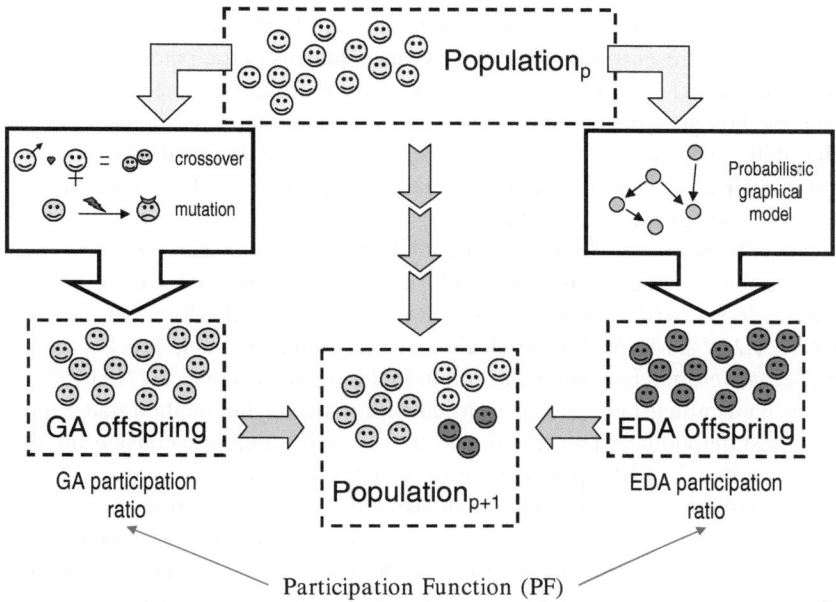

Fig. 1. Hybrid Evolutionary Algorithm Schema

3   Hybrid GA-EDA Algorithm

Hybrid GA-EDA are new algorithms based on both techniques [16,17]. The original objective is to get benefits from both approaches. The main difference from these two evolutionary strategies is how new individuals are generated. These new individuals generated on each generation are called *offspring*. Our new approach generates two groups of offspring individuals, one generated by the GA mechanism and the other by EDA one. On one hand, GAs use crossover and mutation operators as a mechanism to create new individuals from the best individuals of the previous generation. On the other hand, EDAs builds a probabilistic model with the best individuals and then sample the model to generate new ones.

Population$_{p+1}$ is composed by the best overall individuals from (i) the past population (*Population$_p$*), (ii) the GA-evolved offspring, and (iii) EDA-evolved offspring.

The individuals are selected based on their fitness function. This evolutionary schema is quite similar to Steady State GA in which individuals from one population, with better fitness than new individual from the offspring, survive in the next one. In this case we have two offspring pools. Figure 1 shows how this model works.

3.1   Participation Functions

On this approach an additional parameter appears, this parameter has been called *Participation Function*(PF). PF provides a ratio of how many individuals are generated by each mechanism. In other words, the size of GA and EDA offspring sets. The size of these sets also represents how each of these mechanisms participates on the evolution of the population. These ratios are only a proportion for the number of new individuals each method generates, it is not a proportion of individuals in the next population, which is defined by the quality of each particular individual. If a method were better that the other in terms of how it combines the individuals there would be more individuals from this offspring set than the other.

Several alternatives to these Participation Functions were taken into account in previous experiments, being some of them: the Constant Ratio (*x% EDA / y% GA*),the Alternative Ratio (*ALT*), the Incremental Ratio (*EDA++* and *GA++*) or the Dynamic Ratio (*DYNAMIC*). More information about them could be found in [16,17] From all of these alternatives, maybe could be useful to highlight the last one (DYNAMIC), which has a mechanism that increases the participation ratio for the method that happens to generate best individuals. This function evaluates each generation considering the possibility to change the participation criterion as defined by the ratio array.

The DYNAMIC algorithm starts with 50%/50% ratio distribution between the two methods. On each generation the best offspring individuals from each method are compared and the wining method gets a 5% of the ratio of the opposite method (scaled by the amount of relative difference between the methods, dif variable). This mechanism provides a contest-based dynamic function in which methods are competing to get higher ratios as they generate better individuals.

4   The New Range Based Participation Function

In this section we present a new participation function that is based on the first steps of the Mann-Whitney non-parametric test. In this test there is no hypothesis that the initial samples should follow a normal distribution, which is important in this environment.

The new Range Based Participation Function begins by assembling the fitness from GA and EDA populations into a single set of size $N = n_{GA} + n_{EDA}$. These measures are then rank-ordered from lowest ($rank1$) to highest ($rankN$), with tied ranks included where appropriate.

Once they have been sorted out in this fashion, the rankings are then returned to the population, GA or EDA, to which they belong and substituted for the fitness measures that gave rise to them.

The effect of replacing raw measures with ranks is two-fold. The first is that it brings us to focus only on the ordinal relationships among the raw measures ("greater than", "less than" and "equal to") with no illusion or pretense that these raw measures derive from an equal-interval scale. The second is that it transforms the data array into a kind of closed system, many of whose properties can then be known by dint of sheer logic.

Let be,

$T_{GA} = the\_sum\_of\_the\_n_{GA}\_ranks\_in\_group\_GA$

$T_{EDA} = the\_sum\_of\_the\_n_{EDA}\_ranks\_in\_group\_EDA$

Now, we would like to know if GA and EDA do not differ with respect to their effectiveness. If this were true, then the raw measures within fitness in GA and EDA would be about the same, on balance, and the rankings that derive from them would be evenly mixed within fitness in GA and EDA, like cards in a well shuffled deck.

So if this were true, we would expect the separate averages of the GA ranks and the EDA ranks each to approximate the same overall mean value. This entails that the rank-sums of the two groups, T_{GA} and T_{EDA}, would approximate the values,

$Mean_{GA} = n_{GA}(N+1)/2$
$Mean_{EDA} = n_{EDA}(N+1)/2$

Thus we know that:

− The observed value of T_{GA} belongs to a sampling distribution whose mean is equal to $Mean_{GA}$.
− The observed value of T_{EDA} belongs to a sampling distribution whose mean is equal to $Mean_{EDA}$.

Finishing, the effectiveness of GA and EDA will be,

$Effect_{GA} = T_{GA}/Mean_{GA}$
$Effect_{EDA} = T_{GA}/Mean_{EDA}$

Thus, the percentages for the next generation will be,

$Perc_{GA} = Effect_{GA}/Effect_{GA} + Effect_{EDA}$
$Perc_{EDA} = Effect_{EDA}/Effect_{GA} + Effect_{EDA}$

5   Behavior Analysis of DYNAMIC vs. RANGE Participation Functions

The experiments to compare the behavior of DYNAMIC and RANGE Participation Functions have been performed considering five continuous problems:

① *Branin* RCOS function
② *Griewank* function
③ *Rastrigin* function
④ *Schwefel's* problem [8,20]
⑤ A continuous version of the *MaxBit* problem

The hybrid algorithm is composed of the simplest versions of both GA and EDA components. In this sense a real string (real-coded vector) has been used to code all the problems. GA uses *Roulette Wheel* selector, one-point crossover, flip mutation (in this case selecting a random gene, with probability 0.01 and generating a new value using an uniform random distribution) and uniform initializer. EDA uses the continuous version of the Univariate Marginal Distribution Algorithm (UMDA$_c$) [11]. The overall algorithms generate an offspring twice the size of the population. Depending on the ratios provided by the Participation Function, this offspring is then distributed between the two methods. The composition of the new population is defined by a deterministic method, selecting the best fitness scores from the previous population and both offspring sets. The stopping criteria is quite straightforward, we stop when the difference of the

Table 1. Branin

	GA	EDA	DYNAMIC	RANGE
Average fitness	0.4016	0.3987	0.4000	0.3990
Average generation number	19	19	19	19

sum of the fitness values of all individuals in two successive generations is smaller than a predefined value.

After having executed ten consecutive times the experiments, the average of the best fitness values and the average of the number of generations are calculated. Several population sizes have been tested, but in this paper we only present the most representative size. All these experiments have been performed in an 8-nodes cluster of bi-processors with Intel Xeon 2.4GHz with 1GB of RAM and Gigabit network running Linux 2.4.

With the aim of making a good comparison among the results achieved by all the presented algorithms, we have done the Mann-Whitney statistical test to compare them. The fitness values of the best solutions found in the search are used for this purpose.

It is important to highlight that the results presented in this paper depend on the individual representation used for each of the problems.

5.1   Branin RCOS Function

Definition. This problem is a two-variable continuous problem with three global minimum and no local minimum. The problem is defined as follows [3]:

$$f_B(x_1, x_2) = \left(x_2 - \frac{5}{4\pi^2}x_1^2 + \frac{5}{\pi}x_1 - 6 \right)^2 + 10 \left(1 - \frac{1}{8\pi} \right) cos(x_1) + 10$$
$$-5 < x_1 < 10$$
$$0 < x_2 < 15$$

The global optimum for this problem is 0.397887 with the following values (x_1, x_2) $= (-\pi, 12.275), (\pi, 2.275), (9.42478, 2.475)$.

This problem is considered easy not only because of the number of variables, but the small chance to miss the basin of the global minimum in a global optimization procedure. This is due to the probability to reach the global optimum using local optimization methods, started with a small number of random points is quite high.

Results. Branin is a very simple problem where in few generations (approx 19) all the algorithms converge. This problem was solved using a population size of 300 individuals.

As it is possible to appreciate in the table 2, EDA gets better results than GA. However, the hybrid algorithm with the RANGE Participation Function obtains significant better results than GA, EDA and the DYNAMIC Participation Function.

Table 2. Statistical Significance Tests for Branin

Mann-Whitney Test	p-value
EDA better GA	0.1126
RANGE better DYNAMIC	0.2315
RANGE better EDA	0.2697
RANGE better GA	0.0790

5.2 Rastrigin Function

Definition. It is a scalable, continuous, and multimodal function that must be minimized. It's the result of modulating n-dimensional sphere function with $a \cdot cos(\omega x_i)$.

$$f_{Ra5}(\boldsymbol{x}) = a \cdot n + \sum_{i=1}^{n} \left(x_i^2 - a \cdot cos(\omega \cdot x_i) \right)$$

$$a = 10; \omega = 2\pi; n = 5$$

$$-5.12 < x_i < 5.12$$

The global minimum for this problem can be found in the solution $x_i = 0, i = 1, \ldots, n$ with a fitness value of 0.

Results. This problem was solved using a population size of 1000 individuals.

Table 3. Rastrigin

	GA	EDA	DYNAMIC	RANGE
Average fitness	0.11656	4.10471	0.00013	0.00005
Average generation number	27	28	28	28

Although Rastrigin function has no lineal dependency among the variables, the performance of EDAs (with the UMDA approach) is very poor. Nearby the optimum value there are many local optima and EDAs seems to be very sensitive to this characteristic.

Table 4. Statistical Significance Tests for Rastrigin

Mann-Whitney Test	p-value
EDA better GA	0.1126
RANGE better DYNAMIC	0.1601
RANGE better EDA	0
RANGE better GA	0

The table 4 presents the Mann Whitney significance tests for this problem. In this case, EDA is better than GA with a p-value of 0.1126 and the RANGE Participation Function is significantly better than the DYNAMIC Participation Function with a p-value of 0.1601. Moreover, RANGE is better than GAs and EDAs with p-values equal to 0.

Table 5. Schwefel

	GA	EDA	DYNAMIC	RANGE
Average fitness	3.129	1852.836	0.120	0.037
Average generation number	31	26	34	33

5.3   Schwefel's Problem

Definition. Schwefel's function is a continuous multimodal function. It is interesting because it is a separable problem, it means that searching along the coordinate axes gives optimal values for each of the components because function gradient is oriented along the axes. As in the previous case global optimum is surrounded by several local optimum in the neighborhood.

$$f_{S10}(x) = \sum_{i=1}^{n} x_i \cdot sin(\sqrt{|x_i|})$$
$$n = 10$$
$$-500 < x_i < 500$$
$$f_{S10}(x^*) = min(f_{S10}(x))$$

The global minimum for this problem can be found in the solution $x_i = 420.9687$, $i = 1, \dots, n$ with a fitness value of 0.

Results. This problem has been solved with a population of 2000 individuals.

GAs perform very good in this problem because of the separability of the component optimal values. Genetic combination tries to preserve good gene values when generating new individuals. Although (see Table 6) GA is much better than EDA and, one more time, RANGE outperforms DYNAMIC, GAs and EDAs.

Table 6. Statistical Significance Tests for Schwefel

Mann-Whitney Test	**p-value**
EDA better GA	1
GA better EDA	0
RANGE better DYNAMIC	0.0001
RANGE better EDA	0
RANGE better GA	0

5.4   The MaxBit Continuous Problem

Definition. This problem is a redefinition of the binary MaxBit problem previously presented. The aim is to maximize:

$$f_{M12}(\boldsymbol{x}) = \frac{\sum_{i=1}^{n} x_i}{n}$$

$$x_i \in \{0, 1\}; n = 12$$

In the continuous domain this problem is more complex, as the optimum value of the function is located on the boundary of the search space.

Results. This problem has been solved with a population of 250 individuals.

In this experiment (see Table 6), GA is much better than EDA and, one more time, RANGE outperforms DYNAMIC, GAs and EDAs.

In the MaxBit Continuous problem EDA is slightly better than GA (with p-value equal to 0). However, DYNAMIC and RANGE have the same behavior getting the maximum value for all the problem executions.

6   Intensification and Diversification in GAs and EDAs

One interesting issue is to survey the evolution of the DYNAMIC and RANGE Participation Functions in the series of different experiments. These functions, as we have seen, adjust the participation ratio depending on the quality of the individuals each of the method is providing. This measure has been indirectly used to evaluate the quality of each of the methods across the continuous generations of one algorithm.

In Figure 2 the evolution of the two different participation functions is shown. Being the first one associated to the DYNAMIC participation function and the second one to the RANGE participation function. Moreover we have introduced an additional section at the bottom of the figure with the aim of clarifying the progress of diversification and intensification in the optimization process.

DYNAMIC participation function (Figure 2.a) behaves with smooth variation in the rations for each of the evolutionary methods. As diversification features are required in

Table 7. MaxBitCont

	GA	EDA	DYNAMIC	RANGE
Average fitness	0.9940	0,9998	1	1
Average generation number	36	40	33	34

Table 8. Statistical Significance Tests for MaxBitCont

Mann-Whitney Test	**p-value**
EDA better GA	0
RANGE better DYNAMIC	1
DYNAMIC better RANGE	1
RANGE better EDA	0
RANGE better GA	0

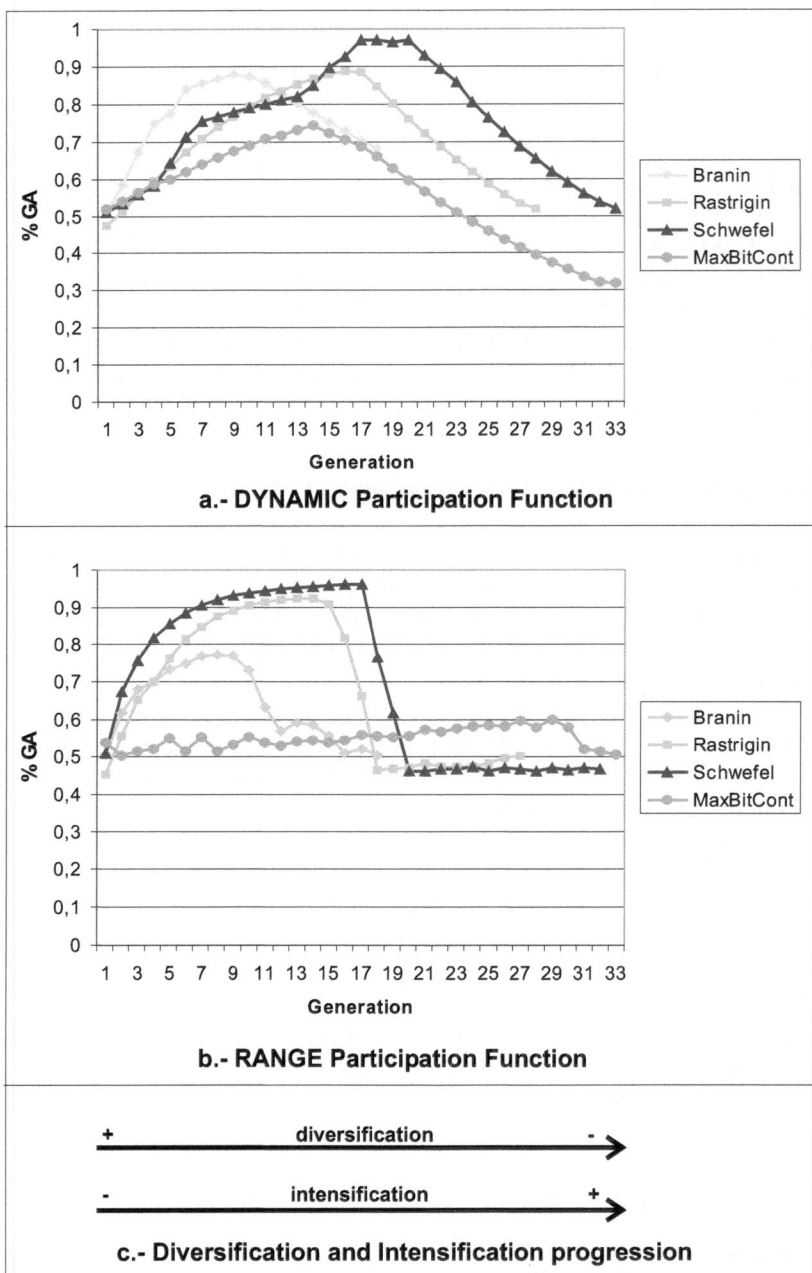

Fig. 2. Evolution of a.- DYNAMIC and b.- RANGE Participation Functions. Progression of Diversification and Intensification during the search

early steps of the process ,during the first generations, genetic algorithms perform better, and therefore their participation ration increases. However, in a second stage, EDAs get profit from their better intensification performance and this characteristic causes that the ration of participation is inverted. The shape of this participation function is similar in all the experiments, and the variations are based on the specific nature of the problem itself.

RANGE participation function (Figure 2.b) presents a similar behavior in general, although (i) there is an abrupt transition between the region in which GAs exploit diversification and the moment in which EDAs are necessary to converge to the optimum value via intensification. (ii) in MaxBitCont problem there are similar proportions of both methods during all the evolution.

7   Conclusions and Future Work

In this contribution a new Participation Function for the hybrid GA-EDA algorithm has been presented. The new function provides a direct adaptability to the results achieved by each of the participating algorithms. This performance seems to fit better at the switching point in which the importance of the diversification decreases and intensification is more required to obtain the optimum value.

Besides, diversification and intensification of both GA and EDA algorithms have been analyzed. This study requires a deeper research to evaluate the theoretical benefits and the quantitative results of these two algorithms according to these concepts.

References

1. V. Bachelet and E. Talbi. Cosearch: A co-evolutionary metaheuritics. In *Proceedings of Congress on Evolutionary Computation CEC2000*, pages 1550–1557, 2000.
2. C. Blum and A. Roli. Metaheuristics in combinatorial optimization: Overview and conceptual comparison. *ACM Computing Surveys*, 35(3):268–308, September 2003.
3. F.K. Branin. A widely convergent method for finding multiple solutions of simultaneous nonlinear equations. *IBM Journal of Research and Development*, pages 504–522, 1972.
4. J. Denzinger and T. Offerman. On cooperation between evolutionary algorithms and other search paradigms. In *Proceedings of Congress on Evolutionary Computation CEC1999*, pages 2317–2324, 1999.
5. F. Foccaci, F. Laburthe, and A. Lodi. Local search and constraint programming. In F. Glover and G. Kochenberger, editors, *Handbook of Metaheuristics*, volume 57 of *International Series in Operations Research and Management Science*, Kluwer Academic Publishers, Norwell, MA.
6. F. Glover and M. Laguna. *Tabu Search*. Kluwer Academic Publishers, 1997.
7. J. Hao, F. Lardeux, and F. Saubion. A hybrid genetic algorithm for the satisfiability problem. In *Proceedings of the First International Workshop on Heuristics*, Beijing, 2002.
8. F. Herrera and M. Lozano. Gradual distributed real-coded genetic algorithms. *IEEE Transactions on Evolutionary Computation*, 4(1), 2000.
9. J.H. Holland. *Adaption in natural and artificial systems*. The University of Michigan Press, Ann Harbor, MI, 1975.
10. S. Kirkpatrick, C.D. Gelatt, and M.P. Vecchi. Optimization by simulated annealing. *Science*, 220(4598):671–680, May 1983.

11. P. Larrañaga, R. Etxeberria, J. A. Lozano, and J. M. Peña. Optimization in continuous domains by learning and simulation of Gaussian networks. In A. S. Wu, editor, *Proceedings of the 2000 Genetic and Evolutionary Computation Conference Workshop Program*, pages 201–204, 2000.
12. P. Larrañaga and J.A. Lozano. *Estimation of Distribution Algorithms. A New Tool for Evolutionary Computation*. Kluwer Academic Publisher, 2001.
13. F.T. Lin, , C.Y. Kao, and C.C. Hsu. Incorporating genetic algorithms into simulated annealing. *Proceedings of the Fourth International Symposium on Artificial Intelligence*, pages 290–297, 1991.
14. O.C. Martin and S.W. Otto. Combining simulated annealing with local search heuristics. *Annals of Operations Research*, 63:57–75, 1996.
15. C.H. Papadimitriou and K. Steiglitz. *Combinatorial Optimization: Algorithms and Complexity*. Prentice-Hall, 1982.
16. J.M. Peña, V. Robles, P. Larrañaga, V. Herves, F. Rosales, and M.S. Pérez. GA-EDA: Hybrid evolutionary algorithm using genetic and estimation of distribution algorithms. In *Innovations in Applied Artificial Intelligence: Proceeding of IEA/AIE 2004*, 2004.
17. V. Robles, J.M. Peña, P. Larrañaga, M.S. Pérez, and V. Herves. GA-EDA: A new hybrid cooperative search evolutionary algorithm. In J.A. Lozano, P. Larrañaga, I. Inza, and E. Bengoetxea, editors, *Towars a New Evolutionary Computation. Advances in Estimation of Distribution Algorithms*. Springer Verlag, 2005. In Press.
18. V. Robles, M.S. Pérez, V. Herves, J.M. Peña, and P. Larrañaga. Parallel stochastic search for protein secondary structure prediction. In *Lecture Notes in Computer Science*, Czestochowa, Poland, 2003.
19. E-G. Talbi. A taxonomy of hybrid metaheuristics. *Journal of Heuristics*, 8(5):541–564, 2002.
20. A. Törn, M. M. Ali, and S. Viitanen. Stochastic global optimization: Problem classes and solution techniques. *Journal of Global Optimization*, 14, 1999.
21. M. Toulouse, T. Crainic, and B. Sansó. An experimental study of the systemic behavior of cooperative search algorithms. In I. Osman S. Voß, S. Martello and C. Roucairol, editors, *In Meta-Heuristics: Advances and Trends in Local Search Paradigms for Optimization*, chapter 26, pages 373–392. Kluwer Academic Publishers, 1999.

Spatial Approach to Pose Variations in Face Verification

Licesio J. Rodríguez-Aragón, Ángel Serrano,
Cristina Conde, and Enrique Cabello

Universidad Rey Juan Carlos, c\ Tulipán, s/n,
E-28933, Móstoles, Madrid, Spain
{licesio.rodriguez.aragon, angel.serrano,
cristina.conde, enrique.cabello}@urjc.es
http://frav.escet.urjc.es

Abstract. Spatial dimension reduction methods called Two Dimensional PCA and Two Dimensional LDA have recently been presented. These variations of traditional PCA and LDA consider images as 2D matrices instead of 1D vectors. The robustness to pose variations of these advances at verification tasks, using SVM as classification algorithm, is here shown.

The new methods endowed with a classification strategy of SVMs, seriously improve, specially for pose variations, the results achieved by the traditional classification of PCA and SVM.

1 Introduction

Some of the fields where biometrics play a relevant role are not only the improvement of security but also the development of smart environments where individuals are able to interact with computers in a human related way [1]. Dimensionality reduction is an important and necessary preprocessing of multidimensional data, as face images. Recent tests to measure the progress recently made towards face recognition show that accuracy on frontal face with indoor lighting goes beyond 90%, which is promising for early stages of recognition tasks [2]. On the other hand, face recognition among different pose or illumination is far from acceptable. Robustness to this changes in facial images is searched in many ways.

Analysis of the effects of pose [3] and illumination [4,5] variations over each face have been studied, searching for invariant characteristics or analyzing the perturbations introduced in the data. A normalization task is aimed by detecting characteristic points and measuring distances [6]. Three dimensional models of facial images are obtained through laser scanners [7], increasing the cost and the complexity of the problem. From our point of view, these methods improve the performance of the classification but traditional methods avoid dealing with an important problem, the spatial structure of the images.

Face recognition is different from classical pattern recognition, since there are many individual classes and only a few images per class. Dimension reduction

A.F. Famili et al. (Eds.): IDA 2005, LNCS 3646, pp. 351–361, 2005.
© Springer-Verlag Berlin Heidelberg 2005

methods commonly used, like Principal Component Analysis (PCA) or Linear Discriminant Analysis (LDA), and Gabor Filters [8], as well as other improved variations, like Independent Component Analysis (ICA) [9] and Kernel Principal Component Analysis (KPCA) [10], obtain a feature set for each image. Classical methods use vectorized representations of the images containing the faces instead of working with data in matrix representation. The main drawbacks of the classical vectorized projection methods is that it is easy to be subjected to gross variations and thus, high sensitive to any changes in pose, illumination etc.

New advances on feature extraction methods called Two-Dimensional Principal Component Analysis [11,12] and Two-Dimensional Linear Discriminant Analysis [13,14] have shortly been presented, and preliminar experiments and junctions of these new methods with SVM are the focus of this work. Experiments are performed over a wide set of subjects, joined in a facial database of images which allow the measurement of the advances of the recognition task to pose variations, specially to rotated faces.

2   Feature Extraction

Traditional feature extraction techniques require that 2D face images are vectorized into a 1D row vector to then perform the dimension reduction [8,9,10]. The resulting image vectors belong to a high-dimensional image vector space where covariance matrices are evaluated with a high associated computational cost.

Recently, a Two-Dimensional PCA method (2DPCA) and Two-Dimensional LDA (2DLDA) have been developed for bidimensional data feature extraction. Both methods are based on 2D matrices rather than 1D vectors, preserving spatial information.

2.1   Principal Component Analysis

Given a set of images $I_1, I_2, \ldots, I_N$ of height h and width w, PCA considers the images as 1D vectors in a $h \cdot w$ dimensional space. The facial images are projected onto the eigenspace spanned by the leading ortonormal eigenvectors, those of higher eigenvalue, from the sample covariance matrix of the training images. Once the set of vectors has been centered, the sample covariance matrix is calculated, resulting a matrix of dimension $h \cdot w \times h \cdot w$. It is widely known that if $N \ll h \cdot w$, there is no need to obtain the eigenvalue decomposition of this matrix, because only N eigenvectors will have a non zero associated eigenvalue [15]. The obtention of these eigenvectors only requires the decomposition of an $N \times N$ matrix, considering as variables the images, instead of the pixels, and therefore considering pixels as individuals.

Once the first d eigenvectors are selected and the proportion of the retained variance fixed (Fig. 1), $\sum_1^d \lambda_i / \sum_1^N \lambda_i$, being $\lambda_1 > \lambda_2 > \cdots > \lambda_N$ the eigenvalues, a projection matrix A is formed with $h \cdot w$ rows and d columns, one for each eigenvector. Then a feature vector $Y_{d \times 1}$ is obtained as a projection of each image $I_{h \cdot w \times 1}$, considered as a 1D vector, onto the new eigenspace.

2.2   Linear Discriminant Analysis

The previous method maximizes the total scatter retained by the fixed dimension. Information provided by the labels of the set of images, $I_1, I_2, \ldots, I_N$, is not used. Linear Discriminant Analysis shapes the scatter in order to make it more reliable for classification. Traditional Linear Discriminant Analysis uses this information to maximize between-class scatter whereas within-class scatter is minimized simplifying the classification process and focusing the problem in a more reliable way.

As images are transformed into a 1D vector, the method faces the difficulty that the within-class scatter matrix, of dimension $h \cdot w \times h \cdot w$, is always singular as the number of images N of the set is usually much lower than the number of pixels in an image. An initial projection using PCA is done to a lower dimensional space so that the within-scatter matrix is non singular. Then applying the standard Fisher Linear Discriminant Analysis, the dimension is finally reduced [16].

2.3   Two-Dimensional Principal Component Analysis

The consideration of images $I_{h \times w}$ as 1D vectors instead as 2D structures is not the right approach to retain spatial information. Pixels are correlated to their neighbours and the transformation of images into vectors produces a loss of information preserving the dimensionality. On the contrary, the main objective of these methods is the reduction of dimensionality and the least loss of information as possible.

The idea recently presented as a variation of traditional PCA, is to project an image $I_{h \times w}$ onto X^{PCA} by the following transformation [11,12],

$$Y_{h \times 1} = I_{h \times w} \cdot X^{PCA}_{w \times 1}. \tag{1}$$

As result, a h dimensional projected vector Y, known as projected feature vector of image I, is obtained. The total covariance matrix S_X over the set of projected feature vectors of training images $I_1, I_2, \ldots, I_N$ is considered. The mean of all the projected vectors, $\overline{Y} = \overline{I} \cdot X^{PCA}$, being $\overline{I}$ the mean image of the training set, is taken into account.

$$S_X = \frac{1}{N} \sum_{i=1}^{N} (Y_i - \overline{Y})(Y_i - \overline{Y})^T \\ = \frac{1}{N} \sum_{i=1}^{N} [(I_i - \overline{I})X][(I_i - \overline{I})X]^T \tag{2}$$

The maximization of the total scatter of projections is chosen as the criterion to select the vector X^{PCA}. The total scatter of the projected samples is characterized by the trace of the covariance matrix of the projected feature vectors. Applying the criterion to (2) the following expression is obtained,

$$J(X) = tr(S_X) = X^T [\frac{1}{N} \sum_{i=1}^{N} (I_i - \overline{I})^T (I_i - \overline{I})] X. \tag{3}$$

What is known as image covariance matrix S defined as a $w \times w$ nonnegative matrix can be directly evaluated using the training samples,

$$S = \frac{1}{N} \sum_{i=1}^{N} (I_i - \overline{I})^T (I_i - \overline{I})]. \tag{4}$$

The optimal projection axis X^{PCA} is the unitary vector that maximizes (3), which corresponds to the eigenvector of S of largest associated eigenvalue.

2.4   Two-Dimensional Linear Discriminant Analysis

The idea presented as 2DPCA, has been upgraded to consider the class information [13,14]. Suppose there are L known pattern clases having M samples for each class, $N = L \cdot M$. The idea is to project each image as in (1), but to obtain X^{LDA} with the information provided by the classes. The covariance over the set of images can be decomposed into between-class and within-class. The mean of projected vectors as in 2DPCA as well as the mean of projected vectors of the same class $\overline{Y^j} = \overline{I^j} \cdot X^{LDA}$, being $\overline{I^j}$ the mean image of the class $j = 1, \ldots, L$, are taken into account.

$$\begin{aligned} S_{XB} &= \sum_{j=1}^{L} M(\overline{Y^j} - \overline{Y})(\overline{Y^j} - \overline{Y})^T \\ &= \sum_{j=1}^{L} M[(\overline{I^j} - \overline{I})X][(\overline{I^j} - \overline{I})X]^T \end{aligned} \tag{5}$$

$$\begin{aligned} S_{XW} &= \sum_{j=1}^{L} \sum_{i=1}^{M} (Y_i^j - \overline{Y^j})(Y_i^j - \overline{Y^j})^T \\ &= \sum_{j=1}^{L} \sum_{i=1}^{M} [(I_i^j - \overline{I^j})X][(I_i^j - \overline{I^j})X]^T \end{aligned} \tag{6}$$

The objective function maximized in this case to select X^{LDA} is considered a class specific linear projection criterion, and can be expressed as

$$J(X) = \frac{tr(S_{XB})}{tr(S_{XW})}. \tag{7}$$

The total between and within covariances are defined as $w \times w$ nonnegative matrices and can be directly evaluated.

$$S_B = \sum_{j=1}^{L} M[(\overline{I^j} - \overline{I})][(\overline{I^j} - \overline{I})]^T; \quad S_W = \sum_{j=1}^{L} \sum_{i=1}^{M} [(I_i^j - \overline{I^j})][(I_i^j - \overline{I^j})]^T \tag{8}$$

Both matrices are formally identical to the corresponding traditional LDA, and by maximizing (7) the within-class scatter is minimized whereas the between-class scatter is maximized, giving as result the maximization of discriminating information. The optimal projection axis X^{LDA} is the unitary vector that maximizes (7), which corresponds to the eigenvector of $S_B \cdot S_W^{-1}$, of largest associated eigenvalue.

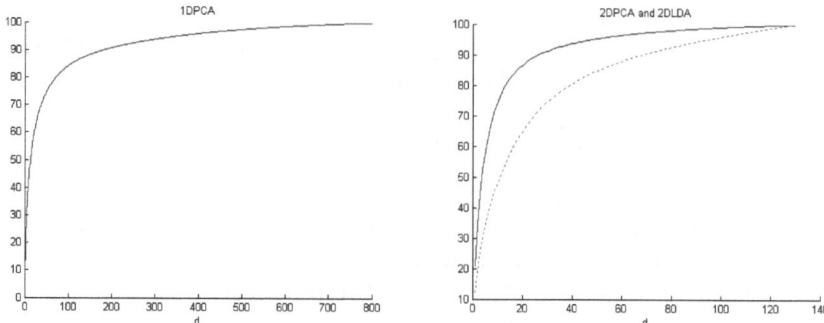

Fig. 1. Evolution of the retained variance percentage for the dimension reduction methods. Left, PCA, for $N = 800$ possible dimensions. Right, 2DPCA in solid line and 2DLDA in dashed line, for $w = 130$ possible dimensions.

3 Projection and Reconstruction

As in traditional PCA, a proportion of retained variance is fixed in 2DPCA and 2DLDA (Fig. 1), $\sum_1^d \lambda_i / \sum_1^w \lambda_i$, where $\lambda_1 > \lambda_2 > \cdots > \lambda_w$ are the eigenvalues and $X_1, X_2, \ldots, X_d$ are the eigenvectors corresponding to the d largest eigenvalues.

Once d is fixed, $X_1, X_2, \ldots, X_d$ are the ortonormal axes used to perform the feature extraction. Let $V = [Y_1, Y_2, \ldots, Y_d]$ and $U = [X_1, X_2, \ldots, X_d]$, then

$$V_{h \times d} = I_{h \times w} \cdot U_{w \times d}. \tag{9}$$

A set of projected vectors, $Y_1, Y_2, \ldots, Y_d$, are obtained for both methods. Each projection over an optimal projection vector is a vector, instead of a scalar as in traditional PCA. A feature matrix $V_{h \times d}$ for each considered dimension reduction method is produced, containing either the most amount of variance, or the most discriminating features of image I.

3.1 Image Reconstruction

In this dimension reduction methods, a reconstruction of the images from the features is possible. An approximation of the original image with the retained information determined by d is obtained.

$$\begin{aligned}
\widetilde{I}_{h \cdot w \times 1} &= A_{h \cdot w \times d} \cdot Y_{d \times 1} \quad \text{PCA image reconstruction.} \\
\widetilde{I}_{h \times w} &= V_{h \times d} \cdot U_{d \times w}^T \quad \text{2DPCA or 2DLDA image reconstruction.}
\end{aligned} \tag{10}$$

4 Classification with SVM

SVM is a method of learning and separating binary classes [17], it is superior in classification performance and is a widely used technique in pattern recognition and especially in face verification tasks [18].

Given a set of features $y_1, y_2, \ldots, y_N$ where $y_i \in \mathbb{R}^n$, and each feature vector associated to a corresponding label $l_1, l_2, \ldots, l_N$ where $l_i \in \{-1, +1\}$, the aim of a SVM is to separate the class label of each feature vector by forming a hyperplane

$$(\omega \cdot y) + b = 0, \quad \omega \in \mathbb{R}^n, b \in \mathbb{R}. \tag{11}$$

The optimal separating hyperplane is determined by giving the largest margin of separation between different classes. This hyperplane is obtained through a minimization process subjected to certain constrains. Theoretical work has solved the existing difficulties of using SVM in practical application [19].

As SVM is a binary classifier, a *one vs. all* scheme is used. For each class, each subject, a binary classifier is generated with positive label associated to feature vectors that correspond to the class, and negative label associated to all the other classes.

4.1   Facial Verification Using SVM

In our experiments a group of images from every subject is selected as the training set and a disjoint group of images is selected as the test set. The training set is used in the feature extraction process through PCA, 2DPCA and 2DLDA. Then, the training images are projected onto the new ortonormal axes and the feature vector (PCA), or vectors (2DPCA,2DLDA), are obtained. For each subject the required SVMs are trained.

Several strategies have been used to train and combine the SVMs. When training and classifying PCA features, each image generates one feature vector $Y_{d \times 1}$ and one SVM is trained for each subject, with its feature vectors labelled as $+1$ and all the other feature vectors as -1.

On the other hand, for feature vectors obtained from 2DPCA and 2DLDA, each image generates a set of projected vectors, $V_{h \times d} = [Y_1, Y_2, \ldots, Y_d]$, and three different strategies have been considered. First strategy generates a unique feature vector through a concatenation of the d projected vectors, then one SMV is trained for each subject as in PCA. The second and third approaches consider the d projected vectors and consequently for each subject d SVMs are trained, one for each feature vector. These d outputs are then combined to produce a final classification output, first through an arithmetic mean and secondly trough a weighted mean.

Once the SVMs are trained, images from the test set are projected onto the eigenspace obtained from the training set. The features of the test set are classified through the SVMs to measure the performance of the generated system.

For the SVM obtained from the PCA and from the concatenation strategy of 2DPCA and 2DLDA feature vectors, the output is compared with the known label of every test image. However, for the ensemble of SVMs obtained from the 2DPCA and 2DLDA feature vectors, the d outputs are combined whether through an arithmetic or a weighted mean. Arithmetic approach combines the d outputs through an arithmetic mean. At weighted approach, every output is weighted with the amount of variance explained by its dimension, that means that each output will be taken in account proportionally to the value of the eigenvalue associated to the corresponding eigenvector: $\lambda_i / \sum_{j=1}^{d} \lambda_j$ is the weight for the i−SVM, $i = 1, 2, \ldots, d$.

To measure the system performance a cross validation procedure is carried out. Results are then described by using Receiver Operating Curve, ROC curve, as there are four possible experiment outcomes: true positive (TP), true negative (TN), false positive (FP) and false negative (FN). The system threshold can then be adjusted to more or less sensitiveness, but in order to achieve fewer errors new and better methods, like 2DPCA and 2DLDA, are required.

5   Design of Experiment

The Face Recognition and Artificial Vision[1] group (FRAV) at the Universidad Rey Juan Carlos, has collected a quite complete set of facial images for 109

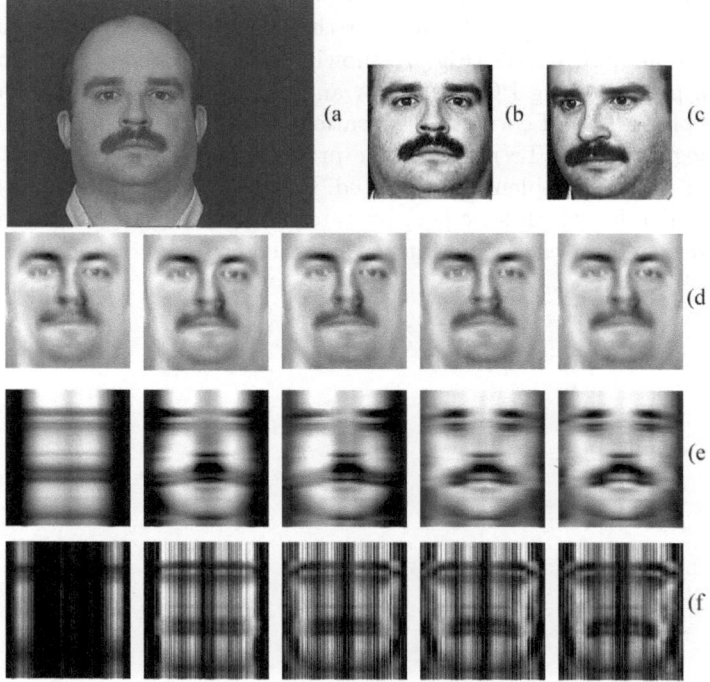

Fig. 2. a) One of the original frontal images in the FRAV2D database. b) Automatically selected window containing the facial expression of the subject in equalized gray scale. c) Sample of a pose variation face, rotated 15°, used to evaluate the performance of the verification. d) From left to right, reconstructed images (10), for $d = 10, 50, 90, 150, 170$, from PCA projection. e) and f) From left to right, reconstructed images (10), for $d = 1, 2, 3, 4, 5$, from 2DPCA and 2DLDA projections respectively.

---

[1] http://frav.escet.urjc.es

subjects. All the images have been taken under controlled conditions of pose and illumination. A partial group of this database is freely available for research purposes.

The images are colored and of size 240 × 320 pixels with homogeneous background color. A window of size 140 × 130 pixels containing the most meaningful part of the face, has been automatically selected in every image and stored in equalized gray scale. That is the information that will be analyzed through the dimension reduction and classification methods (Fig. 2).

The purpose of the following experiments is to confront the robustness to pose variations of the traditional PCA method and classifying strategies to the new proposed 2DPCA and 2DLDA methods in the task of face verification through SVM. Each experiment has been performed for 100 randomly chosen subjects from the whole FRAV2D. In all the experiments, the train set for the extraction of the feature vectors and for the classifiers training is formed by eight frontal images of each subject. Then, the classifiers have been tested over four 15° rotated images to measure the performance of the system at pose variations.

Different tests for the reduced dimension of the projections with different values have been carried out. Results for the best performance of each method are presented as ROC curves (Fig. 3), showing the compared performance of the verification process using PCA, 2DPCA and 2DLDA. True positive rate (TP), that is the proportion of correct classifications to positive verification problems, and true negative rate (TN), that is the proportion of correct classifications to negative verification problems, are plotted. Besides, the equal error rate (EER), that is the value for which false positive rate (FP) is equal to false negative rate (FN), is presented for each experiment that has been undertaken (Fig. 4).

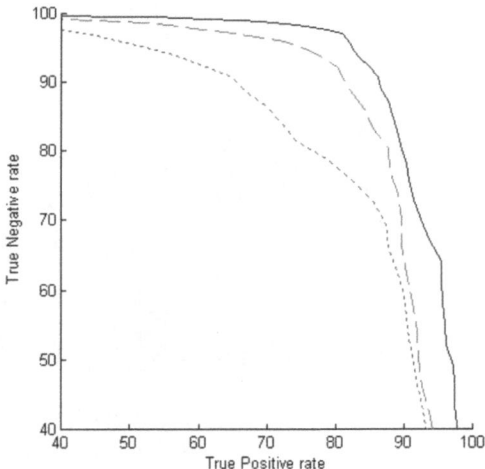

Fig. 3. ROC curves for the best performance of each dimension reduction method, with TP rate in abscises and TN rate in ordinates. The performance of PCA with $d = 170$ in dotted line, 2DLDA with $d = 2$ under concatenated strategy in dashed line and 2DPCA with $d = 1$ in solid line.

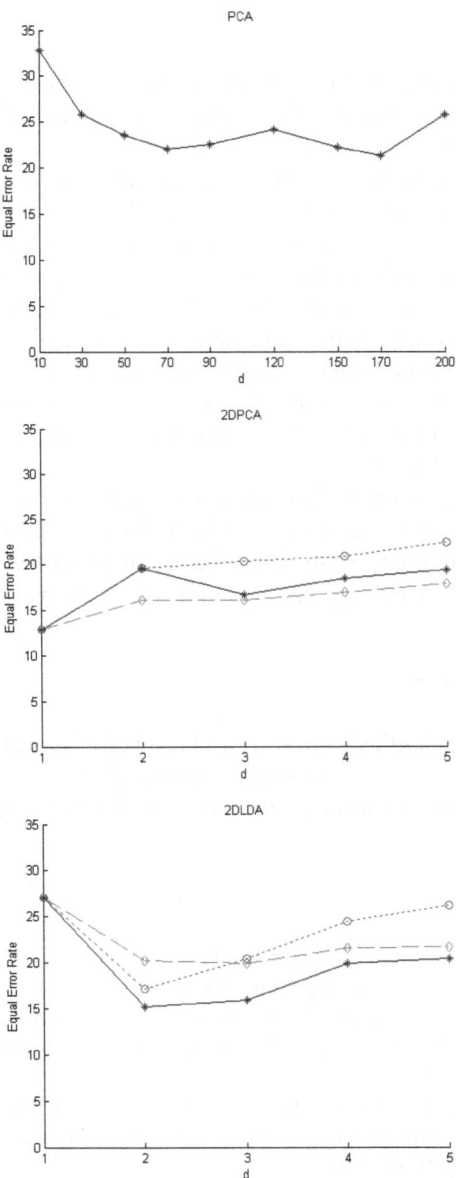

Fig. 4. Top, Equal Error Rate for PCA dimension reduction method for different values of d. Best performance is done for $d = 170$, $EER = 21.33\%$. Center, Equal Error Rate for 2DPCA dimension reduction method for different values of d and the three SVM strategies, concatenated in solid line, arithmetic mean in dotted line and weighted mean in dashed line. Best performance is done for $d = 1$, $EER = 12.89\%$. Bottom, Equal Error Rate for 2DLDA dimension reduction method for different values of d, as in the previous figure the three strategies have been considered. Best performance is done for $d = 2$, $EER = 15.19\%$ with concatenated strategy.

6   Conclusions

Best results are achieved for the spatial reduction method 2DPCA, as presented at the the ROC curves and the EER values for each method (Fig. 3, 4). Improvements are over 8% with respect to PCA.

Both spatial methods improve the performance of traditional PCA but serious differences appear. 2DPCA reaches its maximum accuracy at $d = 1$, while 2DLDA needs $d = 2$ to reach its best performance, both being quite low from $w = 130$ possible dimensions. PCA reaches its best performance at $d = 170$ from $N = 800$ possible dimension. None of the three classifying strategies are able to improve the results while increasing the dimension at 2DPCA. 2DLDA best performance is reached with concatenation strategy, though weighted mean strategy, as in 2DPCA, seems more robust to the increase of dimension. Spatial methods lead to an eigenvector decomposition of matrices with sizes, $w \times w$, much smaller than PCA, $N \times N$.

It is clear that the spatial dimension reduction methods are more reliable for the purpose of face verification, specially for pose variations (Fig. 2), but deeper work has to be done to use all the information provided by the dimension reduction methods in order to achieve a more accurate verification.

Acknowledgments

Authors would like to thank César Morales García for his enthusiastic work. Also thanks must be given to every one that offered his help to join FRAV2D data base. This work has been partially supported by URJC grant GVC$-2004 - 04$.

References

1. Bowyer, K. W.: Face recognition technology: security versus privacy. IEEE Technology and society magazine. **Spring** (2004) 9–20
2. Messer, K., Kittler, J., Sadeghi, M., et al. : Face authentification test on the BANCA database. Proceedings of the International Conference on Pattern Recognition. (2004) 523–532.
3. Gross, R., Jie Yang and Waibel, A.: Growing Gaussian mixture models for pose invariant face recognition. In Proceedings. 15th International Conference on Pattern Recognition. (2000) 1088–1091.
4. Batur, A.U. and Hayes, M.H.I.I.I.: Linear subspaces for illumination robust face recognition. Proceedings of the IEEE Computer Society Conference on Computer Vision and Pattern Recognition. (2001) vol. 2 296–301.
5. Chen, H., Belhumeur, P. and Jacobs, D.: In search of Illumination Invariants. Proceedings of the IEEE Conference on Computer Vision and Pattern Recognition. (2000) 254–261.
6. Conde, C., Cipolla, R., Rodríguez-Aragón, L.J., Serrano Á. and Cabello, E.: 3d facial feature location with spin images. In Proceedings of the 9th International Association for Pattern Recognition Conference on Machine Vision Applications. (2005) 418–421.

7. Lu, X., Colbry, D. and Jain, A. K.: Three-Dimensional Model Based Face Recognition. In Proceedings of International Conference on Pattern Recognition. (2004) 362–366.
8. Pang, S., Kim, D. and Bang, S. Y.: Membership authentication in the dynamic group by face classification using SVM ensemble. Pattern Recognition Letters **24** (2003) 215–225.
9. Kim, T., Kim, H., Hwang, W. and Kittler, J.: Independent Component Analysis in a local facial residue space for face recognition. Pattern Recognition. **37** (2004) 1873–1885.
10. Cao L.J. and Chong W.K.: Feature extraction in support vector machine: a comparsion of PCA, KPCA and ICA. Proceedings of the International Conference on Neural Information Processing. Vol. 2 (2002) 1001–1005.
11. Yang, J. and Yang, J.: From image vector to matrix: a straightforward image projection technique–IMPCA vs. PCA. Pattern Recognition **35** (2002) 1997–1999.
12. Yang, J., Zhang, D., Frangi and F., Yang, J.: Two-Dimmensional PCA: A new approach to apperance-based face representation and recognition. IEEE Transacctions on Pattern Recognition and Machine Intelligence. **26** (2004) 131–137.
13. Li, M., Yuan, B.Z.: A novel statistical linear discriminant analysis for image matrix: two-dimensional fisherfaces. Proceedings of the International Conference on Signal Processing. (2004) 1419–1422.
14. Chen S., Zhu Y., Zhang D. and Yang J.: Feature extraction approaches based on matrix pattern: MatPCA and MatFLDA. Pattern Recognition Letters. In press.
15. Turk, M. and Pentland, A.: Eigenfaces for recognition. Journal of Cognitive Neurosicience. **3** (1991) 71–86.
16. Belhumeur, P.N., Hespanha, J.P., Kriegman, D.J.: Eigenfaces vs. Fisherfaces: recognition using class specific linear projection. IEEE Transactions on Pattern Analysis and Machine Intelligence **19** (1997) 711–720.
17. Cortes, C. and Vapnik, V.: Support vector network. Machine Learning. **20** (1995) 273–297.
18. Fortuna, J. and Capson, D.: Improved support vector classification using PCA and ICA feature space modiffication. Pattern Recognition **37** (2004) 1117–1129
19. Joachims, T.: Making large scale support vector machine learning practical. In: Advances in Kernel Methods: Support Vector Machines. MIT Press, Cambridge, MA.

Analysis of Feature Rankings for Classification

Roberto Ruiz, Jesús S. Aguilar–Ruiz,
José C. Riquelme, and Norberto Díaz–Díaz

Department of Computer Science, University of Seville, Spain
{rruiz, aguilar, riquelme, ndiaz}@lsi.us.es

Abstract. Different ways of contrast generated rankings by feature selection algorithms are presented in this paper, showing several possible
interpretations, depending on the given approach to each study. We begin
from the premise of no existence of only one ideal subset for all cases.
The purpose of these kinds of algorithms is to reduce the data set to
each first attributes without losing prediction against the original data
set. In this paper we propose a method, *feature–ranking performance*,
to compare different feature–ranking methods, based on the Area Under
Feature Ranking Classification Performance Curve (AURC). Conclusions
and trends taken from this paper propose support for the performance
of learning tasks, where some ranking algorithms studied here operate.

1 Introduction

It is a fact that the performance of most practical classifiers improve when
correlated or irrelevant features are removed. Feature selection attempts to select
the minimally sized subset of features according to two criteria: classification
accuracy does not significantly decrease; and resulting class distribution given
only the values for the selected features, is as close as possible to the original class
distribution, given all features. In general, the application of feature selection
helps all phases of the data mining process for successful knowledge discovery.

Feature selection algorithms can be grouped into two categories from the
point of view of a method's output: subset of features or ranking of features.
One category is about choosing a minimum set of features that satisfies an
evaluation criterion; the other is about ranking features according to same evaluation measure. Ideally, feature selection methods search through the subsets
of features and try to find the best one among the competing 2^m candidate
subsets (m: number of whole features), according to some evaluation function.
However, this exhaustive process may be costly and practically prohibitive, even
for a medium–sized feature set size. Other methods based on heuristic or random
search methods attempt to reduce computational complexity by compromising
performance.

When feature selection algorithms are applied as a pre–processing technique
for classification, we are interested in those attributes that better classify new
unseen data. If the feature selection algorithm provides a subset of attributes,
this subset is used to generate the knowledge model that will classify the new

A.F. Famili et al. (Eds.): IDA 2005, LNCS 3646, pp. 362–372, 2005.

data. However, when the algorithm provides a ranking it is not easy to determine how many attributes are necessary to obtain a good classification result.

In this work, we present different ways to compare feature rankings and show the variety of possible interpretations depending on the study approach made. Our intent is to learn if any dependence between classifier and ranking methods exist as well as trying to answer two essential enquiries: What is a good feature ranking? And, how do we value/measure a ranking? To this end, we practise different comparisons using four feature ranking methods: χ^2, Information Gain, ReliefF and SOAP, which are commented on later. We will check the results by calculating the success rate using three classifiers: C4.5, Naïve Bayes and nearest neighbour.

The paper is organized as follows. In Section 2, concepts used throughout the paper are defined. Section 3 reviews related work and the motivation of our approach is presented, feature ranking methods and classification techniques to be used in the experiments are described. The AURC is shown in Section 4, experimental results in Section 5 and finally, in Section 6, the most interesting conclusions are summarized.

2   Definitions

In this section some definitions are given to formally describe the concepts used throughout the paper: feature ranking, classifier, classification accuracy and ranking–based classification accuracy.

Definition 1 (Data). *Let* D *be a set of* N *examples* $e_i = (\overline{x_i}, y_i)$, *where* $\overline{x_i} = (a_1, \ldots, a_m)$ *is a set of input attributes and* y_i *is the output attribute. Each input attribute belongs to the set of attributes* $(a_i \in$ A, *continuous or discrete) and each example belongs to the data* $(e_i \in$ D). *Let* C *be the decision attribute* $(y_i \in$ C), *named class, which will be used to classify the data. For simplicity in the paper,* y_i *means "the class label of the example* e_i*".*

Definition 2 (Feature Ranking). *Let* A $= \{a_1, a_2, \ldots, a_m\}$ *be the set of* m *attributes. Let* r *be a function* $r :$ A$_D \to \mathbb{R}$ *that assigns a value of merit to each attribute* $a \in$ A *from* D. *A feature ranking is a function* F *that assigns a value of merit (relevance) to each attribute* $(a_i \in$ A) *and returns a list of attributes* $(a_i^* \in$ A) *ordered by its relevance, with* $i \in \{1, \ldots, m\}$:
$F(\{a_1, a_2 \ldots, a_m\}) = < a_1^*, a_2^*, \ldots, a_m^* >$ *where* $r(a_1^*) \geq r(a_2^*) \geq \ldots \geq r(a_m^*)$.

By convention, we assume that a high score is indicative of a relevant attribute and that attributes are sorted in decreasing order of $r(a^*)$. We consider ranking criteria defined for individual features, independently of the context of others, and we also limit ourselves to supervised learning criteria.

Definition 3 (Classification). *A classifier is a function* H *that assigns a class label to a new example:* H $:$ A$^p \to$ C, *where* p *is the number of attributes to be used by the classifier,* $1 \leq p \leq m$. *The classification accuracy (CA) is the average success rate provided by the classifier* H *given a set of test examples,* *i.e., the averaged number of times that* H *was able to predict the class of the test examples. Let* $\mathbf{x}$ *be a function that extracts the input attributes from the example* e, $\mathbf{x} : \mathrm{A}^m \times \mathrm{C} \to \mathrm{A}^m$. *For a test example* $e_i^* = (x_i, y_i)$, *if* $\mathrm{H}(\mathbf{x}(e_i^*)) = y_i$ *then* e_i^* *is correctly classified; otherwise misclassified.*

In this paper, to measure the performance of the classifiers only the leaving–one–out method will be used, because it is not dependent on randomness, like k–fold cross–validation or hold out. In the next expression, if $\mathrm{H}(\mathbf{x}(e_i)) = y_i$ then 1 is counted, otherwise 0. $CA = \frac{1}{N} \sum_{i=1}^{N} (\mathrm{H}(\mathbf{x}(e_i)) = y_i)$. As we are interested in rankings, the classification accuracy will be measured with respect to many different subsets of the ranking provided by some feature ranking methods.

Definition 4 (Ranking–based Classification). *Let* S_k^{F} *be a function that returns the subset of the first* k *attributes provided by the feature ranking method* F $(S_k^{\mathrm{F}} : \mathrm{A}^m \to \mathrm{A}^k)$. *The ranking–based classification accuracy of* H *will be as follows:*

$$CA_k(\mathrm{F}, \mathrm{H}) = \frac{1}{N} \sum_{i=1}^{N} \left(\mathrm{H}(S_k^{\mathrm{F}}(\mathbf{x}(e_i))) = y_i \right)$$

Note that S_1^{F} is the first (best) attribute of the ranking provided by F; S_2^{F} are the first two attributes, and thus up to m.

3 Preliminary Study

3.1 Related Work

There are few specific bibliographies where feature ranking comparison is defined. Liu and Motoda [1] comments on the use of learning curves to demonstrate the effect of adding attributes when a list of ordered attributes is provided. There is a paper [2], in which attribute ranking by means of only one subgroup are compared, that one receiving the best classification from all the subgroups needed to obtain the learning curve. But, picking features whose importance is greater than a threshold value [3,4], is more simple and divulged. Irrelevant features (whose values are random) that are used as a threshold in the application of algorithm ranking are inserted in [5].

All the ranking comparison is based on calculate the rankings performance. Two measures currently exist to analyze this; by means of its accuracy or by the area under ROC (Receiver Operating Characteristics) curve [6]. A ROC curve A is said to dominate another ROC curve B if A is always above and to the left of B. In the cases where two ROC curves do not dominate each other in the whole range, or when the class distribution and error costs are unknown, the area under ROC curve (AUC) is a good "summary" for comparing these. So, a curve A dominates to another curve B if $AUC(A) > AUC(B)$, where AUC(A) and AUC(B) denotes the area under ROC curve A or B, respectively, in the ROC space. The main limitation of this measure lies in that it is only easily applicable to problems with two classes. For a problem with c classes,

ROC space is composed of $c * (c - 1)$ dimensions. This fact makes the use of this techniques in problems with a considerable number of classes practically inviable and so, although this measure is better than the previous (based on accuracy), we will not use it. Remember that in this paper we intend to show how the user can choose the best possible method according to what the user is looking for, independently of type of the data set.

In all works of ranking comparison previously mentioned, the measure used to calculate the ranking performance is the exactness obtained by a classifier, with k first features list being different in how the threshold is fixed. This posed the following questions: What exactly is being evaluated, the ranking, or the method to select features? Is this correct? The value which is used in comparison depends on three agents: generated ranking, method of fixing the threshold and learning algorithm. The fact is that the classification model´s exactness can change substantially depending on the features taking part; therefore the way of choosing features seems more important than the order in which they are chosen. Consequentially, we can say that comparisons will be right, but not complete. Our suggestion is to directly value the ranking, without depending on the selection method.

3.2   Description of Methods

We have chosen four criteria to rank attributes (see [7] for review), all of them very different from each other. These feature–ranking methods are briefly described next: χ^2 (CH) was first introduced by Liu an Setiono [8] as a discretization method and later shown to be able to remove redundant and/or irrelevant continuous features; **Information Gain** (IG) is based on the information–theoretical concept of entropy, a measure of the uncertainty of a random variable; **Relief** (RL) algorithm uses an approach based on the nearest-neighbour algorithm to assign a relevance weight to each feature. Relief was originally introduced by Kira and Rendell [9] and later enhanced by Kononenko [10]. Each feature's weight reflects its ability to distinguish among the class values; **Soap** (Selection of Attributes by Projections) evaluation criterion [3] (SP) is based on a unique value called NLC (Number of Label Changes). It relates each attribute with the label used for classification. This value is calculated by projecting data set elements onto the respective axis of the attribute (ordering the examples by this attribute), then crossing the axis from the beginning to the greatest attribute value, and counting the NLC produced.

Once feature rankings are obtained, we check the results calculating the success rate using three classifiers. They are chosen as representatives of different types of classifiers: c4.5 [11] (c4) is a tool that summarizes training data in the form of a decision tree. Along with systems that induce logical rules, decision tree algorithms have proved popular in practice. This is due in part to their robustness and execution speed, and to the fact that explicit concept descriptions are produced, which users can interpret; The naive Bayes [12] (nb) algorithm represents knowledge in the form of probabilistic summaries. It employs a simplified version of Bayes formula to decide which class a novel instances belongs to; Nearest-Neighbour [13] (nn) simply finds the stored instance closest (according to a Euclidean distance metric) to the instance to be classified (we will use only one neighbour, 1NN).

3.3   Motivation

Firstly, we observe the quality of the four feature-ranking methods in respect to the tree classifiers, we will use the Glass2 data set (214 examples, 9 attributes, 2 classes), since it is a representative case to discuss our motivation. Table 1 shows the rankings for χ^2, Information Gain, Relief and SOAP. For each feature–ranking method, the row rk presents the ranking of attributes and, under this row, the classification performance for C4.5, Naïve Bayes and the Nearest Neighbour technique (using only one neighbour), by using the number of attributes from the ranking indicated in the first row, under "Subset". Classification accuracies (using the 9 attributes) from C4.5, Naïve Bayes and 1–NN are very different: 75.5%, 62.0% and 77.3%, respectively. For example, the most relevant attribute for χ^2 and IG was 7, for Relief 3 and SOAP 1. Using only the attribute 7 (CH and IG), C4.5 produced a classification success of 73.6. However, the classification success with attribute 3 was 57.7 (RL) and 77.3 with attribute 1 (SP). The second attribute selected by χ^2 and IG was 1, Relief selected 6 and SOAP, 7. The first three attributes for χ^2, IG and SOAP were the same, so these three classification results are equal. The fourth attribute breaks the tie. Several interesting conclusions can be drawn from the analysis of Table 1: (a) The four feature–ranking methods provide different rankings, what obvi-

Table 1. Feature–rankings for Glass2. FR: Feature–Ranking method (CH: χ^2; IG: Information Gain; RL: Relief; SP: Soap); Cl: Classifier (c4: C4.5; nb: Naïve Bayes; nn: 1–Nearest Neighbour); and rk: ranking of attributes.

FR	Cl	Subset								
		1	2	3	4	5	6	7	8	9
CH	rk	7	1	4	6	3	2	9	8	5
	c4	**73.6**	77.9	**82.2**	78.5	75.5	74.8	73.6	76.1	75.5
	nb	57.1	57.1	66.9	69.9	63.8	63.8	63.2	62.0	62.0
	nn	66.9	**79.7**	75.5	**82.8**	**88.3**	**81.0**	**77.9**	**77.9**	**77.3**
IG	rk	7	1	4	3	6	2	9	8	5
	c4	**73.6**	77.9	**82.2**	82.2	75.5	74.8	73.6	76.1	75.5
	nb	57.1	57.1	66.9	63.8	63.8	63.8	63.2	62.0	62.0
	nn	66.9	**79.7**	75.5	**84.7**	**88.3**	**81.0**	**77.9**	**77.9**	**77.3**
RL	rk	3	6	4	7	1	5	2	8	9
	c4	57.7	67.5	80.4	76.7	75.5	75.5	74.8	77.9	75.5
	nb	**62.0**	62.6	65.0	64.4	63.8	63.8	63.8	62.6	62.0
	nn	58.9	**75.5**	**81.0**	**83.4**	**88.3**	**83.4**	**81.6**	**81.6**	**77.3**
SP	rk	1	7	4	5	2	3	6	9	8
	c4	**77.3**	77.9	**82.8**	**81.6**	**81.6**	**84.1**	74.9	73.0	75.5
	nb	52.2	57.1	66.9	65.6	62.6	63.2	63.8	62.0	62.0
	nn	72.4	**79.7**	75.5	79.8	80.4	82.2	**81.6**	**77.3**	**77.3**

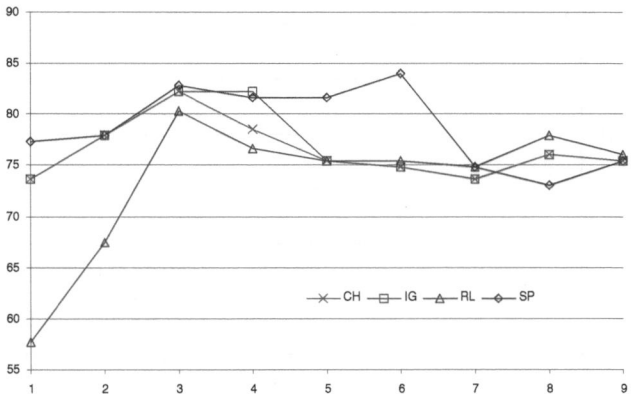

Fig. 1. Accuracy obtained by C4.5 for data set Glass2 (data from Table 1). The number of attributes used to classify are in the abscissa and the success rate in the ordinate.

ously leads to different classification performance. (b) The pair Soap+C4.5 is the only one that provides a classification performance (77.3) using only one attribute (attribute 1) better than using the whole set of attributes (75.5). (c) The sequence of best classification performance is, in principle, arbitrary: (SP+C4, 77.3), (SP+NN, 79.8), (SP+C4, 82.8), (IG+NN, 84.7), ({CH,IG,RL}+NN, 88.3), (SP+NN, 84.1), ({RL+SP}+NN,81.6), (RL+NN,81.6) and the last best value 77.3, with NN. (d) It seems that NN performs very well when the number of attributes is greater than $m/2$. A significant fact is that the best five attributes with 1NN are {1,3,4,6,7}, but the best six attributes are {1,2,3,4,5,7}. Attribute 6 is not that relevant when attributes 2 and 5 are taken into account. In general, a variable that is completely useless by itself can provide a significant performance improvement when it is taken with others.

Figure 1 shows the classification accuracy for C4.5 by using the four feature–ranking methods with the data set Glass2. Although the best subset exactness is similar, SOAP performance is excellent for any feature number and is the only method that in almost all subsets appears above average. In conclusion, we could assert that it is the best ranking of all. The analysis based on the best subset does not exactly show the kindness of features ranking because before or after that subset, the results could be terrible. Taking into account these conclusions, we want to consider the possibility of finding some insight about when one feature–ranking is better than others for a given classifier. Therefore, it would be interesting to explore the ranking method performance along the learning curve described, and extracting conclusions according to the feature proportion used.

Figure 2 shows the possible situations when we compare different rankings for a data subsets. The question posed is: Which ranking is better to classify? The answer would be conditioned by what the user is looking for. This means, if the interest is the ranking identification method that gets the best classified

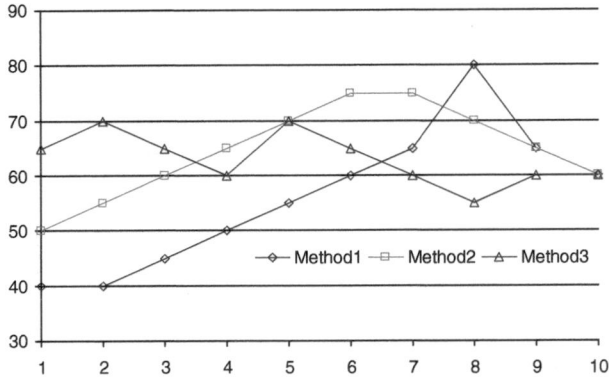

Fig. 2. Fictitious example of three different kind of learning curves

subset for a learning algorithm given, we should choose Method1, remaining conscious of what we need for that eighty percent of features. However, it has been observed that the rest of the curve classification results are almost always under two other methods. If we choose a features number lower than seventy percent, Method1 results will be the worst of the three. If what we are looking for is a best performance method along the whole curve, we must compare the evolution of the three curves point to point. Method2 loses at the beginning (until thirty percent of all features). With Method3, the former is always better than the previous ones, except in the previously commented case (with eighty percent of features). Finally, Method3 is the best, if we want to choose less than thirty percent of the features.

4   Area Under Learning Curve

Comparing subset to subset would be a more complete comparison between two features ranking. Comparing classification results obtained by the first feature of the two lists (the best one), with the two best, and so on successively until m ranked features. We could use this comparison, calculating the average of the obtained results with each list, to compare rankings. The calculation of the area under curve described by previous results would be a very similar study.

Area Under the Curve (AUC) is calculated applying the trapezium formula. In our case, the curve (learning curve) is obtained adding features according to the order assigned by ranking method.

$$\sum_{i=1}^{m-1} (x_{i+1} - x_i) * \frac{(y_{i+1} + y_i)}{2}$$

Definition 5 (AURC). *Given a feature ranking method* F *and a classifier* H, *we can obtain the performance of the classification method regarding the ranking* *provided by the feature ranking by measuring the area under the curve. The curve is drawn by joining every two points* $(CA_k(F,H), CA_{k+1}(F,H))$, *where* $k \in \{1, \ldots, m-1\}$ *and* m *is the number of attributes. The Area Under Ranking Classification Performance Curve* $AURC(F,H)$ *will be calculated as:*

$$AURC(F,H) = \frac{1}{2(m-1)} \sum_{i=1}^{m-1} (CA_i(F,H) + CA_{i+1}(F,H))$$

With this expression, for any pair (F, H), $AURC(F,H) \in [0,1]$ (in Table 3, it appears multiplied by one hundred for a better understanding), which provides us an excellent method to compare the quality of feature rankings with respect to each classification method. Take into account that the best AURC correspond to the best Ranking method.

An interesting property of this curve is that it is not monotone increasing, i.e., for some i, it would be possible that $CA_i(F,H) > CA_{i+1}(F,H)$.

Definition 6 (Feature–Ranking Performance). *The feature–ranking performance is measured as the evolution of the AURC along the ranking of features, with step $\delta\%$. The curve is plotted, for every $\delta\%$ of the attributes as follows:*

$$AURC_\delta(F,H) = \frac{1}{2\delta(m-1)} \sum_{i=1}^{\delta(m-1)} (CA_i(F,H) + CA_{i+1}(F,H))$$

We must consider that the idea concerning every feature selection method (one of them is ranking method) is that it must take the smallest number of features as possible. If we contemplate the possibility that in each learning curve, high and short exactnesses are compensated to the AURC calculation, we must make a study about methods performance using first features and fixed percentages.

5 Experiments

The implementation of induction algorithms and other selectors was done using Weka library [14] and comparison was performed with sixteen data sets from the University of California at Irvine [15] summarized in Table 2. All the experiments were run using leaving one out. The four methods of feature rankings are applied to each data set, and each ranking learning curve is calculated with the three classifiers.

Table 3 shows, for each data set, the Area Under Classification Performance Curve. Boldprint values are the best for the three classifiers, and those underlined are the best for corresponding classifiers. A clear conclusion can not be made, but specific trends can: (a) Results are very similar under each classifier (last line). There are some differences between each one of them. 1–NN is the classifier that offers a better performance with the four feature ranking methods; C4.5 is very close and NB is the last one. (b) If we take into account the best AURC for each data set, 1–NN obtains better results. (c) Most of the RL cases win, so that we could conclude that it is the best ranking method.

Table 2. Data sets used in the experiments

Data set	Id	Instances	Attributes	Classes
anneal	AN	898	38	6
balance	BA	625	4	3
g_credit	GC	1000	20	2
diabetes	DI	768	8	2
glass	GL	214	9	7
glass2	G2	163	9	2
heart–s	HS	270	13	2
ionosphere	IO	351	34	2
iris	IR	150	4	3
kr–vs–kp	KR	3196	36	6
lymphography	LY	148	18	4
segment	SE	2310	19	7
sonar	SO	208	60	2
vehicle	VE	846	18	4
vowel	VW	990	13	11
zoo	ZO	101	16	7

Table 3. AURC value for each ranking–classifier combination

	C4.5				NB				1NN			
DS	CHI2	IG	RLF	SOAP	CHI2	IG	RLF	SOAP	CHI2	IG	RLF	SOAP
an	97.30	97.12	96.90	97.11	85.82	86.30	86.50	86.47	**98.20**	98.09	97.54	97.71
bs	68.61	68.61	68.83	68.61	**75.55**	**75.55**	72.56	**75.55**	72.77	72.77	69.79	72.77
gc	72.39	72.39	71.71	72.31	**74.74**	**74.74**	73.89	74.22	70.16	70.16	70.38	66.83
di	72.85	72.89	73.30	72.52	75.36	**75.73**	75.68	75.15	68.87	69.52	68.09	67.87
gl	64.57	66.09	67.09	67.32	49.15	49.85	47.34	51.37	63.49	67.67	68.17	**71.12**
g2	76.65	77.11	74.35	79.03	63.27	62.50	63.50	62.27	79.41	79.64	**80.37**	78.91
hs	78.23	78.23	77.04	76.54	**83.09**	**83.09**	81.53	80.80	78.43	78.43	75.94	74.34
io	88.69	89.18	**90.03**	86.94	84.52	85.11	85..66	80.23	88.57	88.36	88.57	86.92
ir	95.11	95.11	95.22	95.22	95.56	95.56	95.56	95.56	95.00	95.00	95.56	95.56
kr	95.22	95.13	**96.48**	95.56	87.47	87.47	89..81	86.99	93.97	93.89	95.79	94.63
ly	74.84	75.97	75.66	75.42	78.76	80.25	80.37	80.09	76.61	81.28	**82.31**	80.56
se	92.23	92.15	93.37	92.96	74.63	73.79	78.26	76.77	92.78	93.11	**93.87**	93.26
so	73.92	73.71	76.06	75.15	67.62	67.44	69.57	68.83	84.15	83.94	**84.41**	83.87
ve	64.59	65.79	**68.10**	67.43	41.65	41.54	41.72	41.04	66.09	65.83	65.79	65.69
vw	73.98	74.20	73.96	74.59	61.96	62.46	61.65	62.17	**90.67**	90.66	89.63	90.52
zo	88.18	87.56	86.88	88.27	88.95	88.21	86.42	89.36	**91.34**	90.84	87.69	90.87
Av	79.83	80.08	80.31	80.31	74.25	74.35	74.37	74.18	81.91	82.45	**82.12**	81.96

In order to facilitate the comparison of diverse ranking methods from different points of view, and to extract some conclusions, table 4 is presented. In this table we show a summary of each time a ranking method holds the first position. Different groups of comparisons are set: results obtained by the first features

Table 4. Summary of times each ranking method holds first position. Results assocciated by: first features, percentages and classifiers

Results for	CH	IG	RL	SP
Exactness-1at:	26	28	20	**30**
AURC-2at:	18	22	16	**24**
AURC-3at:	15	17	15	**21**
AURC-4at:	14	15	16	**20**
AURC-5at:	15	**20**	18	17
AURC-25%:	17	**22**	21	17
AURC-50%at:	13	12	**21**	13
AURC-all at:	14	12	**25**	12
C4.5:	39	46	50	**51**
NB:	41	**59**	58	46
NN:	50	43	44	**55**
Total:	130	148	**152**	**152**

are situated in the first block (success rate with the first feature and the AURC with two, three, four and five features are contrasted); the second block shows the comparisons by percentage results (25, 50 with the whole results); and last group is broken down by classifiers.

If we contemplate the tests done by the first features, SOAP ranking method stands out, especially in relation with C4.5 and 1–NN classifiers, the one that offers the best result with NB is IG, using only the first ranking features. IG and RL obtain better results at 25% of ranking (IG: 22, rl: 21 y CH, SP: 17). Partly through a classifier, this position is kept with C4.5 and NB, but not with a NN in first position at 25% for Relief. From here through the whole features set, RL is the one that most frequently holds first position. At a 100% ranking, relief wins with a difference 25 times in comparison to CH, 12 to IG, and 10 to SP, and wins equally at 50% of ranking features. Results are kept with these percentages (50 and 100) for the three classifiers.

If we do the study regarding the entire eight tested by classifiers, there are no large differences. For C4.5 classifier, SP and RL methods stand out with very few differences regarding to IG. IG and RL are those that hold first positions with NB, while with 1NN it is SP. SOAP and Relief, with 152, are the ones which stayed in first position most of the time in all the tests (480); with IG 148, and with chi2 130, following.

We can adhere to the next recommendations due to the results obtained through the last three tests (AURC, AURC´s percentage and AURC with the first features of the arrange list): (I) AURC gives a more complete ranking goodness idea than the exactness obtained by a feature subset. (II) The complete best valued list is generated by the *RL* algorithm. However, if we are going to work with the first features, or with less than 25% of the features, SP and IG methods offer better results in less time. (III) In general, the best classification results are obtained by 1NN, although when the selected features number is smaller (less than the 25%), the performance of C4.5 was better in the four cases than in the rest of the classifiers.

6   Conclusions

Traditional work, where comparisons of feature ranking algorithms are made, mainly evaluate and compare the way of features selection instead of ranking methods. In this paper we present a methodology for evaluating ranking, beginning from the premise of no existence of any singular unique subgroup ideal for every case, and that the best ranking will depend on what the user is looking for.

We can conclude that the Area Under Ranking Classification Performance Curve ($AURC$) shows the complete performance of the orderly features list, globally indicating its predictive power. Based on the analysis of the evolution of AURC, we propose the use of algorithms SP and IG for C4.5 classifier with few features, and the use of RL with classifier 1NN in the rest of cases.

From here, our work aims to confirm if these results can be applied to other larger data sets as well as to study in depth if any relation exists between ranking method and selected classifier. Furthermore, we plan to increase our study with other measures of feature evaluation.

References

1. Liu, H., Motoda, H.: Feature Selection for Knowlegde Discovery and Data Mining. Kluwer Academic Publishers, London, UK (1998)
2. Hall, M., Holmes, G.: Benchmarking attribute selection techniques for discrete class data mining. IEEE Transactions on Knowledge and Data Eng. **15** (2003)
3. Ruiz, R., Riquelme, J., Aguilar-Ruiz, J.: Projection-based measure for efficient feature selection. Journal of Intelligent and Fuzzy System **12** (2002) 175–183
4. Hall, M.: Correlation-based feature selection for discrete and numeric class machine learning. In: 17th Int. Conf. on Machine Learning, Morgan Kaufmann, San Francisco, CA (2000) 359–366
5. Stoppiglia, H., Dreyfus, G., Dubois, R., Oussar, Y.: Ranking a random feature for variable and feature selection. Journal of Machine Learning Research **3** (2003) 1399–1414
6. Huang, J., Ling, C.: Using AUC and accuracy in evaluating learning algorithms. IEEE Transaction on Knowledge and data Engineering **17** (2005) 299–310
7. Yu, L., Liu, H.: Redundancy based feature selection for microarray data. In: 10th ACM SIGKDD Int. Conf. on Knowledge Discovery and Data Mining. (2004)
8. Liu, H., Setiono, R.: Chi2: Feature selection and discretization of numeric attributes. In: 7th IEEE Int. Conf. on Tools with Artificial Intelligence. (1995)
9. Kira, K., Rendell, L.: A practical approach to feature selection. In: 9th Int. Conf. on Machine Learning, Aberdeen, Scotland, Morgan Kaufmann (1992) 249–256
10. Kononenko, I.: Estimating attributes: Analysis and extensions of relief. In: European Conf. on Machine Learning, Vienna, Springer Verlag (1994) 171–182
11. Quinlan, J.R.: C4.5: Programs for machine learning. Morgan Kaufmann, San Mateo, California (1993)
12. Mitchell, T.: Machine Learning. McGraw Hill (1997)
13. Aha, D., Kibler, D., Albert, M.: Instance-based learning algorithms. Machine Learning **6** (1991) 37–66
14. Witten, I., Frank, E.: Data Mining: Practical machine learning tools with Java implementations. Morgan Kaufmann, San Francisco (2000)
15. Blake, C., Merz, E.K.: UCI repository of machine learning databases (1998)

A Mixture Model-Based On-line CEM Algorithm

Allou Samé, Gérard Govaert, and Christophe Ambroise

Université de Technologie de Compiègne,
Département Génie Informatique,HEUDIASYC, UMR CNRS 6599
BP 20529, 60205 Compiègne Cedex
{same, govaert, ambroise}@utc.fr

Abstract. An original on-line mixture model-based clustering algorithm is presented in this paper. The proposed algorithm is a stochastic gradient ascent derived from the Classification EM (CEM) algorithm. It generalizes the on-line k-means algorithm. Using synthetic data sets, the proposed algorithm is compared to CEM and another on-line clustering algorithm. The results show that the proposed method provides a fast and accurate estimation when mixture components are relatively well separated.

1 Introduction

The Classification EM algorithm (CEM) [3], applied using mixture models is a very useful clustering algorithm which generalizes the well known k-means algorithm when assuming specific Gaussian clusters. In a practical point of view, this algorithm is faster that EM algorithm [5] and converges in a few iterations [3]. Many actual applications require massive data sets to be classified in a real-time. In that context, we have applied CEM algorithm in an application dealing with a real-time flaw diagnosis for pressurized containers using acoustic emissions. However, CEM algorithm has not been able to react in real-time when more than 10000 acoustic emissions had to be clustered. Our aim was to develop a faster clustering algorithm without losing the accuracy of the CEM algorithm. For this purpose, we propose in this work an on-line mixture model-based clustering algorithm.

We suppose that data are independent observations $x_1, \ldots, x_n, \ldots$ which are sequentially received and distributed following a mixture density of K components, defined on $\mathbb{R}^p$ by

$$f(x; \Phi) = \sum_{k=1}^{K} \pi_k f_k(x; \theta_k),$$

with $\Phi = (\pi_1, \ldots, \pi_K, \theta_1, \ldots, \theta_K)$ where $\pi_1, \ldots, \pi_K$ denote the proportions of the mixture and $\theta_1, \ldots, \theta_K$ the parameters of each density component. We denote by $z_1, \ldots, z_n, \ldots$ the classes associated to the observations, where $z_n \in \{1, \ldots, K\}$ corresponds to the class of x_n.

A.F. Famili et al. (Eds.): IDA 2005, LNCS 3646, pp. 373–384, 2005.

In the second section we describe the stochastic gradient algorithms in a context of parameter estimation; the third section shows how Titterington derived an on-line clustering algorithm from the EM algorithm [9]; the fourth section expresses the k-means algorithm as a stochastic gradient algorithm; in the fifth section, we propose an on-line clustering algorithm derived from the CEM algorithm; an experimental study is summarized in the sixth section.

2 Stochastic Gradient Algorithms

To estimate the parameter $\boldsymbol{\Phi}$, we choose to use a stochastic gradient algorithm. Generally, stochastic gradient algorithms are used for on-line parameter estimation in signal processing, automatic and pattern recognition for their algorithmic simplicity. They have been shown to be faster than standard algorithms. Using current parameters and new observations, stochastic gradient algorithms update recursively parameters. They allow to maximize the expectation of a criterion [1,2],

$$C(\boldsymbol{\Phi}) = E\left[J(\boldsymbol{x}, \boldsymbol{\Phi})\right].$$

where the criterion $J(\boldsymbol{x}, \boldsymbol{\Phi})$ measures the quality of the parameter $\boldsymbol{\Phi}$ given the observation $\boldsymbol{x}$. The stochastic gradient algorithm aiming to maximize the criterion C is then written

$$\boldsymbol{\Phi}^{(n+1)} = \boldsymbol{\Phi}^{(n)} + \alpha_n \nabla_{\boldsymbol{\Phi}} J(\boldsymbol{x}_{n+1}, \boldsymbol{\Phi}^{(n)}) \tag{1}$$

where the learning rate α_n is a positive scalar or a positive definite matrix such that $\sum |\alpha_n| = \infty$ and $\sum |\alpha_n|^2 < \infty$. Under some regularity conditions on J, Bottou [1,2] shows that this algorithm converges toward a local maximum of $C(\boldsymbol{\Phi})$. In practice, the samples sizes are very large but not infinite. In that case, criterion $C(\boldsymbol{\Phi})$ can be replaced with the empirical mean $\frac{1}{n}\sum_{i=1}^{n} J(\boldsymbol{x}_i; \boldsymbol{\Phi})$ whose maximization is equivalent to the maximization of $\sum_{i=1}^{n} J(\boldsymbol{x}_i; \boldsymbol{\Phi})$.

3 On-line EM Algorithm

This section shows how Titterington [9] has derived a stochastic gradient algorithm from the EM algorithm.

Given the observed data $\mathbf{x}_n = (\boldsymbol{x}_1, \ldots, \boldsymbol{x}_n)$ and some initial parameter $\boldsymbol{\Phi}^{(0)}$, the standard EM algorithm maximizes the log-likelihood $\log p(\mathbf{x}_n; \boldsymbol{\Phi})$ by alternating the two following steps until convergence:

E step (Expectation): computation of the expectation of the complete data conditionally to the available data:

$$Q(\boldsymbol{\Phi}, \boldsymbol{\Phi}^{(q)}) = E[\log p(\mathbf{x}_n, \mathbf{z}_n; \boldsymbol{\Phi})|\mathbf{x}_n, \boldsymbol{\Phi}^{(q)}]$$

$$= \sum_{i=1}^{n} \sum_{k=1}^{K} t_{ik}^{(q)} \log[\pi_k f(\boldsymbol{x}_i; \boldsymbol{\theta}_k)],$$

where $\mathbf{z}_n = (z_1, \ldots, z_n)$ and $t_{ik}^{(q)} = \frac{\pi_k f(\boldsymbol{x}_i; \boldsymbol{\theta}_k)}{\sum_{\ell=1}^{K} \pi_\ell f(\boldsymbol{x}_i; \boldsymbol{\theta}_\ell)}$ is the posterior probability that $\boldsymbol{x}_i$ arises from the kth component of the mixture. This step simply requires the computation of posterior probabilities $t_{ik}^{(q)}$.

M step (Maximization): maximisation of $Q(\boldsymbol{\Phi}, \boldsymbol{\Phi}^{(q)})$ with respect to $\boldsymbol{\Phi}$.

To derive a stochastic algorithm from this formulation, Titterington [9] defined recursively, in the same way as for the EM algorithm, the quantity

$$\begin{cases} Q_1(\boldsymbol{\Phi}, \boldsymbol{\Phi}^{(0)}) & = E[\log p(\boldsymbol{x}_1, z_1; \boldsymbol{\Phi}|\boldsymbol{x}_1; \boldsymbol{\Phi}^{(0)})] \\ Q_{n+1}(\boldsymbol{\Phi}, \boldsymbol{\Phi}^{(n)}) = Q_n(\boldsymbol{\Phi}, \boldsymbol{\Phi}^{(n-1)}) + E[\log p(\boldsymbol{x}_{n+1}, z_{n+1}; \boldsymbol{\Phi})|\boldsymbol{x}_{n+1}; \boldsymbol{\Phi}^{(n)}], \end{cases} \quad (2)$$

where $\boldsymbol{\Phi}^{(n)}$ is the parameter maximizing $Q_n(\boldsymbol{\Phi}, \boldsymbol{\Phi}^{(n-1)})$. The indice n added to letter Q is used to specify that, contrary to the standard EM algorithm, quantity $Q_n(\boldsymbol{\Phi}, \boldsymbol{\Phi}^{(n-1)})$ depends on observations $\mathbf{x}_n = (\boldsymbol{x}_1, \ldots, \boldsymbol{x}_n)$ acquired until the moment n and quantity $Q_{n+1}(\boldsymbol{\Phi}, \boldsymbol{\Phi}^{(n)})$ depends on observations $\mathbf{x}_{n+1} = (\boldsymbol{x}_1, \ldots, \boldsymbol{x}_{n+1})$ acquired until the moment $n + 1$. The maximization of $\frac{1}{n+1}Q_{n+1}(\cdot, \boldsymbol{\Phi}^{(n)})$ using Newton method after approximating the hessian matrix term by its expectation which is the Fisher information matrix

$$I_c(\boldsymbol{\Phi}^{(n)}) = -E[\frac{\partial^2 \log p(\boldsymbol{x}, z; \boldsymbol{\Phi})}{\partial \boldsymbol{\Phi} \partial \boldsymbol{\Phi}^T}]|_{\boldsymbol{\Phi}=\boldsymbol{\Phi}^{(n)}}$$

associated to one complete observation $(\boldsymbol{x}, z)$ results in the algorithm proposed by Titterington:

$$\boldsymbol{\Phi}^{(n+1)} = \boldsymbol{\Phi}^{(n)} + \frac{1}{n+1}[I_c(\boldsymbol{\Phi}^{(n)})]^{-1}\nabla_{\boldsymbol{\Phi}}\log f(\boldsymbol{x}_{n+1}; \boldsymbol{\Phi}^{(n)}). \quad (3)$$

Fisher information matrix $I_c(\boldsymbol{\Phi}^{(n)})$ is positive definite for some density families like the exponential family. In that case, Titterington algorithm has the general form (1) of the stochastic gradient algorithms, which guarantees, under some conditions [1,2], that the criterion maximized by equation (3) is $E[\log f(\boldsymbol{x}; \boldsymbol{\Phi})]$.

4   On-line k-Means Algorithm

This section expresses the well known on-line k-means clustering algorithm [6] as a stochastic gradient algorithm. This algorithm will be generalized by a model-based approach in the next section. It consists in estimating recursively K means $\boldsymbol{\mu}_1, \ldots, \boldsymbol{\mu}_K$ using the algorithm

$$\boldsymbol{\mu}_k^{(n+1)} = \boldsymbol{\mu}_k^{(n)} + \frac{1}{n_k^{(n)} + 1}z_{n+1,k}^{(n)}(\boldsymbol{x}_{n+1} - \boldsymbol{\mu}_k^{(n)}), \quad (4)$$

where $z_{n+1,k}^{(n)}$ equal 1 if k minimizes $(\boldsymbol{x}_{n+1} - \boldsymbol{\mu}_k^{(n)})^T(\boldsymbol{x}_{n+1} - \boldsymbol{\mu}_k^{(n)})$ and 0 otherwise; $n_k^{(n)}$ is the number of observations assigned to component k at the moment n.

This algorithm can also be expressed as a stochastic gradient algorithm with the matrix

$$
\alpha_n = \begin{pmatrix} \frac{1}{n_1^{(n)}+1}I & 0 & \cdots & 0 \\ 0 & \frac{1}{n_2^{(n)}+1}I & \cdots & 0 \\ \vdots & \vdots & & \vdots \\ 0 & 0 & \cdots & \frac{1}{n_K^{(n)}+1}I \end{pmatrix}
$$

as a learning rate, where I is the identity matrix in dimension p. The criterion maximized by this algorithm is the expectation

$$
E\left[\min_{1\leq k\leq K}\frac{1}{2}(\boldsymbol{x}-\boldsymbol{\mu}_k)^T(\boldsymbol{x}-\boldsymbol{\mu}_k)\right] = E[J(\boldsymbol{x},\boldsymbol{\Phi}))] = C(\boldsymbol{\Phi}),
$$

where $\boldsymbol{\Phi} = (\boldsymbol{\mu}_1,\ldots,\boldsymbol{\mu}_K)$.

5   An on Line Clustering Algorithm Derived from CEM Algorithm

This section begins with a recall of the Classification EM (CEM) [3] algorithm in the context of mixture models and then derives a stochastic algorithm from CEM algorithm.

5.1   CEM Algorithm

The Classification EM(CEM) algorithm is an iterative clustering algorithm maximizing, with respect to the components membership vector $\mathbf{z}_n = (z_1,\ldots,z_n)$ and the parameter vector $\boldsymbol{\Phi}$, the classification likelihood criterion

$$
C(\mathbf{z}_n,\boldsymbol{\Phi}) = \log p(\mathbf{x}_n,\mathbf{z}_n;\boldsymbol{\Phi}) = \sum_{i=1}^{n}\sum_{k=1}^{K} z_{ik}\log\pi_k f_k(\boldsymbol{x}_i;\boldsymbol{\theta}_k) \tag{5}
$$

where z_{ik} equal 1 if z_i equal k and 0 otherwise. The classification likelihood criterion is inspired by the criterion

$$
C_1(\mathbf{z}_n,\boldsymbol{\Phi}) = \sum_{i=1}^{n}\sum_{k=1}^{K} z_{ik}\log f_k(\boldsymbol{x}_i;\boldsymbol{\theta}_k)
$$

proposed by Scott an Symons [8] where the sample $\mathbf{x}_n = (\boldsymbol{x}_1,\ldots,\boldsymbol{x}_n)$ is supposed to be formed by separately taking observations of each component of the mixture. The CEM algorithm starts from an initial parameter $\boldsymbol{\Phi}^{(0)}$ and alternates, at qth iteration, the following steps until convergence:

E step (Expectation): computation of the posterior probabilities $t_{ik}^{(q)}$;

C step (Classification): assignation of each observation $\boldsymbol{x}_i$ to the cluster $z_i^{(q)}$ which maximizes $t_{ik}^{(q)}$, $1\leq k\leq K$;

M step (Maximization): maximization of $C(\mathbf{z}_n^{(q)}, \boldsymbol{\Phi})$ with respect to $\boldsymbol{\Phi}$.

Thus, in the mixture model context, the CEM algorithm can be regarded as a classification version of the EM algorithm which incorporates a classification step between the E step and the M step of the EM algorithm. Celeux and Govaert [3] show that each iteration of the CEM algorithm increases the classification likelihood criterion and that convergence is reached in a finite number of iterations.

In order to introduce the on-line CEM algorithm, we should point out that the maximization of the classification likelihood criterion defined by equation (5) is equivalent to the maximization of the criterion

$$L_C(\boldsymbol{\Phi}) = \max_{\mathbf{z}_n}[\log p(\mathbf{x}_n, \mathbf{z}_n; \boldsymbol{\Phi})]$$

$$= \sum_{i=1}^{n} \max_{z_i}[\pi_{z_i} f_{z_i}(\boldsymbol{x}_i; \boldsymbol{\theta}_{z_i})].$$

Each iteration q of the CEM algorithm also consists in maximizing with respect to $\boldsymbol{\Phi}$, the quantity

$$R(\boldsymbol{\Phi}, \boldsymbol{\Phi}^{(q)}) = \log p(\mathbf{x}_n, \mathbf{z}_n^{(q)}; \boldsymbol{\Phi}),$$

where $\mathbf{z}_n^{(q)}$ maximizes $p(\mathbf{x}_n, \mathbf{z}_n; \boldsymbol{\Phi}^{(q)})$.

5.2 On-line CEM Algorithm

To derive from this formulation a stochastic gradient algorithm, we define the quantity R_n as follows:

$$\begin{cases} R_1(\boldsymbol{\Phi}, \boldsymbol{\Phi}^{(0)}) & = \log p(\boldsymbol{x}_1, z_1^{(0)}; \boldsymbol{\Phi}) \\ R_{n+1}(\boldsymbol{\Phi}, \boldsymbol{\Phi}^{(n)}) & = R_n(\boldsymbol{\Phi}, \boldsymbol{\Phi}^{(n-1)}) + \log p(\boldsymbol{x}_{n+1}, z_{n+1}^{(n)}; \boldsymbol{\Phi}), \end{cases} \quad (6)$$

where $\boldsymbol{\Phi}^{(n)}$ maximizes $R_n(\boldsymbol{\Phi}, \boldsymbol{\Phi}^{(n-1)})$ and $z_{n+1}^{(n)}$ maximizes $\log p(\boldsymbol{x}_{n+1}, z_{n+1}; \boldsymbol{\Phi}^{(n)})$. The indice n added to letter R is used to specify, like Titterington approach, that quantity $R_n(\boldsymbol{\Phi}, \boldsymbol{\Phi}^{(n-1)})$ depends on observations $\mathbf{x}_n = (\boldsymbol{x}_1, \ldots, \boldsymbol{x}_n)$ and quantity $R_{n+1}(\boldsymbol{\Phi}, \boldsymbol{\Phi}^{(n)})$ depends on observations $\mathbf{x}_{n+1} = (\boldsymbol{x}_1, \ldots, \boldsymbol{x}_{n+1})$. By maximizing $\frac{1}{n+1}R_{n+1}(\cdot, \boldsymbol{\Phi}^{(n)})$ using Newton method and approximating the hessian matrix term by the Fisher information matrix

$$I_c(\boldsymbol{\Phi}^{(n)}) = -E[\frac{\partial^2 \log p(\boldsymbol{x}, z; \boldsymbol{\Phi})}{\partial \boldsymbol{\Phi} \partial \boldsymbol{\Phi}^T}]|_{\boldsymbol{\Phi} = \boldsymbol{\Phi}^{(n)}}$$

associated to one complete observation $(\boldsymbol{x}, z)$, we get our new algorithm given by the recursive formula

$$\boldsymbol{\Phi}^{(n+1)} = \boldsymbol{\Phi}^{(n)} + \frac{1}{n+1}[I_c(\boldsymbol{\Phi}^{(n)})]^{-1}\nabla_{\boldsymbol{\Phi}}\left[\max_{z_{n+1}} \log p(\boldsymbol{x}_{n+1}, z_{n+1}; \boldsymbol{\Phi}^{(n)})\right] \quad (7)$$

which is recognizable as a stochastic gradient algorithm with the matrix learning rate $\frac{1}{n+1}[I_c(\boldsymbol{\Phi}^{(n)})]^{-1}$. This stochastic gradient algorithm maximizes the expected classification likelihood criterion

$$E[\max_{1 \leq z \leq K} \log p(\boldsymbol{x}, z; \boldsymbol{\Phi})] = E[J(\boldsymbol{x}, \boldsymbol{\Phi})] = C(\boldsymbol{\Phi}).$$

Referring to the definition of stochastic gradient algorithms, this criterion can also be maximized using algorithm (7) where $\frac{1}{n+1}[I_c(\boldsymbol{\Phi}^{(n)})]^{-1}$ is replaced with a scalar learning rate α_n verifying conditions $\sum |\alpha_n| = \infty$ and $\sum |\alpha_n|^2 < \infty$ [7]. Usual scalar learning rates in stochastic approximation take the form $\alpha_n = \frac{1}{an}$ ($a > 0$). In the proposed algorithm, the inverse of the Fisher information matrix may contribute to fast convergence.

Many commonly used mixture models like Gaussian mixtures has their complete data distribution from the exponential family. The next subsection shows that algorithm (7) can be simplified in that situation.

5.3   Exponential Family Model

This part shows how recursion (7) can be simplified if complete data have their distribution from the exponential family.

The complete data $(\boldsymbol{x}, z)$ has its distribution from the exponential family with natural parameter $\boldsymbol{\eta}$ and sufficient statistic $\mathbf{T}(\boldsymbol{x}, z)$ if its distribution can be written

$$p(\boldsymbol{x}, z; \boldsymbol{\eta}) = \exp\left(\boldsymbol{\eta}^T \mathbf{T}(\boldsymbol{x}, z) - a(\boldsymbol{\eta}) + b(\boldsymbol{x}, z)\right).$$

If we re-parameterize the mixture distribution with the expectation parameter $\boldsymbol{\Psi} = E(\mathbf{T}(\boldsymbol{x}, z)|\boldsymbol{\eta})$, the complete data log-likelihood can be derived as followed

$$\frac{\partial \log p(\boldsymbol{x}, z; \boldsymbol{\eta}(\boldsymbol{\Psi}))}{\partial \boldsymbol{\Psi}} = \frac{\partial \boldsymbol{\eta}}{\partial \boldsymbol{\Psi}}\left(\mathbf{T}(\boldsymbol{x}, z) - \frac{\partial a}{\partial \boldsymbol{\eta}}\right)$$

Using the following relations verified by the exponential family:

$$\frac{\partial \boldsymbol{\eta}}{\partial \boldsymbol{\Psi}}, = I_c(\boldsymbol{\Psi})$$

$$\boldsymbol{\Psi} = \frac{\partial a}{\partial \boldsymbol{\eta}},$$

we obtain

$$\frac{\partial \log p(\boldsymbol{x}, z; \boldsymbol{\eta}(\boldsymbol{\Psi}))}{\partial \boldsymbol{\Psi}} = I_c(\boldsymbol{\Psi})(\mathbf{T}(\boldsymbol{x}, z) - \boldsymbol{\Psi}). \quad (8)$$

The derivative of $R_{n+1}(\boldsymbol{\Psi}, \boldsymbol{\Psi}^{(n)})$ with respect to $\boldsymbol{\Psi}$ is then written

$$\frac{\partial R_{n+1}(\boldsymbol{\Psi}, \boldsymbol{\Psi}^{(n)})}{\partial \boldsymbol{\Psi}} = \sum_{i=1}^{n+1} \frac{\partial \log p(\boldsymbol{x}_i, z_i^{(i-1)}; \boldsymbol{\eta}(\boldsymbol{\Psi}))}{\partial \boldsymbol{\Psi}}$$

$$= \sum_{i=1}^{n+1} I_c(\boldsymbol{\Psi}) \big(\mathbf{T}(\boldsymbol{x}_i, z_i^{(i-1)}) - \boldsymbol{\Psi} \big)$$

$$= I_c(\boldsymbol{\Psi}) \Big(\sum_{i=1}^{n+1} \mathbf{T}(\boldsymbol{x}_i, z_i^{(i-1)}) - (n+1)\boldsymbol{\Psi} \Big).$$

The parameter $\boldsymbol{\Psi}^{(n+1)}$ maximizing $R_{n+1}(\boldsymbol{\Psi}, \boldsymbol{\Psi}^{(n)})$ can thus be written

$$\boldsymbol{\Psi}^{(n+1)} = \frac{\sum_{i=1}^{n+1} \mathbf{T}(\boldsymbol{x}_i, z_i^{(i-1)})}{n+1}$$

$$= \frac{\sum_{i=1}^{n} \mathbf{T}(\boldsymbol{x}_i, z_i^{(i-1)}) + \mathbf{T}(\boldsymbol{x}_{n+1}, z_{n+1}^{(n)})}{n+1}$$

$$= \frac{n\boldsymbol{\Psi}^{(n)} + \mathbf{T}(\boldsymbol{x}_{n+1}, z_{n+1}^{(n)})}{n+1}.$$

This parameter is finally written

$$\boldsymbol{\Psi}^{(n+1)} = \boldsymbol{\Psi}^{(n)} + \frac{1}{n+1} \big[\mathbf{T}(\boldsymbol{x}_{n+1}, z_{n+1}^{(n)}) - \boldsymbol{\Psi}^{(n)} \big]. \tag{9}$$

Writing recursive formula (9) as followed:

$$\boldsymbol{\Psi}^{(n+1)} = \boldsymbol{\Psi}^{(n)} + \frac{1}{n+1} I_c(\boldsymbol{\Psi}^{(n)})^{-1} I_c(\boldsymbol{\Psi}^{(n)}) \big[\mathbf{T}(\boldsymbol{x}_{n+1}, z_{n+1}^{(n)}) - \boldsymbol{\Psi}^{(n)} \big]]$$

and using relation (8), it can then be deduced that equation (9) is equivalent to equation (7).

Consequently, in the situation where complete data have their distribution from the exponential family, recursion (7) is obtained exactly (not approximatively) and is written under the simplified form of recursion (9).

Using recursive formula (9), our on-line CEM algorithm for Gaussian mixtures starts with initial proportions $\pi_k^{(0)}$, means vectors $\boldsymbol{\mu}_k^{(0)}$, covariance matrices $\boldsymbol{\Sigma}_k^{(0)}$ and initial number of observations $n_k^{(0)} = 0$ of each cluster k, $1 \leq k \leq K$. The two following steps are then alternated while new observations are received.

Step 1 (iteration $n+1$) assignation of the new observation $\boldsymbol{x}_{n+1}$ to the class k^* which maximizes the posterior probability $t_{n+1k}^{(n)} = \frac{\pi_k^{(n)} f(\boldsymbol{x}_{n+1}; \boldsymbol{\theta}_k^{(n)})}{\sum_{\ell=1}^{K} \pi_\ell^{(n)} f(\boldsymbol{x}_{n+1}; \boldsymbol{\theta}_\ell^{(n)})}$:

$$k^* = \underset{1 \leq k \leq K}{argmax} \Big(\log \pi_k^{(n)} - \frac{1}{2} \log det(\boldsymbol{\Sigma}_k^{(n)}) -$$

$$\frac{1}{2}(\boldsymbol{x}_{n+1} - \boldsymbol{\mu}_k^{(n)})^T \boldsymbol{\Sigma}_k^{(n)^{-1}} (\boldsymbol{x}_{n+1} - \boldsymbol{\mu}_k^{(n)}) \Big);$$

set $z_{n+1,k}^{(n)}$ equals 1 if $k = k^*$ and 0 otherwise.

Step 2 (iteration $n + 1$) updating of the parameters:

$$n_k^{(n+1)} = n_k^{(n)} + z_{n+1,k}^{(n)}$$

$$\pi_k^{(n+1)} = \frac{n_k^{(n+1)}}{n+1}$$

$$\mu_k^{(n+1)} = \mu_k^{(n)} + \frac{z_{n+1,k}^{(n)}}{n_k^{(n+1)}} \cdot (x_{n+1} - \mu_k^{(n)})$$

$$\Sigma_k^{(n+1)} = \Sigma_k^{(n)} + \frac{z_{n+1,k}^{(n)}}{n_k^{(n+1)}} \cdot$$

$$\left(\left(1 - \frac{z_{n+1,k}^{(n)}}{n_k^{(n+1)}}\right)(x_{n+1} - \mu_k^{(n)})(x_{n+1} - \mu_k^{(n)})^T - \Sigma_k^{(n)}\right).$$

Since each new observation x_{n+1} is used only on time, this algorithm does not require a stop condition.

By considering a Gaussian mixture with identical proportions and spherical covariance matrices (equal to the identity matrix), the on-line k-means algorithm is recovered. Thus, this algorithm is a generalization of on-line k-means algorithm.

6   Experiments

This section is designed to evaluate the proposed algorithm in term of precision and computing time. Simulations are restricted to two-dimensional data sets corresponding to a Gaussian mixture with diagonal covariance matrices and equal proportions. We focus on model-based algorithms designed to directly find a partition by optimizing a criterion. An on-line CEM algorithm with a classical scalar learning rate $\alpha_n = \frac{1}{an}$ has been chosen for comparisons. By varying values of a, we observed that the algorithm performs better for values of a between 0.1 and 0.6. Thus, the learning rate $\alpha_n = \frac{1}{0.3n}$ has been used in the current simulations. The three algorithms compared are:

- the CEM algorithm which is our reference algorithm,
- the proposed on-line CEM algorithm with a matrix learning rate,
- the on-line CEM algorithm with the scalar learning rate $\alpha_n = \frac{1}{0.3n}$.

6.1   Evaluation Criteria

The precision of partitions estimated by each algorithm is measured by the misclassification rate between estimated partitions and the true simulated partition. The misclassification rate between two partitions is measured by crossing the classes vectors of the two partitions and then counting the number of misclassified observations. The computing time is given by the CPU running time. We should point out that the processor used for all the simulations is a 2 Ghz Pentium 4 processor.

6.2   Simulation Protocol

The adopted strategy for simulations consists in initially drawing n observations according to a mixture of two bi-dimensional Gaussian distributions, to apply the standard CEM algorithm on a few points (n_0 points) and finally to apply the on-line CEM algorithms on the rest of the points. The standard CEM algorithm is directly applied to the n observations. For each data set, the CEM algorithm starts with 30 different initializations and the solution which provides the greatest likelihood is selected. For each mixture model, 25 random data sets are generated and both misclassification rates and CPU times are averaged.

6.3   Simulation Parameters

We consider mixture parameters corresponding to two kinds of models: a model A with two spherical clusters and a model B with two elliptical clusters. For each model, three overlapping rates were considered, corresponding to 5%, 15% and 25% of Bayes error rate and depending on the distance between Gaussian densities means. The proportions for all models are $\pi_1 = \pi_2 = 1/2$. The covariance matrices are $\Sigma_1 = \Sigma_2 = diag(1,1)$ for model A and $\Sigma_1 = diag(1; 1/8)$, $\Sigma_2 = diag(1/8; 1)$ for model B, where $diag(a, b)$ is the diagonal matrix whose diagonal components vector is (a, b). The Gaussian density means for model A are: $(-2; 0)$, $(1.2; 0)$ for 5% of Bayes error rate, $(-2; 0)$, $(0; 0)$ for 15% of Bayes error rate and $(-2; 0)$, $(-0.7; 0)$ for 25% of Bayes error rate. The Gaussian density centers for model B are: $(3.3; 0)$, $(0; 0)$ for 5% of Bayes error rate, $(1.6; 0)$, $(0; 0)$ for 15% of Bayes error rate and $(0; 0)$, $(0; 0)$ for 25% of Bayes error rate. The sample sizes n varied from 1000 to 20000 by step of 1000 and the number n_0 of observations initially processed with the CEM algorithm is $n_0 = 80$. Figure 1 shows examples of simulation of mixture models A and B for $n = 3000$ with 15% of Bayes error rate.

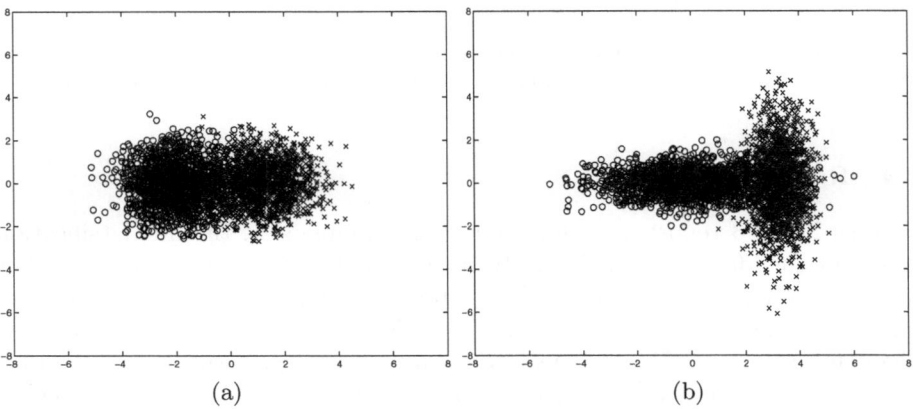

(a)   (b)

Fig. 1. Example of simulation of Mixtures A and B for $n = 3000$ with 15% of Bayes error rate

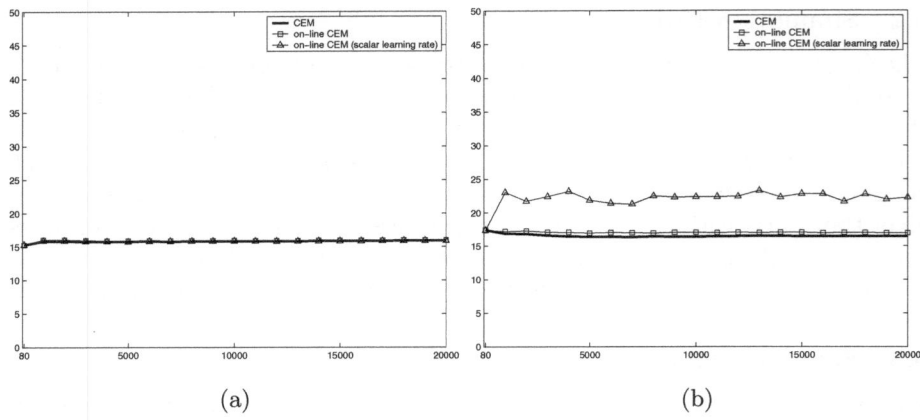

Fig. 2. Misclassification rate as a function of the sample size obtained with the three algorithms, for model A (a) and model B (b), for 15% of Bayes error rate

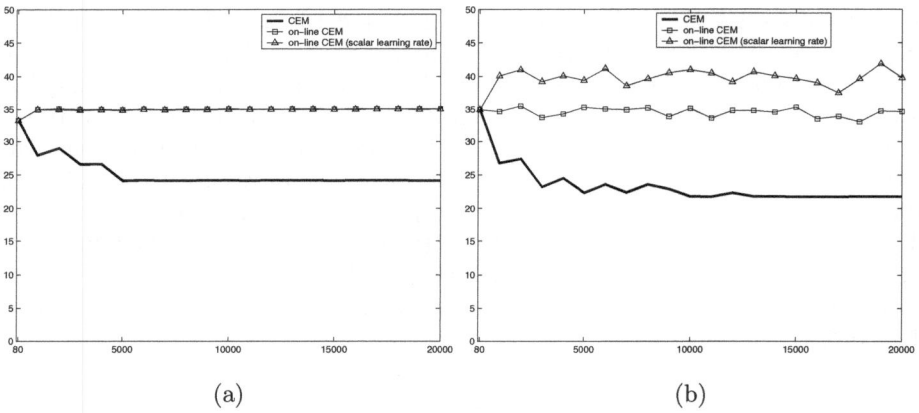

Fig. 3. Misclassification rate as a function of the sample size obtained with the three algorithms, for model A (a) and model B (b), for 25% of Bayes error rate

6.4   Results

Figures 2 and 3 report, as a function of the sample size, the misclassification rates obtained for mixtures A and B, for 15% and 25% of Bayes error rate. For a Bayes error rate leading to 15%, we observe that the two on-line algorithms have the same performances for model A (see figure 2-a) and that our new algorithm (on-line CEM with a matrix learning rate) performs better than his concurrent (on-line CEM using a scalar learning rate) for model B (see figure 2-b). The two algorithms stabilize very quickly and misclassification percentages given by our new algorithm is nearly the same as that of the standard (off-line) CEM algorithm. The same behavior has been observed for 5% of Bayes error rate. Therefore, when the clusters are relatively well separated, our on-

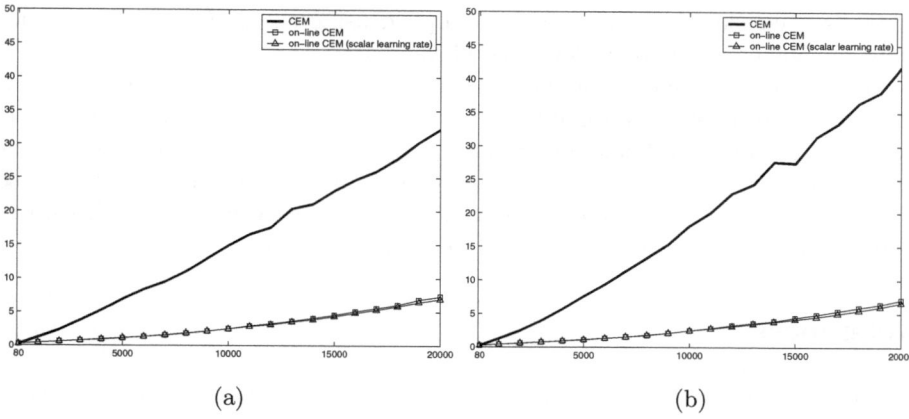

(a)   (b)

Fig. 4. CPU time (in seconds) as a function of the sample size obtained with the three algorithms, for model A (a) and model B (b), for 15% of Bayes error rate

line CEM algorithm provides a faster and better fitting to the data shape and performs better than the on-line CEM algorithm using a scalar learning rate. These performances can be attributed to the matrix learning rate.

When the class overlap is relatively high (25% of Bayes error rate), the two on-line CEM algorithms give poor results for mixture A (see figure 3-a) and miture B (see figure 3-b) but, again, the on-line CEM algorithm with a matrix learning rate performs better than his concurrent for mixture B. This can be attributed to the notoriously poor performances [3] of CEM-type algorithms when clusters are not well separated. This phenomenon seems to be even more pronounced when using on-line algorithms.

Figure 4 reports, as a function of the sample size, the CPU time (in seconds) obtained for mixtures A and B, for 15% of Bayes error rate. It can be observed that CPU times given by the on-line algorithms vary very slowly with sample size. The CPU time for the standard CEM algorithm in fact grows considerably with the sample size. In particular, for 20000 observations, CEM is about six time slower than the two on-line algorithms for mixture B. The same behavior has been observed for the others Bayes error rates (5% and 25%). These experiments clearly show that our proposed on-line CEM algorithm is more efficient than the CEM algorithm in terms of speed.

7   Conclusion

In this paper an on-line clustering algorithm based on mixture models was proposed. This algorithm is a stochastic gradient algorithm with a matrix learning rate. When complete data distribution is from the exponential family, the algorithm is simplified. It provides a generalization of the on-line k-means algorithm introduced by MacQueen [6] when a Gaussian mixture with spherical covariance matrices is considered. It may be also applied using the 28 Gaussian parsimonious models proposed by Celeux and Govaert [4].

Although the proposed method provides reasonably good results, the convergence analysis of the on-line CEM algorithm toward a local maxima of the expected classification likelihood $E[\max_{1 \leq z \leq K} \log p(\boldsymbol{x}, z; \boldsymbol{\Phi})]$ is met only under some conditions [1,2] which are often difficult to prove. The verification of these conditions, at least for some particular models, remains a prospect of this work. The evaluation of the algorithm on real data is in progress.

References

1. Bottou L. *Une approche théorique de l'apprentissage connexioniste; applications à la reconnaissance de la parole*, Thèse de Doctorat, université d'Orsay, 1991.
2. Bottou L. *Online learning and stochastic approximations*. In online learning in neural networks, D. Saad, Ed., Cambridge: Cambridge University Press, 1998.
3. Celeux G. and G. Govaert. *A classification EM algorithm for clustering and two stochastic versions*. Computational Statistics and Data Analysis, 14, 315-332, 1992.
4. Celeux G. and G. Govaert. *Gaussian parsimonious clustering models*. Pattern Recognition, 28(5), 781 – 793, 1995.
5. Dempster A. P., Laird N. M. and D. B. Rubin. *Maximum likelihood from incomplete data via the EM algorithm*. Pattern Recognition, 28(5), Journal of Royal Statistal Society Series B 39(1): 1 – 38, 1977.
6. MacQueen J. *Some methods for classification and analysis of multivariate observations*. In Proceedings of 5th Berkeley Symposium on Mathematics, Statistics and Probability 1, 281 – 298, 1967.
7. Samé A. and Ambroise C. and Govaert G. *A mixture model approach for on-line clustering*. In Proceedings of Computational Statistics, COMPSTAT 2004, Antoch J., ed., Physica-Verlag, 1759 – 1765
8. Scott A. J. and Symons M. J. *Clustering methods based on likelihood ratio criteria*. Biometrics, 27, 387-397, 1971.
9. Titterington D.M. *Recursive parameter estimation using incomplete data*. Journal of Royal Statistal Society Series B 46, 257 – 267, 1984.

Reliable Hierarchical Clustering
with the Self-Organizing Map

Elena V. Samsonova[1,2], Thomas Bäck[2,3],
Joost N. Kok[2], and Ad P. IJzerman[1]

[1] Leiden University, Leiden / Amsterdam Center for Drug Research,
Einsteinweg 55, PO Box 9502, 2300RA Leiden, The Netherlands
`ijzerman@chem.leidenuniv.nl`
[2] Leiden University, Leiden Institute for Advanced Computer Science,
Niels Bohrweg 1, PO Box 9512, 2333CA Leiden, The Netherlands
`{baeck, elena.samsonova, joost}@liacs.nl`
[3] NuTech Solutions GmbH, Martin-Schmeisser-Weg 15,
44227 Dortmund, Germany

Abstract. Clustering problems arise in various domains of science and engineering. A large number of methods have been developed to date. Kohonen self-organizing map (SOM) is a popular tool that maps a high-dimensional space onto a small number of dimensions by placing similar elements close together, forming clusters. Cluster analysis is often left to the user. In this paper we present a method and a set of tools to perform unsupervised SOM cluster analysis, determine cluster confidence and visualize the result as a tree facilitating comparison with existing hierarchical classifiers. We also introduce a distance measure for cluster trees that allows to select a SOM with the most confident clusters.

1 Introduction

Problems of ordering high-dimensional data in a small number of dimensions are frequently encountered. Such data are often noisy or incomplete, but classical clustering methods such as linkage or multidimensional scaling, assume the data to be well defined. Noise may lead to incorrect clusterings, and missing components are often unacceptable. Iterative learning of the data may neutralize these shortcomings, and Kohonen self-organizing map (SOM) [1] is a clustering method that addresses these issues. It is an artificial neural network capable of mapping high-dimensional data onto a low-dimensional grid such that similar data elements are placed close together. Groups of nodes with short distances to each other represent clusters. However, different map initializations and topologies, as well as input order of data elements, may result in different clusterings [2]. Ideally, a large number of SOMs with varying random seed needs to be created, their clusterings analyzed, and only those clusters occurring in a majority of cases should be chosen. This is a lengthy and laborious task, so far not automated. For large data sets it becomes intractable for manual analysis forcing the user to select a single "good" SOM and accept it as the final result omitting

A.F. Famili et al. (Eds.): IDA 2005, LNCS 3646, pp. 385–396, 2005.

tests of confidence. Needless to say that such unverified clusterings may contain faulty conjectures leading to incorrect conclusions.

In this paper we present an unsupervised method for cluster analysis and confidence testing for SOMs. When used for clustering, SOM can be represented as a tree [3] allowing for easy comparisons with the outcomes of hierarchical classifiers widely used in various domains. Moreover, a tree representation allows to solve the problem of cluster confidence testing taking advantage of consensus tree building methods, developed and implemented independently of SOM (e.g. [4], [5]). A *consensus tree* represents an "average" of a set of trees with frequencies of occurrence of its branches compared to the set of all trees representing reliable clusters as subtrees. The exact way of constructing such a tree, as well as the way of measuring the frequencies, depends on the particular method selected. We propose to make one further step in selecting one of the SOMs as the best representative of the consensus. Such combination of a consensus tree providing a cluster hierarchy, and a cluster map revealing spatial ordering of clusters, allows to view the clustering from different perspectives supported by the data.

Other neural network variants allow to overcome sensitivity to topology and initialization by dynamically growing the network, e.g., growing neural gas [6], growing hierarchical self-organizing map [7]. Compared to a standard rectangular SOM, such methods may yield a topology better matching the data, and a more complex final map that is highly data-dependent. This may require complex methods for standardized analysis of such results, especially when comparing with other classification methods. A rectangular SOM, on the other hand, allows for systematic analysis independent of the data. Another approach is to build a tree directly, e.g., with a growing neural tree [8]. This allows for an immediate comparison with other tree-based classification methods but lacks a spatial ordering of data inherent to a map.

We also offer to the scientific community an open source C^{++} implementation *TreeSOM* of the algorithms described in this paper, including the core SOM algorithm. Our implementation is approximately 5.5 times faster than the original version by Kohonen's group [2] due to a number of optimizations. They include replacing linked lists by arrays, computing full distances only if necessary and changing types of numerical values and are fully described in the documentation provided with the source.

The remainder of the paper is organized as follows. First, in section 2 we present the data, models and classification methods used in the running example illustrating *TreeSOM*. Then we outline various aspects of the method: section 3 describes the cluster discovery algorithm, section 4 determines data clustering based on node clustering, section 5 represents SOM clustering as a tree, section 6 examines cluster confidence constructing a consensus cluster tree based on a series of SOMs, and finally section 7 presents an algorithm for finding the most representative SOM for the consensus cluster tree of the series. Section 8 concludes the paper.

2   Protein Data

To illustrate the methods discussed in this paper, we use a running example from protein bioinformatics: clustering of serotonin receptors of several species.

Proteins are composed of amino acids, and primary protein structure is determined by their sequence. The task is to cluster a set of proteins into families not known in advance. By far the most common approach to this problem is to infer phylogeny from a homology model of the protein set resulting in a *phylogenetic tree* where families are represented as subtrees.

When only the amino acid sequence of a protein is available, homology modeling is usually reduced to finding a gaped alignment of the sequences in the data set such that only similar amino acids are aligned with each other [9], [10]. This in itself is already a very complex problem as amino acid similarity as used by the alignment algorithms, appears to depend not only on the chemical properties of the molecules, but on their context as well as on protein class [11].

Inferring phylogeny from an alignment is the process of determining pairwise distances between the proteins in the alignment that are mutually consistent. This step can be performed with a variety of methods resulting in different phylogenies [12]. The choice of the method often depends on the protein class and the data set homogeneity.

The proteins we use in this paper belong to the class of G protein-coupled receptors (GPCRs), cell transmembrane proteins that typically bind hormones and neurotransmitters and convey signals into the cells. Such a mediatory role ensures their importance for medicine, and indeed almost a half of all drugs currently on the market, act through GPCRs. The small set we used to illustrate our method, contains 62 serotonin GPCRs from different species taken from the public GPCR database GPCRDB (*www.gpcr.org*, [13]).

The phylogenetic trees are based on a multiple alignment made with *ClustalW* [14] with increased gap penalties (pwgapopen=60, pwgapext=2, gapopen=70, gapext=2). Phylogenies were inferred according to the probabilities of amino acid change [15] with *PHYLIP* [16]. The trees are constructed with the Neighbor-Joining method [17], their confidence determined with bootstrapping [18] and consensus tree construction [4].

The SOMs are based on the same set of proteins but use a different model. Considering a protein as a string of amino acids, we can analyze its amino acid content, and specifically examine amino acid pairs. Frequencies of occurrence (in percent) of each possible pair comprise a protein model. For GPCRs we looked at third-neighbor pairs rather than at adjacent amino acids, as prompted by the specifics of GPCR structure. GPCRs are believed to have most of the sequence coiled in spiral structures (α-helices) with 3.6 amino acids per turn, such that most amino acids are spatially roughly above their third neighbor suggesting influence or interaction (see also [19]).

With such frequency-based vectors composing the data set, we use a Euclidean SOM, trained in two phases with the following parameters: map size 5x4, Gaussian neighborhood, linear decrease of learning rate and radius; phase 1: starting learning rate 0.2, starting radius 6, 1000 iterations; phase 2: starting learning rate 0.02, starting radius 3, 100,000 iterations. Such training length ensured that cluster tree topology remained unchanged with any additional training. Map size of 20 neurons was selected based on several trials such that confident clusters as well as alternative distributions of data on the map could be achieved. We show in a separate study [20] that smaller SOMs tend to yield more confident clusters, whereas larger SOMs reveal alternative data mappings. To determine confidence, 100 SOMs with different random seeds were produced and their consensus tree constructed with the same method as used for the phylogenetic trees in *PHYLIP* implementation.

Thus, the examples present not only two different classification methods — Euclidean SOM versus phylogeny inference, — but also two different data models — frequencies of third pairs of amino acids in protein sequences versus homology modeling with protein sequence alignment. In our previous work we also used Euclidean SOMs with protein alignment scores [19], and various other combinations are possible such as alignment-based SOMs [21], [22], [23] that are more natural for handling general protein data making use of the standard phylogeny inference models. The tools we present here can be employed for the uniform analysis and visualization of the results, allowing for an easy comparison of various approaches.

3   Cluster Discovery

When self-organizing maps are used for clustering, finding clusters on the SOM becomes a crucial task. Several fairly complex approaches have been developed, e.g. [3], [24]. A node is iteratively updated during training based on the learning vectors such that a well-trained SOM represents a distribution of the input data over a two-dimensional surface preserving topology. In this context we can define a *cluster* as a group of nodes with short distances between them and long distances to the other nodes.

The representation in Figure 1 is similar to the popular *umat* visualization [25]. Using this representation, we can define a cluster as a group of nodes surrounded by an uninterrupted border of a given shade or darker representing distances equal or greater than a given distance threshold. Note that distances between nodes within a cluster may not necessarily be all shorter than the dis-

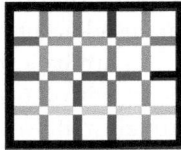

Fig. 1. SOM as a grid of nodes separated by borders with gray shades representing distances between the corresponding nodes. White stands for a zero distance and black for the largest distance between two adjacent nodes found on the map.

Algorithm 1 Cluster discovery algorithm for a given distance threshold

Procedure Cluster-Discovery (for distance threshold T):

- mark all nodes as unvisited
- while there are unvisited nodes, repeat:
  - locate an arbitrary unvisited node N
  - start a new cluster C
  - call procedure *Cluster* for N, C and T

Procedure Cluster (for node N, cluster C and distance threshold T):

- assign N to C
- mark N as visited
- for each unvisited node A adjacent to N such that the distance $|NA| < T$ call procedure *Cluster* for A and C

tance threshold, however every node must be connected to every other node within the cluster along a path consisting exclusively of edges shorter than the distance threshold. Effectively it means that each node within a cluster must be connected to at least one other node within the same cluster with an edge that is shorter than the distance threshold. Based on this observation, we define the cluster discovery algorithm as shown in the algorithm box 1. It may be applied either onto a fully trained SOM to discover the "final" clustering, or to any intermediate SOM snapshot as a monitor of the training progress or even as a part of the termination test. The distance measure is not defined by this algorithm, and is generally the same as the distance measure used during training. However, a different measure may be used if appropriate. Also, since algorithm 1 is independent of the SOM training algorithm as well as of the definition of node neighborhood and adjacency, it is in principle applicable to any SOM variant.

Scaling all distances such that the largest distance between two adjacent nodes equals 1, we can express distance thresholds as values between 0 and

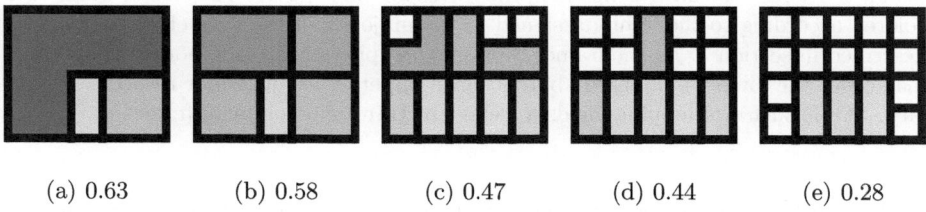

   (a) 0.63   (b) 0.58   (c) 0.47   (d) 0.44   (e) 0.28

Fig. 2. A series of clusterings of the map in figure 1 with the scaled distance thresholds given in the subscripts. The cluster areas are shaded according to the average distance between the nodes in the corresponding cluster.

1 regardless of the actual distances. For each threshold we can find a unique clustering using algorithm 1. Figure 2 shows a series of clusterings of the SOM in figure 1. Here, the cluster borders are always displayed in black, and the cluster areas are shaded according to the same convention as used for the borders in figure 1[1]. However in this case the shade represents the average distance between the nodes *within* the cluster rather than the distance *between* the neighboring nodes. Thus, cluster maps bring forward the clusters by hiding the borders that fall within them, yet provide information on cluster density by shading cluster areas.

A series of cluster maps generated for successive thresholds as shown in figure 2, may be converted into an animated graphical format such as GIF or MPEG using widely available graphical tools. Such a movie gives an insight into cluster formation on a SOM.

4   Calibration

When SOM is used for clustering, it is not the node clustering that is of ultimate interest, but the data clustering, the distribution of training data over the SOM clusters. Mapping of data on a trained SOM is called *calibration*. Each data item is assigned to the node that is most similar to it, so that some nodes may get many data elements, and others none at all.

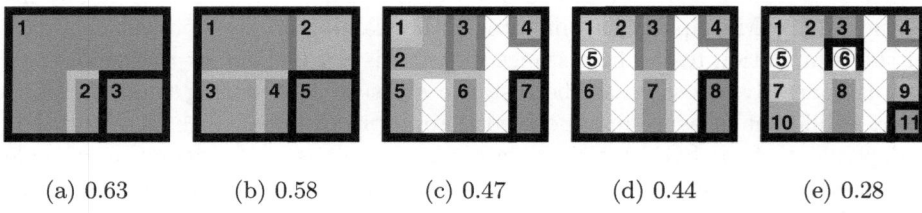

(a) 0.63   (b) 0.58   (c) 0.47   (d) 0.44   (e) 0.28

Fig. 3. A series of calibrated clusterings of the map in figure 1 with the (scaled) distance thresholds given in the subscripts. The cluster areas are shaded according to the average distance between the data elements in the corresponding cluster. The cluster frames are colored according to the family assigned to the majority of the data elements in it, as described in section 5. As the frames overlap, the upper or left-side border of a cluster may cover the lower or right-side border of its upper or left neighbor respectively. On large SOMs such overlapping yields a clearer picture than complete frames.

Figure 3 shows the same series of clusterings as figure 2, but now "filled" with the data. The nodes that contain no data elements are crossed out and are not used in cluster analysis. Here, in contrast to figure 2, the clusters are shaded

---

[1] Shading cluster borders as well as cluster areas yields poorly readable maps. Instead, borders may be colored to display additional information as shown in section 4.

according to the average distance between the data elements rather than nodes within the corresponding cluster, with white indicating zero distance (identical elements) and black the largest distance between any two elements in the data set. Each cluster receives a unique identifier used in the corresponding data clustering list (a file listing each cluster and the elements belonging to it). Since clusters containing a single element represent a special case, they are marked with a circle. Note that box area in such clusters is always white, since the distance of any data element to itself is zero.

Such a visualization allows to immediately distinguish "tight" and "loose" clusters by their shades. Data elements in "tight" clusters have short distances to one another, resulting in a light shade of the cluster area. On the other hand, "loose" clusters contain data elements with large distances between them, resulting in a dark shade of the cluster area. For example, in figure 3b cluster 5 is "looser" than cluster 2, and in figure 3e clusters 1 and 7 are the "tightest" and cluster 11 is the "loosest" on the map.

5   SOM as a Tree

Clustering problems arise in many branches of science and engineering and many areas traditionally use hierarchical clusterings visualizing the result as a tree. Thus when a SOM is used in such a domain, the result needs to be compared with a tree in order to determine how the new attempt differs from the previous ones. In our previous work we addressed the problem of comparing alternative SOM clusterings of protein sequences with phylogenetic trees [19]. Although the method of placing a phylogenetic tree orthogonally to the SOM surface yields a clear graphical representation, it is largely manual and requires tree aggregation. The comparison becomes easier if the SOM itself is represented as a tree, leaving the phylogenetic tree unmodified and reducing the task to comparison of two trees.

The SOM cluster analysis yields a series of nested clusterings that allow to represent cluster development as a tree (see figure 4a). At each threshold in the clusterings series one or more clusters are split into several subclusters that is represented as a node in the tree. The lowest level shows the individual elements found in the corresponding clusters.

In clustering problems it is often required to determine whether some novel data belongs to any of the established families, or whether a novel clustering criterion yields a better result than a previously used one. In such cases some or even all of the data in the training set can be annotated with family information known beforehand. Different families are then represented in color on a SOM cluster map and tree giving an instant overview as in figures 3 and 4a respectively. Each branch on the tree and each cluster frame on the map is assigned the family to which the majority of data elements in the corresponding cluster belong, and colored accordingly. If no majority can be found, the branch is left black.

Phylogenetic trees can be displayed in the same manner allowing for an easy visual comparison (see figure 4b).

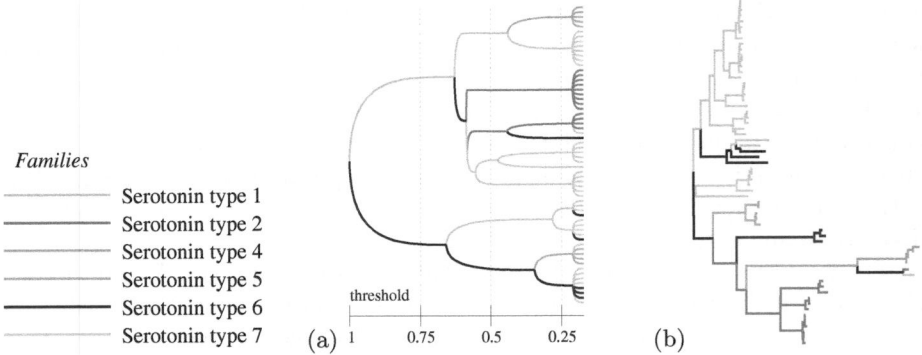

Fig. 4. Tree representations of two different clusterings of serotonin receptors. (a) Cluster tree of a calibrated SOM in figure 3. (b) The corresponding phylogenetic tree.

6   Clustering Confidence

A self-organizing map is initialized randomly and trained with data elements presented in random order, resulting in different SOMs for different random seeds. More specifically, clusterings may be different, both in terms of grouping on the lowest level, as in cluster hierarchies. To determine the "true" clustering a large number of such SOMs has to be analyzed and only the clusters found in the majority of cases, may be included in the final clustering. If each SOM is represented as a tree, this task can be tackled with methods constructing a *consensus tree* from a set of trees. There are many such methods, so that the most appropriate one may be selected for each particular problem. We used method [4] because it is also used in determining confidence of phylogenetic trees.

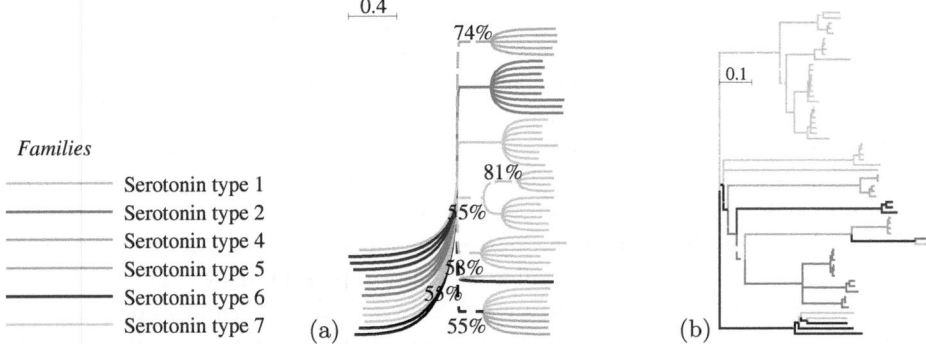

Fig. 5. Consensus trees of two different clusterings of serotonin receptors. (a) Consensus SOM cluster tree. (b) Consensus phylogenetic tree.

Figure 5 shows consensus trees of the SOM clusters and the phylogeny corresponding to the individual trees in figure 4. Confidence of each cluster is shown on the branch leading to the corresponding node. The confidence values in the phylogenetic tree are omitted for the sake of readability. Branch lengths reflect the distances between the corresponding nodes (clusters) and their siblings (nodes with the same parent): assume set A is split into $B_1, \ldots, B_n$, then the branch AB_j has length $|AB_j| = \frac{1}{n-1} \sum_{i \neq j} |B_i B_j|$. The distance between two sets P and Q equals the average distance between each element from P and from Q.

Consensus trees in figure 5 allow to compare two different clusterings of the same data set. Here, not only the clustering methods differ, but also the underlying distances. Thus, the results are not expected to be alike. The difference is reflected in the relative branch lengths in the trees. In this example, the phylogenetic model shows much higher similarity between the proteins grouped in one cluster, than does the statistical model used with the SOM. However, the general clustering on the family level is similar. The difference in clustering methods is visible in the tree topologies: although both the individual SOM cluster trees and the phylogenetic trees are almost fully binary, the consensus phylogenetic tree retained much more of its binary hierarchy with higher confidence than the SOM consensus tree did, suggesting a greater consistency among the phylogenetic trees than among the SOMs.

In spite of using different methods and models for the clusterings in figure 5, the results are notably similar. However, it is the differences that may open up new insights, such as the SOM grouping together receptors of type 4 and 7, whereas the phylogenetic tree clearly separates the two families. Remarkably, these are the only serotonin receptors that exhibit alternative splicing. This result suggests that amino acid composition may contain clues to underlying gene expression. A study of a larger set of proteins is needed to verify this suggestion.

7   The Most Representative SOM

Cluster confidence analysis presented in the previous sections, leads to a final tree converting a spatial ordering of clusters inherent to a SOM, into a hierarchy. Although in many cases it is desirable, in many other cases it is not, as it lacks the information on proximity of clusters to one another. To solve this problem, we propose a distance measure for cluster trees allowing to select an individual tree, and hence a SOM, as the best representative of the consensus. Such a distance measure is defined algorithmically in box 2. Note that this measure is not symmetrical as it is based on the clusters of the reference tree.

In our example, the best SOM is represented by the cluster tree and map in figure 6, and the distance from the consensus tree to it is 0.43. Here clusters of the same family are grouped close together, except serotonin type 2 (clusters 11, 13 and 14) which is split by an empty node. The distance from the consensus tree to the SOM in figures 3 and 4a is 0.7, and here not only type 2 cluster is split, but also type 1 cluster.

Algorithm 2 Distance measure for cluster trees

Procedure Distance-Measure (from reference tree R to tree T)

- initialize score to 0
- for each node N of R repeat:
    - search T for a node N_T equivalent to N_R (use *Node-Equivalence-Test* for nodes N_T and N_R)
    - if found, increment the score
- define similarity of T with respect to R: $S = \frac{score}{|R|}$ where $|R|$ is the number of nodes in R, $S \in [0, 1]$
- define distance from R to T: $D = 1 - S$, $D \in [0, 1]$

Procedure Node-Equivalence-Test (for nodes A and B)

- define a set of leaves of a node N to contain all the leaves directly connected to N and all the leaves connected to any of its descendants
- nodes A and B are equivalent if their sets of leaves are identical

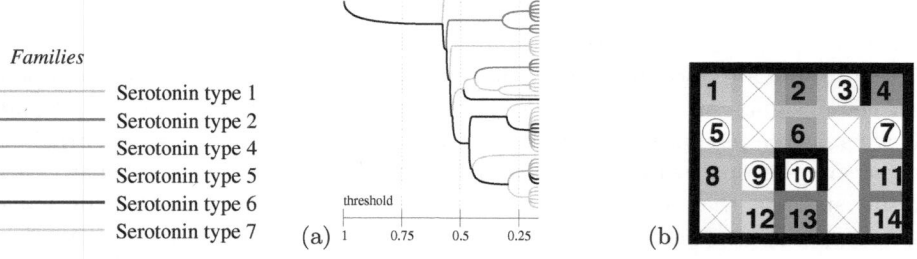

Families

———— Serotonin type 1
———— Serotonin type 2
———— Serotonin type 4
———— Serotonin type 5
———— Serotonin type 6
———— Serotonin type 7

(a) threshold 1 0.75 0.5 0.25

(b)

Fig. 6. Cluster tree (a) and map (b) of the best SOM with respect to the consensus tree in figure 5a

8   Conclusions

In this paper we presented a new look at self-organizing maps improving their applicability to clustering problems and facilitating comparisons of clustering results with those of hierarchical classifiers. We propose an unsupervised cluster analysis method *TreeSOM* dividing up a SOM into nested clusters at different distance thresholds, determining cluster confidence based on many independently trained SOMs, visualizing the resulting clustering as a tree, and indicating a single SOM as the best representative of the final clustering. This method not only reveals the segregation of data, but also a cluster hierarchy. We also make available to the scientific community open source tools implementing all aspects of this method, along with a new, efficient and modular implementation of the core SOM algorithm. Moreover, the visualization tools we present here, may be used with any trees to achieve clear uniform representations aggregating data on different levels, permitting to create detailed diagrams as well as general overviews.

In short, the method and tools that we propose, not only bridge the gap between a map-like SOM structure and well-established tree clusterings, but also analyze and quantify SOM results in a well-defined and uniform manner making it an even more valuable tool for data analysis.

Tools and supplementary material are available from:
http://web.inter.nl.net/users/Elena.Samsonova/resources.shtml#TreeSOM.

Acknowledgments

We thank Bas Groenendijk for his contribution to the implementation and testing of the software and to the anonymous reviewers for insightful comments. We are also grateful to the Netherlands Organisation for Scientific Research (NWO) for the financial support.

References

1. T. Kohonen, *Self-Organizing Maps*, vol. 30 of *Springer Series in Information Sciences*. Springer, second ed., 1997.
2. T. Kohonen, J. Hynninen, J. Kangas, and J. Laaksonen, *SOM_PAK: the self-organizing map program package*, second ed., 1995.
3. J. Himberg, A SOM based cluster visualization and its application for false coloring, in *IEEE-INNS-ENNS International Joint Conference on Neural Networks (IJCNN'00)*, vol. 3, pp. 3587–3592, 2000.
4. T. Margush and F. R. McMorris, Consensus n-trees, Bulletin of Mathematical Biology, vol. 43, pp. 239–244, 1981.
5. E. N. Adams, N-trees as nestings: complexity, similarity, and consensus, Journal of Classification, vol. 3, pp. 299–317, 1986.
6. B. Fritzke, A growing neural gas network learns topologies, in *Advances in Neural Information Processing Systems 7* (G. Tesauro, D. S. Touretzky, and T. K. Leen, eds.), pp. 625–632, Cambridge MA: MIT Press, 1995.
7. M. Dittenbach, A. Rauber, and D. Merkl, Uncovering hierarchical structure in data using the growing hierarchical self-organizing map, Neurocomputing, vol. 48, pp. 199–216, 2002.
8. J. Dopazo and J. M. Carazo, Phylogenetic reconstruction using an unsupervised growing neural network that adopts the topology of a phylogenetic tree, Journal of Molecular Evolution, vol. 44, pp. 226–233, 1997.
9. E. W. Myers and W. Miller, Optimal alignments in linear space, Computer Applications in Biosciences (CABIOS), vol. 4, pp. 11–17, 1988.
10. M. S. Waterman, Parametric and ensemble sequence alignment algorithms, Bulletin of Mathematical Biology, vol. 56, no. 4, pp. 743–767, 1994.
11. S. Henikoff and J. Henikoff, Amino acid substitution matrices from protein blocks, Proceedings of the National Academy of Sciences, USA, vol. 89, pp. 10915–10919, 1992.
12. K. Sjölander, Phylogenomic inference of protein molecular function: advances and challenges, Bioinformatics, vol. 20, no. 2, pp. 170–179, 2004.

13. F. Horn, E. Bettler, L. Oliveira, F. Campagne, F. E. Cohen, and G. Vriend, GPCRDB information system for G protein-coupled receptors, Nucleic Acids Research, vol. 31, no. 1, pp. 294–297, 2003.
14. J. Thompson, D. Higgins, and T. Gibson, CLUSTAL W: improving the sensitivity of progressive multiple sequence alignment through sequence weighting, positions-specific gap penalties and weight matrix choice, Nucleic Acids Research, vol. 22, pp. 4673–4680, 1994.
15. D. T. Jones, W. R. Taylor, and J. M. Thornton, The rapid generation of mutation data matrices from protein sequences, Computer Applications in Biosciences (CABIOS), vol. 8, pp. 275–282, 1992.
16. J. Felsenstein, PHYLIP – phylogeny inference package (version 3.2), Cladistics, vol. 5, pp. 164–166, 1989.
17. N. Saitou and M. Nei, The neighbor-joining method: a new method for reconstructing phylogenetic trees, Mol. Biol. Evol., vol. 4, pp. 406–425, 1987.
18. J. Felsenstein, Confidence limits on phylogenies: an approach using the bootstrap, Evolution, vol. 39, pp. 783–791, 1985.
19. E. V. Samsonova, T. Bäck, M. W. Beukers, A. P. IJzerman, and J. N. Kok, Combining and comparing cluster methods in a receptor database, in *Advances in Intelligent Data Analysis V* (M. R. Berthold, H.-J. Lenz, E. Bradley, R. Kruse, and C. Borgelt, eds.), (Berlin), pp. 341–351, Springer-Verlag, 2003.
20. E. V. Samsonova, J. N. Kok, and A. P. IJzerman, TreeSOM: cluster analysis in the self-organizing map, in *Proceedings of the 5th Workshop On Self-Organizing Maps*, p. in press, 2005.
21. J. Hanke and J. G. Reich, Kohonen map as a visualization tool for the analysis of protein sequences: multiple alignments, domains and segments of secondary structures, Computer Applications in Biosciences (CABIOS), vol. 12, no. 6, pp. 447–454, 1996.
22. T. Kohonen and P. Somervuo, How to make large self-organizing maps for non-vectorial data, Neural Networks, vol. 15, pp. 945–952, 2002.
23. I. Fischer, Similarity-based neural networks for applications in computational molecular biology, in *Advances in Intelligent Data Analysis V* (M. R. Berthold, H.-J. Lenz, E. Bradley, R. Kruse, and C. Borgelt, eds.), (Berlin), pp. 208–218, Springer-Verlag, 2003.
24. J. Vesanto and E. Alhoniemi, Clustering of the self-organizing map, IEEE Transactions on Neural Networks, vol. 11, no. 3, pp. 586–600, 2000.
25. M. A. Kraaijveld, J. Mao, and A. K. Jain, A non-linear projection method based on Kohonen's topology preserving maps, in *Proceedings of the 11th International Conference on Pattern Recognition (11ICPR)*, (Los Alamitos, CA.), pp. 41–45, IEEE Comput. Soc. Press, 1992.

Statistical Recognition of Noun Phrases in Unrestricted Text*

José I. Serrano[1] and Lourdes Araujo[2]

[1] Instituto de Automática, Industrial CSIC, Spain
nachosm@iai.csic.es
[2] Departamento de Sistemas, Informáticos y Programación,
Universidad Complutense de Madrid, Spain
lurdes@sip.ucm.es

Abstract. This paper presents a new model for flexible noun phrase detection, which is able to recognize noun phrases similar enough to the ones given by the inferred noun phrase grammar. To allow this flexibility, we use a very accurate set of probabilities for the transitions between the part-of-speech tag sequence which defines a noun phrase. These accurate probabilities are obtained by means of an evolutionary algorithm, which works with both, positive and negative examples of the language, thus improving the system coverage, while maintaining its precision. We have tested the system on different corpora and compare the results with other systems, what has revealed a clear improvement of the performance.

1 Introduction

There are many Natural Language Processing (NLP) applications for which noun phrase (NP) detection is useful. For instance, NPs can be used in the identification of multiword terms, which are mainly noun phrases, or as a preprocessing step for a subsequent complete syntactic analysis. However, probably the most direct and relevant application nowadays is in information recovering. NPs recover most of the information content of a document, helping to detect the topics it is about. In fact, document indexing by NPs has been shown to improve on other kinds of indexing when retrieval is carried out over a very large corpus [1]. Thus, the distribution of noun phrases can guide search engines in collecting relevant documents according to user queries, or they can be used in text summarizing, in machine translation, etc.

Different approaches have been proposed for the NP identification problem. Some of them rely on linguistic knowledge and use a hand-coded language model. Bourigault [2] uses a handcrafted NP grammar along with some heuristics for identifying NPs of maximal length, and Voutilainen [3] uses a constraint grammar formalism. Other proposals follow a learning approach based on the use of corpora. Church [4] uses a probabilistic model automatically trained on the Brown corpus to detect NPs as well as to assign parts of speech. Ramshaw &

---

* Supported by projects TIC2003-09481-C04 and FIT150500-2003-373.

A.F. Famili et al. (Eds.): IDA 2005, LNCS 3646, pp. 397–408, 2005.

Marcus [5] uses the supervised learning methods proposed by Brill[6] to learn a set transformation rules for the problem. Pla and Prieto [7] use grammatical inference to obtain a FSA which recognizes NPs, what has inspired this work.

This paper presents a new model for a *flexible* identification of basic (non-recursive) NPs in an arbitrary text. The system generates a probabilistic finite-state automaton (FSA) able to recognize the sequences of lexical tags which form an NP. The FSA is generated with the Error Correcting Grammatical Inference (ECGI) algorithm of grammatical inference and initial probabilities are assigned using the collected bigram probabilities. The FSA probabilities provide a method for a flexible recognition of input chains, which are considered to be NPs even if they are not accepted by the FSA but are similar enough to an accepted one. Thus, the system is able to recognize NPs not present in the training examples, what has proven very advantageous for the performance of the system. The FSA probabilities, which are crucial for the flexibility in recognition, are optimized by applying an evolutionary algorithm (EA), what produces a highly robust recognizer. The EA uses both, positive and negative training examples, what contributes to improve the coverage of the system while maintaining a high precision.

Though it is difficult to compare different approaches because they differ in multiple elements, some attempts have been made. Pla [8] gathers results of a number of parses trained on the data set used by Ramshaw [5]: the one proposed by Cardie & Pierce [9], whose technique consists in matching part-of-speech (POS) tag sequences from an initial NP grammar extracted from an annotated corpus and then ranking and pruning the rules according to their achieved precision; the memory-based approach presented in [10], which introduces cascaded chunking, a process in two stages in which the classification of the first stage is used to improve the performance of the second one; the Memory-Based Sequence Learning (MBSL) algorithm [11], which learns sequences of POS tags and brackets, and the hybrid approach of Tjong-Kim-Sang [12], which uses a weighted voting to combine the output of different models. We have also applied our model to the same test set in order to compare the results.

The rest of the paper proceeds as follows: section 2 describes the general scheme of the system, presenting their main elements and its relationships; section 3 is devoted to describe the evolutionary algorithm used to training the FSA; section 4 presents and discusses the experimental results, and section 5 draws the main conclusions of this work.

2   General Scheme of the System

This work presents the design and implementation of a flexible recognizer of basic NPs given as chains of POS tags. Because the system is constructed from examples of objective syntagmas, an algorithm of grammar inference will be used (ECGI). This algorithm extracts from these examples a FSA, which represents the grammar defined by them. In its turn, this algorithm uses the Viterbi algorithm, a dynamic programming one, so that the FSA has an optimal number

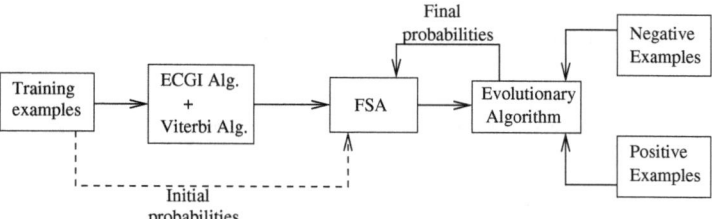

Fig. 1. Scheme of the process to generate the NP flexible recognizer. The FSA is constructed from a set of training examples by applying the ECGI algorithm and using Viterbi to find the most similar chain of tags for a given input. Initial probabilities for the FSA are also obtained from this set of training examples. Then, the probabilities are tuned by applying an evolutionary algorithm which uses new examples, both positive and negative.

of states and paths. In order to extend the recognition capabilities of the FSA beyond the set of training examples, it is necessary to endow the system with a mechanism to recognize also other NPs not appearing in the examples, but very similar to the ones gathered in the grammar. This mechanism to introduce flexibility amounts to allowing the FSA to recognize NPs with a percentage of error. If the number of errors incurred in parsing an NP is below a threshold value, then the new NP is also recognized. An error appears whenever the input tag does not produce any transition from the actual state of the FSA. In order to allow the FSA to find NPs similar to those of the input, their edges are labeled with the probability of each transition between tags within an NP. These probabilities are initially extracted from the NPs examples, and afterwards an evolutionary algorithm is applied to optimize the probabilities. This algorithm uses NP examples different from those used in the FSA construction, as well as negative examples, i.e. syntagmas which must not be recognized as NPs. Figure 1 shows a scheme of the relationships between the different components and phases involved in the flexible recognizer. Let us now describe the different elements of the system.

2.1   Building the FSA for NP Recognition

The technique used to obtain the FSA is an algorithm of Grammar Inference, i.e. a process which, from a set of examples, obtains structural patterns of the language and represents them by means of a grammar. Dupont [13] proposes a general scheme for selecting the most appropriate algorithm of grammar inference under different conditions. According to these ideas, we have chosen the Error Correcting Grammatical Inference (ECGI) algorithm [14] since we are interested in a heuristic construction of a regular grammar such that the final result allows for a flexible recognition. This technique involves a progressive construction of the FSA from positive examples. The FSA is modified in order to correct the errors appearing in the parsing of a new example, obtaining non recursive grammars (a FSA without cycles).

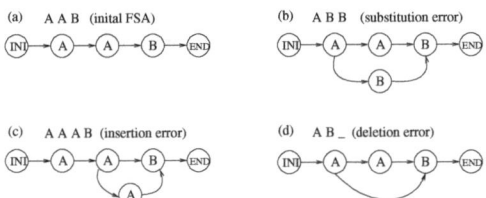

Fig. 2. Examples of application of the ECGI algorithm

The algorithm constructs the grammar by adding in an incremental way the training examples. In order to avoid introducing noise in the recognizer which could drive it to increase the number of false positives (non-nominal syntagmas recognized as nominal), it is necessary to remove this noise by applying a preprocessing step to the training examples. Thus, the typically non-nominal syntagmas which behave as nominal in the training set (the nominalization of a verb, for example) are handily detected and eliminated from the training list. After the preprocessing, the algorithm proceeds as follows. First, a simple FSA is generated which recognizes the first example of the training set. This FSA is then extended with the other examples. To introduce a new example, the Viterbi algorithm is used to find the sequence of states which recognizes the example with less errors. Then, the FSA is extended by adding states and/or transitions which correct the produced errors. In this way, when the process of parsing all the examples finishes, the resulting FSA is able to recognize all of them and does not present cycles.

The errors which can appear between an input chain and the recognized ones are of three types: insertion, substitution and deletion. Figure 2 shows examples of each of them. When the FSA of Figure 2(a) tries to recognize the input chain (A B B) (Figure 2(b)), a substitution error is detected and solved by adding a new state to the FSA with the label of the error. If the new chain is (A A A B) (Figure 2(c)) an insertion error appears, which is solved by adding a new state with the missing label. Finally, if the chain is (A B) (Figure 2(d)), the deletion error is managed by adding a transition between the states neighbor to the deleted one. The result of the algorithm is a regular grammar which generates a finite language, because by construction it has no cycles.

We have introduced a simplification of this algorithm already proposed in the literature [15]. It aims at reducing the complexity of the algorithm to find in the FSA the closest path to the input chain. This simplification amounts to ignoring the deletion errors. This can reduce the recognition complexity below 10%. However, this simplification also reduces the generality of the language, since it discards chains with lengths different from the training examples. Thus, these examples must be sufficiently varied and heterogenous.

The selection of the most similar chain to the input one is carried out by applying the Viterbi algorithm. This algorithm was initially designed to find the most probable path in a Markov chain which produces a given sequence of tags [16]. This algorithm has also been applied to search the minimum path in a multi-stage graph. In our case, the goal is to minimize the number of errors in the FSA for an input chain.

We have modified the Viterbi algorithm to process input chains with a length different from the examples chains. Thus, if the input chain is longer than the longest chain the FSA recognizes, an edge from each state to every other one, including itself, is added and those edges are considered to produce transition errors.

2.2 Flexible Recognition of NPs

The FSA built so far basically recognizes those NPs present in the training examples (some other sequences can also be recognized, though in general they are not NPs). Because we are interested in a recognizer as general as possible and because the training examples usually include only a small sample of the language, it is necessary to endow the system with a mechanism of generalization. This mechanism amounts to allowing the FSA to recognize a chain if it is sufficiently similar to an accepted chain, i.e. if the number of differences is below a certain threshold value. This procedure does not introduces too many chains outside the language, as the experiments have shown. For example, the FSA of Figure 3 does not recognize the chain (D C C). However, if the error threshold value is 1, the chain will be recognized, since the chain (A C C) is in fact recognized. Obviously, the error threshold value is a determining factor for the generalization of the model, and it is object of a detailed study in the experiments.

Now the question is how to parse a chain which presents errors. In this case, the parse arrives at a situation in which no transition is possible, since the next input tag does not matches any of the state for which there is a transition. At this point, we use the statistics derived from the training examples. Each transition is labeled with a real value, which represents its probability in the target language. Then, to process an error tag, the parse follows the most probable transition. The probabilities are relative to each state, in such a way that the values of all the transitions leaving a same state add up to one. Each probability is initially computed as the number of occurrences of a transition relative to the others of the same state according to the training examples. For example, in the FSA of Figure 3, if transition AC appears 7 times in the examples and transition AB

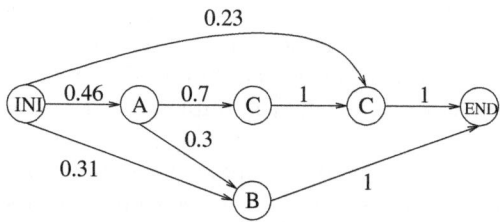

Fig. 3. Probabilistic FSA for flexible NPs recognition

appears 3 times, 10 transitions exist from A to C or B, and thus the probability of the edges that leave state A are 0.7(C) and 0.3(B) (AC = 7/10; AB = 3/10). With the probabilities of the edges so determined, if the chain (D C C) is parsed, as there is not transition from the initial state to D, the parse will take the most probable transition, which leads to A, and afterwards the process continues with the other tags of the input, (C C), which are correct transitions. Thus, the chain (A C C) is obtained with one error with respect to the input. In this way, the parse produces a chain very similar to the input one and which occurs with high probability in the language. Thus, we can expect that a chain which has been only partially recognized, belongs to the language if the error is small.

3   Optimizing the Automata Probabilities: Evolutionary Algorithm

Now the goal is to find a set of probabilities for the edges of the FSA described in the previous section, which allow it to recognize as many NPs as possible, avoiding at the same time recognizing false NPs. Accordingly, the search space is composed of the different sets of probabilities for the edges of the FSA and the algorithm looks for those which optimize the recognition of a set of training examples. This complex search is performed with an evolutionary algorithm.

Systems based on evolutionary algorithms maintain a population of potential solutions and are provided with some selection process based on the fitness of individuals, as natural selection does. The population is renewed by replacing individuals with those obtained by applying "genetic" operators to selected individuals. The most usual "genetic" operators are *crossover* and *mutation*. Crossover obtains new individuals by mixing, in some problem dependent way, two individuals, called parents. Mutation produces a new individual by performing some kind of random change on an individual. The production of new generations continues until resources are exhausted or until some individual in the population is fit enough. Evolutionary algorithms have proven to be very useful search and optimization methods, and have previously been applied to different issues of natural language processing [17], such as text categorization [18], tagging [19] and parsing [20].

In our case, individuals represent probabilistic FSAs, all of them with the same structure but different probabilities. They are implemented as an adjacency matrix, what facilitates the application of the genetic operators. Thus, the probability an edge going from state i to state j is the value which appears in row i column j, A_{ij}. Transitions which do no exist are assigned the value -1. For example, the FSA of Figure 4(a) is implemented as the matrix of Figure 4(b).

The genetic operators applied to renew the population are crossover and mutation. Two variants of crossover have been implemented. The first one is the classic one point crossover, which combines two individuals by combining the first part of one parent up to a crossover point with the second part of the other parent and vice versa. The second variant uses two crossover points and

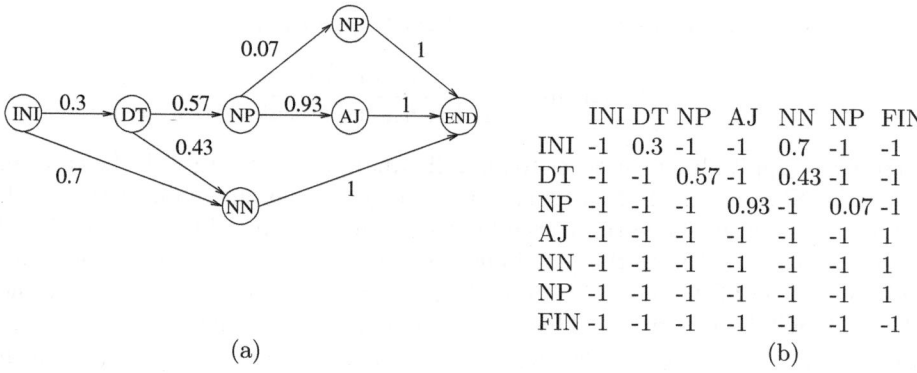

	INI	DT	NP	AJ	NN	NP	FIN
INI	-1	0.3	-1	-1	0.7	-1	-1
DT	-1	-1	0.57	-1	0.43	-1	-1
NP	-1	-1	-1	0.93	-1	0.07	-1
AJ	-1	-1	-1	-1	-1	-1	1
NN	-1	-1	-1	-1	-1	-1	1
NP	-1	-1	-1	-1	-1	-1	1
FIN	-1	-1	-1	-1	-1	-1	-1

(a)   (b)

Fig. 4. Individuals representation

exchanges the bit of the two parents between these points. In both variants, the crossover points are randomly selected.

The mutation operator amounts to choosing an edge at random and varying its probability. This variation may be positive or negative, what is decided at random, and the amount is an input parameter to the algorithm, studied in the experiments. Notice that when an edge is modified the remaining edges of the same state must be updated in order to maintain the value of its addition equal to 1. If the variation produces a value smaller than the zero or greater than one, then the edge is assigned zero or one respectively, and the real variation is computed according to this value.

3.1   The Fitness Function

The fitness function must be a measure of the capability of the FSA to recognize NPs and only them. On the one hand, the fitness function must include a measure of the coverage, or *recall* achieved by the individual, i. e. the number of NPs which have been recognized from the set of proposed NPs:

$$recall = \frac{\text{number of recognized NPs}}{\text{number of proposed NPs}}$$

On the other hand, the fitness function must also take into account the precision achieved by the individual:

$$precision = \frac{\text{number of NPs recognized} + \text{number of non NPs discarded}}{\text{number of proposed syntagmas}}$$

We have considered as fitness function two F-measures which combine these two parameters in different ways:

$$\text{F-measure} = \frac{2 \cdot \text{Precision} \cdot \text{Recall}}{\text{Precision} + \text{Recall}}$$

and F2-measure, which gives a higher weight to precision,

$$\text{F2-measure} = \frac{5 \cdot \text{Precision} \cdot \text{Recall}}{4 \cdot \text{Precision} + \text{Recall}}$$

In order to apply these measures to an individual, we must establish the cases in which an NP is considered recognized. Obviously those input chains corresponding to a path from the initial to the final state are recognized by any individual in the population. But in the remaining cases, the path will depend on the probabilities of the FSA, and only those chains with a number of errors below the threshold will be recognized. This threshold value can be defined as an absolute value or as a percentage of the length of the chain. Both alternatives have been evaluated in the experiments.

3.2   Initial Population

Individuals for the initial population are randomly generated from a seed solution, which helps to guide the search. The seed, which is included in the population as another individual, is the set of probabilities obtained from the training examples used to construct the FSA. The individuals of the remaining population are generated by applying the mutation operator several times to the seed. The variation produced by each mutation in this phase is a parameter studied in the experiments. Notice that a small variation would produce individuals too similar to the seed, which in spite of having a high quality according to the fitness function, do not help to explore other areas of the search space where better solutions could be found, while a great variation can lead to lose the advantages of the seed.

4   Experimental Results

The algorithm has been implemented using C++ language and run on a Pentium III processor. The CPU time spent on generating an automaton from 45 examples, with a maximum length of 7, was 0.3 seconds, and from 156 examples, with 9 as maximum length, 1.6 seconds. The time spent on analyzing a text composed of 1108 different syntagmas, being NPs 570 of them, was 1.1 seconds. For the experiments we have used training and test sets from the Brown corpus portion of the Penn Treebank [21], a database of English sentences manually annotated with syntactic information. After a number of experiments, we have selected a set of default criteria and parameters for the EA, which provide a high performance for different settings of the problem, and which have been used in the experiments described below. The EA uses a two point crossover, F1-measure as fitness function, and elitism[1]. The parameter values are a population size of 100 individuals, a number of 150 generations, a crossover rate of 60%, a mutation

---

[1] By elitism we refer to the technique of retaining in the population the best individuals found so far.

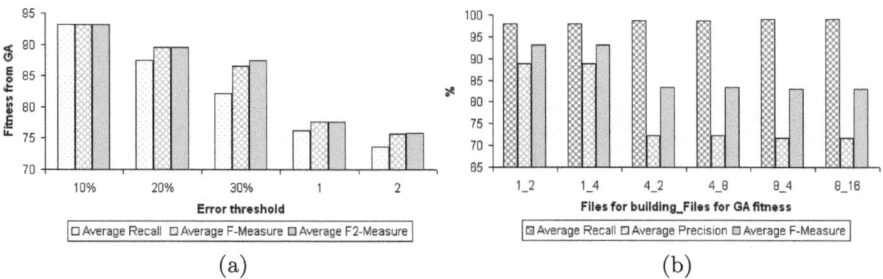

Fig. 5. (a) Error thresholds and fitness function types comparison. (b) F-Measure values for different combinations of number of files for building the recognizer-number of files for EA fitness.

rate of 40%, a variation of FSA probabilities in mutation of 0.5, and a variation of FSA probabilities in the generation of the initial population of 0.7. Each EA has been run five times, and the average and maximum values of these executions are presented for each experiment. Setting the EA parameters to these values, the time spent on ten executions of the EA was over 3 hours and a half, using a test set of 1569 different syntagmas, 543 NPs, for the fitness function calculation.

Before any other empirical trials, we have carried out a test in order to show the benefits of the EA over the recognizer. We have found out that the higher the error threshold the better the improvement. For relative error thresholds of 10%, 20% and 30% the F-Measure over the test increases 4%, 10% and 14%, respectively. For absolute error threshold values of 1 and 2 the increase is about 5% and 12%, respectively.

We also have performed experiments to determine the most appropriate error threshold allowed in the recognition. Figure 5 (a) shows the results obtained with different performance measures. Two different ways of defining the threshold value have been studied: as an absolute number of errors, and as a percentage of the NP length. Best results are obtained with the threshold defined as a percentage and for relative low values, such as 10%. This value has been used in the following experiments.

Another question which has been investigated is the most appropriate size of the training sets for both, the construction of the FSA (C), and for the evaluation of fitness measure of the EA (F). Figure 5 (b) shows the results of the different considered measures for different combination of these values (C-F). Best results for all measures are obtained when using a small number of files for the FSA construction, and a large number of files for the EA. Using a larger training set in the construction of the FSA produces too large FSAs, which recognize too much false NPs and deteriorate the system performance. We can in fact observe that the recall measures are high when using a larger set in the construction of the FSA, at the price of producing very low precision measures.

Since the system also uses training examples to establish the initial probabilities from which the initial population of EA is generated, we have performed a test in order to study the influence of the number of examples used this way

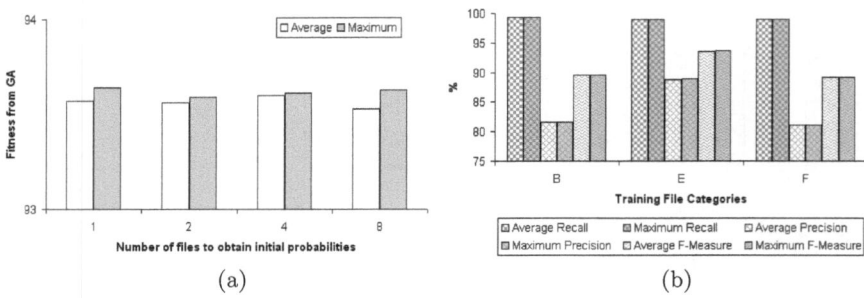

Fig. 6. (a) F-measures values from the GA using different number of files to obtain the initial probabilities or seed. (b) Accuracy values for different categories of the file used to build the recognizer.

in the EA performance. As observed in Figure 6 (a), there exists an improvement the more examples are used, but it is not very significant. However, the decrease in performance when using eight files indicates that it is not convenient to use too many training examples, because the extracted probabilities become overfitted.

We have also performed experiments to determine the impact of using different categories of topics as training sets. Figure 6 (b) shows the results obtained using training files of category B (press editorial), E (skills and hobbies) and F (popular lore). We can observe that the results are quite different depending on the used category, since the kind and frequency of NPs is different in each of them.

In order to compare the overall performance with other related systems, we have applied the system to a standard subset of the Wall Street Journal portion of the Penn Treebank, also used in the other works. Table 1 compares the precision, recall and F-measure values. We can observe that our results improve on those of all the other systems, both in recall and F-measure, and only the Ramshaw95 and Tjong-Kim-Sang00 systems provide a slightly better precision. It proves the usefulness of the mechanism of flexible recognition, since it achieves, as expected, a significant improvement of the performance, maintaining at the same time a high level of precision thanks to the accurate probabilities that the EA provides.

Table 1. Comparison of precision, recall and F-measure values with similar systems over the Ramshaw standard corpus

	Precision(%)	*Recall(%)*	*F-Measure(%)*
[Ramshaw95]	91,8	92,3	92,0
[Cardie98]	90,7	91,1	90,9
[Veenstra98]	89,0	94,3	91,6
[Argamon98]	91,6	91,6	91,6
[Tjong-Kim-Sang00]	93,6	92,9	93,3
[This work]	91,6	97,8	94,5

5   Conclusions

This paper presents the design of a probabilistic FSA for the identification of basic NPs, which can be used for document indexing in information retrieval and other applications.

The FSA probabilities provide a method for a flexible recognition of input chains, which are considered to be NPs if they are similar enough to an accepted one. An evolutionary algorithm is used to optimize the FSA probabilities, and it uses both, positive and negative training examples, what contributes to improve the coverage of the system maintaining at the same time a high precision.

We have performed experiments to tune the system in several ways, such as the EA parameters, the threshold value of the errors allowed in the recognition, and the size of the training sets used in different phases of the process. A comparison with other systems tested on the same set of texts has shown that the mechanism introduced for the flexible recognition of NP clearly improves the coverage of the system, while the adjusted probabilities provided by the EA yield a high level of precision, thus yielding a better overall performance.

A study of the most frequent errors has revealed that those affecting the recall are mainly due to the presence of rare NPs, with a structure very different from the others. However, as they are infrequent, the reduction of the recall is low. The errors which deteriorate the precision are mainly due to some training examples used in the construction of the FSA in which a sequence of tags, which usually does not correspond to an NP, is working as an NP in that particular case, and also to sequences in which many of their tags are not typical NPs tags. According to this observation, we plan to perform a statistical study of the different kinds of errors, in order to introduce mechanisms to filter the training set in some appropriate and automatic manner. We also consider to investigate other improvements, on different aspects of the system: on the application of the EA, with the definition of other genetic operators and fitness functions, on the recognition process, with the consideration of information from negative examples, and on the construction of the FSA, with the possibility of introducing cycles.

References

1. Zhai, C.: Fast statistical parsing of noun phrases for document indexing. In: Proceedings of the Fifth Conference on Applied Natural Language Processing. (1997)
2. Bourigault, D.: Surface grammatical analysis for the extraction of terminological noun phrases. In: Proc. of the Int. Conf. on Computational Linguistics (COLING-92). (1992) 977–981
3. Voutilainen, A.: Nptool, a detector of english noun phrases. In: Proc. of the Worshop on Very Large Corpora (ACL). (1993) 48–57
4. Church, K.W.: A stochastic parts program and noun phrase parser for unrestricted text. In: Proc. of 1st Conference on Applied Natural Language Processing, ANLP. (1988) 136–143

5. Ramshaw, L., Marcus, M.: Text chunking using transformation-based learning. In: Proc. of the third Workshop on Very Large Corpora (ACL). (1995) 82–94
6. Brill, E.: Transformation-based error-driven learning and natural language processing: A case study in part of speech tagging. Computational Linguistics **21** (1995)
7. Pla, F., Molina, A., Prieto, N.: Tagging and chunking with bigrams. In: Proc. of the 17th conference on Computational linguistics. (2000) 614–620
8. Pla, F.: Etiquetado léxico y análisis sintáctico superficial basado en modelos estadísticos (2000)
9. Cardie, C., Pierce, D.: Error-driven pruning of treebank grammars for base noun phrase identification. In: Proc. of COLING-ACL'98. (1998) 218–224
10. Veenstra, J.: Fast np chunking using memory-based learning techniques. In: Proc. of BENELEARN-98: Eighth Belgian-Ducth Conference on Machine Learning. (1998) 71–78
11. Argamon, S., Dagan, I., Krymolowski, Y.: A memory-based approach to learning shallow natural language patterns. In: Proc. of joint International Conference COLING-ACL. (1998) 67–73
12. Tjong-Kim-Sang, E.F.: Noun phrase representation by system combination. In: Proc. of ANLP-NAACL. (2000) 50–55
13. Dupont, P.: Inductive and statistical learning of formal grammars. Technical report, Reseach talk, Departement ingenerie Informatique, Universite Catholique de Louvain (2002)
14. Rulot, H., Vidal, E.: Modelling (sub)string-length-based constraints through a grammatical inference method. In: Pattern Recognition: Theory and Applications. Springer-Verlag (1987) 451–459
15. Torró, F., Vidal, E., , Rulot, H.: Fast and accurate speaker independent speech recognition using structurals models learnt by the ecgi. In: Signal Proccesing V: Theories and Applications. Elsevier Science Publishers B.V. (1990)
16. Forney, G.D.: The viterbi algorithm. Proceedings of The IEEE **61** (1973) 268–278
17. Kool, A.: Literature survey. Technical report, Center for Dutch Language and Speech. University of Antwerp (2000)
18. Serrano, J., Castillo, M.D., Sesmero, M.: Genetic learning of text patterns. In: Proc. of CAEPIA03. (2003) 231–234
19. Araujo, L.: Part-of-speech tagging with evolutionary algorithms. In: Proc. of the Int. Conf. on Intelligent Text Processing and Computational Linguistics (CICLing-2002), Lecture Notes in Computer Science 2276, Springer-Verlag (2002) 230–239
20. Araujo, L.: A probabilistic chart parser implemented with an evolutionary algorithm. In: Proc. of the Int. Conf. on Intelligent Text Processing and Computational Linguistics (CICLing-2004), Lecture Notes in Computer Science 2276, Springer-Verlag (2004) 81–92
21. Marcus, M.P., Santorini, B., Marcinkiewicz, M.A.: Building a large annotated corpus of english: The penn treebank. Computational Linguistics **19** (1994) 313–330

Successive Restrictions Algorithm in Bayesian Networks

Linda Smail and Jean Pierre Raoult

L.A.M.A. Laboratory, Marne-la-Vallée University,
77454 Champs sur Marne, Marne-la-Vallée, France
{linda.smail, jean-pierre.raoult}@univ-mlv.fr
http://www.univ-mlv.fr/lama

Abstract. Given a Bayesian network relative to a set I of discrete random variables, we are interested in computing the probability distribution P_A or the conditional probability distribution $P_{A|B}$, where A and B are two disjoint subsets of I. The general idea of the algorithm of successive restrictions is to manage the succession of summations on all random variables out of the target A in order to keep on it a structure less constraining than the Bayesian network, but which allows saving in memory ; that is the structure of Bayesian Network of Level Two.

1 Introduction

Given a Bayesian network [3] relative to a set $X_I = (X_i)_{i \in I}$ of random variables taking values in finite sets $(\Omega_i)_{i \in I}$, we are interested in computing the joint probability distribution of a subset of random variables $X_A = (X_j)_{j \in A}$ conditionally to another subset (possibly empty) of random variables $X_B = (X_k)_{k \in B}$, where A and B are two disjoint subsets of I. This conditional distribution will be denoted $P_{A/B}$ (or P_A if $B = \emptyset$).

According to Bayes theorem we have

$$P_{A/B}(x_A|x_B) = \frac{P_{A \cup B}(x_A, x_B)}{\sum_{x_A} P_{A \cup B}(x_A, x_B)}.$$

Therefore, to compute this conditional probability, we need to calculate the probability distribution of $X_{A \cup B}$, which requires marginalizing out the set of variables X_i, for $i \in I - (A \cup B) = \overline{A \cup B}$.

The algorithm we propose to solve this problem is called the "Successive Restrictions Algorithm" (SRA) [9]. SRA is a goal-oriented algorithm that tries to find an efficient marginalization (elimination) ordering for an arbitrary joint distribution.

The aim of finding a marginalization (elimination) ordering for an arbitrary target joint distribution is shared by other node elimination algorithms like "variables elimination" [10], "bucket elimination" [2]. The main idea of this goal-oriented approach, common to all these algorithms, is to sum over a set of variables from a list of factors one by one. An ordering of these variables is

A.F. Famili et al. (Eds.): IDA 2005, LNCS 3646, pp. 409–418, 2005.

required as an input and is called an elimination ordering. The computation depends on the ordering elimination: different elimination ordering produce different factors. However, the SRA has two additional objectives:

1. We construct a symbolic probability expression (evaluation tree) representing the elimination ordering regardless the numerical values to be used in the effective numerical evaluation.
2. All intermediate computations are probability distributions instead of simple potentials. In other words, each node of the evaluation tree represents a probability distribution on a subset of variables. This property is very important because each node of the tree may be replaced at runtime by another distribution.

Before detailing the principle of SRA, we introduce the concepts of Bayesian network of level two and close descendants (for more details see [8]).

2　Bayesian Network of Level Two

We consider the probability distribution P_I of a finite family $(X_i)_{i \in I}$ of random variables on a finite space $\Omega_I = \prod_{i \in I} \Omega_i$. Let $\mathcal{I}$ be a partition of I and let us consider a directed acyclic graph $\mathcal{G}$ on $\mathcal{I}$; we say that there is a link from J' to J'' (where J' and J'' are atoms of the partition $\mathcal{I}$) if $(J', J'') \in \mathcal{G}$. If $J \in \mathcal{I}$, we note $p(J)$ the set of parents of J, that is the set of J' such that $(J', J) \in \mathcal{G}$.

Definition 1: *The probability P_I is defined by the Bayesian Network of level two (BN2), on I, $(\mathcal{I}, G, (P_{J|p(J)})_{J \in \mathcal{I}})$, if for each $J \in \mathcal{I}$, we have the conditional probability $P_{J|p(J)}$, in other words the probability of X_J conditioned by $X_{p(J)}$ (which, if $p(J) = \emptyset$, is the marginal probability P_J), so that :*

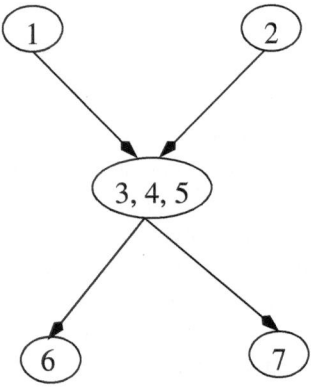

Fig. 1. Example of a Bayesian network of level two

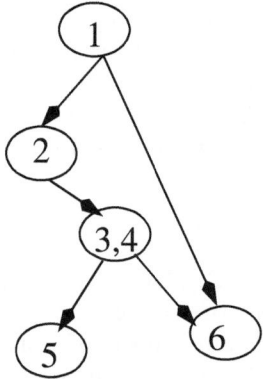

Fig. 2. Bayesian network of level two

$$P_I(x_I) = \prod_{J \in \mathcal{I}} P_{J|p(J)}(x_J|x_{p(J)}).$$

An usual Bayesian network is a particular case of level 2, with the partition of I into singletons.

In some cases, it can be useful to remark that $P_{J/p(J)}$ actually depends only on a subset K of $\cup_{J' \in p(J)} J'$; this is equivalent to the notion of bubble graphs due to Shafer [7].

The probability distribution P_I associated to the Bayesian network of level 2 in figure 1 can be written as

$$P_I(x_1, x_2, x_3, x_4, x_5, x_6, x_7) = P_1(x_1)P_2(x_2)P_{3,4,5/1,2}(x_3, x_4, x_5|x_1, x_2)$$
$$P_{6/3,4,5}(x_6|x_3, x_4, x_5)P_{7/3,4,5}(x_7|x_3, x_4, x_5).$$

2.1   Close Descendants

Definition 2: *Let* $(\mathcal{I}, G, (P_{J|p(J)})_{J \in \mathcal{I}})$ *be a Bayesian network of level two. For each* $J \in \mathcal{J}$, *we define the set of close descendants of* J *(denoted* $cd(J)$*) as the set (possibly empty) of vertices containing the children of* J *and, if there are any out of the children themselves, the vertices located on a path between* J *and one of its children.*

$cd(1) = \{2, \{3, 4\}, 6\}$, which we shall write also: $cd(1) = \{2, 3, 4, 6\}$.

3   Successive Restrictions Algorithm

Let $(I, G, (P_{i|p(i)})_{i \in I})$ be a Bayesian network. Given a subset A of I, known as "target", we consider the problem of computing P_A, the probability distribution of $(X_j)_{j \in A}$.

To solve this problem, we suggest an algorithm called *Successive Restrictions algorithm* (SRA). It is a goal-oriented algorithm that tries to find a marginalization (elimination) ordering for an arbitrary target joint distribution.

The general idea of SRA is to manage the summations that we have to do relative to the random variables X_ℓ for $\ell \in \overline{A}$ ($\overline{A} = I - A$).

3.1   Principle of the Algorithm

Let be given a Bayesian network $(I, G, (P_{i|p(i)})_{i \in I})$ and a subset A of I. For any subset S, let $a(S)$ denote the set of all nodes in S or having descendant in S. For any S, $a(S)$ is naturally embedded within the structure of Bayesian network by restriction to S of the Bayesian network given on I. The main idea of SRA for computation of P_A is to build a sequence of subsets $(I_0, \ldots, I_\ell)$ with $\ell = \mathrm{Card}(a(A)) - \mathrm{Card}(A)$, and, for each $0 \leq s \leq \ell$, a structure of BN2 on I_s, noted $R_s = (\mathcal{I}_s, G_s, (P_{J|p(J)})_{J \in \mathcal{I}_s}$, which defines $P_{\mathcal{I}_s}$, probability distribution of $X_{I_s} = (X_i)_{i \in I_s}$, such that :

1. $\mathcal{I}_0$ is the restriction to $a(A)$ of the initial Bayesian network (so $I_0 = a(A)$).
2. Each element of I_s which contains an element of $\overline{A}$ is a singleton.
3. I_{s+1} contains one element less than I_s, this element is in $\overline{A}$. The node corresponding to this element is called marginalization node. In addition R_{s+1} contains fewer nodes than R_s and a new node, not included in R_s, resulting from the marginalization
4. Once the algorithm is performed, $I_\ell = A$ and the probability distribution of $(X_j)_{j \in A}$ can be computed simply as the product of the conditional probabilities in the Bayesian network of level two obtained on I_ℓ .

Now let's specify the general iteration of this algorithm.

We start with a BN2, on $L \subset I$, $R = (\mathcal{I}, G, (P_{J|p(J)})_{J \in \mathcal{I}})$, defining P_L, where $L = \cup_{J \in \mathcal{I}} J$; we will obtain at the exit a BN2, on $L' = L - \{i\}$, $R' = (\mathcal{I}', G', (P_{J|p(J)})_{J \in \mathcal{I}'})$, defining $P_{L'}$, where i is the selecting marginalization node.

We choose a variable i of $\overline{A} \cap L$; according to what precedes, the singleton i belongs to $\mathcal{I}$; this choice is made in such a way that i has no descendant in $\overline{A}$. This constraint is a sufficient condition to ensure the coherence of the constructed network of each step. However, when several choices are available (several nodes are respecting this constraint) additional criteria in relation with computational cost of the inference task may be applied [8].

Once the marginalization node i is selected, the output BN2 R' admits as vertices the atoms of the partition $\mathcal{I}'$, which are:

1. the set of close descendants of i in R, which we denote E_i ($E_i = cd(i)$),
2. all the atoms of $\mathcal{I}$ other than i and those in E_i .

The graph G' results from the graph G in the following way :

1. we delete the links (for G) relative to i and to the elements of E_i,
2. we keep all other links of G,

3. we introduce to E_i a set of parents, $p'(E_i)$, which includes:
   (a) all parents of i in G ;
   (b) all parents of vertices belonging to the close desendants of i, others than i or those in E_i itself.
4. we introduce as children to E_i all children of vertices of R belonging to E_i, others than those in E_i itself.

The probabilistic data associated to R' can be computed from those associated to R in the following way :

1. we conserve the probability, conditionally to his parents, for each vertex such that the passage from R to R' changes neither itself nor his parents (in other words, each vertex other than i, those in E_i and the children of those in E_i);
2. for each child J of E_i (in G), his probability, conditionally to his parents, is preserved by substitution of E_i to the set of the parents of J (in G) which belongs to E_i, and we conserve the information that only these variables are involved in $p'_{J|E_i}$ as shown in figure 3. Indeed, conditionally to his parents in G, J is independent from the vertices of E_i which are not parents of J. This results from the fact that none of those vertices is a descendant of J.
3. we create the probability of E_i conditionally to $p'(E_i)$, which can be computed using the following formula

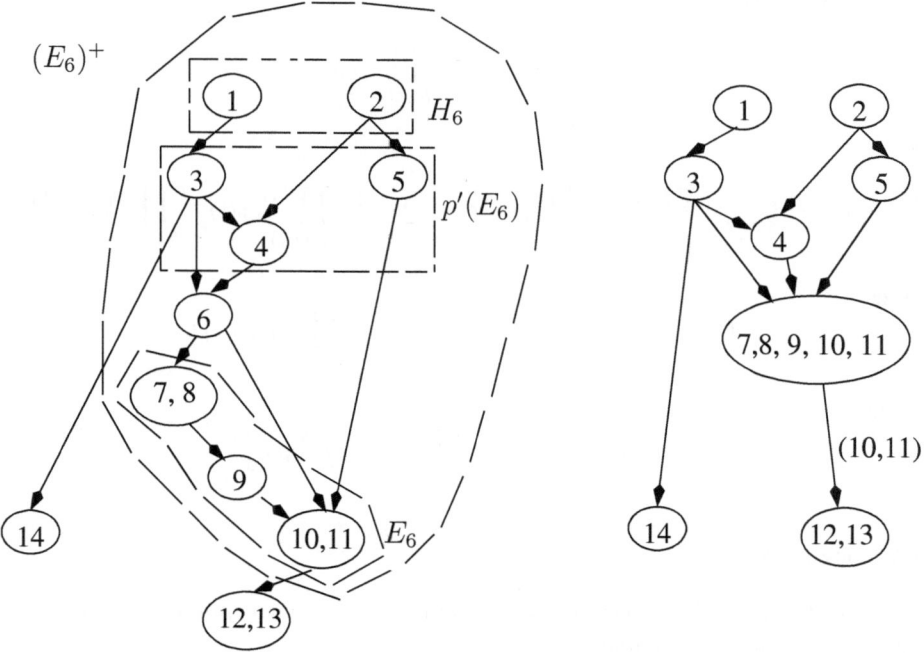

Fig. 3. (a) : Various subsets defining over the node 6. (b) : BN2 resulting after summation over 6 : $E_6 = \{7, 8, 9, 10, 11\}$.

$$P(x_{E_i}|x_{p'(E_i)}) = \sum_{x_i \in \Omega_i} \left[\left(\prod_{J \in E_i} P(x_J|x_{p(J)}) \right) P(x_i|x_{p(i)}) \right].$$

Proof :

To simplify the notation, we will note $\sum_i$ for $\sum_{x_i \in \Omega_i}$ and, for each subset $B = \{b_1, \ldots, b_m\}$ of L, $\sum_B$ for $\sum_{x_{b_1} \in \Omega_{b_1}} \cdots \sum_{x_{b_m} \in \Omega_{b_m}}$; we will omit to write the variables (for example we write $P_{J|p(J)}$ for $P(x_J|x_{p(J)})$).

The various objects intervening in this proof may be visualized on figure 3.

The computation of $P_{E_i|p'(E_i)}$ can be done by using only vertices in $a(E_i)$ which includes $p'(E_i)$ by construction.

We decompose $a(E_i)$ according to the partition $(E_i, \{i\}, p'(E_i), G_i)$, where $a\left(p'(E_i)\right) = p'(E_i) \cup G_i$.

Then

$$P_{E_i \cup p'(E_i)} = \sum_i \sum_{G_i} \left[\left(\prod_{J \in E_i} P_{J|p(J)} \right) P_{i|p(i)} \times \left(\prod_{k \in p'(E_i)} P_{k|p(k)} \right) \left(\prod_{h \in G_i} P_{h|p(h)} \right) \right].$$

We notice that the index i is present only in $\left(\prod_{J \in E_i} P_{J|p(J)} \right) P_{i|p(i)}$, whereas all the ℓ in G_i may be present only in

$$\left(\prod_{k \in p'(E_i)} P_{k|p(k)} \right) \left(\prod_{h \in G_i} P_{h|p(h)} \right);$$

so

$$P_{E_i \cup p'(E_i)} = \left\{ \sum_i \left[\left(\prod_{J \in E_i} P_{J|p(J)} \right) P_{i|p(i)} \right] \right\} \times \left\{ \sum_{G_i} \left[\left(\prod_{k \in p'(E_i)} P_{k|p(k)} \right) \left(\prod_{h \in G_i} P_{h|p(h)} \right) \right] \right\}.$$

Since

$$P_{p'(E_i)} = \sum_{G_i} \left[\left(\prod_{k \in p'(E_i)} P_{k|p(k)} \right) \left(\prod_{h \in G_i} P_{h|p(h)} \right) \right]$$

we get

$$P_{E_i|p'(E_i)} = \sum_i \left[\left(\prod_{J \in E_i} P_{J|p(J)} \right) P_{i|p(i)} \right].$$

3.2   Example

Consider the Bayesian network given in figure 4. The corresponding joint distribution is given by the equation :

$$P_I(x_I) = P_1(x_1)P_2(x_2)P_3(x_3)P_{4/1}(x_4|x_1)P_{5/2,4}(x_5|x_2,x_4)P_{6/5}(x_6|x_5)$$

$$P_{7/3,5}(x_7|x_3,x_5)P_{8/1,6}(x_8|x_1,x_6)P_{9/2,6}(x_9|x_2,x_6)P_{10/2,7}(x_{10}|x_2,x_7)$$

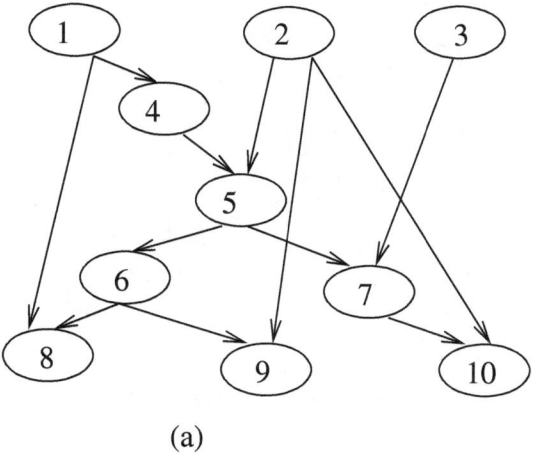

(a)

Fig. 4. Example of a Bayesian network

Suppose we are interested in computing the marginal distribution P_A, where $A = \{1, 3, 5, 7, 8, 9, 10\}$.

This computation requires marginalizing out the variables with indexes in $\overline{A} = \{2, 4, 6\}$ (i.e X_2, X_4, and X_6). For each step of the algorithm, we have to choose a variable in $\overline{A}$ to marginalize out.

Step 1.
For the first step, note that only X_6 has no descendants in $\overline{A}$. By marginalizing out X_6 we get:

$$\sum_{x_6} P_{8/1,6}(x_8|x_1, x_6) P_{9/2,6}(x_9|x_2, x_6) P_{6/5}(x_6|x_5) = P_{8,9/1,2,5}(x_8, x_9|x_1, x_2, x_5).$$

The resulting graph is given in figure 5. It has the structure of a Bayesian network of level 2 and is constructed as follows :

1. The node 6 is suppressed.
2. A new node $E_6 = \{8, 9\}$ is created. Its parents are 1, 2, and 5 (i.e. the initial parents of 6 and the parents of its close descendants 8 and 9).

The joint distribution corresponding to this new BN2 can be written as :

$$P_{I-\{6\}}(x_{I-\{6\}}) = P_1(x_1)P_2(x_2)P_3(x_3)P_{4/1}(x_4|x_1)P_{5/2,4}(x_5|x_2, x_4)$$

$$P_{E_6/1,2,5}(x_{E_6}|x_1, x_2, x_5)P_{7/3,5}(x_7|x_3, x_5)P_{10/2,7}(x_{10}|x_2, x_7).$$

Step 2.
For the second step of the algorithm, note that both X_2 and X_4 have no descendants in $\overline{A}$. So, they are both candidates to be marginalized out. In order

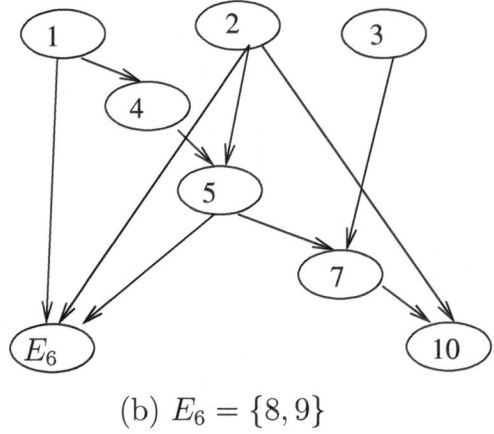

(b) $E_6 = \{8, 9\}$

Fig. 5. BN2 resulting after summation over 6

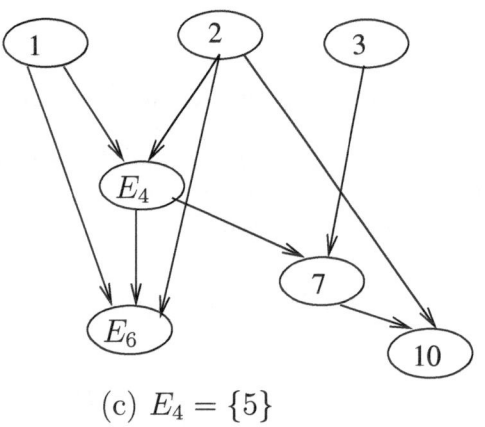

(c) $E_4 = \{5\}$

Fig. 6. BN2 resulting after summation over 4

to decide which variable is to be selected, additional criteria in relation with the computational cost induced by each choice can be introduced. Suppose we choose the node 4 (X_4). By marginalizing out X_4 we get:

$$\sum_{x_4} P_{4/1}(x_4|x_1) P_{5/2,4}(x_5|x_2, x_4) = P_{E_4/1,2}(x_{E_4}|x_1, x_2).$$

The resulting graph is given in figure 6 and is constructed as follows :

1. The node 4 is suppressed.
2. A new node $E_4 = \{5\}$ is created. Its parents are 1 and 2 (i.e. the initial parents of 4 and the parents of its close descendants 5).

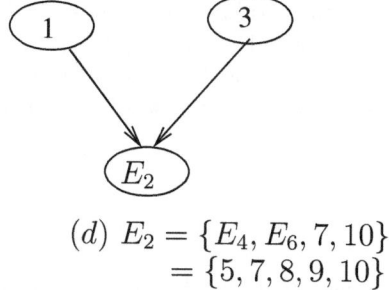

$$(d)\ E_2 = \{E_4, E_6, 7, 10\}$$
$$= \{5, 7, 8, 9, 10\}$$

Fig. 7. BN2 resulting after summation over 2

The joint distribution corresponding to this new BN2 can be written as :

$$P_{I-\{6,4\}}(x_{I-\{6,4\}}) = P_1(x_1)P_2(x_2)P_3(x_3)P_{E_4/1,2}(x_{E_4}|x_1, x_2)$$
$$P_{7/3,5}(x_7|x_3, x_5)P_{10/2,7}(x_{10}|x_2, x_7)P_{E_6/1,2,5}(x_{E_6}|x_1, x_2, x_5).$$

Step 3.

Finally, we have to marginalize out X_2 and we get :

$$\left[\sum_{x_2} P_{E_6/1,2,E_4}(x_{E_6}|x_1, x_2, x_{E_4})P_{E_4/1,2}(x_{E_4}|x_1, x_2)P_{10/2,7}(x_{10}|x_2, x_7)\right.$$
$$\left. P_2(x_2)\right] \times P_{7/3,5}(x_7|x_3, x_5) = P_{E_6,E_4,7,10/1,3}(x_{E_6}, x_{E_4}, x_7, x_{10}|x_1, x_3),$$

The resulting graph is given in figure 7 and is constructed as follows :

1. The node 2 is suppressed.
2. A new node $E_2 = \{E_6, E_4, 7, 10\}$ is created. Its parents are 1 and 3.

The joint distribution corresponding to this new BN2 can be written as :

$$P_{I-\{6,4,2\}}(x_{I-\{6,4,2\}}) = P_1(x_1)P_3(x_3)P_{E_2/1,3}(x_{E_2}|x_1, x_3).$$

This joint distribution corresponds to our target distribution $P(x_A)$. So

$$P_A(x_A) = P_1(x_1)P_3(x_3)P_{E_2/1,3}(x_{E_2}|x_1, x_3).$$

4   Conclusion

We have introduced an algorithm which makes possible the computation of the probability distribution over a subset of random variables $(X_a)_{a \in A}$ of the initial graph. It would be possible to compute the probability distribution of a subset of variables X_A conditionally to another subset X_B ($P_{A|B}$) using SRA.

This algorithm aims to construct a symbolic representation of the target distribution by finding a marginalization ordering that takes into account the computational constraints of the application.

It may happen that, in certain simple cases, the SRA would be less powerful than the traditional methods [1], [4], [5], [6], [7], [10], but it has the advantages of adapting to any subset on nodes of the initial graph, and also to present in each stage interpretable result in terms of conditional probabilities, and thus technically usable.

References

1. Roberto G. Cowell and A. Philip Dawid and Stefenl L. Lauritzen and David.J. Spiegelhalter : Probabilistic Networks and Expert Systems. Springer (1999)
2. Dechter, R. : Bucket elimination : A unifying framework for probabilistic inference. In E. Horvits and F. Jensen, editor, Proc. Twelthth Conf. on Uncertainty in Artificial Intelligence. (1996) 211-219
3. F. V. Jensen : An introduction to Bayesian Networks. UCL Press (1999)
4. Richard E. Neapolitan : Probabilistic Reasoning in Expert Systems. A Wiley-Interscience Publication, John Wiley and Sons, Inc. (1990)
5. Judea, Pearl and Verma, T. : Influence Diagrams and d-Separation UCLA Cognitive Systems Laboratory, Technical Report 880052" (1988)
6. Petr Hájek and Tomás Havránek and Radim Jirouśek : Information Processing in Expert Systems", CRC Press (1992)
7. G. Shafer : Probabilistic Expert System. CBMS-NSF Regional Conference Series in Applied Mathematics 67, SIAM (1996)
8. L. Smail : Algorithmic for Bayesian Networks and their Extensions. Ph.D. Thesis, Laboratory of Applied Mathematics, University of Marne-la-Vallée, France (2004)
9. L. Smail : Successive Restrictions Algorithm. Fourth International Conference on Mathematical Methods in Reliability Methodology and Practice. June 21-25, Santa Fe, New Mexico (2004)
10. N. L. Zhang and D. Poole : A simple approach to bayesian network computations. In Proc. of the Tenth Canadian Conference on Artificial Intelligence, pp. 171-178 (1994)

Modelling the Relationship Between Streamflow and Electrical Conductivity in Hollin Creek, Southeastern Australia

Jess Spate

Australian National University, Institute of Mathematical Sciences
and Integrated Catchment Assessment and Management Centre
jessica.spate@anu.edu.au
http://icam.anu.edu.au/

Abstract. The relationship between streamflow q and electrical conductivity k is explored in this paper, using data from Hollin Cave Spring in New South Wales, Australia. A temporal rule extraction algorithm is used to identify frequent patterns in each time series. The frequent patterns are then refined using the concept of profile convexity, and parametrised for compactness of representation, before the coupling between flow and conductivity is examined. Results show that two frequent peak patterns occur in flow and two troughs in electrical conductivity, and that the shapes of all these can be characterised with a single magnitude parameter. The coupling between events in the two series is investigated, and reveals that the depth of k troughs depend heavily on the initial state of k, and more weakly on the magnitude of the flow peak.

1 Introduction

Electrical conductivity k provides an estimate of dissolved solids in-stream, as the relationship between salinity and electrical conductivity in water is well established. Compared to most other water quality indicators such as suspended sediment load, it is easy to measure. Where some contaminant concentrations rise after rainfall due to increased input, k will fall as the salt load delivered by groundwater is diluted by a greater volume of streamflow or discharge q.

Hollin Cave catchment studied here lies in the Snowy Mountains of South Eastern Australia, entirely within the Yarrangobilly Limestone. The catchment is small (less than 500 Ha) and very steep. More than 300 caves and numerous hot and warm springs have been identified in the area [5]. Hollin Cave Creek is connected to other streams underground in Eagle's Nest Cave, one of the deepest cave complexes on the Australian mainland. Data were collected from this system in the hope of better understanding the karst hydrology regimes.

The Hollin Cave Spring dataset consists of 31341 paired data points for q in litres/second, and k in micro-Siemens/cm at half-hour intervals with only a few breaks over a three year record. Almost all of these breaks are short, of the order of a few hours. The apparatus is discussed at length in [6]. Electrical conductivity

A.F. Famili et al. (Eds.): IDA 2005, LNCS 3646, pp. 419–428, 2005.

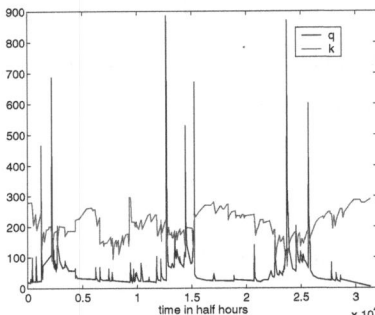 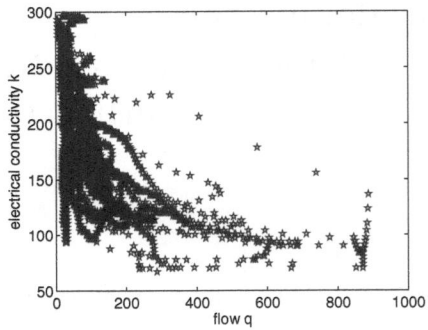

Fig. 1. The k and q time series, and k vs q

and flow are plotted over time and then against one another in Figure 1, which demonstrates that the behaviour of k is not solely controlled by volume of flow. However, as a starting point we shall assume that it is and create a linear model to use as a baseline for comparison with subsequent schemes. Linear regression yields $k = -0.4119q + 230.2105$. Using ten-fold cross-validation, we find that the average mean absolute error in k is 30.8929 micro-Siemens per cm, or 81.3717%.

The obvious next choice of model is of the form $k_i = \alpha k_{i-1} + \beta q_i + \gamma q_{i-1} + \delta$ where α, β, γ, and δ are constant parameters. This format is common in hydrology, and is used in many models involving flow, such as [4]. With a resolution of half an hour, best fit is achieved with $\alpha = 1, \beta = 0, \gamma = 0, \delta = 0$, or in other words $k_i = k_{i-1}$. Resampling to a resolution of 24 hours yields $k_i = 0.938k_{i-1} - 0.135q_i + 0.120q_{i-1} + 13.757$, but this model is still following the observed k with a one-timestep lag. Consequently, error is highly correlated with gradient to the left of k_i and the model tells us little about the system we could not have found by inspection of Figure 1. In the remainder of this paper we develop an alternative to such classical time series modelling. We use data mining techniques to build a model explaining the behaviour of k and q in Sections 3 and 4, and in Section 6 investigate the relationship between the two quantities.

2   Data Structure

Our first task is to build a compact representation of the data. The fine temporal resolution of the record is an asset, but as the difference between adjacent points is so small, there is redundancy in the 31341 point series. A simple discretisation method was applied to both q and k series, breaking them into sections of near-constant gradient, a process akin to piecewise linear approximation. The data are now in the form of a record of points where slope changes significantly. The k or q value of these points is compared to the original dataset in Figure 2 below. The length of the k series is now 231 points and the length of the q series 248 points, a very significant reduction from the original 31341.

Clearly, some loss of information occurs in deriving this compact representation, but it captures the essential behaviour of both flow and electrical con-

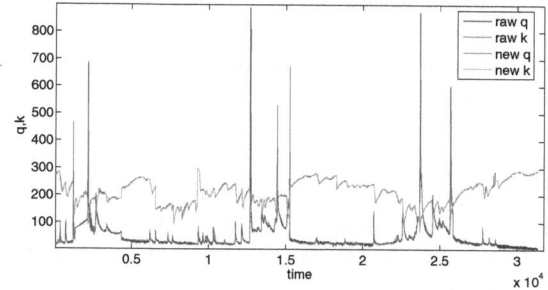

Fig. 2. Discretised and raw flow and electrical conductivity time series

ductivity series very well. Now that we have a simplified but hopefully adequate representation, we proceed to the investigatory stage.

3   Rule Extraction

Two factors that determine the value of any given point in the k series are the antecedent electrical conductivity, and the current flow behaviour. For the flow series, the only relationship available for scrutiny is that with antecedent flow conditions, as the prior assumption that k does not affect q is sufficiently fundamental to the above challenge. A rainfall record could be added to the analysis at this point, but a complete record is unavailable.

Assuming that we know little else about the system, a simple way of stating our aim is to say that if the flow behaves as Q, the behaviour K of the electrical conductivity should become known. Here the capitalised letters indicate some event in the flow or electrical conductivity series. This characterisation of the problem leads naturally towards rule extraction methods. Recent applications of temporal data mining techniques for a diverse range of problems include [3], [8], [9]. A discussion of temporal rule extraction as a method can be found in [2]. As noted there, the basic rule $A \Rightarrow B$ will need to be extended to $A \Rightarrow^T B$, meaning if A happens in the flow series, B will follow within time T. B may belong to either the q or k series. We shall apply a rule extraction algorithm to each series, and use the output frequent subsequences as archetypes to describe the events that make up the record. Then, the connection between q and k can be investigated in a simple way.

Complications arise when considering real-valued series and rule extraction, a technique usually applied to nominal or discrete data. Our sections of constant gradient, however, are easily divisible into three groups: positive gradient, negative gradient, and zero or near-zero gradient. These categories shall be denoted by the letters U for up, D for down, and 0 zero (flat). The magnitude of the rise or fall and the length of the sections shall be quantified later. To create instances for processing with a rule extraction routine, we use a sliding window that selects a subsequence of the input series. All non-flat points are allowed to act as the head of a subsequence (this follows from the assumption that flatness represents a ground state). The subsequence is cut off when one of the following conditions is met: 1. time between points exceeds a threshold t_{lim}, 2. the number of points exceeds a threshold l_{lim}, or 3. the system returns to the ground (flat) state. After some trial and error the threshold magnitude, below which gradient is considered flat, was set to 0.05.

The classic Apriori algorithm [1], which is known to be fast and robust, was modified to suit our purposes. Chief among the changes was the preservation of order within the instances, so that [U D] is distinct from [D U]. Apriori extracts frequent itemsets (or in this case subsequences) using the property that no subgroup of a candidate set C can occur frequently if C itself is not frequent, and therefore only the frequent candidates need to be searched for more complex frequent patterns. Frequency can be defined as having a support s (or rate of occurrence) greater than a chosen threshold. Support is most often simply the fraction of the total data pool where the given pattern occurs.

For example, if we have a set of items [**milk, cheese, tofu, coffee**] from which purchases may be made, and only **milk**, **coffee**, and **cheese** have a support greater than the threshold, no combination that contains tofu can be frequent. From the reduced pool [**milk, cheese, coffee**], we search the more complex patterns. Of these, perhaps only **milk and coffee** will be frequent. Thus we could state that if a person buys milk, it is likely that they will also buy coffee and vice-versa. We shall ignore which item or items are the predicate (cause) and which the consequent (effect) in the rule relationship and simply find the patterns that occur frequently. Another relevant quantity is confidence c of a rule, here defined as the ratio of the number of instances where the sequence of length l is complete to the total number of joint occurrences of the first $l - 1$ parts of the sequence.

In addressing the issue of capturing gradient magnitudes it is evident that a wide range of possible values exist. Rather than discretise the likely range of gradients, we note that for shape characteristion purposes another distinction exists. Where a string of rising or falling sections occur together, convexity or concavity of subsequences can be established simply by enquiring whether each gradient is larger or smaller than the last. Each instance with a run of U or D can be treated by appending a marker to all but the first value that records whether it is greater or lesser in magnitude than the previous value. Of course, no pair of adjacent gradients will be the same. The notation shall take the form of an appended minus if the magnitude of the gradient reduces, or a subscripted plus otherwise. By the Apriori property [1], no convex, concave, or mixed profile may be frequent unless the basic sequence of U, D, and 0 values is also frequent. Therefore, the comparison between adjacent gradients need be made only rarely.

4   Frequent Subsequences

Initially, the time threshold t_{lim} was set at 24 hours for both q and k. After experimentation it was increased to three days for k, because only small numbers of events with more than two elements were being discovered. There was little difference in the generated dataset when t_{lim} was increased to four days, and for the q series the same applied when t_{lim} was increased to two days. The upper limit on the number of elements l_{lim} was generously defined as 10. With those settings, plenty of interesting candidate subsequences were generated. It is possible that a characteristic length could have been extracted from the data and used to generate an estimate for t_{lim}, and in further work investigations will be made in this direction. For the moment, however, the subjective trial and error method is deemed to suffice.

The minimum support for the modified Apriori algorithm was set at $s_{min} = 0.10$. Note that this was calculated as the proportion of the dataset accounted for by the subsequence divided by the total number of elements, not as the number of occurrences of the subsequence divided by the total number of points. In addition, the frequent subsequences were treated as rules with the last value on the right as consequent. Confidences were calculated for the rules and any rule with confidence less than 0.3 (a rough and naive estimate of chance) deleted. The remaining frequent patterns are listed below, along with support values listed as percentages.

For q: [U D] s = 28   [D D-] s = 15   [U D D-] s = 12

For k: [D U] s = 21   [U U-] s = 14   [D D-] s = 11   [D U U-] s = 10

In q, we can see that the first two frequent patterns are subpatterns of the last, although the long pattern does not occur as frequently, either because [D U] ends with a 0 value or is interrupted by another event. The distinction is physically important, so minor changes were made to the algorithm so that the confidences for sequences constructed by taking the frequent patterns and appending a zero were also calculated. In this way, a measure of how often the sequence terminates as opposed to being followed by another event was obtained. The probability of [U D] being followed by a zero was only 0.35, which rose to 0.76 for [D D 0] and 0.75 for [U D D 0]. So most of the time [U D] was followed by another action, which we would expect as [U D D] is also frequent. From this we can conclude that while [U D] is a part of the longer [U D D], it also terminates independently without being interrupted. In contrast, the support for [U D D] is only three percent less than the support for [D D], and it is fair to conclude that [D D] does not usually occur except where [U D D] occurs.

Convexity/concavity requirements are almost never violated. Therefore the support values effectively remain the same between the raw output from the rule learner and the post-processing convexity check. An exception occurs for the sequence [D D] in the k dataset where profiles were divided almost evenly between concave and convex. Apart from [D D], the frequent sequences are again all subsequences of the longest frequent pattern. The pattern [D U] terminates in a zero with probability 0.46, and [D U U] has $c = 0.49$, so the probability of [D U] being an interrupted [D U D] is very low. Both [U U] and [D U U] have greater than 0.5 probability of being followed by a zero. The level of support for [U U] is quite close to that of [D U U], implying that most of the occurrences of [U U] are a part of a [D U U] pattern.

While the two-part patterns are not necessarily interrupted three-part patterns, the difference between the two may arise through the method used to find sections of constant or near-constant gradient. It is possible that an additional distinction exists within one section of a two-part sequence, but was too fine to be captured. In summary, for the q series we see two characteristic peak shapes: [U D] and [U D D-]. Similarly for k, [D U] and [D U U-] are characteristic trough shapes.

Before moving onto consideration of the quantitative description of these shapes, we shall consider how much of each series is covered by them. Support values tell us the percentage of the time each sequence is active, but some fraction of each series is settled in the ground state (zero gradient). Roughly two fifths for k and one quarter for q are flat in terms of number of sections, and considerably more in terms of time. With this data and s values for the characteristic shapes, it can be calculated that the two peak shapes together cover just over 40% of the non-zero q series, and the two trough shapes about 55% of the non-zero k series.

5   Quantifying Characteristic Shapes

Now that we have a preliminary idea of the shapes that typically occur in each series, it is necessary to quantify them in some way. The left side of Figure 3 is a three dimensional plot of flow peaks following the pattern [U D D-]. The axes are the breakpoints i evenly spaced and normalised to begin at zero (a substitute for a time axis), peak magnitude from initial point to maximum, and the flow value again normalised to be initially zero. The magnitude axis is included to show that while the peak shape does not remain the same over all scales (ie peaks of different magnitude are not congruent) it can be parameterised by magnitude. Therefore a single magnitude value uniquely, albeit approximately, defines peak shape. Similar trends exist for the other frequent subsequences (frequent shapes), but for brevity we shall confine discussion to the [U D D-] pattern.

Consider the height of the shape at each of the defining points as $q_i, i = 1, 2, 3, 4$ where the index identifies the point on the time axis. Flow units are Litres per second Ls_{-1}. We shall assume that each shape has been normalised so that $q_1 = 0$. Similarly, let t_i denote the normalised time at these points. Modelling each of the other q_i with the magnitude (mag) of the peak, which is equal to q_2, least squares fitting of degree one yields the following.

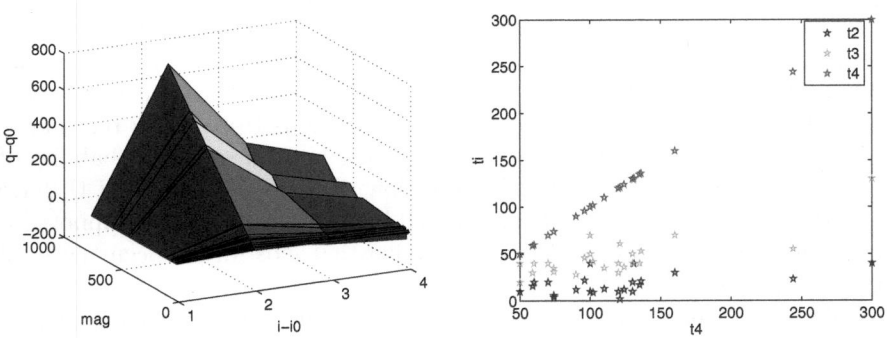

Fig. 3. [U D D-] flow patterns, without and with time scale

$q_3 = 0.278mag + 0.912$ with mean error $= 16.94Ls^{-1}$
$q_4 = 0.125mag - 10.815$ with mean error $= 16.80Ls^{-1}$

The time scales of each shape on the left side of Figure 3 have been deformed, as the component sections are defined by constant gradient rather than cut off after a set time interval. When proper time scales are added, the plot becomes far more complicated, and the neat parameterisation by magnitude is no longer obvious. However, the length in time of each segment can be parameterised in terms of one quantity. The right side of Figure 3 is a plot of break of gradient times t_i against the total time t_{tot}. All times are normalised to begin at zero, so t_1 is uniformly zero and so is not shown. The times t_2 where the first section ends and t_3 where the second section ends can be roughly modelled as a function of the total time. In the case of [U D D-] total time $t_{tot} = t_4$. Least squares linear models for each unknown t_i are given below, along with the mean error in half-hour time periods and hours.

$t_2 = 0.086t_{tot} + 7.719$ with mean error $= 7.3$ (3.7 hours)
$t_3 = 0.316t_{tot} + 13.002$ with mean error $= 13.0$ (6.5 hours)

Similar parameterisations can be found for all the other frequently occurring peak and trough shapes. We now have all the necessary time and flow magnitude information representing a [U D D-] peak encoded in two numbers, mag and t_{tot}. Of course, to recreate the time series we do still need to know the position of the peak in time and the starting flow value.

6   Two-Dimensional Relationships

Now that characteristics of each series individually have been investigated, we can return to the original problem of discovering the relationship between flow and electrical conductivity. If a large number of different characteristic shapes had been identified as frequent in either series, we could proceed by building new instances for rule learning where each q event was followed by any k event that occurred within a certain time. However, the behaviour of each series has been decomposed into only two main patterns, and we shall simply test the hypothesis that a peak in flow causes either a [D U] or [D U U-] event in k.

Setting the time limit between events at a generous two days between the start of the flow peak and the start of the k trough, we find 22 occurrences of a trough close to a peak. Recalling support values, we find that 34 troughs and 35 peaks were identified. Some of these were small in size, and examination of the magnitude values of generating and generated shapes reveals that they are mostly relatively large. Small peaks do not typically cause detectable changes in k and, correspondingly, many small k events cannot be traced back to an event in q.

Curiously, in a handful of cases the trough actually appeared to begin before the peak. These points were checked in detail, and there often appeared to be no other possible generative peak in streamflow to the left. Therefore we allow the condition, that the k response follow the q event, be relaxed to simple occurrence within an interval around the start of the peak. This behaviour may seem unphysical, but it is in keeping with our policy of making as few assumptions as possible. Most of the troughs, however, fall within 12 hours after the streamflow peak begins.

In the last few months of the record, several events are recorded in k, but no matching q peaks were identified. This anomalous behaviour can be observed in Figure 1 as well. The researchers originally responsible for collecting the data revealed that the flow record in this region was in fact questionable, due to the gradual degradation of the streamflow recording mechanism [5]. This resolves the inconsistency in the hypothesis that a q event implies a k event.

Inspection reveals that while there is a mild tendency towards [U D] peaks that are not subsequences of [U D D] producing troughs of the shorter form and [U D D-] the longer form, it is not at all pronounced. Support information yielded the observation that while some of the two-part patterns are the beginnings of three-part patterns, this is not always true. Some may be interrupted, but others terminate. Ideally, all three cases would be treated separately along with the three-part pattern, but analysis at this level of detail demands more data than we have available.

The problem of finding a relationship between magnitude of events in the two variables remains. Simply plotting the two against one another yields very little information. However, when the initial state of k is introduced into the plot as an additional axis, useful order begins to appear. Figure 4 (left) below shows one view of a plot with magnitude of flow peak $qmag$, initial k state $kstate$, and the minimum point $kmin$ to which k drops. When combined with $kstate$, this last variable contains the same information as the magnitude of the trough in k, but produces a clearer trend which suggests a surface on which $kmin$ is proportional to $kstate$ with gradient decreasing with increasing $qmag$. This relationship is quite surprising, especially the relatively weak coupling between $qmag$ and $kmag$. However, it is useful. In the dense region $qmag < 200$ or so, the relationship between $kstate$ and $kmin$ is fairly well defined, but becomes a matter of hypothesis in the sparser region where fewer events have been observed.

The simple hypothesis that the total time of the k event increases with the total time of the q event can be proven false by a simple scatter plot (Figure 4, right). Note from this plot that there appears to be a characteristic time scale for q events, but a wide variation in total times for k events. Clearly, other

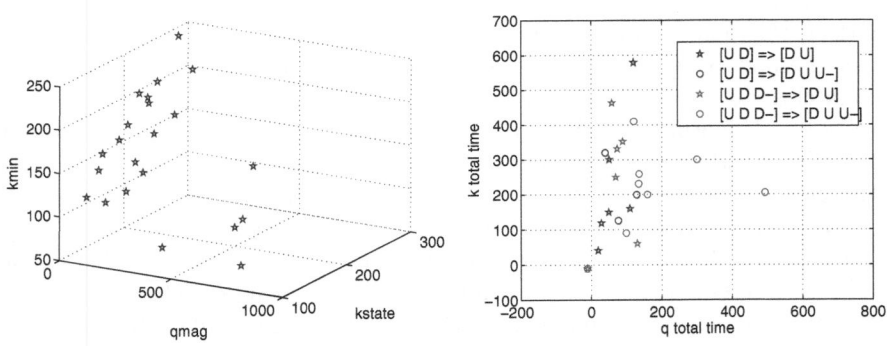

Fig. 4. q magnitude vs k state vs k min, and k total time vs q total time

factors are at work. Experiments with multivariate linear regression did not reveal a sufficiently strong relationship with either the q time parameter (total flow event time) or any other variable to determine time scale of the k event.

7   Conclusions

This exploration of the q and k series of record from Hollin Cave and the relationship between them revealed strong regularities in both series. Characteristic peak and trough shapes, easily identifiable by cursory inspection of Figure 1, were extracted by the modified Apriori algorithm. It also revealed that the two-part patterns [U D] and [D U] existed independent from the subsequences [U D D] and [D U U], although the difference between the short and long patterns may be an artefact of the method used to break the series into lengths of near-homogenous gradient. In both q and k, the three-part sequences proved to be of almost uniform convexity and were thus refined to [U D D-] and [D U U-], thus removing unusual and probably unhelpful data from the shape analysis. The examination of shape convexity is to the author's knowledge novel, and is expected to be useful in the future analysis of larger and more unwieldy datasets. Figure 3 illustrates the [U D D-] patterns, and shows that the peaks are not congruent to one another, but can be parameterised by magnitude and total time. Similar results were obtained for the other frequently occurring shapes.

The hypothesis that events were linked in time was then verified. A relationship between flow peak size, initial k state and the point to which k falls was established for the dense region $q < 200Ls^{-1}$ and postulated for $q \geq 200Ls^{-1}$. The initial state of the conductivity at the start of a trough, $kstate$, proved to be far more important than the size of the precipitating flow event. The problem of determining the time scale of k events remains, although we have established that flow peak magnitude does not have a significant effect, and there is some reliance on $kmin$, the minimum conductivity attained. When the nature of total event time t_{tot} can be established for k, a complete model will have been obtained and we will be able to reconstruct both time series. Given a starting value of k, the series may be recoverable to a reasonable accuracy.

These two points form the main foci for work on this problem in the near future. Other approaches to the time issue, such as uniform sampling, are also under consideration. The methods developed will also be applied to other, longer records, and datasets with a greater number of interconnected variables. Unfortunately, the short length of record prevents us from testing on an independent sample and so we cannot obtain real measures of how well the system dynamics can be reproduced, but far larger water quality datasets have been obtained and will be processed using the techniques discussed here, and independent testing will certainly be possible with this new data.

Although we do not yet have a complete picture of the dynamics at work in the flow/electrical conductivity system, our qualitative and quantitative knowledge has been advanced considerably over that discussed in the Introduction. Qualitatively, both series are made of characteristic peak and trough events. The depth of the troughs in k is determined largely by the starting point but also by the magnitude of the corresponding q peak. Quantitatively, the relations controlling event shapes have been extracted, and for most flow magnitudes the relationship between peak and trough magnitudes can be determined. The nature of this relationship is unexpected, with surprisingly weak dependence on flow peak magnitude for most flows, and correspondingly strong dependence on the initial state of k.

In terms of the dynamics of Hollin Creek system, the fact that electrical conductivity on some occasions begins to fall before any increase in flow shows that Hollin Creek is certainly connected to another system underground. Unlike most surface streams, the Hollin Creek flows through restricted spaces, and some passages may only be used at high flows. When presented with the above results, speleological experts [5] suggested that as water levels in the cave begin to rise, a branch of Hollin creek takes a different path through the cave, collecting less salts or allowing more to be precipitated out as calcium carbonate (calcite). This could result in a small quantity of very low conductivity water entering the system before the main peak in flow, as we see.

It has also been suggested that the flow peaks are delayed in some way by cave passages too small to accommodate the full flow. The system may contain a bottleneck where water accumulates until the water level rises to a second passage where it can overflow. Complex three dimensional structures of this kind are common in many caves. Either or both of these dynamics may be at work in Hollin Cave, and may also explain why a stable relationship between q and k time scales remains so elusive.

References

1. Agrawal, R., Srikant, R.: Fast Algorithms for Mining Association Rules. VLDB 1994: Proceedings of the 20th international conference on Very Large DataBases. (1994) 487–499
2. Antunes, CM., Oliveira, AL.: Temporal Data Mining: An Overview. KDD 2001: Proceedings of the 7th ACM SIGKDD International Conference on Knowledge Discovery and Data Mining. (2001)
3. Chen, J., He, H., Williams, GJ., Jin, H.: Temporal Sequence Associations for Rare Events. PAKDD2004: Advances in Knowledge Discovery and Data Mining, 8th Pacific-Asia Conference. (2004) 239–239
4. Hipel, KW., McLeod AI., Lennox, WC.: Advances in Box-Jenkins modelling, 1, model construction. Water Resources Research. **13** (1977) 567-575
5. Optimal Karst Managment.: Personal Correspondence. (2004,2005)
6. Spate, AP., Jennings, JN., Smith, DI., James, JM.: A Triple Dye Tracing Experiment at Yarrangobilly. Helictite: Journal of Australasian Cave Research. **14** (1976) 27–48
7. Spate, JM.: A Detailed Analysis of Electrical Conductivity in Hollin Cave Spring. In preparation. (2005)
8. Su, F., Zhou, C., Lyne, V., Du, Y., Shi, W.: A Data-Mining Approach to Determine the Spatio-Temporal Relationship Between Environmental Factors and Fish Distribution. Ecological Modelling. **174** (2004) 421–431
9. Winarko, E., Roddick, JF.: Relative Temporal Association Rule Mining. Proceedings of the 2nd Australasian Data Mining Workshop. (2003) 121–142

Biological Cluster Validity Indices
Based on the Gene Ontology

Nora Speer, Christian Spieth, and Andreas Zell

Centre for Bioinformatics Tübingen (ZBIT),
University of Tübingen, Sand 1, D-72076 Tübingen, Germany
nspeer@informatik.uni-tuebingen.de

Abstract. With the invention of biotechnological high throughput
methods like DNA microarrays and the analysis of the resulting huge
amounts of biological data, clustering algorithms gain new popularity.
In practice the question arises, which clustering algorithm as well as
which parameter set generates the most promising results. Little work
is addressed to the question of evaluating and comparing the clustering results, especially according to their biological relevance, as well on
distinguishing biologically interesting clusters from less interesting ones.
This paper presents two cluster validity indices intended to evaluate clusterings of gene expression data in a biological manner.

1   Introduction

In an attempt to understand complex biological regulatory mechanisms of a
cell, biologists tend to use large scale techniques to collect huge amounts of gene
expression data. Thus, DNA microarrays became a popular tool in the past few
years. A problem inherent in the use of DNA arrays is the tremendous amount
of data produced, whose analysis itself constitutes a challenge. Data mining
techniques like cluster algorithms are utilized to extract gene expression patterns
inherent in the data and thus find potentially co-regulated genes [14]. Various
methods have been applied, such as Self-Organizing-Maps (SOMs) [22], K-Means
[23], Hierarchical Clustering [7] as well as Evolutionary Algorithms [13,20].

Since different cluster algorithms or different runs of the same algorithm
generate different solutions given the same data set, in practice, biologists are
faced with the problem of choosing an appropriate algorithm with appropriate
parameters for the data set. The evaluation of cluster results is a process known
as cluster validity and is an important task in cluster analysis.

Several cluster validity indices are known in literature, such as Dunn's Index
[6], Rand Index [15], Figure of Merit [25], Silhouette Index [18] or Davies-Bouldin
Index [5] and many of them have already been used with gene expression data
[1,3,25]. All these indices evaluate the mathematical properties of a clustering,
but especially for gene expression data, the biological cluster quality plays an
important role, too [17,19]. Some attempts in this direction were based on text
mining methods for literature abstracts [16]. Others simply count Gene Ontology
annotations per cluster [2,17,19], but in contrast to our approach, none of them

A.F. Famili et al. (Eds.): IDA 2005, LNCS 3646, pp. 429–439, 2005.

relies on biological distances between genes, an advantage that enables the use of established cluster indices.

The paper is organized as follows: a brief introduction to the Gene Ontology is given in section 2. Section 3 explains our method in detail. The performance on real world data sets is shown in section 4. Finally, in section 5, we conclude.

2   The Gene Ontology

The Gene Ontology (GO) is one of the most important ontologies within the bioinformatics community and is developed by the Gene Ontology Consortium [24]. It is specifically intended for annotating gene products with a consistent, controlled and structured vocabulary. Gene products are for instance sequences in databases as well as measured expression profiles. The GO is independent from any biological species and is rapidly growing. Additionally, new ontologies covering other biological or medical aspects are being developed.

The GO represents terms in a Directed Acyclic Graph (DAG), covering three orthogonal taxonomies or "aspects": *molecular function, biological process* and *cellular component*. The GO-graph consists of over 18.000 terms, represented as nodes within the DAG, connected by relationships, represented as edges. Terms are allowed to have multiple parents as well as multiple children. Two different kinds of relationship exist: the "is-a" relationship (*photoreceptor cell differentiation* is, for example, a child of *cell differentiation*) and the "part-of" relationship that describes, for instance, that *regulation of cell differentiation* is part of *cell differentiation*.

By providing a standard vocabulary across any biological resources, the GO enables researchers to use this information for automatic data analysis done by computers and not by humans.

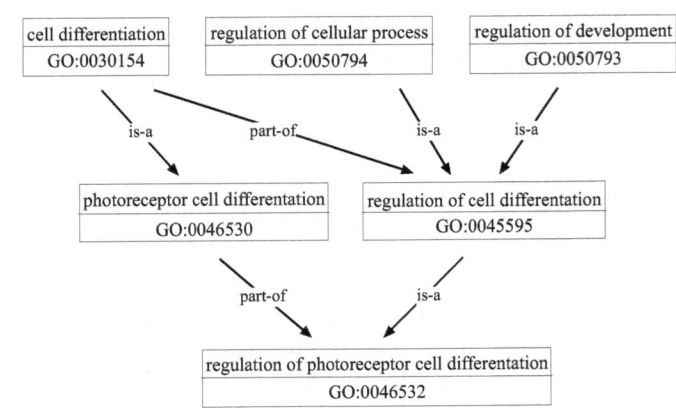

Fig. 1. Relations in the Gene Ontology. Each node is annotated with a unique accession number.

3 Methods

3.1 Mapping Genes to the Gene Ontology

To properly evaluate a clustering result with GO information, a mapping M that relates the clustered genes to the nodes in the GO graph is required. For eucaryotic genes the common biological databases (e.g. TrEMBL or GenBank) provide GO annotation for their entries and also biotech companies like Affymetrix provide GO mappings for their DNA microarrays. Such a mapping is not one-to-one, which means that there are genes annotated with more than one GO term as well as genes without a GO annotation. The first point will be discussed later in this section, the latter reduces the number of genes that can take part in such an analysis.

3.2 Distances Within the Gene Ontology

To calculate biological distances within the GO, we rely on a technique that was originally developed for other taxonomies like WordNet to measure semantic distances between words [11]. The distance measure is based on the information content of a GO term. Following the notation in information theory, the information content (IC) of a term t can be quantified as follows:

$$IC(t) = -\ln P(t) \tag{1}$$

where $P(t)$ is the probability of encountering an instance of term t.

In the case of a hierarchical structure, such as the GO, where a term in the hierarchy subsumes those lower in the hierarchy, this implies that $P(t)$ is monotonic as one moves towards the root node. As the node's probability increases, its information content or its informativeness decreases. The root node has a probability of 1, hence its information content is 0. As the three aspects of the GO are disconnected subgraphs, this is still true if we ignore the root node "Gene Ontology" and take, for example, "biological process" as our root node instead.

To compute a similarity between two terms one can compute the IC of their common ancestor. As the GO allows multiple parents for each term, two terms can share ancestors by multiple paths. We take the minimum $P(t)$, if there is more than one ancestor. This is called P_{ms}, for *probability of the minimum subsumer* [12]:

$$P_{\text{ms}}(t_i, t_j) = \min_{t \in S(t_i, t_j)} P(t) \tag{2}$$

where $S(t_i, t_j)$ is the set of parental terms shared by both t_i and t_j. Based on Eq. 1 and 2, Jiang and Conrath developed the following distance measure [11]:

$$d(t_i, t_j) = 2\ln P_{ms}(t_i, t_j) - (\ln P(t_i) + \ln P(t_j)) \tag{3}$$

Since genes can have more than one function and are therefore often annotated with more than one GO term, multiple functional distances can be computed between two genes. Since, we don't know which of these functions play a role in the underlying biological experiment, we assume the best and use the smallest distance between two genes during the calculation of cluster validities.

3.3   Cluster Validities

A good cluster validity index should be independent of the number of clusters, thus allowing to compare two clusterings with different number of clusters. At the same time, it is desirable that genes in one cluster have minimum possible distance to each other and maximum distance to the genes in other clusters, in other words, we seek clusters that are compact and well separated. Two cluster validity measures that fulfill these criteria are the Silhouette and the Davies-Bouldin index [18,5].

Given a set of genes $G = \{g_1, g_2, \ldots, g_n\}$ and a clustering of G in $C = \{C_1, C_2, \ldots, C_k\}$, the Silhouette index is defined as follows [18]: for each gene g_i of cluster C_j, a confidence measure, the Silhouette width $s(g_i)$, is calculated that indicates if gene g_i belongs to cluster C_j. The Silhouette width $s(g_i)$ is defined as follows:

$$s(g_i) = \frac{\min(\bar{d}_B(g_i)) - \bar{d}_W(g_i)}{\max\{\bar{d}_W(g_i), \min(\bar{d}_B(g_i))\}} \tag{4}$$

where $\bar{d}_W(g_i)$ is the average distance from g_i to all other genes of the cluster to which g_i is assigned and $\bar{d}_B(g_i)$ is the average distance between g_i and all other genes assigned to the clusters C_l with $l = 1, \cdots, k \wedge j \neq l$. Observations with a large $s(g_i)$ (almost 1) are very well clustered, a small $s(g_i)$ (around 0) means that the observation lies between two clusters, and observations with a negative $s(g_i)$ are probably placed in the wrong cluster. Thus, for each cluster C_j, a mean Silhouette index

$$S_j(C_j) = \frac{1}{|C_j|} \sum_{i=1}^{|C_j|} s(g_i) \tag{5}$$

can be computed. $|C_j|$ denotes the number of genes included in cluster C_j. The index ranges between 1 (for a perfect cluster/clustering) and -1. Thus, the overall quality of a clustering C can be measured using:

$$S(C) = \frac{1}{n} \sum_{i=1}^{n} s(g_i), \tag{6}$$

Given the same notation as above, the Davies-Bouldin index has been defined in [5] as:

$$DB_j(C_j) = \max_{i \neq j} \left\{ \frac{\Delta(C_i) + \Delta(C_j)}{\delta(C_i, C_j)} \right\} \tag{7}$$

where $\Delta(C_i)$ and $\Delta(C_j)$ represent the inner cluster distance of cluster C_i and C_j and $\delta(C_i, C_j)$ denotes the distance between the clusters C_i and C_j. Usually $\Delta(C_i)$ and $\delta(C_i, C_j)$ are calculated as the sum of distances to the respective cluster center and the distance between the centers of two clusters. Since means are not defined in a DAG, we use the average diameter of a cluster as $\Delta(C_i)$ and the average linkage between two clusters as $\delta(C_i, C_j)$:

$$\Delta(C_i) = \frac{1}{|C_i|(|C_i - 1|)} \sum_{g_i, g_j \in C_i, g_i \neq g_j} d(g_i, g_j) \tag{8}$$

$$\delta(C_i, C_j) = \frac{1}{|C_i||C_j|} \sum_{g_i \in C_i, g_j \in C_j} d(g_i, g_j) \tag{9}$$

where $d(g_i, g_j)$ defines the distance between the genes g_i and g_j. It is clear from the above definition, that $DB_j(C_j)$ is the average similarity between cluster C_j, and its most similar one. It is desirable for the clusters to have minimum possible similarity to each other. Therefore, we seek clusterings that minimize $DB_j(C_j)$. The index for the whole clustering can be computed as:

$$DB(C) = \frac{1}{k} \sum_{j=1}^{k} DB_j(C_j). \tag{10}$$

4   Results

4.1   Data Sets

The performance of the cluster validity indices are discussed on two real world data sets. For our work, we only use the taxonomy *biological process*, because we are mostly interested in gene function. However, our method can be applied in the same way for the other two taxonomies.

The authors of the first data set examined the response of human fibroblasts to serum on cDNA microarrays in order to study growth control and cell cycle progression. They found 517 genes whose expression levels varied significantly, for details see [10]. We used these 517 genes for which the authors provide NCBI accession numbers. The GO mapping was done using GeneLynx [8]. After mapping to the GO, 238 genes showed one or more mappings to *biological process* or a child term of *biological process*. These 238 genes were used for the clustering. We selected 14 clusters as indicated in our previous publication [21].

In order to study gene regulation during eukaryotic mitosis, the authors of the second data set examined the transcriptional profiling of human fibroblasts during cell cycle using microarrays [4]. Duplicate experiments were carried out at 13 different time points ranging from 0 to 24 hours. Cho *et al.* found 388 genes whose expression levels varied significantly [4]. In [9] Hvidsten *et al.* provide a mapping of the data set to the GO. 233 of the 388 genes showed at least one mapping to the *biological process* taxonomy and were thus used for clustering. We selected 10 clusters as indicated in our previous publication [21].

4.2   Computational Experiments

If our proposed cluster indices are able to distinguish biologically meaningful clusterings from less meaningful ones, a functional clustering according to the GO annotations should show better validity index values than a clustering that was produced according to the normalized expression vectors of the genes.

Therefore, in our experiments, we used a clustering algorithm based on an Evolutionary Algorithm from earlier publications [20,21] to produce these two

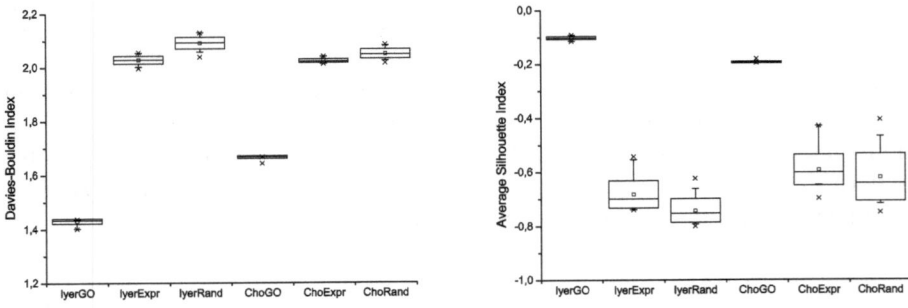

Fig. 2. Davies-Bouldin index (left, small values indicate good clusterings) and Silhouette index (right, large values indicate good clusterings) averaged over 25 runs. Maximum and minimum values are indicated by a cross, the mean by the rectangle, the standard deviation is indicated by the box, the error bars indicate the 5-95 confidence intervals.

different clusterings for each data set: an expression based clustering and a functional clustering. In principle, any cluster algorithm could be used in place that does not rely on mean calculation (this is important for the functional clustering, since we cannot compute means in the GO as mentioned earlier). The only reason why we use this algorithm is that we got good results compared to other non-mean based methods like Average Linkage clustering [20,21]. While producing these two clusterings, all parameters of the algorithm were fixed (200 generations, population size of 40 and 40% mutation and recombination rate), except the distance function used: for the functional clustering, we used the GO distance (Eq. 3) and for the expression based clustering, we used the Euclidean distance of the normalized expression vectors of each gene. The normalization was performed as described in [23]. We also compared the clusterings to a random partition. For the random partition, one result corresponds to the best partition out of 8000 (200 generations * 40 individuals) tries. All results are averaged over 25 runs.

Fig. 2 shows the Davies-Bouldin index (left) and the Silhouette index (right) of the expression based and functional clusterings and for the random partition for both data sets. Maximum and minimum values are indicated by a cross, the mean by the rectangle, the standard deviation is indicated by the box, the error bars indicate the 5-95 confidence intervals. For both indices and both data sets, the GO based clustering obtains significant better values than a gene expression based clustering. These results were of course expected since we used a biological clustering method to produce this clustering. But nevertheless, it indicates that our validity measures are able to detect biological meaningful clusterings. Beside that, it is notable that the expression based clustering is only slightly better than random concerning its biological similarity, which emphasizes the need for methods that can distinguish between biologically interesting and less interesting clusterings.

Table 1. Cluster validity values for the individual clusters for a GO based clustering. A low value of the Davies-Bouldin and a high value for the Silhouette index indicate good clusters. A good and a bad cluster are marked in bold.

Cluster	Davies-Bouldin Index	Silhouette Index
1	1.49	-0.67
2	1.76	-0.55
3	1.32	-0.09
4	1.29	0.24
5	1.55	0.16
6	1.73	-0.20
7	1.39	0.21
8	**1.76**	**-0.40**
9	1.57	-0.22
10	1.29	-0.21
11	1.32	-0.26
12	1.28	-0.16
13	**1.07**	**0.49**
14	1.29	0.05

Furthermore, the presented cluster validity measures can not only be used to distinguish between whole clusterings but also to validate individual clusters and thus find interesting clusters that contain genes that are biologically closely related and already known to be involved in the same pathway. Such a cluster would indicate that a whole biological process might be switched on or off under the given experimental condition, e.g. that cells leave the G_O- phase and enter cell proliferation. Tab. 1 shows the individual cluster validity values for the overall best clustering. As an example, we show two extreme clusters in more detail.

For both cluster validity measures, cluster 13 has a good quality, whereas cluster 8 is much more functionally diverse. The GO annotations of cluster 13 are displayed in Tab. 2 and those of cluster 8 are shown in Tab. 3. The genes of the good cluster are mostly closely related to DNA replication and repair, which is a defined and separated process in biology. So cluster 13 is a small and functionally compact cluster that was also indicated by the validity values. Instead, the other example is larger and much more diverse. Genes in that cluster are related to cell adhesion, cell motility, inter- and intra-cellular signal transduction, metabolism, nervous system development and pregnancy. All theses functions are quite different biological processes, which was already indicated by the validity measures.

We showed that our two biological cluster indices are able to distinguish biologically more homogeneous clusters from less homogeneous ones, a fact that can be used to find those clusters in a clustering that contain genes that are not only co-expressed, but also related to the same biological process. Additionally, we showed that one can use these indices to measure the biological quality of

Table 2. Example of the GO annotation of a good functional cluster (cluster 13)

Probeset Id	GO Term Name
H63374	DNA repair
	pyrimidine-dimer repair, DNA damage excision
N22858	chromosome organization and biogenesis (sensu Eukarya)
	DNA methylation
	DNA recombination
	DNA repair
N68268	DNA replication
	DNA replication, priming
W93122	DNA dependent DNA replication
	DNA replication
N93479	DNA replication
H29274	DNA repair
	DNA replication
	double-strand break repair
	UV protection
AA053076	DNA replication
AA031961	cell cycle
	regulation of cell cycle
	cell proliferation
	DNA repair
	regulation of CDK activity

a whole clustering and therefore find biologically meaningful clusterings out of a bunch of given clusterings. Thus, our two presented biological cluster validity indices can be used to evaluate clusterings and single clusters of genes in a biological manner.

5   Conclusion

In this paper, we presented two biological cluster validity indices that are based on the Gene Ontology. We showed that they can be utilized to detect clusters of genes that share similar functions. This is especially important, because such clusters indicate that a whole regulatory pathway might be affected under the given conditions, which leads to an information gain about the underlying regulatory mechanisms of a cell. The fact that a clustering due to gene expression profiles does not always implicate a biological clustering as shown by our results even emphasizes the need of a tool like the presented biological cluster indices.

The advantage of our method compared to other approaches is, that it is based on biological distances, which enable the usage of established cluster validity measures including the knowledge of their weaknesses and advantages. Beside that, the utilized GO annotation is easy to obtain from biological databases.

One problem of our method is, of course, that for each gene at least one Gene Ontology annotation is needed. In most of the cases the GO annotation is available in public databases. Nevertheless, there are still some genes that do

Table 3. Example of the GO annotation of a bad functional cluster (cluster 8)

Probeset Id	GO Term Name
W89002	peroxidase reaction
H63779	central nervous system development
	epidermal differentiation
	lipid metabolism
	peripheral nervous system development
N79778	cell-matrix adhesion
N67806	respiratory gaseous exchange
R37986	pregnancy
AA029995	pregnancy
W86618	DNA metabolism
	intracellular protein transport
	G2 phase of mitotic cell cycle
	NLS-bearing substrate-nucleus import
	regulation of DNA recombination
	spindle pole body and microtubule cycle (sensu Saccharomyces)
T70079	chemotaxis
	G-protein coupled receptor protein signaling pathway
	inflammatory response
T62835	cell adhesion
N22383	cell adhesion
	cell-matrix adhesion
	cell-substrate junction assembly
	integrin-mediated signaling pathway
AA056401	cellular morphogenesis
	epidermal differentiation
N63308	cell adhesion
	neuronal cell recognition
AA037351	cell adhesion
	neuronal cell recognition
AA045473	cell adhesion
N93476	cell adhesion
	G-protein coupled receptor protein signaling
W49619	cell adhesion
R80217	cell motility
	inflammatory response
	peroxidase reaction
	physiological processes
	prostaglandin metabolism
AA044993	cell adhesion
	cell growth and/or maintenance
	cell motility
	DNA metabolism
	epidermal differentiation

not have that kind of annotation. One way to solve this problem might be to use all genes for clustering, but calculate the validity index only with those that can be annotated. In this case, one might additionally think of giving a score to each cluster, indicating how many genes participate in the validity index. We will address this point in future work.

Acknowledgment

This work was supported by the National Genome Research Network (NGFN) of the Federal Ministry of Education and Research in Germany under contract number 0313323.

References

1. F. Azuaje. A cluster validity framework for genome expression data. *Bioinformatics*, 18(2):319–320, 2001.
2. T. Beißbarth and T. Speed. GOstat: find statistically overexpressed Gene Ontologies within groups of genes. *Bioinformatics*, 20(9):1464–1465, 2004.
3. N. Bolshakova, F. Azuaje, and P. Cunningham. An integrated tool for microarray data clustering and cluster validity assessment. *Bioinformatics*, 21(4):451–455, 2004.
4. R.J. Cho, M. Huang, M.J. Campbell, H. Dong, L. Steinmetz, L. Sapinoso, G. Hampton, S.J. Elledge, R.W. Davis, and D.J. Lockhart. Transcriptional regulation and function during the human cell cycle. *Nature Genetics*, 27(1):48–54, 2001.
5. J.L. Davies and D.W. Bouldin. A cluster separation measure. *IEEE Transactions on Pattern Analysis and Machine Intelligence*, 1:224–227, 1979.
6. J.C. Dunn. Well separated clusters and optimal fuzzy partitions. *Journal of Cybernetics*, 4:95–104, 1974.
7. M. Eisen, P. Spellman, D. Botstein, and P. Brown. Cluster analysis and display of genome-wide expression patterns. In *Proceedings of the National Academy of Sciences, USA*, volume 95, pages 14863–14867, 1998.
8. Gene Lynx. http://www.genelynx.org, 2004.
9. T.R. Hvidsten, A. Laegreid, and J. Komorowski. Learning rule-based models of biological process from gene expression time profiles using Gene Ontology. *Bioinformatics*, 19(9):1116–1123, 2003.
10. V.R. Iyer, M.B. Eisen, D.T. Ross, G. Schuler, T. Moore, J.C.F. Lee, J.M. Trent, L.M. Staudt, J Hudson Jr, M.S. Boguski, D. Lashkari, D. Shalon, D. Botstein, and P.O. Brown. The transcriptional program in response of human fibroblasts to serum. *Science*, 283:83–87, 1999.
11. J.J. Jiang and D.W. Conrath. Semantic similarity based on corpus statistics and lexical taxonomy. In *Proceedings of the International Conference on Research in Computational Linguistics*, Taiwan, 1998. ROCLING X.
12. P.W. Lord, R.D. Stevens, A. Brass, and C.A. Goble. Semantic similarity measures as tools for exploring the gene ontology. In *Proceedings of the Pacific Symposium on Biocomputing*, pages 601–612, 2003.
13. Peter Merz. Clustering gene expression profiles with memetic algorithms. In *Proceedings of the 7th International Conference on Parallel Problem Solving from Nature, PPSN VII*, pages 811–820. Lecture Notes in Computer Science, Springer, Berlin, Heidelberg, 2002.

14. J. Quackenbush. Computational analysis of microarray data. *Nature Reviews Genetics*, 2(6):418–427, 2001.
15. W. M. Rand. Objective criteria for the evaluation of clustering methods. *Journal of the American Statistical Association*, 66:846–850, 1971.
16. S. Raychaudhuri and R.B. Altman. A literature-based method for assessing the functional coherence of a gene group. *Bioinformatics*, 19(3):396–401, 2003.
17. P.N Robinson, A. Wollstein, U. Böhme, and B. Beattie. Ontologizing gene-expression microarray data: characterizing clusters with gene ontology. *Bioinformatics*, 20(6):979–981, 2003.
18. P.J. Rousseeuw. Silhouettes: a graphical aid to the interpretation and validation of cluster analysis. *Journal of Computational Applications in Math*, 20:53–65, 1987.
19. N.H. Shah and N.V. Fedoroff. CLENCH: a program for calculating Cluster ENriCHment using Gene Ontology. *Bioinformatics*, 20(7):1196–1197, 2004.
20. N. Speer, P. Merz, C. Spieth, and A. Zell. Clustering gene expression data with memetic algorithms based on minimum spanning trees. In *Proceedings of the IEEE Congress on Evolutionary Computation (CEC 2003)*, volume 3, pages 1848–1855. IEEE Press, 2003.
21. N. Speer, C. Spieth, and A. Zell. A memetic clustering algorithm for the functional partition of genes based on the Gene Ontology. In *Proceedings of the 2004 IEEE Symposium on Computational Intelligence in Bioinformatics and Computational Biology (CIBCB 2004)*, pages 252–259. IEEE Press, 2004.
22. P. Tamayo, D. Slonim, J. Mesirov, Q. Zhu, S. Kitareewan, E. Dmitrovsky, E.S. Lander, and T.R. Golub. Interpreting patterns of gene expression with self-organizing maps: Methods and application to hematopoietic differentiation. In *Proceedings of the National Academy of Sciences, USA*, volume 96, pages 2907–2912, 1999.
23. S. Tavazoie, J.D. Hughes, M.J. Campbell, R.J. Cho, and G.M. Church. Systematic determination of genetic network architecture. *Nature Genetics*, 22:281–285, 1999.
24. The Gene Ontology Consortium. The gene ontology (GO) database and informatics resource. *Nucleic Acids Research*, 32:D258–D261, 2004.
25. K.Y. Yeung, D.R. Haynor, and W.L. Ruzzo. Validating clustering for gene expression data. *Bioinformatics*, 17:309–318, 2001.

An Evaluation of Filter and Wrapper Methods for Feature Selection in Categorical Clustering

Luis Talavera

Dept. Llenguatges i Sistemes Informàtics,
Universitat Politècnica de Catalunya,
Jordi Girona 1-3 08034 Barcelona, Spain
talavera@lsi.upc.edu

Abstract. Feature selection for clustering is a problem rarely addressed in the literature. Although recently there has been some work on the area, there is a lack of extensive empirical evaluation to assess the potential of each method. In this paper, we propose a new implementation of a wrapper and adapt an existing filter method to perform experiments over several data sets and compare both approaches. Results confirm the utility of feature selection for clustering and the theoretical superiority of wrapper methods. However, it raises some problems that arise from using greedy search procedures and also suggest evidence that filters are a reasonably alternative with limited computational cost.

1 Introduction

It is widely recognized that a large number of, possibly irrelevant, features can adversely affect the performance of inductive learning algorithms, and clustering is not an exception. However, while there exists a large body of literature devoted to this problem for supervised learning tasks [9,1], feature selection for clustering has been rarely addressed. The problem appears to be a difficult one given that it inherits all the uncertainties that surround this type of inductive learning. Particularly, that there is not a single performance measure widely accepted for this task and the lack of supervision available (e.g. class labels).

Although recently there has been a growing interest in feature selection for clustering, a number of questions still remain open. Wrappers for feature selection have been recently proposed with some success. However, they exhibit some limitations. The first, and probably on of the most important deficits is the lack of a more extensive empirical evaluation of the methods and, in particular, a comparison between filters and wrappers. A second shortcoming is that many of these approaches are focused on numerical clustering, and there is no theoretical or experimental evidence related to their behavior on categorical data.

In this paper we present a first attempt to fill these gaps by comparing the performance of wrapper and filter methods over several data sets. We propose a new wrapper implementation and use a filter technique based upon previous work for the experiments.

A.F. Famili et al. (Eds.): IDA 2005, LNCS 3646, pp. 440–451, 2005.

2   Feature Selection for Clustering

In supervised learning, feature selection is often viewed as a search problem in a space of feature subsets. To carry out this search we must specify a starting point, a strategy to traverse the space of subsets, an evaluation function and a stopping criterion. Although this formulation allows a variety of solutions to be developed, usually two families of methods are considered, namely *filter* and *wrapper* methods [9]. On one hand, filter methods use an evaluation function that relies solely on properties of the data, thus is independent on any particular algorithm. On the other hand, wrappers use the inductive algorithm to estimate the value of a given subset.

Wrapper methods are widely recognized as a superior alternative in supervised learning problems, since by employing the inductive algorithm to evaluate alternatives they have into account the particular biases of the algorithm. However, even for algorithms that exhibit a moderate complexity, the number of executions required by the search process results in a high computational cost, especially as more complex search strategies are used.

Implementing a wrapper is a straightforward task in supervised learning, since there always some external validation measure available. Typically, one executes a classifier and obtains an estimation of the accuracy in predicting a class label that is known. Although class label prediction can be used as an external measure to assess the validity of a clustering in rediscovering a known structure, labels are not available during the learning process, so they cannot be used in a wrapper implemention for clustering.

A solution is to assume that the goal of clustering is to optimize some objective function which helps to obtain 'good' clusters and use this function to estimate the quality of different feature subsets. Despite the unavailability of class labels, this approach seems to be more reasonable that requiring clustering algorithms to maximize accuracy over a piece of information which they do not have access to. Actually, we can view the objective function as the "accuracy" of clustering algorithms. When a given algorithm is used, there is an implicit assumption that the higher (lower) the value of its objective function the better are the properties that the groups discovered exhibit.

When an objective function is used to evaluate feature subsets in a search, it must applied to clusterings obtained with subsets of different cardinality. Since we need to compare these results, the function must be defined in a way that is not biased with respect to the number of features, that is, it should not be monotonically increasing or decreasing as a function of the dimensionality of the data. For example, as reported in [5] the scatter separability and the maximum likelihood criteria suffer this drawback.

Filter methods appear to be a, probably less optimal, but reasonable compromise for feature selection problems. But then again, for clustering tasks this turns out to be a hard problem since we need to decide what is going to be relevant to discover a structure that we do not know in advance. Similarly to wrapper methods, existing supervised approaches for filtering rely mainly in properties and relationships between the data and a predefined class label.

A particularly optimal implementation of filters are methods that employ some criterion to score each feature and provide a ranking. From this ordering, several feature subsets can be chosen, either manually of setting a threshold. This special case of the filter approach, that will be refer to as *rankers*, can be extremely efficient because is a one step process without any search involved. In practice, the efficiency depends on the computational complexity of the scoring procedure.

3   EM Clustering with Feature Selection

In this work, we adopt a commonly used and simple probabilistic framework for clustering assuming that the data comes from a multinomial mixture model with k sources corresponding to the number of clusters ([11]). This model is closely related to the Naive Bayes model for classification as it relies on the assumption that all features are rendered mutually independent by the cluster variable.

We use the EM algorithm to estimate the maximum likelihood (ML) parameters and the posterior cluster probabilities for each data point. Briefly, this algorithm is an iterative procedure that alternates between two steps: the Expectation step (E) and the Maximization step (M). In the E step we use the current parameters to compute the partial assignment (weights) to the k clusters for each data point. In the M step, we reestimate the parameters as the ML assignment given these weights.

There are not as many clustering algorithms for categorical data as for numerical data, but still there are other possible approaches, notably COBWEB [6]. However, we made the choice of EM because it produces flat clusterings as opposed to COBWEB, which builds cluster hierarchies. We think that for adequately assessing feature selection methods, the representational bias is an important factor that should be fixed, and, currently, flat clustering algorithms are more representative. Nevertheless, most categorical clustering algorithms rely on counting and computing frequencies, so that our results within the EM framework have a good chance to generalize to other algorithms.

3.1   An EM Wrapper

As previously noted, the ML criterion for cluster quality has a bias of increasing as the number of features decreases, so that it cannot be used to define a wrapper. We propose a solution that assumes that the goal of feature selection is to obtain a clustering with a reduced set of features of similar or better quality as that obtained by using all the features. Intuitively, if we build a clustering with a reduced feature set, then compute the objective function adding the rest of features and find that the resulting score is as good as the one that is obtained by using all the features, this is an indicator that the non-selected features were not relevant. Therefore, the full set log-likelihood can be used to both assess the resulting clusterings and guide the search for the wrapper approach. Note that this method of evaluation can be potentially applied to any objective function, not only likelihood-based approaches. An equivalent proposition has been made in the context of feature selection for unsupervised learning of conditional Gaussian networks [13].

In our probabilistic framework, we can run the EM algorithm for a given subset and estimate the model parameters, and then compute the log-likehood that these parameters yield using the full feature set. We can estimate this score in a simple manner by running an additional M step of the *EM* algorithm in which the parameters for the removed features are estimated from the weights obtained using only the selected subset. A subsequent E step would provide the full feature set likelihood estimation.

Since using exhaustive search strategies is prohibitive, wrapper methods often resort to heuristic methods and, particularly, greedy approaches. A commonly used procedure is *sequential stepwise selection* that adds or removes a single feature at each step of the search. We can start from the full set of features and use a removal operator (*backward elimination*) or start from the empty set and add one feature at a time (*forward selection*). Since repeatedly using the clustering algorithm is already a costly solution, in this paper we resort to an implementation that combines EM with forward selection because is significantly cheaper that backward elimination. We call this implementation EM-WFS (EM wrapper with forward search).

With these assumptions we have defined an starting point, a search strategy and an evaluation function, but we also need a stopping criterion. Usually we would continue the process until no improvement on the evaluation function is found. However, we have noted that, at certain points, the change of the function scores is very small. Because of that, in our implementation we stop if the relative change of the score is less than a fixed threshold.

3.2   A (Dependency-Based) EM Ranker

One view about the relevance of features conjectures that features that are not highly correlated with other features are not likely to play an important role in the clustering process and can be deemed as irrelevant [15]. This conjecture can be explained from two points of view.

The first view argues that a general principle common to most clustering systems is to form clusters having most feature values common to their members (*cohesion*) and few values common to members of other clusters (*distinctiveness*). These properties can be expressed in the form of the conditional probabilities $P(F_i = V_{ij} \mid C_k)$ and $P(C_k \mid F_i = V_{ij})$, where F_i is a given feature, V_{ij} is some value of this feature and C_k is a cluster. By rewarding clusterings that simultaneously maximize both probabilities for given values, at the same time, clusters formed around feature correlations are favored (see [15] for examples). Therefore, features that exhibit low dependencies with other features, are not good candidates to obtain cohesive and distinct groups and, hence, irrelevant.

A second approach stems from considering the clustering problem as *mixture modeling* in which the data is assumed as being generated from a mixture of several distributions. This approach can be encoded as a Bayesian network which contains a hidden variable corresponding the clusters in the data. A commonly used simplification assumes that all the features are conditionally independent of every other feature given the cluster variable, so that the underlying dependency model is a Naive Bayes model. The Bayesian interpretation of this approach is that the hidden variable explains or captures the dependencies of the rest of features. Thus, the resulting clusters will be most influenced by the strongest feature dependencies in the data. Hence again, features that are least correlated with other features are likely to be good candidates to eliminate.

Formulated in either way, the assumption that feature dependences are important to determine their importance for clustering tasks is independent of any labeling of the data. Therefore, it can be employed as a foundation in designing filters for feature selection for clustering. Still, this is a very general formulation that does not indicate nor how to model these dependencies neither how to employ this information in the feature selection process.

The previous assumption relating dependency and irrelevance of features provides a guide to design filter methods in feature selection for clustering, We can score each feature with a measure reflecting the degree in which this feature is dependant of other features in the data. With such a measure, we can implement a feature selection method by constructing a rank of features and selecting the best k, where k is a user given parameter.

We will assume that we can capture feature dependencies via pairwise interactions. For instance, using a mutual information measure, we can define the score of a feature F_i to be:

$$score(F_i) = \sum_{j=1, j \neq i}^{n} I(F_i; F_j) \tag{1}$$

where $I(F_i; F_j)$ stands for the usual definition of the mutual information between two variables x and y:

$$I(x; y) = \sum_x \sum_y p(x, y) log \frac{p(x, y)}{p(x)p(y)} \tag{2}$$

A simple method can be implemented by using this measure to order the features and obtain a ranking with a $O(nm^2)$ cost, where n is the number of instances and m is the number of features. We will refer to this method as EM-PWDR (EM pairwise dependency ranker).

The straightforward implementation of a ranker lefts up to the user the task of decide the number of features selected. To provide some help in this task, we added an additional step that builds a clustering with each of the feature subsets that result from the ordering (with one feature, two features, three features and so on) and then perform a single iteration to obtain the log-likelihood over the full feature set, as explained before. This figure can be used to conjecture the behavior or different subsets, although being obtained from training data, can be somewhat optimistic.

4 Empirical Evaluation

In order to compare the performance of the EM-WFS and EM-PWDR methods, we performed experiments on ten data sets from the UCI Repository. The data sets and their characteristics are listed in Table 1. Data sets including numerical features were previously discretized and missings were removed by substituting those values by the mode.

As previously described, performance is estimated by computing the log-likelihood of the obtained clustering over the full feature set at the end of the process. To avoid an optimistic estimation, we applied a ten-fold cross validation procedure in order to apply the feature selection procedure over a training set and compute the log-likelihood over a separate test set.The same folds were used for each of the methods.

Since the EM algorithm can be trapped in at a local maximum, both the wrapper and the ranker used at each run of EM the best of 5 runs starting with different random weight assignments. Additionally, we made the algorithm to stop when the relative difference between the likelihoods computed in two consecutive iterations did not change by 0.0001. This constrain is justified by the fact that this algorithm tends to converge asymptotically.

Figure 1 shows the log-likelihood averaged over the training and testing sets when a fixed number of features is selected for each fold. A first trend that can be observed is that feature selection does not tend to decrease the quality of the clusterings with respect to the original score using all the features. Obviously, selecting the smaller subsets drops cluster quality, but the rest of combinations consistently equal or improve the full feature set results. It appears that in some data sets using the full set of features hinders the capability of the EM algorithm to converge to a good model. This result suggest that feature selection might be even more important in clustering that in supervised learning, which makes sense, since clustering algorithms must consider a large number of possible relationships between the features.

A second, possibly surprising, trend that some data sets exhibit is that performance on training data is a good predictor of performance on unseen test

Table 1. Characteristics of the data sets used in the experiments

Dataset	Instances	Attributes
vote	435	16
mushroom	8124	22
LED+17	5000	24
WDBC	569	30
ionosphere	351	34
spambase	4601	57
sonar	208	60
splice	3186	60
yeast	208	79
musk	6598	166

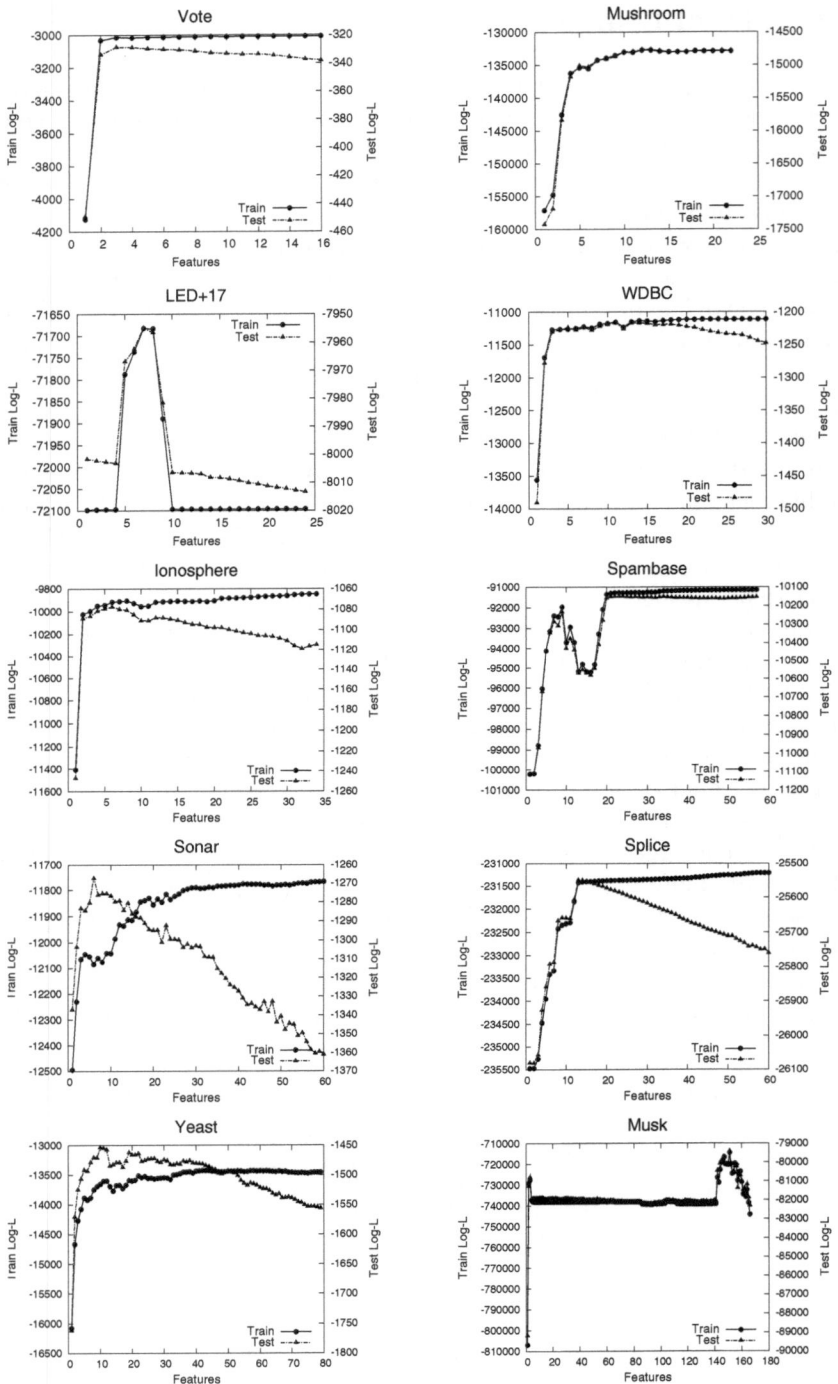

Fig. 1. Average log-likelihood over training and testing sets of EM-PWDR over different number of features

Table 2. Average test log-likelihood for different stopping criteria of EM-WFS, EM-PWDR with heuristic selection of the number of features and the best result of EM-PWDR

Dataset	EM-WFS-0.001 Log-L	Feat.	EM-WFS-0.0001 Log-L	Feat.	EM-PWDR 1% Log-L	Feat.	EM-PWDR-best Log-L	Feat.
vote	-330.12	4.1	-331.40	6.1	-332.52	2	-327.77	5.3
mushroom	-15908.62	2.5	-14896.97	4.5	14907.50	8	-14750.90	15.3
LED+17	-8001.47	1	-8001.47	1	-7954.89	7	-7953.23	3
wdbc	-1215.62	4.6	-1217.71	8.4	-1218.27	7.9	1209.70	11.10
ionosphere	-1089.65	5.5	-1092.57	12.4	-1079.29	5	-1075.87	5.7
spambase	-10619.14	3.8	-10138.22	15.8	-10233.10	9	-10143.42	29.8
sonar	-1247.05	4.5	-1255.77	12.1	-1287.79	16	-1255.40	11.4
splice	-26080.95	1	-26080.95	1	-25791.47	6	-25543.38	11.1
yeast	-1437.85	9	-1437.40	14.3	-1473.25	25	-1440.11	13.9
musk	-88381.39	1	-88185.38	1.9	-79399.90	151	-77563.88	92.6

data. Particularly on the vote, mushroom, LED+17, spambase and musk data sets the overlapping is close to perfect. And in most of the rest, even differing to some extend, training performance still can be used as a guide to select a reasonably good subset. Note that if, instead of selecting the subset with maximum training quality, we allow a deviation from the maximum, we still can obtain impressive results even with those data sets.

Table 2 shows the results for the EM-WFS method with two different stopping thresholds, namely 0.001 and 0.0001. Additionally, results for a manual selection method for EM-PWDR that chooses a number of features based on the maximum likelihood over training data is also shown. To avoid overfitting, we allow a 1% deviation from the maximum quality observed in the curve. The final column lists the best possible selection that could be made for the EM-PWDR. As expected, the wrapper performs well and somewhat better in general than the ranker. There are times where EM-PWDR could obtain a similar result but at the expense of selecting more features. However, most of the times the quality decreases by a relative factor under 1%. Moreover, in three data sets EM-WFS gets trapped in a local maximum, selecting too few features and producing unsatisfactory results.

As we could expect, wrapper methods are significantly more expensive than filter ones. In order to develop a machine independent measure of complexity, we will consider a more abstract measure than running times based upon the number of required feature comparisons. The EM algorithm exhibits a complexity $O(mnk)$ in each iteration for n instances, m features and k clusters. Therefore, we assume that a single execution of the algorithm performs $mnkI$ feature comparisons. On the other hand, the ranking method requires to compute the mutual information $(m(m-1)n)/2$ times. Note than in order to simulate a manual selection of the number of features by plotting the curve of the likelihood on the full feature set additional runs of the EM algorithm over each subset is required with the added cost. Figure 2 shows the computational complexity for each method on each data set. With the exception of the cases in which the wrapper is trapped on local maxima, the computational cost is always more expensive than filter methods, especially as the number of features increases. Note

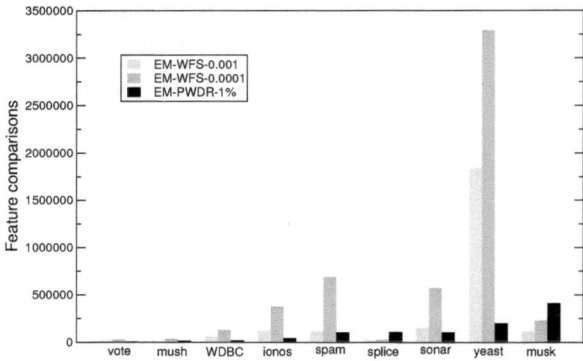

Fig. 2. Relative computational cost of the FS methods as a function of the number of feature comparisons

that the repeated execution of the clustering algorithm is likely to be always an expensive procedure, since, unlike some lazy or semi-lazy supervised approaches (e.g. Naive Bayes) most batch clustering methods rely on some form of iterative optimization.

The most important advantage of using wrappers lies in the fact that, in some cases, they are able to achieve the same performance than filters with a more reduced subset. The most probable reason is that the dependency-based ranker is sensitive to redundant features. Even though we aim to find correlated features to be the core of the discovered clusters, there will be cases in which some features will not provide any improvement over the selected subset.

Summing up, experimental evidence suggests that the ranker is able to perform a reasonably good job given the limited information that uses and its significant lower complexity respect to the wrapper. The wrapper has the potential to make an accurate selection but experiments suggest evidence that it is too prone to get trapped in local maxima, a well known problem for forward search strategies. A more conservative backward search method or different search strategies, such as best first search [9], could be used to overcome this problem but at the price of increasing the already high complexity of the wrapper solution.

5   Related Work

Although recently several works studying the problem of feature selection for clustering have appeared in the literature, filter based approaches are still uncommon. A notable exception is a proposal which develops an unsupervised entropy measure for ranking features [2,3,4]. Although several data sets are used in the evaluation, different assessment measures are employed in these works making difficult a direct comparison.

The dependency-based ranker presented in this paper has been previously used with success with hierarchical clusterings with alternative evaluation measures. In [15] the method is evaluated by comparing cluster predictions with ground truth labels, while in [14] the average predictive power over all the features (flexible prediction) is employed. A variant of the dependency assumption for continuous features has been presented in [13] for feature selection in learning conditional Gaussian networks.

An alternative to filter methods is to embed the feature selection task into the clustering process itself. The model based paradigm offers a natural way of achieving this goal by modeling feature relevance as parameters of the model. Examples of this approach are found in [10] and [17]. Results are, again, difficult to compare since the former work makes a very limited empirical evaluation using error rates and only numerical data, while the latter is focused on document clustering.

Early work in embedding feature selection into the clustering process traces back to work by Gennari [7] that implemented a wrapper over the CLASSIT hierarchical clustering system, although at that time there was a limited availability of data for evaluation. The work is based upon selecting the features that most contribute to the clustering objective function, an idea that is also used in a filter proposed in [16] also for hierarchical clusterings.

Finally, wrapper approaches are found in [5] and [8]. The experimental evidence in these papers tend to focus on investigating the particular issues of the presented methods rather than on exploring the performance on a wide range of data sets. As it is the general case, evaluation is performed basically on numerical data.

6   Concluding Remarks

In this work we have presented, to our knowledge, the first extensive empirical comparison between filter and wrapper methods of feature selection for clustering for categorical data. As it is the case with supervised learning approaches, feature selection can increase the quality of the results while reducing the complexity of the learning task.

As widely reported in the literature, wrapper methods tend to be superior to filters, and it appears that clustering is not an exception. However, the forward selection mechanism used in this work has not proved to be reliable enough, being too prone to stop in local maxima. This is an interesting result not mentioned in other papers using wrappers in feature selection for clustering. Although this could be a byproduct of our particular evaluation function, we think that the lack of references in other works to this undesirable behavior is the limited variety of data sets used.

Our results confirm previous work in that dependency based filters are a reasonably feature selection alternative. Interestingly, most often than not training quality has shown to be a good indicator of performance so that the resulting curves could be used as a guide to select the appropriate number of features. Our evaluation function appears to be intuitive and can be generalized to any objective function. However, future work could pursue a comparison with alternative approaches, such as the one presented in [5].

The computation of pairwise dependencies used in this work relies on the implicit assumption that all the features are independent given each other. This may not be the case, but supervised methods such as Naive Bayes that make the same assumption have been successfully used in a variety of learning tasks. Moreover, this is actually the same assumption that is made by the simple probabilistic model used in our implementation of the EM algorithm. It remains to be seen whether performance can be improving by using methods that do not assume that all features are independent of each other. More complex dependencies involving several features might exist but not be correctly reflected by these scores. In some cases, we could expect that by summing across all the features, some spurious dependencies might amplify the score thus producing a less accurate ranking. Future work could study more elaborated methods to score the dependence between features.

The previous issue might be connected with a limitation of the ranker method, its inadequacy to detect redundant features. The score computed cannot differentiate between required correlations that lead to good clusters and those that do not provide improvements on the light of the already selected features. This might not be a trivial problem to solve using filters, since the characterization of when a feature has to be considered redundant in clustering problems remains still an open issue.

An additional problem that could hinder the capabilities of filter methods is the existence of different good feature subsets that may lead to different clusterings of equivalent quality. In such a case, the feature ranking could be mixing features that are relevant in different contexts, thus yielding an suboptimal ordering. This assumption makes an interesting connection to a different area of research, subspace clustering [12] that could be worth to pursue.

Finally, it would be interesting to perform additional comparisons employing alternative filter approaches. Although there is almost no work on this area, the method suggested in [2] appears to be a good candidate.

References

1. A. L. Blum and P. Langley. Selection of relevant features and examples in machine learning. *Artificial Intelligence*, 97:245–271, 1997.
2. M. Dash, K. Choi, P. Scheuermann, and H. Liu. Feature selection for clustering - a filter solution. In *Proceedings of the 2002 IEEE International Conference on Data Mining (ICDM 2002)*, pages 115–122, Maebashi City, Japan, 2002. IEEE Computer Society.
3. M. Dash and H. Liu. Feature selection for clustering. In *Knowledge Discovery and Data Mining, Current Issues and New Applications, 4th Pacific-Asia Conference, PADKK 2000*, volume 1805 of *Lecture Notes in Computer Science*, pages 110–121, Kyoto, Japan, 2000. Springer.
4. M. Dash, H. Liu, and J. Yao. Dimensionality reduction for unsupervised data. In *Ninth IEEE International Conference on Tools with AI, ICTAI'97*, 1997.
5. J. G. Dy and C. E. Brodley. Feature selection for unsupervised learning. *Journal of Machine Learning Research*, 5:845–889, 2004.

6. D. H. Fisher. Knowledge acquisition via incremental conceptual clustering. *Machine Learning*, 2:139–172, 1987.
7. J. H. Gennari. Concept formation and attention. In *Proceedings of the Seventh Annual Conference of the Cognitive Science Society*, pages 724–728, Irvine,CA, 1991. Lawrence Erlbaum Associates.
8. Y. Kim, W. N. Street, and F. Menczer. Evolutionary model selection in unsupervised learning. *Intelligent Data Analysis*, 6(6):531–556, 2002.
9. R. Kohavi and G. H. John. Wrappers for feature subset selection. *Artificial Intelligence*, 97:273–324, 1997.
10. M. H. C. Law, M. A. T. Figueiredo, and A. K. Jain. Simultaneous feature selection and clustering using mixture models. *IEEE Transactions on Pattern Analysis and Machine Intelligence*, 26(9):1154–1166, 2004.
11. M. Meila and D. Heckerman. An experimental comparison of model-based clustering methods. *Machine Learning*, 42(1/2):9–29, 2001.
12. L. Parsons, E. Haque, and H. Liu. Subspace clustering for high dimensional data: a review. *SIGKDD Explorations*, 6(1):90–105, 2004.
13. J. M. Peña, J. A. Lozano, P. Larrañaga, and I. Inza. Dimensionality reduction in unsupervised learning of conditional gaussian networks. *IEEE Transactions on Pattern Analysis and Machine Intelligence*, 23(6):590–603, 2001.
14. L. Talavera. Feature selection as a preprocessing step for hierarchical clustering. In *Proceedings of the Sixteenth International Conference on Machine Learning*, pages 389–397, Bled, Slovenia, 1999. Morgan Kaufmann.
15. L. Talavera. Dependency-based feature selection for symbolic clustering. *Intelligent Data Analysis*, 4(1), 2000.
16. L. Talavera. Feature selection and incremental learning of probabilistic conc ept hierarchies. In *Proceedings of the Seventeenth International Conference on Machine Learning*, pages 951–958, Stanford, CA, 2000. Morgan Kaufmann.
17. S. Vaithyanathan and B. Dom. Model selection in unsupervised learning with applications to do cument clustering. In *Proceedings of the Sixteenth International Conference on Machine Learning*, pages 433–443, Bled, Slovenia, 1999. Morgan Kaufmann.

Dealing with Data Corruption
in Remote Sensing*

Choh Man Teng

Institute for Human and Machine Cognition,
40 South Alcaniz Street, Pensacola FL 32502, USA
cmteng@ihmc.us

Abstract. Remote sensing has resulted in repositories of data that grow at a pace much faster than can be readily analyzed. One of the obstacles in dealing with remotely sensed data and others is the variable quality of the data. Instrument failures can result in entire missing observation cycles, while cloud cover frequently results in missing or distorted values. We investigated the use of several methods that automatically deal with corruptions in the data. These include robust measures which avoid overfitting, filtering which discards the corrupted instances, and polishing by which the corrupted elements are fitted with more appropriate values. We applied such methods to a data set of vegetation indices and land cover type assembled from NASA's Moderate Resolution Imaging Spectroradiometer (MODIS) data collection.

1 Introduction

Except for data from highly constrained environments, it is almost unavoidable that data repositories and streaming data contain imperfections. Remote sensing data products suffer various kinds of corruption during the process of acquisition, processing, analysis, and mapping of the data [9]. Sources of imperfections include instrument failures, less than ideal observation conditions, recording and formatting anomalies, and transmission errors. In the case of satellite observations, an obscuring cloud may result in missing data pixels, while the shadow of this cloud may produce data points with distorted values.

These imperfections in many cases are difficult to avoid due to resource constraints, such as instrument capabilities, and external factors, such as weather conditions. Repeated measurements to improve the data are sometimes not only undesirable but practically impossible in tasks such as reanalyzing historical data collected using different instruments and at a bygone time.

In this paper we examine several methods from the machine learning and data mining communities for dealing with data corruption. We compare the performance of these methods using a data set of vegetation indices and land cover type assembled from the MODIS collection.

---

* This work was supported by NASA NCC2-1239, NNA04CK88A and ONR N00014-03-1-0516.

A.F. Famili et al. (Eds.): IDA 2005, LNCS 3646, pp. 452–463, 2005.

2   Approaches to Handling Data Corruption

Broadly conceived the data corruption problem is pervasive, and different communities focus on solving different parts of the problem. For instance, one of the prime concerns in the database community is the reconciliation of inconsistent records and the pruning of duplicate ones [10]. In the signal processing community there has been much work on identifying outliers and smoothing modulating signals [15]. In earth science, geostatistical techniques are employed for interpolating spatially correlated quantities [4].

Each of these methods tackles a characteristic kind of data problem. We have identified three general approaches to handling imperfections that are most relevant to the problem of rectifying corruptions in data with independently derived or loosely coupled instances: We may leave the imperfections to individual applications that make use of the data, discard the corrupted portions of the data by using a filter, or correct the values that are in error.

In the machine learning and data mining communities, the standard approach to coping with imperfections is to delegate the burden to the theory builder. This is typically accomplished by avoiding overfitting, so that the theory does not develop overly complicated substructures just to fit the noise [2,11]. The corrupted instances are retained in the data set, and each algorithm has to institute its own corruption handling routine to ensure robustness, duplicating the effort required even when the same data set is used in each case. In addition, this noise tolerance can interfere with the quality of the results obtained from the theory builder. For instance, the predictive accuracy may suffer and the representation of the theory thus built may be less compact.

Another approach is to eliminate from the data set instances that are suspected of being corrupted according to certain evaluation mechanisms [1,6]. A theory is then constructed using only the remaining instances. Similar ideas can be found in robust regression and outlier detection techniques in statistics [12]. In filtering out the corrupted instances, there is an obvious tradeoff between the amount of corruption removed and the amount of data retained; the more corrupted instances we remove, the less data is available for meaningful analysis. In the extreme case where every instance is in some way less than perfect, the whole data set may get discarded and we are left with nothing to analyze. Thus, filtering does not make efficient use of the data.

A third approach orrects the corrupted instances rather than eliminating them [5,13]. Possibly incorrect elements in an instance are identified, but instead of discarding the whole instance as in the case of filtering, the corrupted elements are repaired and the repaired instances are reintroduced into the data set. Ideally, the resulting data set would preserve and recover the maximal information available in the data and better approximate the noise-free situation. A theory built from this corrected data should have a higher predictive power and a more streamlined representation. We have developed a method of data correction, called *polishing*, which will be described in more detail in the next section.

3   Polishing

In polishing, the corrupted instances are not only identified but also corrected. To accomplish this we exploit the interdependency between the components of a data set. For example, in the context of classification, the task is to predict the target concept by examining the feature values. The basic assumption here is that the feature and target concept values are related. Rather than utilizing the features only to predict the target concept, we can in addition turn the process around and utilize the target together with selected features to predict the value of another feature. This provides a means for identifying corrupted elements together with their correct values. Note that except for totally irrelevant elements, each feature would be at least related to some extent to the target concept, even if not to any other features. The problem we need to address is how to harness effectively this not always obvious relationship for data correction.

The basic algorithm of polishing consists of two phases: *prediction* and *adjustment*. In the prediction phase, elements in the data that are suspect are identified together with a nominated replacement value. In the adjustment phase, we selectively incorporate the nominated changes into the data set.

3.1   Prediction

In the first phase, the predictions are carried out by systematically swapping the target and particular features of the data set and performing a ten-fold cross validation using a chosen classification algorithm for the prediction of the feature values. The data set is partitioned in the ten-fold trials in a way to achieve a balanced distribution of target values in each fold. In each trial nine parts of the data are used for training a classifier which is then applied to predict the feature values of the instances in the remaining one part.

An instance can be represented as a tuple consisting of the values of features $F_1, \ldots, F_n$ and a class variable C. The classification task then can be broadly described as using the features values to predict the class value:

$$\langle F_1, F_2, \ldots, F_n \rangle \leadsto C.$$

In polishing, for each feature F_i a classifier is built to predict the value of F_i in each instance using the remaining features together with the original class variable:

$$\langle F_1, \ldots, F_{i-1}, C, F_{i+1}, \ldots, F_n \rangle \leadsto F_i.$$

The same learning algorithm used for classifying C can be applied to learn to classify F_i. This procedure effectively exchanges the roles of the feature F_i and the class C in the classification task.

If the predicted value of a feature in an instance is different from the stated value in the data set, the location of the discrepancy is flagged and recorded together with the predicted value. This information is passed on to the next phase, where we institute the actual adjustments.

3.2   Adjustment

Since the polishing process itself is based on imperfect data, the predictions obtained in the first phase can contain errors as well. We should not indiscriminately incorporate all the nominated changes. Rather, in the second phase, the adjustment phase, we selectively adopt appropriate changes from those predicted in the first phase, using a number of strategies to identify the best combination of changes that would improve the fitness of a datum. We perform a ten-fold cross validation on the data, and the instances that are classified incorrectly are selected for adjustment. A set of changes to a datum is deemed acceptable if it leads to a correct prediction of the target concept by the classifiers obtained from the cross validation process.

The number of possible replacement values and possible combinations of changes we can make to a datum grows exponentially with the dimensionality of the data, that is, the number of features and the number or range of feature values. We have adopted a number of heuristics for selecting an appropriate set of changes, as well as for keeping the search space manageable. For example, the features are sorted according to their classification accuracy in the prediction phase, and more reliable features are given higher priority in the adjustment phase. This is based on the assumption that their nominated replacement values, when available, are more reliable due to a higher predictive accuracy.

Another heuristic is to make as few changes as possible to the data. Thus, we only make adjustments when a datum cannot be classified correctly, and changes involving fewer features are preferred over changes involving many features, when both sets of changes can result in a correct classification of a previously misclassified instance. In general the number of features alone may not be a good indicator of the amount of change involved. A more sophisticated measure will in addition consider factors such as the importance of the feature and the type of change we are making to the feature. In addition, an estimated corruption level can give an indication of the number of corrupted features we can expect in an instance.

After making the suggested adjustments, we obtain a polished data set which can be used for further analysis.

4   Remote Sensing Data from MODIS and Data Preparation

4.1   Vegetation Indices and Land Cover

We assembled a data set compiled from the MODIS vegetation indices and land cover products. Two vegetation indices were selected: the Normalized Difference Vegetation Index (NDVI) and the Enhanced Vegetation Index (EVI). Both indices are concerned with the proportion of photosynthetically absorbed and reflected radiation, and have been extensively used as remotely sensed indicators of the density of vegetation growth [14,8, for example]. NDVI is calculated as a function of red and infrared bands, while EVI in addition employs the blue band to correct for background clutter and atmospheric influences.

Each instance in our data set consists of one year of NDVI and EVI values at 16-day intervals, starting from October 2000. The class label to be predicted is the land cover type of the instance according to the International Geosphere-Biosphere Programme classification scheme. The data pixels are sampled uniformly and globally, and in order to minimize spatial dependencies, the pixels selected are at least 100km apart.

Intuitively, the type of land cover represented in a pixel can be inferred from its vegetation index signatures. For instance, an evergreen forest is green year round, while a deciduous forest is greener in the summer than in the winter. Similarly, cropland may have intervals of vegetation activities part of the year, each followed by a period of bare soil after harvesting.

We excluded the easier to discriminate classes. For instance, barren land (rock, sand, etc.) and ocean have no vegetation and can be readily identified by their relatively constant and low vegetation index values throughout the year; these pixels are not included in the experimental data set.

4.2   Seasonal Changes and Missing Observation Cycles

There was one more problem we needed to tackle before we fed the data to the imperfection handling mechanisms we were going to compare.

In June 2001, the MODIS instrument experienced an "anomaly", resulting in two 16-day periods with no vegetation indices produced for any data pixel. This translates to two completely uninstantiated attributes in our data set. These empty attributes affect the three approaches we will consider in various ways:

- Robust algorithms (or any other classification method for that matter) have effectively two fewer attributes.
- Filtering, in an idealized sense where all partially corrupted instances are discarded, will remove all instances as they all have two missing attributes.
- Polishing, in the absence of any values for these attributes, cannot be used to identify correct values for them since the prediction of the values depend on the pattern of relationship between attributes in instances with instantiated values.

We observed that the growth cycle of vegetation follows the change of seasons, and the seasons differ between the northern and southern hemispheres. Deciduous trees in the northern hemisphere can be expected to be the greenest in June, while those in the southern hemisphere can be expected to peak in December. We therefore re-aligned the data attributes according to seasons rather than according to calendar dates. This should give us a more consistent seasonal pattern of vegetation indices. In addition, it provides us a way to avoid the totally empty attributes. After the re-alignment, instead of two attributes with absolutely no values in any instance, we have four attributes each with some values according to whether a pixel represents a location in the northern or southern hemisphere.

Table 1. Corruption Characteristics: Percentages of items corrupted and percentages of elements corrupted, without and with a nearest neighbor model. An item is considered corrupted if any one of its elements (attributes or class) has been corrupted.

Corruption Level	Original		with Nearest Neighbor	
	% Instance	% Elements	% Instance	% Elements
0; 0	0.0	0.0	0.0	0.0
500; 10	100.0	38.4	100.0	36.4
1000; 20	100.0	61.7	100.0	54.4
1500; 30	100.0	73.1	100.0	62.5
2000; 40	100.0	79.0	100.0	67.1

5 Experiments

5.1 Simulating Data Corruption

In remote sensing, typically optical systems (for example MODIS) are corrupted by additive, normally distributed random processes, while radar systems in addition are subject to multiplicative noise [3]. There are further sources contributing errors to every stage of the data collection and analysis process, but here we will model the corruption as an additive Gaussian process with mean 0 and a specified standard deviation, simulating optical disturbances.

The training data were artificially corrupted by introducing random noise into both the attributes and the class. A corruption level of $x; y$ denotes that

- A value generated from a Gaussian with mean 0 and standard deviation x is added to each vegetation index attribute.
- Each class label (the land cover type) is assigned a random value $y\%$ of the time, with each alternative class value being equally likely to be selected.

Table 1 shows the amount of change induced in the data at various corruption levels. The columns under "Original" are actual differences between the original data and the corrupted data. (The columns under "with Nearest Neighbor" will be discussed in Section 7.3.) An element (attribute or class) is considered corrupted if it is different from its corresponding value in the original data set. An instance is considered corrupted if any one of its elements is corrupted.

Note that at nearly every corruption level considered, 100% of the instances have at least one corrupted element. Thus, if we were to use an idealized filtering technique, that is, one that can identify and discard every corrupted instance, all the data would be eliminated. The filtering methods we use in practice are typically not this effective; nonetheless this is one argument for attempting to repair the corruption instead of discarding corrupted instances wholesale.

5.2 Mechanisms Evaluated

The basic learning algorithm we used is c4.5 [11] the decision tree builder, using the information gain ration criterion as the basis for choosing splitting attributes.

One of the reasons for adopting decision trees is that the MODIS land cover classification has been carried out partly also using decision trees.

The three corruption handling mechanisms evaluated in this study are as follows:

Robust : c4.5, with its built in mechanisms for avoiding overfitting. These include, for instance, post-pruning, and stop conditions that prevent further splitting of a leaf node.

Filtering : Instances that have been misclassified by the decision tree built by c4.5 are discarded, and a new tree is built using the remaining data. This is similar to the approach taken in [7].

Polishing : Instances that have been misclassified by the decision tree built by c4.5 are polished, and a new tree is built using the polished data, according to the mechanism described in Section 3.

We performed ten-fold cross validation on the data set, using the above three methods (robust, filtering, and polishing) in turn to obtain the classifiers. In each trial, nine parts of the data were used for training, and the remaining one part was held for testing. The performance results obtained from the ten-fold cross validation trials were compared using a number of evaluation metrics. First we will look at the more traditional classifier specific metrics, namely classification accuracy and decision tree size, and then we will concentrate on data correction, and examine some classifier independent metrics, which directly compare the three versions available of each data set: the original (supposedly noise-free), the artificially corrupted (noisy), and the treated (cleaned) versions.

6   Classifier Specific Comparisons

We compared the classification accuracy and size of the decision trees built. The results are summarized in Tables 2 and 3.

Table 2 shows the classification accuracy of trees obtained using the three methods. We compared the methods in pairs (robust vs. filtering; robust vs. polishing; filtering vs. polishing), and differences that are significant at the 0.05

Table 2. Classification accuracy. An "*" indicates a significant improvement of the latter method over the former at the 0.05 level.

Corruption Level	Robust	Filtering (Percentage)	Polishing	Robust/ Filtering	Robust/ Polishing	Filtering/ Polishing
0; 0	72.4	73.6	74.4	*	*	*
500; 10	69.1	72.8	74.2	*	*	*
1000; 20	56.0	65.9	66.8	*	*	
1500; 30	49.3	58.6	59.3	*	*	*
2000; 40	39.5	51.3	52.6	*	*	*

Table 3. Tree Size

Corruption Level	Robust	Filtering	Polishing
0; 0	2435.8	1492.6	1594.6
500; 10	3550.6	1592.2	1738.6
1000; 20	3945.4	1852.6	2141.8
1500; 30	3898.6	1601.8	1885.0
2000; 40	3521.8	1862.2	2225.8

level using a paired t-test are marked with an $*$. (An "$*$" indicates the latter method performed better than the former in the pair being compared.)

In almost every case considered, polishing outperformed filtering, which outperformed c4.5 (the robust algorithm). Both filtering and polishing achieved a relatively large improvement over c4.5. The difference in accuracy between filtering and polishing is relatively much smaller but statistically significant. The comparison between c4.5 and filtering also confirmed the results reported in [7], in which trees built from the filtered data were in many cases more accurate than trees built from the unfiltered data.

Note that even at the 0;0 corruption level—that is, no artificial corruption has been added—filtering and polishing gave rise to more accurate classifiers than c4.5. It is almost certain that there is corruption inherent in the original data, and the performance of the three methods at the 0;0 corruption level may serve as an indication of their ability to deal with the existing "real" corruption in the data.

Table 3 shows the size of the trees built using the three methods. The trend is that the robust algorithm resulted in the largest trees, followed by polishing, and filtering resulted in the smallest trees. By discarding corrupted instances (filtering) or repairing the corrupted elements (polishing), the data set became more uniform and therefore could be represented by a smaller tree. In the case of filtering, the data set in addition became smaller after the corrupted instances have been discarded; this reduction in training data size may have also contributed to a reduction in the resulting tree size.

7   Classifier Independent Comparisons (Proximity Metrics)

As mentioned before, we considered two classes of evaluation metrics, namely, classifier specific metrics and classifier independent metrics. The performance measures discussed in the previous section are classifier specific. We judge how well an imperfection handling mechanism fares by looking at the quality of the classifier built from the treated data set. In this section we consider a couple of more general classifier independent metrics. These metrics assess the quality of the correction directly from the data. They would be more appropriate than classifier dependent metrics when the corrected data is intended for purposes beyond classification. They would also provide more direct evidence as to whether the corruption in the data has indeed been successfully repaired. Note that these metrics are only applicable to data correction methods, since neither robust algorithms nor filtering makes changes to the retained instances in the data set.

Consider the i-th instance in a data set. Let us call this instance in the original noise-free data set the *root instance* x_i; the (possibly) corrupted version of it in the noisy data set the *noisy instance* y_i; and the version in the polished data set the *polished instance* z_i. The question we want to answer is: have we been able to repair the corruption in the i-th instance, and if so, to what extent? In other words, is the polished instance z_i any less noisy than the noisy instance y_i?

One simple (but not quite satisfactory, as we will see) interpretation of this question is: is z_i any better than y_i in approximating the root instance x_i? We might count the differences between z_i and x_i, and compare that to the number of differences between y_i and x_i. However, there are a number of problems associated with quantifying proximity this way. Consider the following scenarios.

- Suppose one of the attributes is a dummy attribute whose values are entirely random. It is unreasonable to expect a noise correction mechanism to be able to correct the "noise" in an attribute with random values.
- Suppose it just so happens that an instance is corrupted in a way so that it matches exactly another noise-free instance. The "noise" in the instance exists only with respect to the original root instance, but in the broader picture, the instance has been changed from one noise-free instance to another also noise-free instance, and in this sense should not be counted as noisy.

We can generalize these two scenarios to the case of irrelevant attributes and the case of adjustment towards an alternative.

7.1   Irrelevant Attributes

In a not as extreme case, some attributes may not be totally random, but they may not be very relevant to the classification of the target concept either. To get a clearer picture of how much noise we have corrected, we should exclude these irrelevant attributes from the tally or weigh the attributes according to their respective relevance.

To accomplish this we need to identify the (ir)relevant attributes. The approach we have adopted is to consider as relevant only those attributes used in building the original classifiers from the corrupted data. If an attribute is not part of a classifier, it is irrelevant at least with respect to this classifier and the classification task at hand, regardless of whether it can be important in other aspects. This gives us an operational definition of relevance which is simple to use.

Note that the irrelevant attributes do not have to be removed from the data set at any stage; they are just skipped over in our count.

7.2   Adjustment Towards an Alternative

Another difficulty with measuring proximity is that the noise added and the adjustments we made may move an instance towards an alternative instance (that is not the root instance) in the original noise-free data set. The root instance may not be the noise-free instance in the original data set that is closest to the noisy instance. Fewer changes may be required of the corrupted instance to match an alternative clean instance than to revert back to the root instance.

In some sense these adjustments towards an alternative instance should be considered correct with respect to the alternative, yet they would be counted as incorrect with respect to the root instance, as the polished instance is moved (usually) even farther away from the root instance in these cases. Thus, a straightforward comparison between the root, noisy, and polished versions of an instance would miscount these adjustments as undesirable, and penalize any deviation away from the root instance towards an alternative, even if the instance has been repaired to perfectly match the alternative clean instance.

7.3   Relevant Nearest Neighbor

We developed the notion of a relevant nearest neighbor to circumvent the two problems discussed above. The relevant attributes are taken to be the ones that have been used to build the original classifiers. The distance $d(z, x)$ between two instances z and x is measured by the number of differences between their relevant attributes plus the class. A *relevant nearest neighbor* of an instance z is then an instance z^* in the original noise-free data set, where $d(z, z^*)$ is minimal.

We should also note that a noisy instance and its polished version might not have the same nearest neighbor, as each may be changed so that they move closer to different clean instances in the original data set. To obtain a fair comparison, we need to take this into account. Rather than utilizing a single fixed nearest neighbor for both the noisy and polished versions of an instance, we quantify the amount of noise present by the distances relative to their respective (relevant) nearest neighbors.

We formulated two proximity metrics based on the idea of a relevant nearest neighbor: the net percentage reduction in overall noise, and the percentage of correct adjustment.

Let y_i and z_i be the noisy and polished versions of the i-th instance in the data set, and y_i^* and z_i^* be their respective relevant nearest neighbors. The distance $d(y_i, y_i^*)$ denotes the difference between the two instances. Let n_i and m_i be the number of correct and incorrect adjustments made to the i-th instance (with respect to a relevant nearest neighbor). We then have

$$\text{Net Reduction (NR):} \quad \frac{\sum_i [d(y_i, y_i^*) - d(z_i, z_i^*)]}{\sum_i d(y_i, y_i^*)} \quad ;$$

$$\text{Correct Adjustment (CA):} \quad \frac{\sum_i n_i}{\sum_i (n_i + m_i)} \quad .$$

Table 4. Proximity Metrics: Net Reduction (NR) and Correct Adjustment (CA), without and with a nearest neighbor model

Corruption Level	Original		with Nearest Neighbor	
	NR	CA	NR	CA
0; 0	N/A	0.0	N/A	38.3
500; 10	0.7	53.1	2.9	83.8
1000; 20	1.9	61.1	3.8	91.5
1500; 30	1.6	61.5	3.5	91.6
2000; 40	2.3	71.7	3.4	94.4

Net Reduction indicates how much less (or more) noise exists in a data set after polishing, and Correct Adjustment reports what proportion of the adjustments made is considered correct.

The proximity metric values for our data set, corrupted to various corruption levels and then polished, are shown in Table 4. The amount of corruption added at the different levels, as measured using the nearest neighbor model, is shown in the columns under "with Nearest Neighbor" in Table 1.

While a fairly low percentage of noise was repaired (NR values in Table 4), almost all the adjustments made were effectively correct (CA values). Both the NR and CA scores increased when the comparisons were made with respect to the nearest neighbors rather than with respect to the original root instances. These results are encouraging as it is of vast importance that a data correction mechanism be reliable—false positives, or incorrect repairs, should be minimized as much as possible.

Even though polishing decreased the overall corruption only a little, the adjustments were mostly "correct", and they contributed to an improved classification performance (as shown in Table 2). This may be an indication that our Net Reduction metric needs further fine tuning, as it does not reflect the extent to which the repairs have improved the quality of the classification.

Note that although the *percentage* of noise reduction decreased as the noise level increased, this does not automatically translate to a decrease in the net *amount* of noise repaired. At a higher noise level, a larger amount of noise is present, and therefore even a smaller percentage may still amount to a considerable quantity.

8   Concluding Remarks

Automated methods for handling data imperfections are especially useful when the volume of data precludes us from manually inspecting the data in any meaningful way. We studied three such methods in the context of remote sensing. In our experiments both filtering and polishing were able to improve on the results achieved by the robust algorithm c4.5, with polishing being slightly better than filtering.

In addition to classification accuracy and classifier size, we in particular examined metrics to evaluate data correction methods. These nearest neighbor metrics are not specific to polishing, but apply in general to any method that manipulates the data. Establishing an accurate evaluation metric is a first and essential step in safeguarding data integrity. An accurate model of the data and imperfection distribution can be used to achieve a good understanding of the tradeoff between the sensitivity of polishing and the reliability of the repairs.

References

1. Carla E. Brodley and Mark A. Friedl. Identifying mislabeled training data. *Journal of Artificial Intelligence Research*, 11:131–167, 1999.
2. P. Clark and T. Niblett. The CN2 induction algorithm. *Machine Learning*, 3(4):261–283, 1989.
3. B. R. Corner, R. M. Narayanan, and S. E. Reichenbach. Noise estimation in remote sensing imagery using data masking. *International Journal of Remote Sensing*, 24(4):689–702, 2003.
4. Noel A. C. Cressie. *Statistics for Spatial Data*. Wiley, revised edition, 1993.
5. George Drastal. Informed pruning in constructive induction. In *Proceedings of the Eighth International Workshop on Machine Learning*, pages 132–136, 1991.
6. Dragan Gamberger, Nada Lavrač, and Ciril Grošelj. Experiments with noise filtering in a medical domain. In *Proceedings of the Sixteenth International Conference on Machine Learning*, pages 143–151, 1999.
7. George H. John. Robust decision trees: Removing outliers from databases. In *Proceedings of the First International Conference on Knowledge Discovery and Data Mining*, pages 174–179, 1995.
8. H. Q. Liu and A. R. Huete. A feedback based modification of the NDVI to minimize canopy background and atmospheric noise. *IEEE Transactions on Geoscience and Remote Sensing*, 33:457–465, 1995.
9. R. S. Lunetta, R. G. Congalton, L. K. Fenstermaker, J. R. Jensen, K. C. McGwire, and L. R. Tinney. Remote sensing and geographic information system data integration: Error sources and research issues. *Photogrammetric Engineering and Remote Sensing*, 57(6):677–687, 1991.
10. J. Maletic and A. Marcus. Data cleansing: Beyond integrity analysis. In *Proceedings of the Conference on Information Quality*, pages 200–209, 2000.
11. J. Ross Quinlan. *C4.5: Programs for Machine Learning*. Morgan Kaufmann, 1993.
12. Peter J. Rousseeuw and Annick M. Leroy. *Robust Regression and Outlier Detection*. John Wiley & Sons, 1987.
13. Choh Man Teng. Correcting noisy data. In *Proceedings of the Sixteenth International Conference on Machine Learning*, pages 239–248, 1999.
14. C. J. Tucker, W. W. Newcomb, S. O. Los, and S. D. Prince. Mean and inter-year variation of growing-season normalized difference vegetation index for the Sahel 1981–1989. *International Journal of Remote Sensing*, 12:1113–1115, 1991.
15. Saeed V. Vaseghi. *Advanced Digital Signal Processing and Noise Reduction*. Wiley, second edition, 2000.

Regularized Least-Squares for Parse Ranking

Evgeni Tsivtsivadze, Tapio Pahikkala, Sampo Pyysalo, Jorma Boberg,
Aleksandr Mylläri, and Tapio Salakoski

Turku Centre for Computer Science (TUCS),
Department of Information Technology, University of Turku,
Lemminkäisenkatu 14 A, FIN-20520 Turku, Finland
firstname.lastname@it.utu.fi

Abstract. We present an adaptation of the Regularized Least-Squares
algorithm for the rank learning problem and an application of the method
to reranking of the parses produced by the Link Grammar (LG) dependency parser. We study the use of several grammatically motivated features extracted from parses and evaluate the ranker with individual features and the combination of all features on a set of biomedical sentences
annotated for syntactic dependencies. Using a parse goodness function
based on the F-score, we demonstrate that our method produces a statistically significant increase in rank correlation from 0.18 to 0.42 compared
to the built-in ranking heuristics of the LG parser. Further, we analyze
the performance of our ranker with respect to the number of sentences
and parses per sentence used for training and illustrate that the method
is applicable to sparse datasets, showing improved performance with as
few as 100 training sentences.

1 Introduction

Ranking, or ordinal regression, has many applications in Natural Language Processing (NLP) and has recently received significant attention in the context of
parse ranking [1]. In this paper, we study parse reranking in the domain of
biomedical texts. The Link Grammar (LG) parser [2] used in our research is
a full dependency parser based on a broad-coverage hand-written grammar.
The LG parser generates all parses allowed by its grammar and applies a set
of built-in heuristics to rank the parses. However, the ranking performance of
the heuristics has been found to be poor when applied to biomedical text [3].
Therefore, a primary motivation for this work is to present a machine learning
approach for the parse reranking task in order to improve the applicability of
the parser to the domain. We propose a method based on the Regularized Least-
Squares (RLS) algorithm (see e.g. [4]), which is closely related to Support Vector
Machines (SVM) (see e.g. [5]). We combine the algorithm and rank correlation
measure with grammatically motivated features, which convey the most relevant
information about parses.

Several applications of SVM-related machine-learning methods to ranking
have been described in literature. Herbrich et al. [6] introduced SVM ordinal

A.F. Famili et al. (Eds.): IDA 2005, LNCS 3646, pp. 464–474, 2005.

regression algorithm based on a loss function between rank pairs. Joachims [7] proposed a related SVM ranking approach for optimizing the retrieval quality of search engines. SVM-based algorithms have also been applied to parse reranking, see, for example, [8]. For a recent evaluation of several parse reranking methods, see [9].

One of the aspects of the method introduced in this paper is its applicability to cases where only small amounts of data are available. The annotation of data for supervised learning is often resource-intensive, and in many domains, large annotated corpora are not available. This is especially true in the biomedical domain. In this study, we use the Biomedical Dependency Bank (BDB) dependency corpus[1] which contains 1100 annotated sentences.

The task of rank learning using the RLS-based regression method, termed here Regularized Least-Squares ranking (RLS ranking), can be applied as an machine learning approach alternative to the built-in heuristics of the LG parser. We address several aspects of parse ranking in the domain. We introduce an F-score based parse goodness function, where parses generated by the LG parser are evaluated by comparing the linkage structure to the annotated data from BDB. For evaluating ranking performance, we apply the commonly used rank correlation coefficient introduced by Kendall [10] and adopt his approach for addressing the issues of tied ranks. An application of the method to the parse rank learning task is presented, and an extensive comparison of the performance of the built-in LG parser heuristics to RLS ranking is undertaken. We demonstrate that our method produces a statistically significant increase in rank correlation from 0.18 to 0.42 compared to the built-in ranking heuristics of the LG parser.

The paper is organized as follows: in Section 2, we describe a set of grammatically motivated features for ranking dependency parses; in Section 3, we introduce a parse goodness function; in Section 4, we discuss the Regularized Least-Squares algorithm; in Section 5, we provide the performance measure applied to parse ranking and discuss the problem of tied ranks; in Section 6, we evaluate the applicability of the ranker to the task and benchmark it with respect to dataset size and the number of parses used in training; we conclude this paper in Section 7.

2   Features for Dependency Parse Ranking

The features used by a learning machine are essential to its performance, and in the problem considered in this paper, particular attention to the extracted features is required due to the sparseness of the data. We propose features that are grammatically relevant and applicable even when relatively few training examples are available. The output of the LG parser contains the following information for each input sentence: the linkage consisting of pairwise dependencies between pairs of words termed links, the link types (the grammatical roles assigned to the links) and the part-of-speech (POS) tags of the words. As LG does not perform any morphological analysis, the POS tagset used by LG is limited, consisting

---

[1] http://www.it.utu.fi/~BDB

mostly of generic verb, noun and adjective categories. Different parses of a single sentence have a different combination of these elements. Each of the features we use are described below.

Grammatical bigram. This feature is defined as a pair of words connected by a link. In the example linkage of Figure 1, the extracted grammatical bigrams are *absence—of, of—alpha-syntrophin, absence—leads*, etc. These grammatical bigrams can be considered a lower-order model related to the grammatical trigrams proposed as the basis of a probabilistic model of LG in [11]. Grammatical bigram features allow the learning machine to identify words that are commonly linked, such as *leads—to* and *binds—to*. Further, as erroneous parses are provided in training, the learning machine also has the opportunity to learn to avoid links between words that should not be linked.

Word & POS tag. This feature contains the word with the POS tag assigned to the word by LG. In the example, the extracted word & POS features are *absence.n, alpha-syntrophin.n, leads.v*, etc. Note that if LG does not assign POS to a word, no word & POS feature is extracted for that word. These features allow the ranker to learn preferences for word classes; for example, that "binds" occurs much more frequently as a verb than as a noun in the domain.

Link type. In addition to the linkage structure and POS tags, the parses contain information about the link types used to connect word pairs. The link types present in the example are *Mp, Js, Ss*, etc. The link types carry information about the grammatical structures used in the sentence and allow the ranker to learn to favor some structures over others.

Word & Link type. This feature combines each word in the sentence with the type of each link connected to the word, for example, *absence—Mp, absence—Ss, of—Js*, etc. The word & link type feature can be considered as an intermediate between grammatical unigram and bigram features, and offers a possibility for addressing potential sparseness issues of grammatical bigrams while still allowing a distinction between different linkages, unlike unigrams. This feature can also allow the ranker to learn partial selectional preferences of words, for example, that "binds" prefers to link directly to a preposition.

Link length. This feature represents the number of words that a link in the sentence spans. In Figure 1, the extracted features of this type are *1, 1, 3*, etc. This feature allows the ranker to learn the distinction between parses, which have different link length. The total summed link length is also used as a part of LG ordering heuristics, on the intuition that linkages with shorter link lengths are preferred [2].

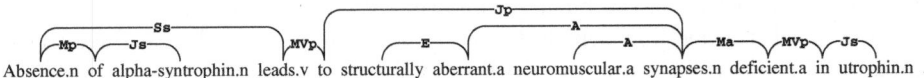

Fig. 1. Example of parsed sentence

Link length & Link type. This feature combines the type of the link in the sentence with the number of words it spans. In Figure 1, the extracted features of this type are *1—Mp, 1—Js, 3—Ss, 1—MVp*, etc. The feature is also related to the total length property applied by the LG parser heuristics, which always favor linkages with shorter total link length. However, the link length & link type feature allows finer distinctions to be made by the ranker, for example, favoring short links overall but not penalizing long links to prepositions as much as other long links.

Link bigram. The link bigram features extracted for each word of the sentence are all the possible combinations of two links connected to the word, ordered leftmost link first. In the example, link bigrams are *Mp—Ss, Mp—Js, Ss—MVp*, etc.

3 F-Score Based Goodness Function for Parses

The corpus BDB is a set of manually annotated sentences, that is, for each sentence of BDB, we have a manually annotated correct parse. Let P be the set of parses produced by the LG parser when applied to the sentences of BDB. We define a parse goodness function as

$$f^* : P \mapsto \mathbb{R}_+$$

which measures the similarity of the parse $p \in P$ with respect to its correct parse p^*. We propose an F-score based goodness function that assigns a goodness value to each parse based on information about the correct linkage structure. This function becomes the target output value that we try to predict with the RLS algorithm.

Let $L(p)$ denote the set of links with link types of a parse p. The functions calculating numbers of true positives (TP), false positives (FP) and false negatives (FN) links with link types are defined as follows:

$$TP(p) = | L(p) \cap L(p^*) | \tag{1}$$

$$FP(p) = | L(p) \setminus L(p^*) | \tag{2}$$

$$FN(p) = | L(p^*) \setminus L(p) | \tag{3}$$

The links are considered to be equal if and only if they have the same link type and the indices of the words connected with the links are the same in the sentence in question. We adopt one exception in (2) because of the characteristics of the corpus annotation. Namely the corpus annotation does not have all links, which the corresponding LG linkage would have: for example, punctuation is not linked in the corpus. As a consequence, links in $L(p)$ having one end connected to a token without links in $L(p^*)$, are not considered in (2). The parse goodness function is defined as an F-score

$$f^*(p) = \frac{2TP(p)}{2TP(p) + FP(p) + FN(p)}. \tag{4}$$

High values of (4) indicate that a parse contains a small number of errors, and therefore, the bigger $f^*(p)$ is, the better is parse p.

Next we consider the Regularized Least-Squares algorithm by which the measure f^* can be predicted.

4   Regularized Least-Squares Algorithm

Let $\{(x_1, y_1), \ldots, (x_m, y_m)\}$, where $x_i \in P, y_i \in \mathbb{R}$, be the set of training examples. We consider the Regularized Least-Squares (RLS) algorithm as a special case of the following regularization problem known as Tikhonov regularization (for a more comprehensive introduction, see e.g. [4]):

$$\min_f \sum_{i=1}^{m} l(f(x_i), y_i) + \lambda \|f\|_k^2, \tag{5}$$

where l is the loss function used by the learning machine, $f : P \to \mathbb{R}$ is a function, $\lambda \in \mathbb{R}_+$ is a regularization parameter, and $\|\cdot\|_k$ is a norm in a Reproducing Kernel Hilbert Space defined by a positive definite kernel function k. Here P can be any set, but in our problem, P is a set of parses of the sentences of the BDB corpus. The target output value y_i is calculated by a parse goodness function, that is $y_i = f^*(x_i)$, and is the one which we predict with RLS algorithm. The second term in (5) is called a regularizer. The loss function used with RLS for regression problems is called least squares loss and is defined as

$$l(f(x), y) = (y - f(x))^2.$$

By the Representer Theorem (see e.g. [12]), the minimizer of equation (5) has the following form:

$$f(x) = \sum_{i=1}^{m} a_i k(x, x_i),$$

where $a_i \in \mathbb{R}$ and k is the kernel function associated with the Reproducing Kernel Hilbert Space mentioned above.

Kernel functions are similarity measures of data points in the input space P, and they correspond to the inner product in a feature space H to which the input space data points are mapped. Formally, kernel functions are defined as

$$k(x, x') = \langle \Phi(x), \Phi(x') \rangle,$$

where $\Phi : P \to H$.

5   Performance Measure for Ranking

In this section, we present the performance measures used to evaluate the parse ranking methods. We follow Kendall's definition of rank correlation coefficient [10] and measure the degree of correspondence between the true ranking and the ranking output by an evaluated ranking method. If two rankings are equal, then correlation is $+1$, and on the other hand, if one ranking is the inverse of the other, correlation is -1.

The problem of the parse ranking can be formalized as follows. Let s be a sentence of BDB, and let $P_s = \{p_1, \ldots, p_n\} \subseteq P$ be the set of all parses of s produced by the LG parser. We apply the parse goodness function f^* to provide the target output variables for the parses by defining the following preference function

$$R_{f^*}(p_i, p_j) = \begin{cases} 1 & \text{if } f^*(p_i) > f^*(p_j) \\ -1 & \text{if } f^*(p_i) < f^*(p_j) \\ 0 & \text{otherwise} \end{cases}$$

which determines the ranking of the parses $p_i, p_j \in P_s$. We also define a preference function $R_f(p_i, p_j)$ in a similar way for the regression function f learned by the RLS algorithm. In order to measure how well the ranking R_f is correlated with the target ranking R_{f^*}, we adopt Kendall's commonly used rank correlation measure τ. Let us define the score S_{ij} of a pair p_i and p_j to be the product

$$S_{ij} = R_f(p_i, p_j) R_{f^*}(p_i, p_j).$$

If score is $+1$, then the rankings agree on the ordering of p_i and p_j, otherwise score is -1. The total score is defined as

$$S = \sum_{i < j \leq n} S_{ij}.$$

The number of all different pairwise comparisons of the parses of P_s that can be made is $\binom{n}{2} = \frac{1}{2} \cdot n (n - 1)$. This corresponds to the maximum value of the total score, when agreement between the rankings is perfect. The correlation coefficient τ_a defined by Kendall is:

$$\tau_a = \frac{S}{\frac{1}{2} \cdot n (n - 1)}.$$

While τ_a is well applicable in many cases, there is an important issue that is not fully addressed by this coefficient—tied ranks, that is, $f^*(p_i) = f^*(p_j)$ or $f(p_i) = f(p_j)$ for some i, j. To take into account possible occurrences of tied ranks, Kendall proposes an alternative correlation coefficient

$$\tau_b = \frac{S}{\frac{1}{2} \sqrt{\sum R_{f^*}(p_i, p_j)^2 \cdot \sum R_f(p_i, p_j)^2}}.$$

With tied ranks the usage of τ_b is more justified than τ_a. For example, if both rankings are tied except the last member, then $\tau_b = 1$ indicating complete agreement between two rankings, while $\tau_a = \frac{2}{n}$. Then for large values of n this measure is very close to 0, and therefore inappropriate. Due to many ties in the data, we use the correlation coefficient τ_b to evaluate performance of our ranker.

6   Experiments

In the experiments, BDB consisting of 1100 sentences was split into two datasets. First 500 were used for parameter estimation and feature evaluation using 10-fold cross-validation, and the rest were reserved for final validation. Each of the sentences has a variable amount of associated parses generated by LG. To address the computational complexity, we limited the number of considered parses per sentence to 5 in the training and to 20 in testing dataset. We also considered the effect of varying the number of parses per sentence used in training (Section 6.3). When more parses than the limit were available, we sampled the desired number of parses from these. When fewer were available, all parses were used.

We conducted several experiments to evaluate the performance of the method with respect to different features and learning ability of the ranker. The RLS algorithm has a regularization parameter λ which controls the tradeoff between the minimization of training errors and the complexity of the regression function. The optimal value of this parameter was determined independently by grid search in each of the experiments.

6.1   Evaluation of Features

In Section 2, we described features that were used to convey information about parses to the ranker. To measure the influence of individual feature, we conducted an experiment where features were introduced to the ranker one by one. Performance is measured using τ_b coefficient with respect to the correct ranking based on the parse goodness function f^*. As a baseline we considered the correlation between LG ranking and the correct ranking, which is 0.16. We observed that most of the features alone perform above or close to the baseline,

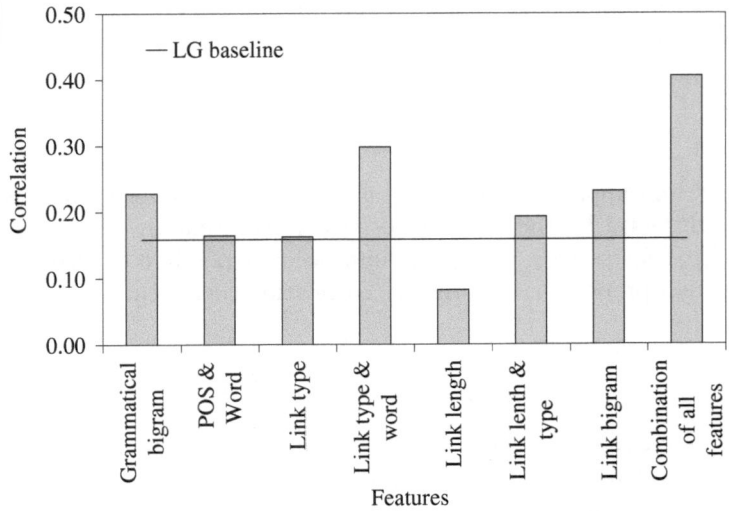

Fig. 2. RLS ranking performance with different features separately and combined

and performance of the RLS ranker with all the seven features combined on evaluation set is 0.42, supporting the relevance of proposed grammatical features. The figure 2 shows the performance of RLS ranking with respect to the different features separately and combined.

6.2 Final Validation

To test the statistical significance of the proposed method, we used the robust 5×2 cross-validation test [13]. The test avoids the problem of dependence between folds in N-fold cross-validation schemes and results in more realistic estimate than, for example, t-test. The performance of RLS ranker with all the seven features combined on validation set is 0.42 and the improvement is statistically significant ($p < 0.01$) when compared to 0.18 obtained by the LG parser. We also measured the performance of our ranker with respect to the parse ranked first. The average F-score of the true best parse in the corpus is 73.2%. The average F-score value obtained by the parses ranked first by the RLS ranker was found to be 67.6%, and the corresponding value for the LG parser heuristics was 64.2%, showing therefore better performance of our method also for this measure. Note that average F-scores of the highest ranked parses were obtained from the downsampled 20 parses per sentence.

6.3 Learning Ability of the Ranker with Respect to the Training Data

To address the issue of applicability of the proposed method to very sparse datasets, we measured performance of the RLS ranker with respect to two main criteria: the number of sentences and the number of parses per sentence used for training. In these experiments all grammatical features were used.

Number of Sentences. The training dataset of 500 sentences was divided into several parts and for testing a separate set of 500 sentences was used. The validation procedure was applied for each of the parts, representing sets of sizes 50, 100,..., 500 sentences. The number of parses used per sentence for training was 5 and for testing 20. We observed that even with a very sparse dataset our method gives a relatively good performance of 0.37 while the learning set size remains as small as 100 sentences. The learning procedure reflected expected tendency of the increased ranker performance with increased number of sentences, reaching 0.42 with 500 sentences.

Number of Parses. We measured performance of the RLS ranker based on the number of parses per sentence used for training with dataset size fixed to 150 sentences. Number of parses per sentence in training was selected to be 5, 10,..., 50 for each validation run. Test dataset consisted of 500 sentences each containing 20 parses. We observed that major improvement in ranker performance occurs while using only 10 or 20 parses per sentence for training corresponding to 0.41 and 0.43 performance respectively. When using 50 parses per sentence, performance was 0.45, indicating a small positive difference compared to results obtained with less number of parses.

Parse-Sentence Tradeoff. In this experiment, we fixed the number of training examples, representing number of sentences multiplied by number of available parses per sentence, to be approximately one thousand. Datasets of 20, 30, 50, 100, 200, 300, 500 sentences with number of parses per sentence 50, 30, 20, 10, 5, 3, 2, respectively, were validated with 500 sentences each containing 20 parses. The results of these experiments are presented in Table 1. We found that the best performance of ranker was achieved using 100 sentences and 10 parses per each sentence for training, corresponding to 0.41 correlation. The decrease in performance was observed when either having large number of parses with small amount of sentences or vice versa.

Table 1. Ranking performance with different number of sentences and parses

# Sentences	# Parses	Correlation	Difference in correlation
20	50	0.3303	0.0841
30	30	0.3529	0.0615
50	20	0.3788	0.0357
100	*10*	*0.4145*	*0.0000*
200	5	0.3798	0.0347
300	3	0.3809	0.0335
500	2	0.3659	0.0485

7   Discussion and Conclusions

In this study, we proposed a method for parse ranking based on Regularized Least-Squares algorithm coupled with rank correlation measure and grammatically motivated features. We introduce an F-score based parse goodness function. To convey the most important information about parse structure to the ranker, we apply features such as grammatical bigrams, link types, a combination of link length and link type, part-of-speech information, and others. When evaluating the ranker with respect to each feature separately and all features combined, we observed that most of them let the ranker to outperform Link Grammar parser built-in heuristics. For example, grammatical bigram (pair of words connected by a link) and link bigram (pair of links related by words) underline importance of link dependency structure for ranking. Another feature yielding good performance is link type & word, representing an alternative grammatical structure and providing additional information in case of similar parses. We observed that link length feature, which is related to LG heuristics, leads to poor, below the baseline, performance, whereas other features appear to have more positive effect.

We performed several experiments to estimate learning abilities of the ranker, and demonstrate that the method is applicable for sparse datasets. A tradeoff spot between number of parses and sentences used for training demonstrates that maximum performance is obtained at 100 sentences and 10 parses per sentence supporting our claim for applicability of the ranker to small datasets. Experimental results suggest that for practical reasons the use of 10 to 20 parses per sentence for training is sufficient. We compared RLS ranking to the built-in heuristics of LG parser and a statistically significant improvement in performance from 0.18 to 0.42 was observed.

In the future, we plan to address the issue of RLS algorithm adaptation for ranking by applying and developing kernel functions, which would use domain knowledge about parse structure. Several preliminary experiments with multiple output regression seemed promising and are worth exploring in more detail. In addition, we plan to incorporate RLS ranking into the LG parser as an alternative ranking possibility to its built-in heuristics.

Acknowledgments

This work has been supported by Tekes, the Finnish National Technology Agency and we also thank CSC, the Finnish IT center for science for providing us extensive computing resources.

References

1. Collins, M.: Discriminative reranking for natural language parsing. In Langley, P., ed.: Proceedings of the Seventeenth International Conference on Machine Learning, San Francisco, CA, Morgan Kaufmann (2000) 175–182
2. Sleator, D.D., Temperley, D.: Parsing english with a link grammar. Technical Report CMU-CS-91-196, Department of Computer Science, Carnegie Mellon University, Pittsburgh, PA (1991)
3. Pyysalo, S., Ginter, F., Pahikkala, T., Boberg, J., Järvinen, J., Salakoski, T., Koivula, J.: Analysis of link grammar on biomedical dependency corpus targeted at protein-protein interactions. In Collier, N., Ruch, P., Nazarenko, A., eds.: Proceedings of the JNLPBA workshop at COLING'04, Geneva. (2004) 15–21
4. Poggio, T., Smale, S.: The mathematics of learning: Dealing with data. Amer. Math. Soc. Notice **50** (2003) 537–544
5. Vapnik, V.N.: The nature of statistical learning theory. Springer-Verlag New York, Inc. (1995)
6. Herbrich, R., Graepel, T., Obermayer, K.: Support vector learning for ordinal regression. In: Proceedings of the Ninth International Conference on Artificial Neural Networks, London, UK, IEE (1999) 97–102
7. Joachims, T.: Optimizing search engines using clickthrough data. In: Proceedings of the ACM Conference on Knowledge Discovery and Data Mining, New York, NY, USA, ACM Press (2002) 133–142
8. Shen, L., Joshi, A.K.: An svm-based voting algorithm with application to parse reranking. In Daelemans, W., Osborne, M., eds.: Proceedings of CoNLL-2003. (2003) 9–16
9. Collins, M., Koo, T.: Discriminative reranking for natural language parsing (2004) To appear in Computational Linguistics, available at http://people.csail.mit.edu/people/mcollins/papers/collinskoo.ps.

10. Kendall, M.G.: Rank Correlation Methods. 4. edn. Griffin, London (1970)
11. Lafferty, J., Sleator, D., Temperley, D.: Grammatical trigrams: A probabilistic model of link grammar. In: Proceedings of the AAAI Conference on Probabilistic Approaches to Natural Language, Menlo Park, CA, AAAI Press (1992) 89–97
12. Schölkopf, B., Herbrich, R., Smola, A.J.: A generalized representer theorem. In Helmbold, D., Williamson, R., eds.: Proceedings of the 14th Annual Conference on Computational Learning Theory and and 5th European Conference on Computational Learning Theory, Berlin, Germany, Springer-Verlag (2001) 416–426
13. Alpaydin, E.: Combined 5×2 cv F-test for comparing supervised classification learning algorithms. Neural Computation **11** (1999) 1885–1892

Bayesian Network Classifiers for Time-Series Microarray Data

Allan Tucker[1], Veronica Vinciotti[1], Peter A.C. 't Hoen[2], and Xiaohui Liu[1,3]

[1] School of Information Systems Computing and Maths,
Brunel University, Uxbridge UB8 3PH, UK
[2] Center for Human and Clinical Genetics, Leiden Genome Technology Center,
Leiden University Medical Center, Wassenaarseweg 72, 2333 AL Leiden, Netherlands
[3] Leiden Institute of Advanced Computer Science,
Leiden University P.O. Box 9512, 2300 RA Leiden, Netherlands

Abstract. Microarray data from time-series experiments, where gene expression profiles are measured over the course of the experiment, require specialised algorithms. In this paper we introduce new architectures of Bayesian classifiers that highlight how both relative and absolute temporal relationships can be captured in order to understand how biological mechanisms differ. We show that these classifiers improve the classification of microarray data and at the same time ensure that the models can easily be analysed by biologists by incorporating time transparently. In this paper we focus on data that has been generated to explore different types of muscular dystrophy.

1 Introduction

The analysis of microarray data has previously focussed on the clustering of genes into groups of similar expression profiles. This has amongst other things allowed biologists to infer the functions of previously unknown genes. More recently, methods to learn gene networks from such data have been explored with the aim of trying to investigate more than just pairwise relationships and understand the interactions between genes in more detail [5]. Another research problem that has arisen from microarray data is the classification of different samples of data into categories such as diseased and control groups. Many microarray datasets contain thousands of genes and the number of samples are usually very small. Therefore methods such as feature selection are required to prevent over-fitting. Previously we have developed a method for classifying this sort of data that uses simple models, sampling and global feature selection algorithms [12].

Microarray data from time-series experiments, where gene expression profiles are measured over the course of the experiment, require specialised algorithms. Recently, papers have documented using time-series models to capture the temporal relationships between genes [14]. Time-series in microarray data contain two types of temporal information: *relative temporal relationships* and *absolute temporal relationships*. Considering the former, a point in a time-series can be classified based upon the changes that occur between time points. In other words

A.F. Famili et al. (Eds.): IDA 2005, LNCS 3646, pp. 475–485, 2005.
© Springer-Verlag Berlin Heidelberg 2005

it is the dynamics of the data that are used to classify the series. Ramoni et al. [8] exploit these relationships in time-series models for clustering. Previously, we have investigated temporal Bayesian classifiers for modelling relationships between different variables over time to classify visual field data [9]. However, another method can be used to model temporal relationships using absolute time which is relative only to a fixed point (or *reference*), say the birth of an organism or the onset of a medical condition, where any feature over time is measured from that point. Here, it is not the dynamics of the data that are used to classify but a combination of the data values and the length of time from the reference. For example, if a variable can be used to determine medication for a condition that has been diagnosed previously then the decision is dependent on the time since the diagnosis (the reference). In this case, the relative time between features in the data are irrelevant as it is the time since diagnosis that will be important in the decision to give medication. Friedman et al. [5] model temporal relationships in gene expression data by adding a node representing the phase of a yeast's cell cycle whilst numerous papers use dynamic Bayesian networks to exploit relative temporal relationships [14]. It should be noted that a dynamic Bayesian network should in theory be able to model absolute temporal relationships if it is of a suitably high order. However, much more data is required to parameterise such models than is typically available in microarray datasets.

In this paper we introduce new models that highlight how both relative and absolute temporal relationships can be captured in order to understand how biological mechanisms differ. We aim to improve the classification of time-series microarray data using new forms of Bayesian classifiers, whilst at the same time ensuring that the models can easily be analysed by biologists by using models that incorporate time transparently. These classifiers are described in section 2. We focus on microarray data that has been generated in order to explore the different types of muscular dystrophy based upon three previously identified biological pathways. The data and the experiments carried out are also described in section 2. Section 3 documents the results and analyses the results whilst section 4 concludes with implications and lines for future work.

2   Methods

2.1   Bayesian Classifiers

Bayesian Networks (BNs) [7] are probabilistic models that can be used to model data transparently. This means that it is relatively easy to explain to non-statisticians how the data are being modelled unlike other 'black box' methods. A BN is a directed acyclic graph consisting of links between nodes that represent variables in the domain. Links are directed from a parent node to a child node, and with each node there is an associated set of conditional probability distributions.

Bayesian classifiers are a special form of Bayesian network where one node represents some classification of the data. The simplest Bayesian classifier is the Naïve Bayes. This classifier has been used with surprising success, given its simplicity, on a number of different applications. It consists of a set of probability distributions for each variable given the class. The assumption behind this is that each variable is independent of one another given the class. A more general Bayesian classifier includes links between the predictor variables [4]. This requires learning a network structure between variables using some scoring metric coupled with a heuristic search. An example of the structure of each classifier is illustrated in Figure 1.

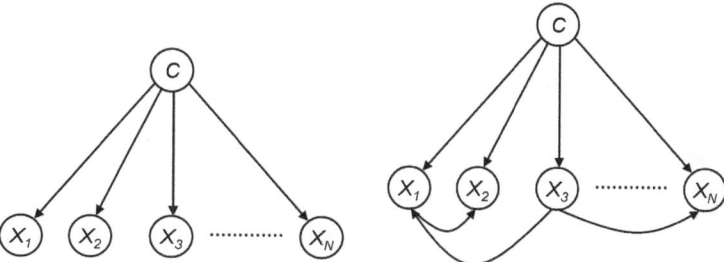

Fig. 1. Common Bayesian network classifier architectures: Naïve Bayes Classifier (NBC), Bayesian Network Classifier (BNC)

Many datasets involve measurements of variables over time and the dynamic Bayesian network (DBN) [3] is an extension of the BN to handle the sort of relationships found in time-series. A DBN is a BN where the N nodes represent variables at differing time slices. Therefore links occur between nodes over time and within the same time lag. Inference in DBNs is very similar to standard inference in static BNs. In this paper, we use a form of stochastic simulation called logic sampling [7] because of its speed and its intuitive appeal.

As opposed to DBNs, BNs can also incorporate time by including temporal nodes into the Bayesian network structures. For example, Friedman et al. [5] used this method when modelling yeast cell-cycle data where the temporal node was made the parent of every gene node.

We now discuss how these methods of incorporating time can be used to analyse and classify gene expression data with respect to absolute and relative temporal relationships.

2.2 Incorporating Time into Bayesian Classifiers

Figure 2 illustrates the two novel classifier architectures that we explore in this paper. From now on we refer to them as the Temporal BN Classifier (TBNC) and the DBN Classifier (DBNC). The TBNC allows genes to be conditioned upon a node that represents the time from some reference as well as other genes and the class node. Therefore the classification will take into account the time from the reference as well as the log ratio of the gene. The DBNC on the other hand, allows genes to be conditioned on genes from previous time points (including themselves as auto-regressive links) as well as the class node so that the change over time can assist the classification. We will also explore a classifier that is a hybrid of the two, the Temporal Dynamic BN Classifier (TDBNC), where a temporal node is included as well as dynamic links to see whether both types of temporal information can be discovered within the microarray data.

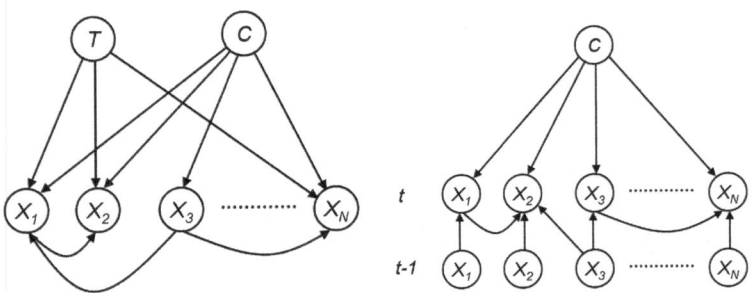

Fig. 2. Proposed architectures for incorporating time: Temporal BNC (TBNC) and Dynamic BNC (DBNC)

2.3   Learning the Classifiers

For all classifiers the links between the class node and every gene are automatically inserted and fixed during the search. Links between genes and between the time node and genes are explored using a simulated annealing approach similar to one we used in [12] but that minimises the Mimimum Description Length (MDL) of the network [6]. This global optimisation search was chosen with the aim of avoiding local optima, which many greedy searches suffer from. The main idea behind our method is to make small changes to the classifier structure and then score the network. The changes involve using three operators, **add**, **delete** and **swap**, which randomly add a link, remove a link and swap a link, respectively.

The optimisation algorithm is documented fully below, where D represents the input data, the initial annealing temperature is denoted by t_0, the cooling parameter for the temperature by c, the maximum number of scoring function calls by **maxfc** and the score of a network by **score(bn)**, computed by the MDL. $R(0, 1)$ is a uniform random number generator with limits 0 and 1. For all our experiments, we set t_0 to 1. This was based upon the initial scores when applied to the dataset investigated in this paper (we have generally found that a good starting temperature is similar to the changes in score in the early iterations). **maxfc** was set to 10000 as this was found through empirical analysis to ensure that convergence has occurred on the dataset explored. c was set to 0.999, calculated to ensure that the temperature after **maxfc** iterations was suitably close to zero.

```
Input: t_0, maxfc, D
       fc = 0, t = t_0
       Initialise bn to a Bayesian
       classifier with no inter-gene links
       result = bn
       oldscore = score(bn)
       While fc ≤ maxfc do
           For each operator do
                   Apply operator to bn
                   newscore = score(bn)
                   fc = fc + 1
                   dscore = newscore − oldscore
                   If dscore < 0 then
                       result = bn
                   Else If R(0,1) < e^{dscore/t} Then
                       Undo the operator
                   End If
           End For
           t = t × c
       End While
Output: result
```

Algorithm 1: Simulated annealing for building Bayesian networks

2.4 Muscular Dystrophy Data

Muscular dystrophies are a heterogeneous group of inherited disorders characterized by progressive muscle wasting and weakness. The genetic defects underlying
many muscular dystrophies have been elucidated [1,2]. A particular subset of
muscular dystrophies is caused by mutations in genes coding for constituents of
the dystrophin-associated glycoprotein complex (DGC). Mutations in the dystrophin gene cause Duchenne muscular dystrophy, whereas mutations in sarcoglycan genes are responsible for Limb-Girdle Muscular Dystrophies. Large-scale
gene expression profiling of mouse models known to recapitulate different human
muscular dystrophies was performed to delineate the molecular mechanisms underlying the shared and distinct phenotypic characteristics [10,11].

The MDX mouse is a mouse model for Duchenne muscular dystrophy, beta-
sarcoglyan-deficient (BSG) and gamma-sarcoglycan-deficient (GSG) mice are
mouse models for Limb-Girdle Muscular Dystrophies 2E and 2C. Expression
profiles were generated from two individual mice (two biological replicates) at
different ages: 1, 2.5, 4, 6, 8, 10, 12, 14, 20 weeks. There were four technical
replicates in the experiment: the arrays were spotted in duplicate and the samples
were hybridized twice (dye-swapped). The arrays used were spotted 7.5K 65-mer
oligonucleotide arrays (Sigma-Genosys mouse library).

A temporal loop hybridization design was applied in which consecutive time
points were hybridized to the same array. For this paper, we focused on the

analysis of a subset of 28 genes, belonging to 3 different pathways, which are related to muscle regeneration. These pathways have been identified in MDX mice, where effective muscle regeneration accounts for regression of the pathology. The first pathway is the Notch-Delta signaling pathway, the second the bone morphogenetic protein pathway, the third the neuregulin pathway. Together, these pathways appear to regulate the proliferation and differentiation of muscle precursor cells, which gives rise to the formation of new muscle fibres and consequently muscle repair. It is not clear yet if these pathways are equally active in the beta- and gamma-sarcoglycan-deficient mice that seem to suffer from a more progressive muscle pathology than the MDX mice.

As there are two independent biological samples for each class of muscular dystrophy, we have decided to perform two-fold cross validation where one experiment involves training from data based solely on one biological sample and tested on the other. This approach avoids testing on data that are highly correlated with the training set (as a higher correlation is expected between technical repeats). Furthermore, for each fold, we repeat the network search 10 times, due to the stochastic nature of our simulated annealing algorithm. For each of these runs, the frequency count is maintained for each link in all networks generated on the training data for the corresponding fold and the classifiers tested on the portion of data taken out. In this way we are able to produce a confidence measure for each link in the network based on different training samples. This is similar to the method used by [5], where the confidence measure on links in a Bayesian network is achieved by bootstrapping the data.

The data were normalised using the `all.norm` function from the smida R-library. The method essentially corrects for spatial, dye and across-array effects. These normalization procedures are applied in a sequential manner, starting with local corrections and proceeding towards more global corrections like across-array normalization. More details about the methods can be found in [13]. The log-ratios of the gene expression at a particular time point with respect to the first time point were then considered for the study. These were estimated from the raw log-ratios using a simple linear model [13]. The data were then discretised into two states using a frequency-based policy whereby the resultant genes appear in each state with equal probability.

3   Results

First of all we investigate the accuracy of the classifiers when determining whether the disease is not present (the wild-type) or, if it is, which form the disease takes - a four class classification problem. Figure 3 (top) shows the accuracy of the classifiers when applied to each of the two folds (i.e trained on one biological sample and tested on the other). It is evident that the simple NBC performs poorly at around 60% (note that a totally random classification output would result in 25% accuracy) with very large variation between the samples. Surprisingly, the BNC which models the dependencies between genes results in a small decrease in the mean accuracy but considerably reduces the variation. When the

DBNC is used to classify the data, a small improvement is seen over the BNC. The TBNC performs considerably better with a 64% accuracy and the TDBNC hybrid (with both dynamic links and a time node) records a further albeit small improvement still at 65%. It seems that the variation between folds is greater when the time node is used (in TBNC and TDBNC compared to BNC and DBNC). This could be due to the increasing complexity and also the fact that the time node only has a small number of examples in the data for each instantiation.

We now look at the accuracy of the methods for each of the four classes in Figure 3 (bottom). It appears that the wild-type (WT) and the MDX form of muscular dystrophy are more easily classified than the BSG and GSG forms. The genes used in this study are known to be associated with the MDX form and were thought likely to be involved in the other two forms. However, the obtained result implies that the three pathways investigated here are more closely associated with dystrophin deficiency (MDX) than with sarcoglycan deficiency

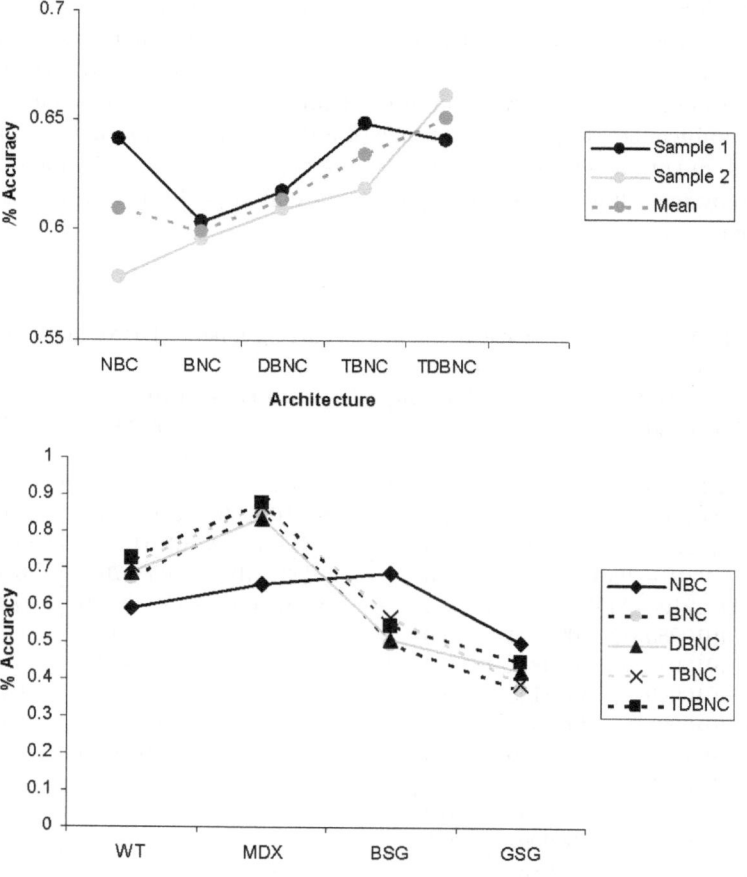

Fig. 3. Top: Comparison of classification accuracy for each biological sample (fold). Bottom: Accuracy for each disease type

(BSG and GSG), which seems reasonable given the regeneration pathways being more active in the MDX mice than in the BSG and GSG mice.

Due to the transparent nature of Bayesian network classifiers we can explore some of the discovered relationships between genes including the temporal aspect in DBNC, TBNC and TDBNC. We turn firstly to the resulting networks generated from the DBNC.

The first observation to make is that only 4% of the links found in the DBNC are dynamic (i.e. spanning two timepoints). Table 1 reports the most commonly occurring links during learning the DBNCs. A single line arrow denotes a normal link whereas a double lined arrow denotes a dynamic link.

Four out of 19 links found by DBNC with highest confidence were associated with genes functioning in the same Notch-Delta signaling pathway. When we consider only the links found with consistently high confidence in the two folds, the percentage of links from genes within the Notch-Delta signaling pathway is even higher (3 out of 5). This is in agreement with our expectation that, of the three studied signaling pathways, the Notch-Delta pathway shows the most coherent regulation. The other links were mapping to genes functioning in different pathways. Since we expect the three different pathways to act synergistically in muscle regeneration, co-expression or co-regulation of these genes may have biological significance, despite the absence of a direct interaction between the gene products. Future assessment of the false positive rate in the identified links, for example by including genes in completely unrelated pathways, will be necessary to further evaluate their biological relevance.

Table 1. Table of the most frequently occurring links in DBNC and TBNC

DBNC Link	Percent	TBNC Link	Percent
NM010950→NM009861	0.5	Time→NM008380	0.7
NM007866→NM009758	0.35	Time→NM013871	0.5
D32210→D90156	0.35	Time→NM008284	0.5
NM007866→NM008734	0.3	NM010950→NM009861	0.45
D32210→NM010091	0.3	Time→NM010423	0.4
D32210→NM021877	0.3	D32210→AF059176	0.4
D32210→AF059176	0.3	NM010949→NM008110	0.4
NM010950→NM008734	0.3	Time→NM010091	0.35
NM010949→NM013871	0.3	NM010949→NM008163	0.35
NM010949→NM008163	0.3	NM010949→NM008734	0.35
NM010091→NM008380	0.3	NM010091→D32210	0.35
NM021877→NM008380	0.3	NM009861→NM010866	0.35
NM010423→NM008163	0.3		
NM010496→NM008540	0.3		
NM008110→NM009097	0.3		
NM009757→NM009861	0.3		
NM009861→NM010950	0.3		
NM008284→NM007888	0.3		
NM013871⇒NM009758	0.2		

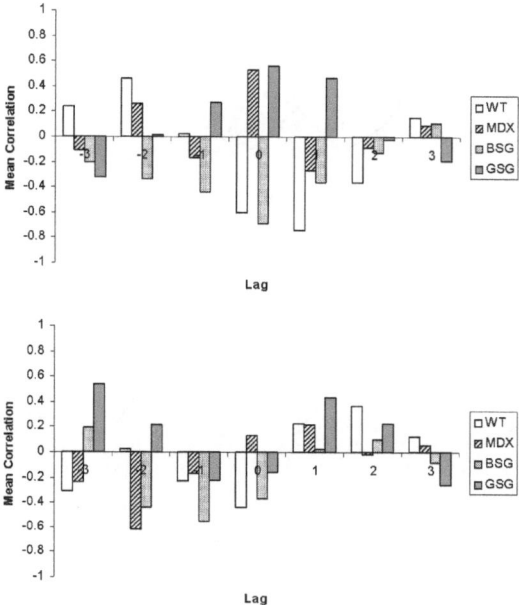

Fig. 4. Comparing cross correlations of discovered dynamic links: NM013871→NM009758 (top) found in 20% and NM008110→NM008380 (bottom) found in 10%

Figure 4 plots the cross-correlation of two of the dynamic links that were discovered in the highest proportion of networks during the 10 runs over the two folds. This is the correlation between gene profiles over varying time lags and for each of the classes, where the class profile is averaged across the biological and technical repeats. It is evident that for many of these plots the most significant correlation is at time lags of one or greater implying a relationship that spans time. Interestingly, for some genes it appears that the highest correlation is at a time lag of one but only in certain classes. For example, in the top cross correlation plot the most significant correlation was at a lag of zero for BSG , MDX and GSG but at a lag of one for wild type (WT).

For the TBNC classifiers, we plot the genes that were discovered to be most commonly associated with the time node. Note that the plots in figure 5 have peaks and troughs for all classes rather than having a high or low log-ratio throughout. It is the *shape* of the plot that differentiates them. In other words, you must take the time when a gene is expressed into account. For example, the gene NM008380 is over-expressed at the first and seventh timepoint for the wild-type. On the other hand, GSG is over-expressed at all time points except the seventh and BSG appears under-expressed at the seventh timepoint. Table 1 reports the most commonly occurring links during learning the TBNCs where *Time* denotes the temporal node.

Most of the identified links between genes are between genes functioning in different regenerative pathways. This suggests that the regenerative pathways are si-

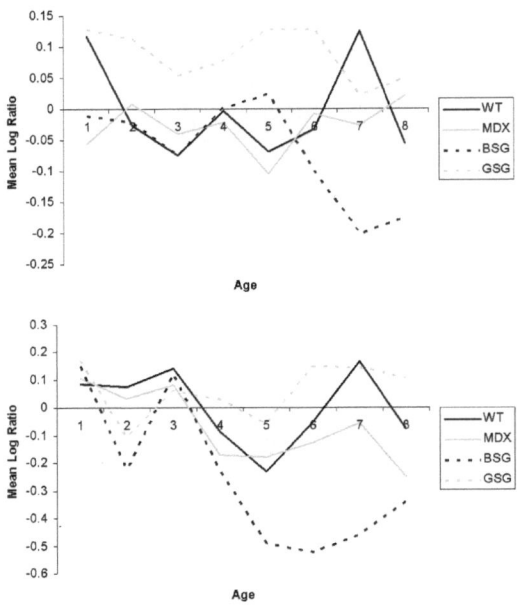

Fig. 5. Examples of genes found associated with the time node. NM008380 and NM008284 were found in 70% and 50% of runs, respectively

multaneously activated and potentially cooperate, while there is less evidence for synchronized activation of members of a specific pathway. Notch2 signalling maybe a notable exception, since DBNC links Notch2 (D32210) to several downstream targets: Dvl1 (NM010091), Hr (NM021877) Myog (D90156). A link between Numb-like protein (NM010950) and the low molecular weight GTPase Cdc42 (NM009861) was found by both DBNC and TNBC with high confidence. These genes seem to have similar temporal profiles of expression for BSG mice, discriminating them from other mouse models. Although a biological interaction between these proteins has not been identified yet, it may be rewarding to study this further.

4   Conclusions

In this paper we have investigated different Bayesian classifier architectures to classify time-series data with the aim of explaining the underlying structure of the data. The methods have been applied and tested on real-world microarray data in order to classify forms of muscular dystrophy. In addition, the classifiers have enabled us to explore the interactions between genes responsible for differentiating between the classes. We have found that incorporating both absolute temporal information which refers to some reference time (e.g. the birth of an organism) and relative temporal information in the form of dynamic links between variables results in more accurate classifiers.

In [12], we explored the use of simple Bayesian classifiers for selecting genes that differentiate between different classes. Due to the small sample sizes of our datasets, we only explored classifiers that assumed independence between genes. However, more interesting features could be discovered if we took into account the relationships between genes including the temporal ones. We intend to combine our work presented in this paper with our feature selection methods in order to identify combinations of genes that work together over time that determine the class of the gene profile in question. We also intend to incorporate expert knowledge by hard-wiring certain key relationships into the networks and experiment with a number of different biological datasets.

Acknowledgements

This work was supported in part by the BBSRC (grants EGM17735 and BBC5062641).

References

1. R.D. Cohn and K.P. Campbell. Molecular basis of muscular dystrophies. *Muscle Nerve*, 23:1456–1471, 2000.
2. I. Dalkilic and L.M. Kunkel. Muscular dystrophies: genes to pathogenesis. *Curr.Opin.Genet.Dev.*, 13:213–238, 2003.
3. N. Friedman. Learning the structure of dynamic probabilistic networks. In *Proceedings of the 14th Annual Conference on Uncertainty in AI*, pages 139–147, 1998.
4. N. Friedman, D. Geiger, and M. Goldszmidt. Bayesian network classifiers. *Machine Learning*, 29:131–163, 1997.
5. N. Friedman, M. Linial, I. Nachman, and D. Pe'er. Using Bayesian networks to analyze expression data. *Journal of Computational Biology,*, 7:601–620, 2000.
6. W. Lam and F. Bacchus. Learning Bayesian belief networks: an approach based on the MDL principle. *Computational Intelligence*, 10(4):269–293, 1994.
7. J. Pearl. *Probabilistic Reasoning in Intelligent Systems: Networks of Plausible Inference*. Morgan Kaufmann, 1988.
8. M. Ramoni, P. Sebstiani, and P. Cohen. Bayesian clustering by dynamics. *Machine Learning*, 47(1):91–121, 2002.
9. A. Tucker, V. Vinciotti, D. Garway-Heath, and X. Liu. A spatio-temporal bayesian network classifier for understanding visual field deterioration. *Artificial Intelligence in Medicine*, To Appear, 2005.
10. R. Turk, E. Sterrenburg, E.J. de Meijer, G.J.B. van Ommen, J.T. den Dunnen, and P.A.C. 't Hoen. Muscle regeneration in dystrophin-deficient mdx mice studied by gene expression profiling. *submitted*.
11. R. Turk, E. Sterrenburg, C.G.C. van der Wees, E.J. de Meijer, S. Groh, K. Campbell, S. Noguchi, G.J.B. van Ommen, J.T. den Dunnen, and P.A.C. 't Hoen. Common pathological mechanisms in mouse models for muscular dystrophies. *submitted*.
12. V. Vinciotti, A. Tucker, X. Liu, E. Panteris, and P. Kellam. Identifying genes with high confidence from small samples. In *Workshop on Data Mining in Functional Genomics and Proteomics, European Conference in Artificial Intelligence*, 2004.
13. E. C. Wit and J. D. McClure. *Statistics for Microarrays: Design, Analysis and Inference*. John Wiley & Sons, 2004.
14. M. Zou and S.D. Conzen. A new dynamic bayesian network approach for identifying gene regulatory networks from time course microarray data. *Bioinformatics*, 21:71–79, 2005.

Feature Discovery in Classification Problems

Manuel del Valle[1], Beatriz Sánchez[1,2],
Luis F. Lago-Fernández[1,3], and Fernando J. Corbacho[1,3]

[1] Escuela Politécnica Superior,
Universidad Autónoma de Madrid, 28049 Madrid, Spain
[2] Telefónica Investigación y Desarrollo,
C/ Emilio Vargas 6, 28043 Madrid, Spain
[3] Cognodata Consulting, C/ Caracas 23, 28010 Madrid, Spain

Abstract. In most problems of Knowledge Discovery the human analyst previously constructs a new set of features, derived from the initial problem input attributes, based on a priori knowledge of the problem structure. These different features are constructed from different transformations which must be selected by the analyst. This paper provides a first step towards a methodology that allows the search for near-optimal representations in classification problems by allowing the automatic selection and composition of feature transformations from an initial set of basis functions. In many cases, the original representation for the problem data is not the most appropriate, and the search for a new representation space that is closer to the structure of the problem to be solved is critical for the successful solution of the problem. On the other hand, once this optimal representation is found, most of the problems may be solved by a linear classification method. As a proof of concept we present two classification problems where the class distributions have a very intricate overlap on the space of original attributes. For these problems, the proposed methodology is able to construct representations based on function compositions from the trigonometric and polynomial bases that provide a solution where some of the classical learning methods, e.g. multilayer perceptrons and decision trees, fail. The methodology consists of a discrete search within the space of compositions of the basis functions and a linear mapping performed by a Fisher discriminant. We play special emphasis on the first part. Finding the optimal composition of basis functions is a difficult problem because of its nongradient nature and the large number of possible combinations. We rely on the global search capabilities of a genetic algorithm to scan the space of function compositions.

1 Introduction

Knowledge discovery from large data sets has become an increasingly important field of research due to the large potential to be tapped from many commercial, scientific and industrial databases. Nevertheless, in most cases the knowledge discovery processes are still quite costly due to the number of iterations that the human analysts have to perform over the discovery loop [3]. To reduce

A.F. Famili et al. (Eds.): IDA 2005, LNCS 3646, pp. 486–496, 2005.

this cost a number of tasks within the knowledge discovery loop can be partially automated. This is the case for feature selection and feature construction [4,5,6,7,8,11,14,15,17,19]. In most problems of knowledge discovery the human analyst previously constructs a new set of features, derived from the initial problem input attributes, based on a priori knowledge of the problem structure. These different features are constructed from different transformations which must be selected by the analyst.

For each new feature, a subset of input attributes must be selected (attribute selection) and a transformation to be applied to those attributes must be also selected (transformation selection). Both processes can be viewed as a search process and hence both can be automated to some degree by heuristic search. In this regard, domain knowledge can be introduced by choosing a set of bases that include transformations closer to the problem structure and heuristics that guide/bias the search process. The methodology described in this paper intertwines attribute selection and transformation selection in an overall search process implemented by a genetic algorithm. We have introduced the bias by means of the set of basis functions included. The different bases provide with different transformation properties, for instance the trigonometric basis introduces periodicity in an explicit manner. Furthermore, the architecture presented in this paper allows for basis function composition. Function composition enriches the expressive power by allowing the construction of features that have combined properties from the selected bases while giving rise to more compact representations. This is so since a basis closer to the problem structure gives rise to a more compact representation of the problem solution.

Classical methods for pattern classification are based on the existence of statistical differences among the distributions of the different classes. The best possible situation is perfect knowledge of these distributions. In such a case, Bayes classification rule gives the recipe to obtain the best possible solution. In real problems, however, class distributions are rarely available because the number of patterns is generally small compared with the dimensionality of the feature space. To tackle this problem many techniques of density estimation have been developed, both parametric and non-parametric [2]. When density estimation becomes too difficult, there is a variety of supervised learning algorithms, such as neural networks [1] or support vector machines [18], that try to find a non-linear projection of the original attribute space on to a new space where a simple linear discriminant is able to find an acceptable solution.

Let us assume a particular classification problem in which, when looking at the original attribute space, we observe an almost complete overlap among the class distributions. Following the Bayes rule, we see that for any point in this attribute space, the probabilities of belonging to any of the classes are all equal. We could be tempted to conclude that there is no solution to the problem better than choosing the class randomly. However, it could be that the overlapping is due to a bad representation of input data, and that there exists a transformation that separates the classes. We hypothesize that if such a transformation exists, there must exist a suitable basis in which it has a simple and compact expression. So solving such a problem can be reduced to finding the most appropriate basis or representation for the input data (with respect to the classification target). Once this representation is found, a linear discriminant will suffice to find a simple and compact solution. We propose an expansion of the Evolutionary Functional Link Networks (EFLNs) of Sierra et al. [16]. They used a genetic algorithm to construct polynomial combinations of the input attributes, which in turn constituted the input for a linear network. They obtained very compact solutions on problems from public databases. Here we incorporate other bases apart from the polynomial one. We use a genetic algorithm to perform both variable selection and search in the transformation space, and a Fisher discriminant that performs the final linear projection. We show that this approach is able to solve problems where other methods fail to find a solution, even when the overlap is so large that there are no apparent statistical differences among the classes. This overlap may be due simply to the fact that the original representation of data is not well suited to the problem. Actually, it is well known that many classification problems are solved only after the application of some "intelligent" transformations provided by a domain "expert". Here we want to go a step closer into the automatic selection of these intelligent transformations, by allowing the algorithm to search for the optimal basis.

2   Methodology

When facing a two-class classification problem, our starting point is the assumption that there exists a non-linear function that projects the input data onto a unidimensional space where a linear separator is able to discriminate among the two classes. This function must have a simple and compact form in some basis, so finding an appropriate set of basis functions will strongly contribute to the simplification of the problem: the final projection may be constructed as a linear combination of these non-linear transformations. Here we propose to explore jointly the Taylor and Fourier bases, as well as compositions of both. We use a genetic algorithm (GA) to construct the non-linear transformations that operate on the raw input data, and a Fisher discriminant to perform the linear projection on the transformed attributes. The separating threshold is selected to minimize the classification error. In this regard our approach follows on the work developed by [16] with the EFLN algorithm, introducing two main differences: first, we do not limit the transformations to polynomials, but we expand the representation capabilities by adding trigonometric functions; and second, the linear projection is performed by a Fisher discriminant, instead of a linear neural network. Both methods are equivalent and they do not suffer from local minima, but the Fisher projection provides the solution avoiding gradient descent optimization.

The proposed algorithm includes feature construction as well as feature selection. For the first task, it combines different bases of transformation (e.g. polynomial and trigonometric) to generate the input for the linear classifier. Feature selection is performed by the application of the genetic algorithm, which selects

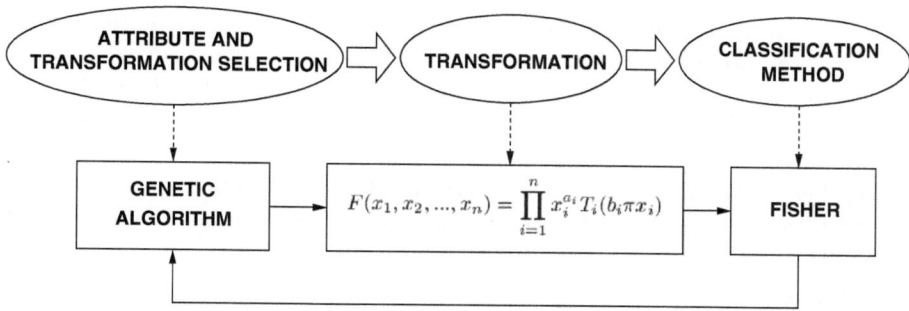

Fig. 1. Schematics of the overall methodology. The genetic algorithm evolves individuals consisting of different sets of transformations that operate on the input data. The transformed attributes are then fed into a Fisher discriminant whose error rate determines the fitness of the individual, used by the genetic algorithm to compute the next generation of transformation sets.

the best subsets of transformed variables by using the linear classifier error rate as the fitness criterion. Consequently our algorithm can be viewed as a wrapper method [7]. The general form for the input transformations operating in a n-dimensional feature space is given by the expression:

$$F(x_1, x_2, ..., x_n) = \prod_{i=1}^{n} y_i^{a_i} T_i(b_i \pi y_i) \tag{1}$$

where the x_i represent the original input variables, a_i and b_i are integer coefficients, and T_i is a trigonometric function (a sine or a cosine). Each y_i is either equal to x_i or to a new $F(x_1, x_2, ..., x_n)$. In this way compositions of polynomials and trigonometric functions can be constructed. Function composition strongly increments computational complexity and is not always allowed.

Our algorithm starts by generating K different function sets, each one composed of m functions as that of equation 1:

$$S_i = \{F_1^i, F_2^i, ..., F_m^i\}, i = 1, 2, ..., K \tag{2}$$

Each of these sets S_i corresponds to an individual in the initial population which will be evolved by the genetic algorithm. The fitness of the individual S_i is calculated as the classification error of a Fisher linear discriminant operating on the transformed attributes $F_1^i(\mathbf{x}), F_2^i(\mathbf{x}), ..., F_m^i(\mathbf{x})$. Note that an exhaustive search over the space of input transformations would be computationally too expensive and would not scale properly on the number of input variables. This fact, together with the absence of gradient information, makes the use of an evolutionary approach very appropriate. In figure 1 we show a scheme of the algorithm, which is briefly described below:

1. *Initialize the first population of individuals randomly, and set the parameters for the GA, such as the number of iterations and the mutation probability.*
2. *For each evolution iteration:*
   (a) For each individual:

      i. *Generate the new features applying the input transformations to the original attributes.*

     ii. *Evaluate its fitness value as the classification error of the Fisher Linear Discriminant applied to the transformed features on the training and validation data sets.*

  (b) *Select the lowest error individuals for the next iteration.*

  (c) *Generate a new population applying genetic operators and the individuals selected in (b).*

3. *Evaluate the most accurate individual on the test data set.*

3   Test Cases

We have applied the previous methodology to two different synthetic data sets. Both of them consist of two classes, A and B, in a two-dimensional input space, given by the attributes x and y. The two problems present the following properties: (i) there exists an appropriate non-linear transformation that is able to separate the classes with no error; and (ii) in the original input space the classes present a very high overlap and, given the number of examples, seem to follow the same distribution. This last fact makes the problems particularly difficult to solve. For both problems we present the results of our algorithm in comparison with the results obtained with other classification methods, namely multilayer perceptrons trained with backpropagation, decision trees trained with the C4.5 algorithm, and evolutionary FLNs that use the polynomial basis. The backpropagation algorithm was tested using networks of one single hidden layer, with different number of hidden units (ranging from 3 to 10) with a sigmoidal activation function. Different values for the learning rate between 0.01 and 0.3 were tried. For the decision trees, we used Quinlan's C4.5 algorithm [13] with probabilistic thresholds for continuous attributes, windowing, a gain ratio criterion to select tests and an iterative mode with ten trials. Finally, the evolutionary FLN was trained as described in [16], with polynomials of up to degree 3.

3.1   Case 1

The first test case we consider consists of 2000 patterns, 1000 of class A and 1000 of class B, defined in the interval $[0 \le x \le 100, 0 \le y \le 100]$. Class A patterns are defined in the following way:

$$(x, y) \in A \longleftrightarrow mod(int(x), 2) = mod(int(y), 2) \tag{3}$$

where $int(x)$ is the integer part of x and $mod(x, 2)$ is the remainder of $x/2$. Class B patterns are those that do not satisfy the equality in eq. 3. We can imagine the input space as a big chess board where class A patterns occupy black squares and class B patterns occupy white ones. In principle, this problem seems not particularly difficult to solve. However, as far as the number of patterns is

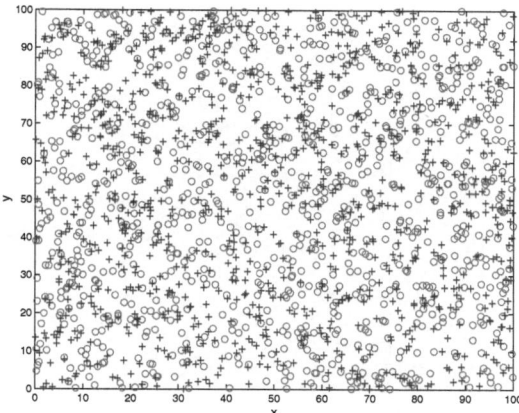

Fig. 2. Input patterns for test case 1, consisting of two classes, A and B, and two attributes, x and y. The problem data consist of 1000 patterns of class A (circles) and 1000 patterns of class B (crosses). Apparently the two classes follow the same (uniform) distribution in the considered interval, and classical methods will have great difficulties to deal with this problem.

very small compared with the number of squares, it becomes more complicated to discover the hidden structure. Here we have forced this situation, and the two classes appear to follow the same (uniform) statistical distribution in the considered interval (see figure 2).

In table 1 we show the results obtained by the different tested classification methods when trying to solve this problem. None of the tested strategies achieves a successful result. All of them achieve error rates close to a 50% on the test set. This means that they are not performing much better than selecting the class randomly. The difficulty these traditional methods are confronting is due to the high overlap between the two classes. Note that for an absolute class overlap, even the best (Bayes) class estimator fails. However Bayesian decision theory assumes perfect knowledge of class distributions, which is not the present case. In fact, we know that below the apparent class mixing there is a hidden structure that the tested methods are not able to discover when just focusing on the original input space.

Table 1. Comparison of performances of various classification methods on the problem of test case 1

Algorithm	Train Error %	Test Error %
Backpropagation	51.1	48.7
C4.5	49.1	51.4
EFLN	46.1	43.8

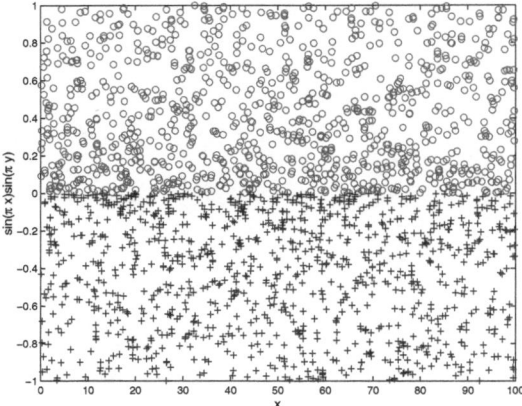

Fig. 3. Plot of $sin(\pi x)sin(\pi y)$ vs x for the patterns of the problem of test case 1. The final transformation discovered by the proposed algorithm allows for a linear separation of the two classes.

We applied the algorithm of section 2 to this problem, using populations of up to 50 individuals, each one consisting of a set of $m = 3$ input transformations. Function compositions were not allowed. The optimization was performed using a standard GA package [10]. The different trials we ran converged quickly (in no more than 50 GA iterations) to optimal solutions with a perfect class separation. As an example we show the outcome of one of the trials, for which the best individual corresponded to the following set of input transformations:

$$\begin{pmatrix} 0 \\ x \sin(2\pi x) \sin(\pi y) \\ \sin(\pi x) \sin(\pi y) \end{pmatrix}$$

These transformations constitute the input for the Fisher discriminant. The resulting Fisher projection is given by the vector:

$$\begin{pmatrix} 0 & 0 & -9.33 \end{pmatrix}$$

So the final transformation reached by the algorithm is $-9.33 sin(\pi x) sin(\pi y)$. Note that the Fisher projection is ignoring all the terms except the third one. As shown in figure 3, this transformation allows a perfect linear separation of the two classes.

3.2   Case 2

Let us consider a second test case constructed in a similar manner as test case 1. As before, it consists of 2000 patterns in a two-dimensional input space, defined in the interval $[0 \le x \le 100, 0 \le y \le 100]$. We select 1000 patterns of each class. The patterns of class A are defined as:

$$(x, y) \in A \longleftrightarrow mod(int(x^2 y^2), 2) = mod(int(y), 2) \tag{4}$$

where $int(x)$ and $mod(x, 2)$ are as before. Class B patterns are those that do not satisfy the equality in eq. 4. As with the previous example, there exists a non-linear transformation that solves this problem with 0 error. However in this case the required transformation involves a composition of polynomial and trigonometric functions. If we plot all the patterns in the original space we obtain the result shown in figure 4. As before, in spite of the deterministic nature of the problem, there appears to be an absolute class mixing. This is due to the relatively small number of patterns.

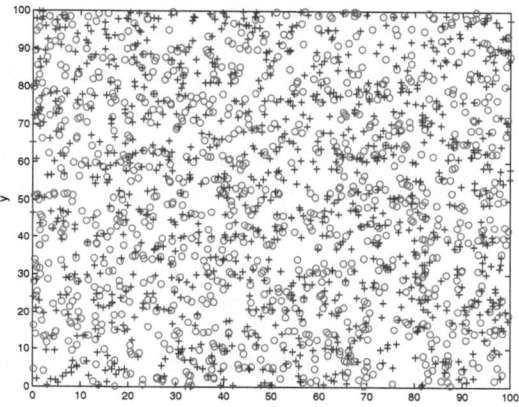

Fig. 4. Input patterns for test case 2, consisting of two classes, A and B, and two attributes, x and y. The problem data consist of 1000 patterns of class A (circles) and 1000 patterns of class B (crosses).

We applied the same three traditional methods to this new problem, obtaining the results shown in table 2. Classification error rates are in all cases close to 50%, which indicates that no improvement with respect to random class selection is achieved.

Table 2. Comparison of performances of various classification methods on the problem of test case 2

Algorithm	Train Error %	Test Error %
Backpropagation	50.4	49.5
C4.5	46.4	50.2
EFLN	44.6	46.4

Finally we tested our algorithm. We used the same experimental conditions as for test case 1, but using sets of $m = 6$ input transformations and allowing function composition. All the trials we ran converged to the optimal solution, the outcome of one of them is shown below:

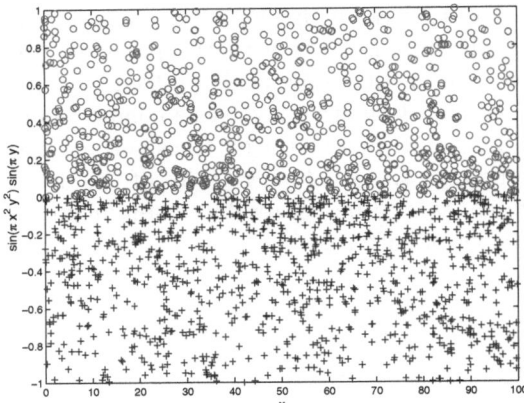

Fig. 5. Plot of $sin(\pi x^2 y^2)sin(\pi y)$ vs x for the patterns of the problem of test case 2. The final transformation discovered by the proposed algorithm allows for a linear separation of the two classes.

$$
\begin{pmatrix}
\sin(\pi x^2 y^2)\sin(\pi y) \\
0 \\
x^2\sin(3\pi x^2 y^2)y\sin(2\pi x^2 y^2\sin(3\pi x)\sin(\pi y)) \\
0 \\
0 \\
\cos(3\pi x^2 y^2)
\end{pmatrix}
$$

The corresponding Fisher projection is given by:

$$
\begin{pmatrix} -9.53 & 0 & 0 & 0 & 0 & 0 \end{pmatrix}
$$

Which produces the final transformation $-9.53\sin(\pi x^2 y^2)\sin(\pi y)$ that separates the two classes with no error (see figure 5).

4 Conclusions

This paper presents a proof of concept for the construction of near-optimal problem representations in classification problems, based on the combination of functions selected from an initial family of transformations. The selection of an appropriate transformation allows the solution of complex nonlinear problems by a simple linear discriminant in the newly transformed space of attributes. The proposed approach has been tested using two complex synthetic problems that are not properly solved by other standard classification algorithms. A few trials on classification problems from the UCI repository [12] have also been performed, where our methodology seems to reach error rates similar to other methods. However the solutions obtained are highly complex and difficult to interpret. Note that we have not included any mechanisms to control the complexity of the solutions.

Work in progress includes the introduction of a more extensive family of basis functions that will allow for the construction of a wider repertoire of problem representations. Additionally, mechanisms to control the combinatorial explosion in the space of representations and the complexity of solutions will be analyzed. Additional work in progress also includes information/statistical measures that allow to uncover the structural/statistical properties of the input attributes and this in turn provides additional heuristics over which transformations to select.

Other advantages of the proposed method are that a closer, more compact problem representation usually allows for easier model interpretation [16], and, hence, a deeper understanding of the structure and mechanisms underlaying the problem under study. Related work on the extraction of hidden causes [9], which provide the generative alphabet, will be farther explored.

Acknowledgments

We want to thank P. Hoelgaard, M. Sánchez-Montañés and A. Sierra for very interesting comments and discussions, and Ministerio de Ciencia y Tecnología for financial support (BFI2003-07276).

References

1. Bishop, C.M.: Neural Networks for Pattern Recognition. Oxford Univ. Press (1995)
2. Duda, R.O., Hart, P.E., Stork, D.G.: Pattern Classification. John Wiley and Sons (2001) 84–214
3. Fayyad, U.M., Piatetsky-Shapiro, G., Smyth, P.: From data mining to knowledge discovery: an overview. In: Advances in Knowledge Discovery and Data Mining, U.M. Fayyad, G. Piatetsky-Shapiro, P. Smyth, and R. Uthurusamy (eds.). Menlo Park, CA, AAAI Press (1996) 1–34
4. Flach, P.A., Lavrac, N.: The role of Feature Construction in Inductive Rule Learning. ICML (2000) 1–11
5. Guyon, I., Elisseeff, A.: An Introduction to Variable and Feature Selection. Machine Learning (2003) 1157–1182
6. Kramer, S.: Demand-Driven Construction of Structural Features in ILP. ILP **2157** (2001) 132–141
7. Kohavi, R., John, G.H.: Wrappers for Feature Subset Selection. Artificial Intelligence **97** (1–2) (1997) 273–324
8. Kudenko, D., Hirsh, H.: Feature Generation for Sequence Categorization. American Association for Artificial Intelligence (1998) 733–738
9. Lago-Fernández, L.F., Corbacho, F.J.: Optimal Extraction of Hidden Causes. LNCS **2415** (2002) 631–636
10. Levine, D.: Users Guide to the PGAPack Parallel Genetic Algorithm Library. T.R.ANL-95/18 (1996)
11. Pagallo, G.: Boolean Feature Discovery in Empirical Learning. Machine Learning **5** (1) (1990) 71–99
12. Prechelt, L.: Proben1: A set of Neural Network Benchmark Problems and Benchmarking Rules. Tech. Rep. 21/94, Fakultät für Informatik, Univ. Karlsruhe, Karlsruhe, Germany (1994)

13. Quinlan, J.R.: C4.5: Programs for Machine Learning. Morgan Kaufmann Publishers Inc. (1992)
14. Ragavan, H., Rendell, M.: Lookahead Feature Construction for Learning Hard Concepts. ICML (1993) 252–259
15. Rennie, J.D.M., Jaakkola, T.: Automatic Feature Induction for Text Classification. MIT Artificial Intelligence Laboratory Abstract Book (2002)
16. Sierra, A., Macías, J.A., Corbacho, F.: Evolution of Functional Link Networks. IEEE Trans. Evol. Comp. **5** (1) (2001) 54–65
17. Utgoff, P.E., Precup, D.: Constructive Function Approximation. In: Feature Extraction, Construction and Selection: a Data Mining Perspective, H. Liu and H. Motoda (eds.). Boston, Kluwer Academic (1998) 219-235
18. Vapnik, V.N.: Statistical Learning Theory. John Wiley and Sons (1998)
19. Zucker, J.D., Ganascia, J.G.: Representation Changes for Efficient Learning in Structural Domains. ICML (1996) 543–551

A New Hybrid NM Method and Particle Swarm Algorithm for Multimodal Function Optimization

Fang Wang, Yuhui Qiu, and Yun Bai

Intelligent Software and Software Engineering Laboratory,
Southwest-China Normal University, Chongqing, 400715, China
{teresa78, yhqiu, baiyun}@swnu.edu.cn

Abstract. In this paper, we introduce a hybrid technique based on particle swarm optimization (PSO) algorithm combined with the nonlinear simplex search method. This approach is applied to multimodal function optimizing tasks. To evaluate its reliability and efficiency, we empirically compare the performance of two variants of the Particle Swarm Optimizer with our hybrid algorithm. The computational results obtained in experiments on large variety of test functions indicate that the hybrid algorithm is competitive with other techniques, and can be successfully applied to more demanding problem domains.

1 Introduction

Particle Swarm Optimization (PSO) [1], which is inspired by the analogy of social behavior of insects and animals, is a recently proposed meta-heuristic algorithm that can be used to find approximate solutions to difficult continuous function optimization tasks. Since its introduction in 1995, the Particle Swarm Optimization paradigm has undergone various modifications and improvements, and has been successfully applied to wide range of industrial fields [2], [3], [4]. It is known from the literature that the convergence rate of PSO is typically slower than those of local search techniques [5]. As an evolutionary computation technique, PSO is severely limited by high computational cost in solving multimodal optimization problems.

The nonlinear simplex search method proposed by Nelder and Mead (NM method) is a simple direct search technique that has been widely used in various unconstrained optimization problems [6]. It is very easy to implement and does not need any derivative information of the objective function. But the NM method is very sensitive to the initial points and is not guaranteed to obtain global optima [7].

Hybrid PSO algorithms with the NM method are proved to be superior to the original two techniques and have many advantages over other heuristic algorithms [8], such as hybrid GA, continuous GA, simulated annealing (SA), and tabu search (TS). Generating initial swarm by the simplex method (SM) might improve, but is not satisfying for multimodal function optimizing tasks [9]. Developing the simplex search as an operator during the optimization may increase the computational complex considerably [8].

In this paper, the nonlinear simplex method is adopted at late stage of the Canonical PSO algorithm when particles hunt quite close to the extrema. In the hybrid ap-

A.F. Famili et al. (Eds.): IDA 2005, LNCS 3646, pp. 497–508, 2005.

proach, PSO contributes to ensure that the search is not likely to be immersed in local optima, while the simplex search makes the search converge faster than pure PSO procedures. Experimental results on several famous test functions show that this is a very promising way to increase both the convergence speed and the success rate significantly of multimodal function optimization.

We briefly introduce the NM method and the PSO algorithm in section 2 and 3; in section 4, the proposed hybrid algorithm and experimental design is described; correlative results of experiments are exhibited in section 5. The paper comes to the end with terse conclusions and some ideas for further work.

2   The NM Method

The basic simplex method was presented by Spendley et al in 1962, and then improved by Nelder and Mead [6], to what is called the nonlinear simplex method. Since its publication in 1965, the NM simplex algorithm has become one of the most widely used local direct search methods for nonlinear unconstrained optimization.

The NM method attempts to minimize a scalar-valued nonlinear function of D real variables using only function values, without any derivative information. At each iteration, the simplex-based direct search method begins with a simplex, specified by its $D+1$ vertices and the values of objective function. One or more test points are computed, along with their function values, and the iteration terminates with bounded level sets.

A D-dimensional simplex is a geometrical figure consisting of $D+1$ vertices (D-dimensional points) and all their interconnecting segments, polygonal faces etc. We consider only simplexes that are non-degenerated, i.e., that enclose a finite inner D-dimensional volume.

The nonlinear simplex search procedure starts with an initial simplex which is generated using the found minimum as one of its vertices and generating the rest D points randomly. Then it takes a series of steps to rescale the simplex: first, it finds the points where the objective function is highest (the least favorable trial W) and lowest (the most favorable trial B); then it reflects the simplex around the high point to point R. If the solution is better, it tries an expansion in that direction; else if the solution is worse than the second-highest (next-to-the worst) point S, it tries an intermediate point. When the method reaches a "valley floor", the simplex is contracted in the transverse direction in order to ooze down the valley or it can be contracted in all directions, pulling itself in around its lowest point, and started again.

At each step, the rejected trial W is replaced by one of the following trials on conditions that:

$$R = \overline{C} + \alpha(\overline{C} - W), \ \ if \ \ f_B < f_R < f_S$$
$$E = \overline{C} + \gamma(\overline{C} - W), \ \ if \ \ f_R < f_B$$
$$C+ = \overline{C} + \beta^+(\overline{C} - W), \ \ if \ \ f_S < f_R < f_M$$
$$C- = \overline{C} - \beta^-(\overline{C} - W), \ \ if \ \ f_R > f_M$$

The different test points of a simplex are shown in Figure 1.

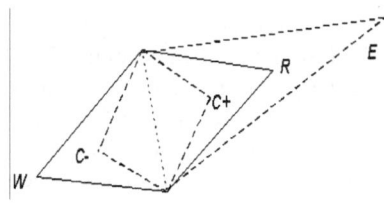

Fig. 1. Different simplex test points from the rejected trial condition in two dimension space. W= the rejected trial, R = reflection, E = expansion, C+ = positive contraction, C- = negative contraction.

Where $\overline{C}$ is the centroid of the remaining vertices; α, γ, β^+ and β^- is coefficients of reflection, expansion, positive contraction and negative contraction; f_B, f_S, f_W and f_R are the values of object function on point B, S, W and R respectively.

3 The Particle Swarm Algorithm

Particle Swarm Optimization (PSO) algorithm proposed by Kennedy and Eberhart is one of the latest evolutionary techniques for continuous function optimization tasks [10]. In comparison with previous population-based evolutionary approaches, such as genetic algorithm (GA) and evolutionary strategy (ES), PSO does not implement the filtering operation, instead, a simulated social behavior, where members of a group tend to follow the lead of the best of the group, is adopted by PSO [11]. Generally speaking, PSO is applicable to most optimization problems which can be transferred into a general form of continuous global optimization problems.

The theory of PSO describes that each particle flies through the multidimensional search space while the particle's velocity and position are updated based on the best previous performance of the particle and of the particle's neighbors, as well as the best performance of particles in the entire population so that each particle can benefit from the current search results of other particles.

In the original PSO formulae, particle i is denoted as $X_i=(x_{i1},x_{i2},...,x_{iD})$, which represents a potential solution to a problem in D-dimensional space. Each particle maintains a memory of its previous best position, $P_i=(p_{i1},p_{i2},...,p_{iD})$, and a velocity along each dimension, represented as $V_i=(v_{i1},v_{i2},...,v_{iD})$. At each iteration, the P vector of the particle with the best fitness in the local neighborhood, designated g, and the P vector of the current particle are combined to adjust the velocity along each dimension, and that velocity is then used to compute a new position for the particle.

The evolutionary equations of the swarm are:

$$v_{id} = w*v_{id}+c1*rand()*(p_{id} - x_{id})+c2*Rand()*(p_{gd} - x_{id}) \tag{1}$$
$$x_{id} = x_{id}+v_{id} \tag{2}$$

Constants c1 and c2 determine the relative influence of the social and cognition components (learning rates), which often both are set to the same value to give each component equal weight. Rand() and rand() are random values in the rang (0,1). A constant, V_{max}, was used to limit the velocities of the particles. The parameter w, which was introduced as an inertia factor, can dynamically adjust the velocity over time, gradually focusing the PSO into a local search [12], [13], [14].

We can see that only a very few parameters are need to adjust in PSO, which makes it very attractive in the literature of meta-heuristic algorithms. Maurice Clerc has derived a constriction coefficient K, a modification of the PSO that runs without V_{max}, reducing some undesirable explosive feedback effects. The constriction factor is computed as [15]:

$$K = \frac{2}{\left|2 - \varphi - \sqrt{\varphi^2 - 4\varphi}\right|}, \quad \varphi = c_1 + c_2, \quad \varphi > 4 \tag{3}$$

With the constriction factor K, the PSO formula for computing the new velocity is:

$$v_{id} = K*(v_{id}+c1*rand()*(p_{id} - x_{id})+c2*Rand()*(p_{gd} - x_{id})) \tag{4}$$

Carlisle and Doziert investigated the influence of different parameters in PSO, selected c1=2.8, c2=1.3, population size as 30, and proposed the Canonical PSO [16].

4   The Proposed Algorithm and Experimental Design

At late stage of PSO running, promising regions of solutions have been located. Applying the nonlinear simplex method to enhance exploitation search at this stage is capable

Table 1. Parameters for each test function

Function	GM	Error Goal	Xmin	Xmax
F1: Branin$_2$	0.397887	10^{-6}	-5	15
F2: Easom$_2$	-1	10^{-8}	-100	100
F3: Shubert$_2$	-186.7309	10^{-7}	-10	10
F4: Zakharov$_2$	0	10^{-6}	-5	10
F5: Griewank$_2$	1	10^{-7}	-300	600
F6: H$_{3,4}$	-3.86343	10^{-7}	0	1
F7: S$_{4,10}$	-10.53641	10^{-5}	0	10
F8: H$_{6,4}$	-3.32237	10^{-5}	0	1
F9: Rastrigin$_{10}$	0	10	-5.12	5.12
F10: Griewank$_{10}$	9	10^{-1}	-300	600
F11: Zakharov$_{10}$	0	10^{-1}	-5	10

*Note: The subscript of each function name denotes its dimension.

Table 2. Pseudo code of the proposed hybrid algorithm

Randomly generate a swarm of size PopSize on the range [Xmin..Xmax], set p_i to x_i, set v_i to random values.

Calculate fitness of all particles, set pbesti to fitnessi, and find particle gbest with the best pbest.

For(iter=1:iter$_{max}$) do

    If(8) algorithm successfully terminate

    End if

    For each particle i with fitness<DRadius

        Generate an initial simplex randomly with the mean of x_i and standard deviation of DRadius

        For(NMiter=1:NMitermax) do

            Apply the nonlinear simplex search operator to the simplex

            Replace particle i with the update

            if (8) algorithm successfully terminate

            End if

        End for

        Update pbest$_i$, and gbest of the swarm

    End for

    Update v_i, x_i

    Calculate fitness of all particles

    Update pbest of all particles, and gbest of the swarm

End for

---

of improving the solution quality and convergence rate, utilizing the merits of NM method's accurate exploitation abilities and PSO algorithm's finer exploration abilities.

We propose a hybrid NM Method PSO, which isolates a particle and apply the nonlinear simplex search to it when it reaches quite close to the extrema (within the diversion radius). If the particle "lands" within a specified precision of a goal solution (error goal) during the simplex search procedure, a PSO process is considered to be successful, otherwise it may be laid back to the swarm and start the next PSO iteration.

The diversion radius is computed as:

$$DRadius = ErrorGoal + \delta \tag{6}$$

$$\delta = \begin{cases} 100 * ErrorGoal, & if \quad ErrorGoal <= 10^{-4} \\ 0.01 * ErrorGoal, & otherwise \end{cases} \tag{7}$$

In a SM process, an initial simplex is consists of the isolated particle i and other D vertices randomly generated with the mean of X_i and standard deviation of DRadius.

The stopping criterion is defined as:

$$|f(gbest) - GM| < ErrorGoal \tag{8}$$

Where GM (theoretic global minimum) and Error Goal (search accuracy) for each test function is defined in Table 1. In order to get quicker convergence, we set maximums of iterations in all experiments as a second stopping criterion. In the later case, we consider the search process to be failed.

The pseudo code of the proposed hybrid algorithm is shown in Table 2.

The benchmark functions on which the proposed algorithm has been tested and compared to other methods in the literature, as well as the equation of each one and the corresponding parameters are listed in the appendix [8], [9], [17]. To avoid the influence of different initial swarms and make unbiased comparisons, we implement 200 experiments for each test, and demonstrate the statistical results in Section 5. The maximum number of PSO iterations is set to be 500 (iter$_{max}$), swarm size is 30 (PopSize). Parameters used in the NM method are: α=1.0, γ=2.0, $\beta^+ = \beta^-$=0.5, NMitermax=50. We used unsymmetrical search space, as shown in Table 1. All algorithms are programmed in Matlab 7.0 and the simulations are executed on a Pentium IV 2.8G with memory capacity of 512 MB under Windows2000 Professional Operating System.

5 Experimental Results

We have conducted large variety of experiments on over 20 test functions, among which most functions are multimodal, abnormal or computational time consuming, and can hardly get favorable results by current optimizers. Comparisons to several other published methods are also investigated, among which we only select NS-PSO and Canonical PSO to list in the paper. NS-PSO is another SM hybrid PSO proposed by Parsopoulos and Vrahatis, which can improve the overall performance of PSO algorithm by generating initial swarm with the simplex method [9]. The Canonical PSO [16] exhibits quite promising on most optimization problems in the literature, which has been used most to accomplish comparison experiments in this research field.

The rate of success, mean function evaluations, average optima and total CPU time for each test are listed in table 3 and table 4, from which we can see that the overall performance of HNMPSO algorithm is apparently superior to other 2 algorithms taken from the literature in terms of success rate, solution quality and convergence speed as well.

The averages of the objective functions evaluation numbers of HNMPSO for lower dimension test functions are all considerably less than those of the pure PSO algorithm (CPSO), indicating that the efficiency of PSO can be improved by the hybrid method.

Table 3. The rate of success and mean function evaluations for each test function

	Rate of success			Mean function evaluations		
	HNMPSO	NS-PSO	CPSO	HNMPSO	NS-PSO	CPSO
F1	1	0.89	1	1466.1	2749.5	1686.5
F2	1	0.98	1	2599.3	3823.1	2951.3
F3	0.995	0.01	0.025	42019	14903	14727
F4	1	0.995	1	1213.9	1089.8	1389
F5	0.775	0.73	0.79	7952.4	8831.9	8047.5
F6	1	0.125	0.055	3889.1	13338	14284
F7	0.62	0.355	0.56	8348.3	10633	8683.5
F8	0.525	0.69	0.5	9082.1	6135	8591.7
F9	0.945	0.84	0.96	5687.1	6012.9	5418.8
F10	0.85	0.84	0.825	22552	7001.7	7483.6
F11	1	0.875	0.985	22036	10026	9619.6

*Note:
HNMPSO: the proposed algorithm
NS-PSO: another SM hybrid PSO proposed by Parsopoulos and Vrahatis [9]
CPSO: the Canonical PSO, Carlisle A [16]

For function Shubert, which has 760 local minima and 18 global minima, HNMPSO possesses absolute predominance over other alternatives. As to high dimension function optimizing, the numbers of evaluations and total CPU time are larger than the other methods due to its high computational expense, but the ratios of success and the solution qualities are absolutely perfect.

From the results of computational experiments and the analyses above, it can be anticipated that the proposed HNMPSO approach remains quite competitive as compared to the other published methods.

6   Conclusions and Future Work

In this paper, we thoroughly investigate a new hybrid Particle Swarm Optimization algorithm, which applies the nonlinear simplex method at late stage of PSO running when the most promising regions of solutions have been located. We implement wide variety of experiments on well-known benchmark functions to test the proposed algorithm. The results compared to other competitive methods that this method is great potential in solving continuous multimodal functions.

Table 4. The average optima and total CPU time for each test function

	Average optima			Total CPU time		
	HNMPSO	NS-PSO	CPSO	HNMPSO	NS-PSO	CPSO
F1	0.39789	0.62566	0.39789	6.1406	10.281	5.75
F2	-1	-0.98489	-1	9.6563	13.609	10.109
F3	-186.73	-174.02	-186.731	173.58	68.781	67.7344
F4	4.567e-7	0.14469	4.789e-7	5.625	5.375	5.5625
F5	1.0015	1.0022	1.0014	40.438	44.953	40.047
F6	-3.8634	-3.8486	-3.8634	24.875	69.078	72.516
F7	-8.0113	-5.6708	-7.642	150.3	191.91	155.06
F8	-3.265	-3.2772	-3.2598	49.188	32.797	44.922
F9	9.7901	10.194	9.6313	31.25	33.078	29.281
F10	9.0971	9.0994	9.0993	104.58	41.766	43.875
F11	9.1628e-7	4.0497	0.44228	98.36	43.75	41.25

Future work may focus on investigating the influence of scaling behavior, accelerating the convergence for high dimension problems, extending the approach to constrained multi-objective optimization, and developing parallel algorithm of this hybrid technique.

References

1. Kennedy, J. and Eberhart, R. C.: Particle swarm optimization. Proceedings of IEEE International Conference on Neural Networks, Piscataway, NJ (1995) 1942-1948
2. Yoshida, H., Kawata, K., Fukuyama, Y. and Nakanishi, Y.: A particle swarm optimization for reactive power and voltage control considering voltage stability. Proceedings of International Conference on Intelligent System Application to Power Systems, Rio de Janeiro, Brazil (1999) 117–121
3. Parsopoulos, K. E. and Vrahatis, M. N.: Recent approaches to global optimization problems through particle swarm optimization. Natural Computing, Vol. 1 (2002) 235-306
4. Hu, X., Eberhart, R. C., and Shi, Y. H.: Engineering optimization with particle swarm. Proceedings of the IEEE Swarm Intelligence Symposium 2003, Indianapolis, Indiana, USA (2003) 53-57
5. Clerc, M. and Kennedy, J.: The particle swarm-explosion, stability, and convergence in a multidimensional complex space. IEEE Transactions on Evolutionary Computation, Vol. 6, No. 1 (2002) 58-73

6. Nelder, J. and Mead, R: A simplex method for function minimization. Computer Journal, Vol. 7 (1965) 308-313.
7. Shi, Y. H. and Eberhart, R. C.): Empirical study of particle swarm optimization. Proceedings of the IEEE Congress on Evolutionary Computation, Piscataway, NJ (1999) 1945-1950
8. Shu-Kai S. Fan, Yun-Chia Liang, and Erwie Zahara: Hybrid Simplex Search and Particle Swarm Optimization for the Global Optimization of Multimodal Functions. Engineering Optimization, Vol. 36, No. 4 (2004) 401-418
9. Parsopoulos, K. E. and Vrahatis, M. N.: Initializing the particle swarm optimizer using the nonlinear simplex Method. In Grmela, A. and Mastorakis, N. E. (eds.) Advances in Intelligent Systems, Fuzzy Systems, Evolutionary Computation, WSEAS Press (2002) 216-221
10. J. Kennedy and R. Eberhart: Swarm Intelligence. Morgan Kaufmann Publishers (2001)
11. E. Bonabeau, M. Dorigo, and G. Theraulaz: Swarm Intelligence: From Natural to Artificial Systems. Oxford Press (1999)
12. Shi, Y. H. and Eberhart, R. C.: Parameter selection in particle swarm optimization. Evolutionary Programming VII: Proceedings of the Seventh Annual Conference on Evolutionary Programming, New York (1998) 591-600
13. Shi, Y. H. and Eberhart, R. C.: A modified particle swarm optimizer. Proceedings of the IEEE Congress on Evolutionary Computation, Piscataway, NJ (1998) 69-73
14. Lagarias, J.C., J. A. Reeds, M. H. Wright, and P. E. Wright: Convergence Properties of the Nelder-Mead Simplex Method in Low Dimensions. SIAM Journal of Optimization, Vol. 9, No. 1 (1998) 112-147
15. Clerc, M.: The swarm and the queen: towards a deterministic and adaptive particle swarm optimization. Proceedings of the IEEE Congress on Evolutionary Computation (1999) 1951-1957
16. Carlisle, A. and Doziert, G.: An off-the-shelf PSO. Proceedings of the Workshop on Particle Swarm Optimization, Indianapolis (2001)
17. Levy A, Montalvo A, Gomez S, and et al.: Topics in Global Optimization. Springer-Verlag, New York (1981)
18. Birge, B.: PSOt: a particle swarm optimization toolbox for use with MATLAB. Proceedings of the IEEE Swarm Intelligence Symposium 2003, Indianapolis, Indiana, USA (2003) 182-186

Appendix: List of Test Functions

Branin

$$f(x) = (x_2 - \frac{5}{4\pi^2}x_1^2 + \frac{5}{\pi}x_1 - 6)^2 + 10(1 - \frac{1}{8\pi})\cos(x_1) + 10$$

Dimensions: 2
X1: [-5, 10]
X2: [0, 15]
3 global minima
Theoretic optimum: 0.397887
Error Goal: 10^{-5}

Easom

$$f(x) = -\cos(x_1)\cos(x_2)e^{-((x_1-\pi)^2 + (x_2-\pi)^2)}$$

Dimensions: 2
X: [-100, 100]
1 global minimum
Several local minima
Theoretic optimum: -1
Error Goal: 10^{-6}

Shubert

$$f(x) = \sum_{i=1}^{5}(i\cos((i+1)x_1+i))\sum_{j=1}^{5}(j\cos((j+1)x_2+j))$$

Dimensions: 2
X: [-10, 10]
18 global minima
760 local minima
Theoretic optimum: -186.7309
Error Goal: 10^{-6}

Zakharov

$$f(x) = \sum_{i=1}^{n} x_i^2 + (\sum_{i=1}^{n} 0.5ix_i)^2 + (\sum_{i=1}^{n} 0.5ix_i)^4$$

Dimensions: 2, 10
X: [-5, 10]
1 global minimum
Several local minima
Theoretic optimum: 0
Error Goal: 10^{-5}

Griewank

$$f(x) = \sum_{i=1}^{n} \frac{x_i^2}{4000} - \prod_{i=1}^{n} \cos(x_i/\sqrt{i}) + 1$$

Dimensions: 2, 10
X: [-300, 600]
1 global minimum
Several local minima
Theoretic optimum: n-1
Error Goal: 10^{-7}, 10^{-1}(for 10 dimension)

Rastrigin

$$f(x) = \sum_{i=1}^{n}(x_i^2 - 10\cos(2\pi x_i) + 10)$$

Dimensions: 10
X: [-5.12, 5.12]
1 global minimum
More than 50 local minima for 2 dimension
Theoretic optimum: 0
Error Goal: 10

Hartmann ($H_{3,4}$)

$$f(x) = -\sum_{i=1}^{4} c_i \exp(-\sum_{j=1}^{3} a_{ij}(x_j - p_{ij})^2)$$

Dimensions: 3
X: [0, 1]
1 global minimum
4 local minima (pi)
Theoretic optimum: -3.86343
Error Goal: 10^{-6}

i	a_{ij}			c_i	p_{ij}		
1	3.0	10.0	30.0	1.0	0.3689	0.1170	0.2673
2	0.1	10.0	35.0	1.2	0.4699	0.4387	0.7470
3	3.0	10.0	30.0	3.0	0.1091	0.8732	0.5547
4	0.1	10.0	35.0	3.2	0.0381	0.5743	0.8827

Hartmann ($H_{6,4}$)

$$f(x) = -\sum_{i=1}^{4} c_i \exp(-\sum_{j=1}^{6} a_{ij}(x_j - p_{ij})^2), \quad c = (1.0,\ 1.2\ ,3.0,\ 3.2)$$

Dimensions: 6
X: [0, 1]
1 global minimum
4 local minima (pi)
Theoretic optimum: -3.32237
Error Goal: 10^{-4}

j	a_{ij}				p_{ij}			
1	10.0	0.05	3.00	17.0	0.1312	0.2329	0.2384	0.4047
2	3.0	10.0	3.50	8.00	0.1696	0.4135	0.1451	0.8828
3	71.0	17.0	1.70	0.05	0.5569	0.8307	0.3522	0.8732
4	3.50	0.10	10.0	10.0	0.0124	0.3736	0.2883	0.5743
5	1.70	8.00	17.0	0.10	0.8283	0.1004	0.3047	0.1091
6	8.00	14.0	8.00	14.0	0.5886	0.9991	0.6650	0.0381

Shekel ($S_{4,n}$)

$$f(x) = -\sum_{i=1}^{n} ((x-a_i)^T (x-a_i) + c_i)^{-1}$$

Dimensions: 4
X: [0, 10]
1 global minimum
n local minima
Theoretic optimum: -10.53641 (for n=10)
Error Goal: 10^{-4}

i		a_i^T			c_i
1	4.0	4.0	4.0	4.0	0.1
2	1.0	1.0	1.0	1.0	0.2
3	8.0	8.0	8.0	8.0	0.2
4	6.0	6.0	6.0	6.0	0.4
5	3.0	7.0	3.0	7.0	0.4
6	2.0	9.0	2.0	9.0	0.6
7	5.0	5.0	3.0	3.0	0.3
8	8.0	1.0	8.0	1.0	0.7
9	6.0	2.0	6.0	2.0	0.5
10	7.0	3.6	7.0	3.6	0.5

Detecting Groups of Anomalously Similar Objects in Large Data Sets

Zhicheng Zhang* and David J. Hand

Department of Mathematics, Imperial College London,
180 Queens Gate, Huxley Building, London, SW7 2AZ, UK
{zhzhang, d.j.hand}@imperial.ac.uk

Abstract. Pattern discovery is a facet of data mining concerned with the detection of "small local" structures in large data sets. In high dimensions this is typically difficult because of the computational work involved in searching over the data space. In this paper we outline a tool called PEAKER which can detect patterns efficiently in high dimensions. We approach the subject through the two aspects of pattern discovery, detection and verification. We demonstrate various ways of using PEAKER as well as its various inherent properties, emphasizing the exploratory nature of the tool.

1 Introduction

We define patterns in data as local structures which deviate in some way from the expected structure (see [1] and [2]). The search for patterns in large data sets constitutes a major component of data mining, with the need for such searches arising in a wide variety of settings including banking (e.g. fraud detection), astronomy (e.g. detection of anomalous radio signals), customer relationship management (e.g. market basket analysis) and pharmaceuticals (e.g. adverse-event detection) and the Internet (e.g. text-mining).

In this paper we focus on the area of *unsupervised pattern discovery*, in which the structure being sought is not specified a priori. Typically one is faced with a high-dimensional data set and one of the major problems is how to search over such a data space for patterns. We describe a pattern-search tool called *PEAKER* which is efficient in searching over high-dimensional space. We illustrate the use of PEAKER for both aspects of pattern discovery: the problem of *detecting* potential patterns in the first place and the problem of *verifying* the reality of these detected patterns.

The layout of the paper is as follows. Section 2 introduces the PEAKER algorithm. Section 3 discusses various ways of detecting and verifying significant peaks. Finally, some illustrative examples, both simulated and real, are given, demonstrating some of the properties of PEAKER.

---

* The work of the first author was supported by an EPSRC CASE award in conjunction with GlaxoSmithKline. We are particularly grateful for the support given by Dr Steven Barrett.

A.F. Famili et al. (Eds.): IDA 2005, LNCS 3646, pp. 509–519, 2005.
© Springer-Verlag Berlin Heidelberg 2005

2   The PEAKER Algorithm

The basic PEAKER algorithm is defined as follow: Given a data set of n points, $(x_i, i = 1, \ldots, n)$ of some dimension d, we calculate the probability density estimates $\hat{f}(x_i)$ for all x_i. For each point x_i, we say that it is a peak with $M(x_i) = m$ iff $\hat{f}(x_i) > \hat{f}(x_j), \forall x_j \in N_m(x_i)$ but $\hat{f}(x_i) \leq \hat{f}(x^{(m+1)})$, where $N_m(x_i)$ is the set of the m nearest neighbours to x_i (defined by some distance function $d(x_i, x_j)$) and $x^{(k)}$ is the (k)th nearest neighbour to x_i.

PEAKER essentially estimates the pdf at each data point and performs nearest neighbour search on all data points, identifying those points which have a higher pdf estimate than their surrounding points (we will call these local peaks). Clearly peaks arise as a result of a local region of anomalously high concentration of data points. A local mode of the underlying true density will thus tend to lead to a peak. We can thus regard peaks as approximations to local modes in the underlying true density. We say approximations since peaks (a) derive from density *estimates* rather than the true underlying density, and (b) are restricted to the locations of actual data points, which are unlikely to be in exactly the same position as the local pdf modes. Note that peaks can also arise, by random variation, even when there is no underlying local pdf mode. The essence of verification, discussed below, is determining when peaks do reflect real underlying modes, and when they are merely chance occurrences. This random occurrence of peaks which do not reflect underlying modes can occur (a) because of a chance grouping of data points, or (b) because of an interaction between the estimated pdfs and the locations of the data points. We illustrate this last situation below.

One attractive property of PEAKER is its computational efficiency in high-dimensional spaces which follows from restricting the search to the actual data points only. This avoids the potentially huge search required (in, for example scan statistics) when moving a window over a high-dimensional space.

The algorithm is not uniquely defined above. Two of the most important aspects to be fixed are the choice of distance metric and the method of estimating the probability density. As far as choice of metric is concerned, this is crucial in determining how points are configured in the data space: the metric determines the commensurability of the different dimensions. As far as the density estimation method is concerned, this determines what can be detected as a peak: over-smoothing will smooth away the localised small peaks while under-smoothing will give many peaks. The traditional bump-hunting literature uses some form of optimal density estimation. In kernel estimation, for instance, one would typically use an optimal bandwidth obtained through the minimization of some criterion function (see [3]). In our context, however, one must be careful to strike a balance between the theoretical nicety of optimal estimates and the practical aspect of being able to detect small localized peaks.

3   Significant Peaks

This section describes the criteria by which we can detect peaks and the methods with which we can verify the significance of a peak once detected (that is, to determine whether a detected peak represents a true anomaly of the underlying mechanism or whether it is a chance occurrence). Similarly we should consider whether we are more interested in detecting sharp peaks or peaks that occupy a large proportion of the density space (which we will refer to simply as large peaks). It can be argued that large peaks represent the more obvious structures within a data set, which would be picked up by standard modeling strategies, and that we are mainly concerned with only the small sharp peaks.

All the above considerations (as well as the choices of distance metrics and density estimation) lead to different variants of the PEAKER algorithm. As such, there is no single best algorithm which can answer all the questions posed. PEAKER is an exploratory tool, and any of its different variants might be useful in detecting interesting and useful local structures.

Before we move on, we must make an important distinction. Any data point can be one of the following

1. A mode (a maximum) - one which is a maximum in the density surface. To decide whether a point is a mode one needs to evaluate the entire density surface around the point.
2. A maximum peak - a data point with higher pdf estimate than its nearest neighbours and which is also an approximation to a maximum. This is of course the essence of PEAKER, since we only make use of density estimates at the data points and not the entire surface and so can only look for approximations.
3. A non-maximum peak - This is simply a point which has a larger pdf estimate than some neighbourhood of data points. Such a point does not need to approximate a maximum and arises due to a chance configuration of the neighbouring points (in one dimension, for instance, the nearest points with lower pdf estimates lying to one side of the peak).
4. A non-peak - A data point which has lower pdf estimate than its nearest neighbour.

Intuitively one would like to detect maxima or approximations to maxima. This is rather difficult since it involves evaluation of the density surface. We will describe a method in the next section by which we can achieve this probabilistically.

3.1   Setting M

The parameter m, defined above, determines the area of the peaks, in the sense that it gives us the number of neighbouring points which have smaller estimated pdf than that at the point in question. Based on this, an intuitively attractive criterion for detecting peaks is to set a minimum neighbourhood size M, requiring that all points flagged as potential peaks should have $m \geq M$. In the extreme case, of course, we can choose to look at all peaks, which is equivalent to setting $M = 1$. Alternatively we can set M beforehand at a sensible size, say $M = 50$ or some proportion of the data size. This is clearly rather arbitrary and subject to personal taste - but, once again, this is legitimate in an exploratory tool. Choosing M small means that there is a greater chance of detecting minor fluctuations in the estimated pdf as apparent peaks, while choosing M large means that only the largest fluctuations will be detected. In the limit, setting M equal to $n - 1$, using the entire data set, will mean that a single peak will be detected - that corresponding to the point which has the highest pdf estimate.

An extension to this idea, and one which sits with the idea of an exploratory tool, is to let M range between 1 and $n-1$. As M decreases in size, so more peaks will be detected. Figures 1 to 3 shows the increase in the number peaks as M decreases. Here the underlying distribution is a mixture of a uniform distribution on the square $(-2, 2), (-2, 2)$ and three bivariate normal distributions with means located at the three peaks indicated in Fig. 1. We can see the increase in the number of peaks from the three true peaks in Fig. 1 to the spurious peaks as M decreases.

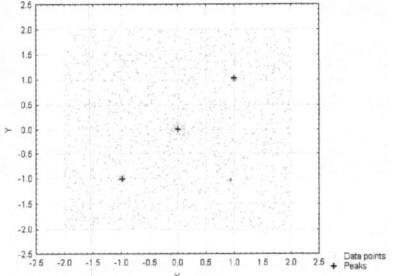
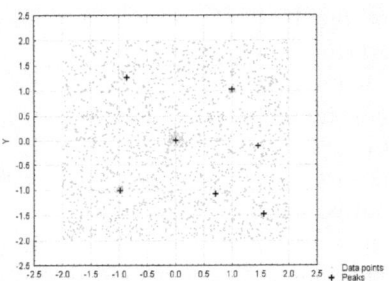

Fig. 1. Peaks Detected With M = 500   **Fig. 2.** Peaks Detected With M = 50

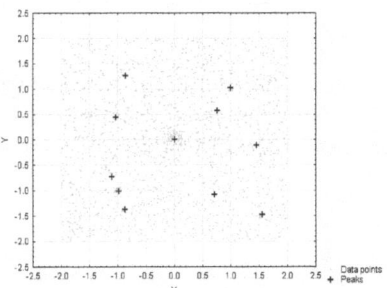

Fig. 3. Peaks Detected With M = 10

Note that anomalies arising from the interaction between data point configurations and estimated pdfs can occur, though we believe they will be unimportant. Alternatively, to introduce some objectivity, we can venture the following method. It seems sensible to decide that we only want to detect modes, and so would like to exclude peaks which do not constitute a local maximum. A peak is not a mode when a specific configuration of the nearest points arises. It is easy to see this in one dimension. Let a set of points $(x_1, \ldots, x_n)$ be generated from some underlying probability density f. For any three consecutive points x_{i-1}, x_i, x_{i+1}, with estimated pdf $\hat{f}(x_{i-1}) > \hat{f}(x_i) > \hat{f}(x_{i+1})$, then x_i is a peak with $M(x_i) > 0$ iff $d(x_{i-1}, x_i) > d(x_i, x_{i+1})$. In this instance, we have x_i being a peak but not a mode. We can extend this further, so that for a set of consecutive points $x_{i-1}, x_i, \ldots, x_{i+m}$ with $\hat{f}(x_{i-1}) > \hat{f}(x_i) > \ldots > \hat{f}(x_{i+m})$, x_i is a peak (but not a mode) with $M(x_i) \geq m$. This particular configuration of points (in one dimension) is the only way for which a point can be a peak but not a mode. Thus we can make a binomial argument and say that the probability of a point being a peak with $M(x_i) = m$ but not a mode is simply $R_m = R^m(1 - R)$, where R is the probability of a point being generated closer to x_i on the right than on the left. We can show that this probability converges to $\frac{1}{2}$ for the one dimensional case. For the higher dimensions R converges to values below one half, but here we must keep in mind that there are points which cannot be peaks (the two end points in the one-dimensional case, for instance) and such points distort the convergent value of R_m. Clearly this gives us a way of setting a reasonable value for a minimum M so as to exclude non-modal peaks.

3.2   Sharpness of a Peak

Detecting peaks, as potential modes of the underlying pdf, is one thing, but we need to be reasonably confident that they are genuine, and not merely due to random fluctuation. We can explore this using ideas of significance tests, and of the probability mass of a peak. We now describe some methods of verifying the significance of detected peaks.

A simple approach is to calculate the ratio of the pdf estimate at a peak to the average of the pdf estimates at the points in a surrounding neighbourhood. This will be a measure of sharpness of the peak (which we will call T_{sharp}). The general form is given by

$$T_{sharp} = \frac{N_s \hat{f}(x)}{\sum_{i=1}^{N_s} \hat{f}(x_i)} \tag{1}$$

where N_s is the number of nearest neighbours from which we take the average . For simplicity we can set $N_s = \frac{m}{2}$. Of course this is again rather arbitrary and moreover prone to distortion by specific arrangements of peaks. For instance, the sharpness of one peak can be very much affected by the shape of the neighbouring peaks (if a small sharp peak is close to a large flat peak, it is likely to have a smaller value of T_{sharp}).

There are more sophisticated methods of measuring the sharpness of peaks. [4] describes two such tests (which they refer to as tests for leptokurtosis). The idea is to assume a null hypothesis in which the points are locally uniformly distributed in a d-dimensional hypersphere of radius R. Denoting the distance of each point to the centre as X, then the transformed variate $Y = (\frac{X}{R})^d$ is distributed as $U(0, 1)$. The two tests are based on this transformed variate Y.

The first test uses the test statistic

$$T_{lepto1} = \frac{1}{n} \sum_{i=1}^{n} \Phi^{-1}(Y_i) \tag{2}$$

where Φ^{-1} is the inverse of the cumulative normal distribution. T_{lepto1} is distributed as a $N(0, \frac{1}{n})$ distribution. This test is based on the fact that a local dense region will have an abundance of short distances and a large negative value of T_{lepto1}.

The second test uses the rth order statistic $Y_{(r)} = (\frac{X_{(r)}}{R})^d$ and $Y_{(r)}$ follows a $Beta(r, n - r + 1)$. Here $Y_{(r)}$ will be smaller if there is a region of high density.

3.3   Probability Mass Contained in Peaks

An alternative to testing for the sharpness of a peak, is to derive a measure of the probability mass contained in a peak. This section follows the approaches contained in [5] and [6], which are used to test for modes in the one-dimensional case. Both assume the use of gaussian kernel estimates. The gaussian kernel has the property that the number of modes found in the data is non-decreasing with decreasing bandwidth h, which facilitate the testing procedures. In [5], to test the null hypothesis that the underlying density g has k modes, the test proposed uses the test statistic

$$h_{crit} = \inf(h; \hat{f}(x, h) \text{ has at most k-modes}). \tag{3}$$

Here h_{crit} is the smallest value at which the data remains k-modal. With the property of the gaussian kernel, one can perform a binary search to calculate h_{crit}. To obtain the significance level $P(h_{crit} > h_0)$, [5] proposes sampling for from g if g is known or sampling from a bootstrap density g_0 obtained from the data if g is unknown.

[6] extends the above to investigate each mode in turn. For a single mode v_i, he defined $h_{test,i}$ as the smallest h at which the mode remains a single mode, and proposed the test statistic

$$M_i = \int_{u_{i-1}}^{u_{i+1}} [\hat{f}(x) - max(\hat{f}(u_{i-1}), \hat{f}(u_{i+1}))]dx \tag{4}$$

where u_{i-1} and u_{i+1} are the two anti-modes either side of mode v_i. In effect M_i is the area of probability mass above the higher of the two surrounding anti-modes. To obtain a p-value for M_i, the approach of [5] is followed and points are sampled from a representative density of the region with no modes (possibly a uniform density).

Higher-dimensional extensions to the above ideas are difficult to implement since it is computationally expensive to either (a) look for anti-modes in high-dimensional space and (b) obtain accurate significance levels by re-sampling.

For our case, we will simply calculate the ratio between an estimate of the density in the local region, and the density that this region would have if the local region was part of a larger region in which the density was uniform. Such a measure would have the generic form

$$T_{density} = \frac{\sum_{N_s} \hat{f}(x_i)/N_s}{\sum_M \hat{f}(x_j)/M} \tag{5}$$

where $\sum_{N_s} \hat{f}(x_i)/N_s$ is an estimate of the density contained in the local region, defined as the region containing the nearest N_s points, and $\sum_M \hat{f}(x_j)/M$ is an estimate of the density over the larger region, containing the nearest M points. There is, of course, some arbitrariness about the choices of M and N_s.

3.4   Size of the Local Neighbourhood

Much of the discussion in the previous sections are based on rather arbitrary selections of the neighbourhood size that we use to calculate various measures. This includes the neighbourhood size N_s used in T_{sharp} and $T_{density}$ and the size of the uniform hyper-sphere used in the two tests by [4]. The problem in all these cases is that these measures work well when the null model of global uniformity holds. When this is not the case, there are various ways for these measures to be distorted. For instance, if we have two identical peaks, one against a uniform background of higher pdf estimates than the other, then the measure T_{sharp} would return very different values for the two peaks, similarly for the two tests contained in [4]. In [6], such problems are resolved by only testing for uniformity of a mode in the local region bounded by the two nearest anti-modes. In our case, the neighbourhood M extends all the way to the next peak.

In the case of PEAKER we cannot evaluate the anti-modes since we only work with the data points themselves. The most intuitive analogy to the anti-mode is the anti-peak (ie, a point which has a lower pdf estimate than its nearest points). We can make use of this in deciding on the neighbourhood sizes. Clearly there are also problems associated with this, since anti-peaks may be false and so make the neighbourhood sizes again rather arbitrary.

Having said this, we must not forget the exploratory nature of PEAKER and as such it is important to experiment with differing neighbourhood sizes to look for otherwise undetected peaks. Therefore we are quite happy in using rather arbitrary parameters in the PEAKER algorithm knowing that one needs to explore these parameters to obtain peaks of different properties.

4   Examples

Our first example demonstrates the use of PEAKER as a high-dimensional exploratory tool. The data set consists of 5000 data points in 10-dimensional space drawn from a mixture of background uniform and three normal distributions. The normals have covariance matrices $0.025I, 0.025I, 0.1I$ with mixing probabilities $0.0025, 0.0025, 0.02$. The background uniform is drawn in the hyper-box

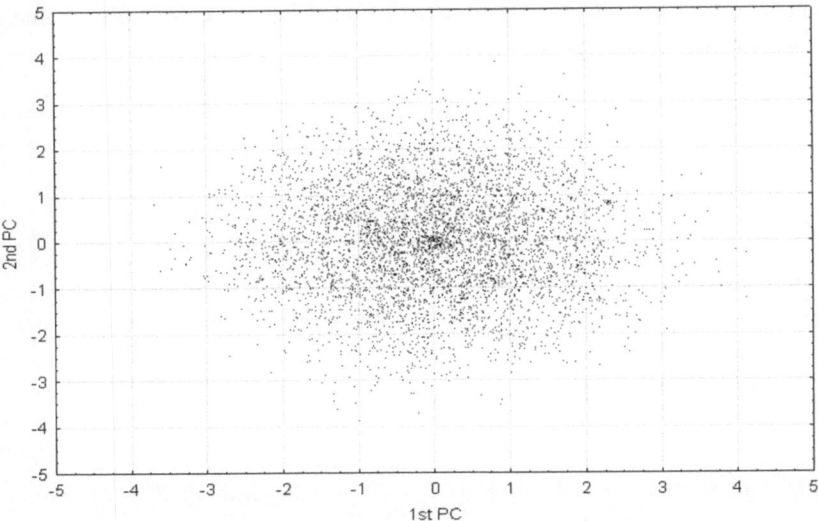

Fig. 4. First Two Principal Components of 10-dimensional Data With Three Modes

$(-2, 2), \ldots, (-2, 2)$. For visualization we have reduced the data set down to two dimensions by plotting the points on the first two principal components.

The three normal means are at $(0, 0), (2.32, 0.81), (-2.32, -0.81)$ in Fig.4. The peak at (0.0) is detectable visually (but this is distorted by the fact that we are plotting the first two principal components, which pushes the points towards the center. It is not nearly as clear if we simply plot any two random dimensions) , but the other two peaks are not at all clear by visualization (since the mixing probabilities are so low). It is interesting to note that here many peaks are thrown up for $M = 200$ say, but the three normals are clearly detected when one uses the measure T_{sharp} (in that they give much higher values of T_{sharp} than other peaks). Of course PEAKER is well designed for detecting such peaks otherwise undetectable by eye (since at high dimensions points are sparse and a small but dense region would be easily detected by PEAKER).

The second example shows the properties of PEAKER when applied to simulated data sets with changing parameters. Here, again, we simulated data sets of mixtures of background noise and normal distributions. The parameters we change are 1) dimensionality, 2) sample size, 3) covariance matrices of the normals, 4) mixing probabilities of the normals and 5) separation of the normal means. To illustrate the use of PEAKER, we plot a histogram of how close the detected peaks are to the true normal means.

Figure 5 shows the histogram with the distance of a detected peak to a true peak on the horizontal axis. We can see that the majority of the peaks detected are close to the true normal means. We must keep in mind that some deviation are unavoidable since we are only looking at points approximating the true means. This demonstrates that PEAKER is rather robust in detecting different

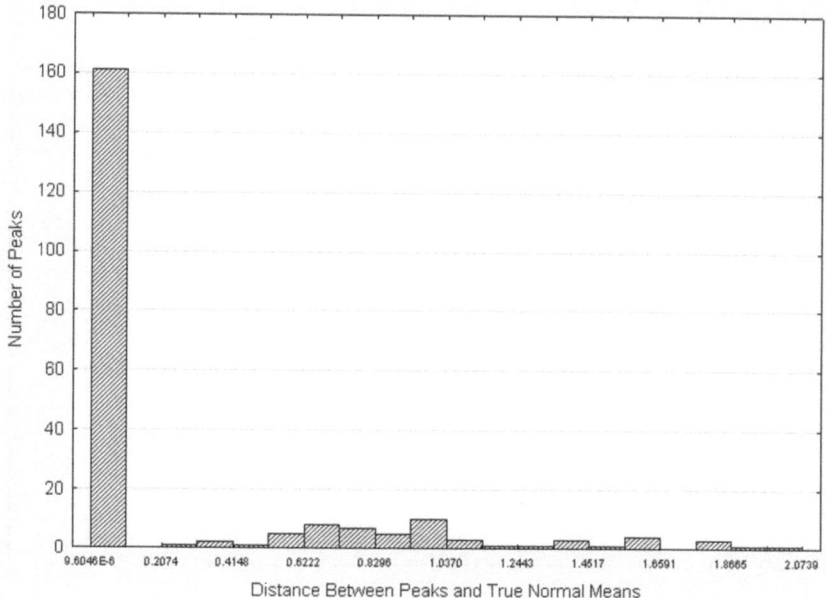

Fig. 5. Histogram of Distances Between Peaks and Normal Means

types of peaks (in our example here normal means of different dimensionality, size, sharpness and separation within a uniform background). The peaks arising at around 0.7 and 1.0 represent those which are merged from two normal distributions and are consequently only detected as one peak. This is a common problem in bump-hunting whereby it can be difficult to separate modes close to one another.

Lastly we take an example from the pharmaceutical industry. The data set we have comprises of 931 drugs. We have a distance matrix derived from using the Jaccard coefficient on the chemical properties of the drugs (fingerprints of the presence of molecules). We ran PEAKER to find dense regions of drugs which may have similar properties. Figure 6 shows the configuration of the drugs in 2-dimensional space obtained via Principal Components Analysis.

We have indicated various peaks in Fig.6 which represent some interesting patterns. All these peaks contain drugs almost exclusively of individual types. With a smooth bandwidth and $M = 20$, we discover mostly peaks of the more common drugs (including anti-inflammatory, antineoplastic, anti-bacterial, anti-convulsant, anesthetic, anti-depressant and analgesic drugs) as shown in Fig. 6. It is surprising that we have managed to find patterns of such purity (in terms of homogeneous drugs).

Figure 7 shows the peaks found using a smaller bandwidth and $M = 5$. Some of the peaks found with $M = 5$ are not immediately obvious in similarity and it is only with further investigation that they are revealed as interestingly similar drugs. Certain drug peaks found have very low incidence rate in the data set

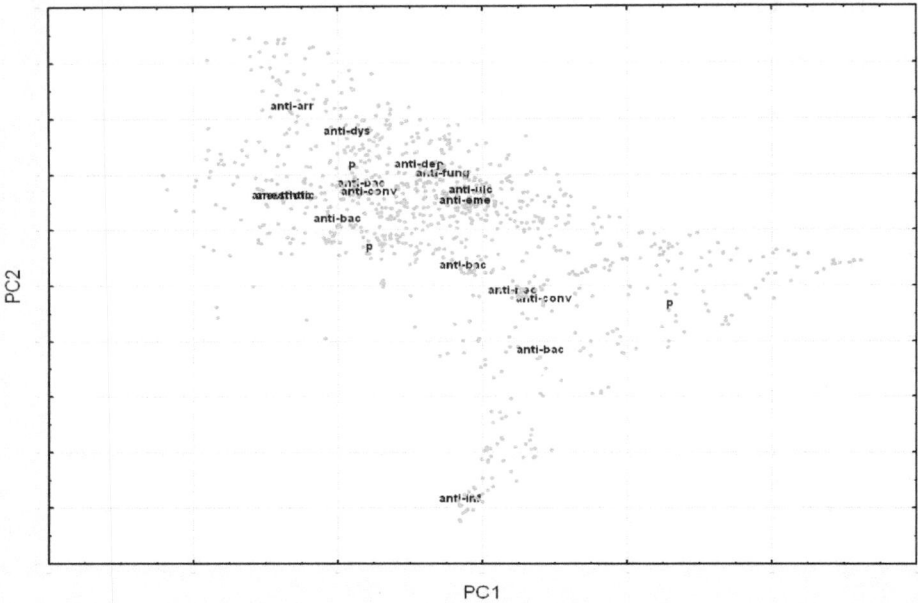

Fig. 6. First Two Principal Components Drugs Data Set Showing Peaks of the Common Drugs

(sometimes only 3 or 4 drugs). Overall there are not very many false peaks and most of the peaks found consist of interestingly similar groups of drugs, but this may be due to the closeness of chemical similarity to general behaviour of a drug. The general trend seems to be that sharper peaks tend to show more homogeneous drug clusters. Indeed it does seem that PEAKER can be applied to search for the rarer drugs that constitute interesting patterns or associations.

5 Conclusion

In this paper we have focused our attention on a pattern discovery algorithm called PEAKER. We have explored some of the properties of PEAKER and demonstrated its value as an exploratory tool, particularly for high dimensional data due to its computational efficiency and ability to detect peaks otherwise difficult to identify. As discussed, high-dimensional data is typically difficult to explore efficiently and PEAKER achieves this by restricting attention only to the data points so that we no longer have to explore the entire space. On the other hand, what we gain in terms of efficiency and simplicity we lose out somewhat when it comes to an accurate evaluation of the data space. As such, we can only approximate the true modes in the data. However, we believe that given the exploratory nature of PEAKER such approximations are justified.

Moreover, we have discussed various ways of using PEAKER (in terms of both detection and evaluation of peaks), depending on personal judgement of

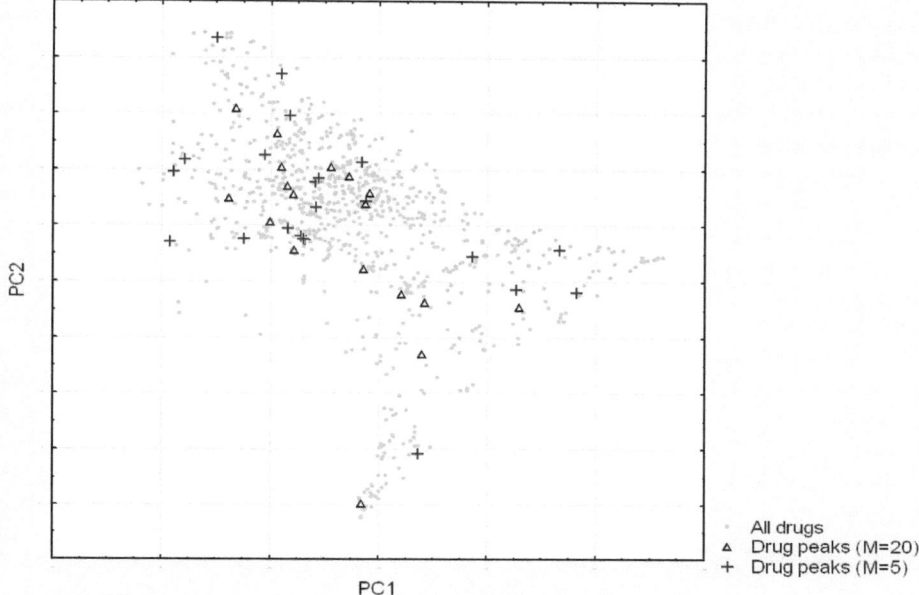

Fig. 7. First Two Principal Components Drugs Data Set Showing All Peaks

the types of peaks that one prefers to detect. Of course, the true essence of PEAKER relies on the user to experiment with the various parameters inherent in the tool (density estimation, distance metric, M, neighbourhood size, etc) so as to detect peaks with different properties.

Work on PEAKER is still ongoing and we hope to explore deeper the various properties offered by the algorithm.

References

1. Hand, D., Bolton, R.: Pattern discovery and detection: a unified statistical methodology. Journal of Applied Statistics **31** (2004) 885–924
2. Hand, D., Mannila, H., Smyth, P.: Principles of Data Mining. MIT Press, Cambridge, Mass (2001)
3. Sain, S., Baggerly, K., Scott, D.: Cross-validation of multivariate densities. Journal of the American Statistical Association **89** (1994) 807–817
4. Bolton, R., Hand, D., Crowder, M.: Significance tests for unsupervised pattern discovery in large continuous multivariate data sets. Computational Statistics and Data Analysis **46** (2004) 57–79
5. Silverman, B.: Using kernel density estimates to investigate multimodality. Journal of the Royal Statistical Association. Series B **43** (1981) 97–99
6. Minnotte, M.: Nonparametric testing of existing modes. The Annals of Statistics **25** (1997) 1646–1660

Author Index

Aguilar-Ruiz, Jesús S.   362
Allende, Héctor   272
Ambroise, Christophe   373
Andreoli, Jean-Marc   1
Angiulli, Fabrizio   12
Araujo, Lourdes   397
Aue, Anthony   121

Bäck, Thomas   385
Bai, Yun   497
Batista, Gustavo E.A.P.A.   24
Beringer, Jürgen   168
Berthold, Michael R.   97
Boberg, Jorma   464
Bonnaire, Xavier   328
Bouchard, Guillaume   1
Boulicaut, Jean-François   293

Cabello, Enrique   351
Carmona-Saez, Pedro   74
Chen, Zhihua   86
Cobb, Barry R.   36
Combarro, E.F.   239
Conde, Cristina   351
Corbacho, Fernando J.   486
Corston-Oliver, Simon   121
Couto, Julia   46
Cubo, Óscar   57, 339

Díaz, I.   239
Díaz-Díaz, Norberto   362
Dopazo, Esther   66

Edgerton, Mary   86

Famili, A. Fazel   74
Fisher, Doug   86
Flach, Peter A.   145
Frey, Lewis   86
Fürnkranz, Johannes   180

Gabriel, Thomas R.   97
Gamberoni, Giacomo   109
Gamon, Michael   121
González-Pachón, Jacinto   66

Govaert, Gérard   249, 373
Gunetti, Daniele   133
Gyftodimos, Elias   145

Hand, David J.   509
Herrero, P.   339
Hoen, Peter A.C. 't   475
Horman, Yoav   157
Hüllermeier, Eyke   168, 180

IJzerman, Ad P.   385

Kaminka, Gal A.   157
Kegelmeyer, W. Philip   192
Klawonn, Frank   316
Koegler, Wendy S.   192
Kok, Joost N.   385
Kokol, Peter   305
Kruse, Rudolf   316

Lago-Fernández, Luis F.   486
Lamma, Evelina   109
Lehtimäki, Pasi   204
Liu, Jian   216
Liu, Xiaohui   475
Liu, Ziying   74

Martín-Merino, Manuel   228
Menasalvas, Ernestina   57
Monard, Maria C.   24
Montañés, E.   239
Moossen, Michael   328
Moraga, Claudio   272
Muñoz, Alberto   228
Mullick, Alaka   74
Mumcuoğlu, Ü. Erkan   283
Mylläri, Aleksandr   464

Nadif, Mohamed   249
Nam, Mi Young   260
Ñanculef, Ricardo   272

Oğul, Hasan   283

Pahikkala, Tapio   464
Peña, J.M.   339
Pensa, Ruggero G.   293

Pérez, M.S.   339
Picardi, Claudia   133
Pintilie, A. Simona   97
Povalej, Petra   305
Prati, Ronaldo C.   24
Pyysalo, Sampo   464

Qiu, Yuhui   497

Raivio, Kimmo   204
Ranilla, J.   239
Raoult, Jean Pierre   409
Rehm, Frank   316
Rhee, Phill Kyu   260
Riff, María-Cristina   328
Riguzzi, Fabrizio   109
Ringger, Eric   121
Riquelme, José C.   362
Robles, Juan   66
Robles, Víctor   57, 339
Rodríguez-Aragón, Licesio J.   351
Ruffo, Giancarlo   133
Ruiz, Roberto   362
Rumí, Rafael   36

Salakoski, Tapio   464
Salmerón, Antonio   36
Samé, Allou   373
Samsonova, Elena V.   385
Sánchez, Beatriz   486

Segovia, Javier   57
Serrano, Ángel   351
Serrano, José I.   397
Smail, Linda   409
Spate, Jess   419
Speer, Nora   429
Spieth, Christian   429
Stiglic, Bruno   305
Storari, Sergio   109

Talavera, Luis   440
Tang, Lianhong   86
Teng, Choh Man   452
Tsivtsivadze, Evgeni   464
Tucker, Allan   475

Valle, Carlos   272
Valle, Manuel del   486
Vinciotti, Veronica   475
Volinia, Stefano   109

Wang, Fang   497
Welzer Družovec, Tatjana   305
Wu, Gengfeng   216

Yao, Jianxin   216

Zell, Andreas   429
Zhang, Zhicheng   509

Lecture Notes in Computer Science

For information about Vols. 1–3569

please contact your bookseller or Springer

Vol. 3697: W. Duch, J. Kacprzyk, E. Oja, S. Zadrożny (Eds.), Artificial Neural Networks: Formal Models and Their Applications - ICANN 2005, Part II. XXXII, 1045 pages. 2005.

Vol. 3696: W. Duch, J. Kacprzyk, E. Oja, S. Zadrożny (Eds.), Artificial Neural Networks: Biological Inspirations - ICANN 2005, Part I. XXXI, 703 pages. 2005.

Vol. 3687: S. Singh, M. Singh, C. Apte, P. Perner (Eds.), Pattern Recognition and Image Analysis, Part II. XXV, 809 pages. 2005.

Vol. 3686: S. Singh, M. Singh, C. Apte, P. Perner (Eds.), Pattern Recognition and Data Mining, Part I. XXVI, 689 pages. 2005.

Vol. 3674: W. Jonker, M. Petković (Eds.), Secure Data Management. X, 241 pages. 2005.

Vol. 3672: C. Hankin, I. Siveroni (Eds.), Static Analysis. X, 369 pages. 2005.

Vol. 3671: S. Bressan, S. Ceri, E. Hunt, Z.G. Ives, Z. Bellahsène, M. Rys, R. Unland (Eds.), Database and XML Technologies. X, 239 pages. 2005.

Vol. 3670: M. Bravetti, L. Kloul, G. Zavattaro (Eds.), Formal Techniques for Computer Systems and Business Processes. XIII, 349 pages. 2005.

Vol. 3664: C. Türker, M. Agosti, H.-J. Schek (Eds.), Peer-to-Peer, Grid, and Service-Orientation in Digital Library Architectures. X, 261 pages. 2005.

Vol. 3663: W. Kropatsch, R. Sablatnig, A. Hanbury (Eds.), Pattern Recognition. XIV, 512 pages. 2005.

Vol. 3662: C. Baral, G. Greco, N. Leone, G. Terracina (Eds.), Logic Programming and Nonmonotonic Reasoning. XIII, 454 pages. 2005. (Subseries LNAI).

Vol. 3660: M. Beigl, S. Intille, J. Rekimoto, H. Tokuda (Eds.), UbiComp 2005: Ubiquitous Computing. XVII, 394 pages. 2005.

Vol. 3659: J.R. Rao, B. Sunar (Eds.), Cryptographic Hardware and Embedded Systems – CHES 2005. XIV, 458 pages. 2005.

Vol. 3658: V. Matoušek, P. Mautner, T. Pavelka (Eds.), Text, Speech and Dialogue. XV, 460 pages. 2005. (Subseries LNAI).

Vol. 3654: S. Jajodia, D. Wijesekera (Eds.), Data and Applications Security XIX. X, 353 pages. 2005.

Vol. 3653: M. Abadi, L.d. Alfaro (Eds.), CONCUR 2005 – Concurrency Theory. XIV, 578 pages. 2005.

Vol. 3649: W.M.P. van der Aalst, B. Benatallah, F. Casati, F. Curbera (Eds.), Business Process Management. XII, 472 pages. 2005.

Vol. 3648: J.C. Cunha, P.D. Medeiros (Eds.), Euro-Par 2005 Parallel Processing. XXXVI, 1299 pages. 2005.

Vol. 3646: A. F. Famili, J.N. Kok, J.M. Peña, A. Siebes, A. Feelders (Eds.), Advances in Intelligent Data Analysis VI. XIV, 522 pages. 2005.

Vol. 3645: D.-S. Huang, X.-P. Zhang, G.-B. Huang (Eds.), Advances in Intelligent Computing, Part II. XIII, 1010 pages. 2005.

Vol. 3644: D.-S. Huang, X.-P. Zhang, G.-B. Huang (Eds.), Advances in Intelligent Computing, Part I. XXVII, 1101 pages. 2005.

Vol. 3642: D. Ślezak, J. Yao, J.F. Peters, W. Ziarko, X. Hu (Eds.), Rough Sets, Fuzzy Sets, Data Mining, and Granular Computing, Part II. XXIV, 738 pages. 2005. (Subseries LNAI).

Vol. 3641: D. Ślezak, G. Wang, M.S. Szczuka, I. Düntsch, Y. Yao (Eds.), Rough Sets, Fuzzy Sets, Data Mining, and Granular Computing, Part I. XXIV, 742 pages. 2005. (Subseries LNAI).

Vol. 3639: P. Godefroid (Ed.), Model Checking Software. XI, 289 pages. 2005.

Vol. 3638: A. Butz, B. Fisher, A. Krüger, P. Olivier (Eds.), Smart Graphics. XI, 269 pages. 2005.

Vol. 3637: J. M. Moreno, J. Madrenas, J. Cosp (Eds.), Evolvable Systems: From Biology to Hardware. XI, 227 pages. 2005.

Vol. 3636: M.J. Blesa, C. Blum, A. Roli, M. Sampels (Eds.), Hybrid Metaheuristics. XII, 155 pages. 2005.

Vol. 3634: L. Ong (Ed.), Computer Science Logic. XI, 567 pages. 2005.

Vol. 3633: C. Bauzer Medeiros, M. Egenhofer, E. Bertino (Eds.), Advances in Spatial and Temporal Databases. XIII, 433 pages. 2005.

Vol. 3632: R. Nieuwenhuis (Ed.), Automated Deduction – CADE-20. XIII, 459 pages. 2005. (Subseries LNAI).

Vol. 3629: J.L. Fiadeiro, N. Harman, M. Roggenbach, J. Rutten (Eds.), Algebra and Coalgebra in Computer Science. XI, 457 pages. 2005.

Vol. 3628: T. Gschwind, U. Aßmann, O. Nierstrasz (Eds.), Software Composition. X, 199 pages. 2005.

Vol. 3627: C. Jacob, M.L. Pilat, P.J. Bentley, J. Timmis (Eds.), Artificial Immune Systems. XII, 500 pages. 2005.

Vol. 3626: B. Ganter, G. Stumme, R. Wille (Eds.), Formal Concept Analysis. X, 349 pages. 2005. (Subseries LNAI).

Vol. 3625: S. Kramer, B. Pfahringer (Eds.), Inductive Logic Programming. XIII, 427 pages. 2005. (Subseries LNAI).

Vol. 3624: C. Chekuri, K. Jansen, J.D.P. Rolim, L. Trevisan (Eds.), Approximation, Randomization and Combinatorial Optimization. XI, 495 pages. 2005.

Vol. 3623: M. Liśkiewicz, R. Reischuk (Eds.), Fundamentals of Computation Theory. XV, 576 pages. 2005.

Vol. 3621: V. Shoup (Ed.), Advances in Cryptology – CRYPTO 2005. XI, 568 pages. 2005.

Vol. 3620: H. Muñoz-Avila, F. Ricci (Eds.), Case-Based Reasoning Research and Development. XV, 654 pages. 2005. (Subseries LNAI).

Vol. 3619: X. Lu, W. Zhao (Eds.), Networking and Mobile Computing. XXIV, 1299 pages. 2005.

Vol. 3618: J. Jedrzejowicz, A. Szepietowski (Eds.), Mathematical Foundations of Computer Science 2005. XVI, 814 pages. 2005.

Vol. 3617: F. Roli, S. Vitulano (Eds.), Image Analysis and Processing – ICIAP 2005. XXIV, 1219 pages. 2005.

Vol. 3615: B. Ludäscher, L. Raschid (Eds.), Data Integration in the Life Sciences. XII, 344 pages. 2005. (Subseries LNBI).

Vol. 3614: L. Wang, Y. Jin (Eds.), Fuzzy Systems and Knowledge Discovery, Part II. XLI, 1314 pages. 2005. (Subseries LNAI).

Vol. 3613: L. Wang, Y. Jin (Eds.), Fuzzy Systems and Knowledge Discovery, Part I. XLI, 1334 pages. 2005. (Subseries LNAI).

Vol. 3612: L. Wang, K. Chen, Y. S. Ong (Eds.), Advances in Natural Computation, Part III. LXI, 1326 pages. 2005.

Vol. 3611: L. Wang, K. Chen, Y. S. Ong (Eds.), Advances in Natural Computation, Part II. LXI, 1292 pages. 2005.

Vol. 3610: L. Wang, K. Chen, Y. S. Ong (Eds.), Advances in Natural Computation, Part I. LXI, 1302 pages. 2005.

Vol. 3608: F. Dehne, A. López-Ortiz, J.-R. Sack (Eds.), Algorithms and Data Structures. XIV, 446 pages. 2005.

Vol. 3607: J.-D. Zucker, L. Saitta (Eds.), Abstraction, Reformulation and Approximation. XII, 376 pages. 2005. (Subseries LNAI).

Vol. 3606: V. Malyshkin (Ed.), Parallel Computing Technologies. XII, 470 pages. 2005.

Vol. 3604: R. Martin, H. Bez, M. Sabin (Eds.), Mathematics of Surfaces XI. IX, 473 pages. 2005.

Vol. 3603: J. Hurd, T. Melham (Eds.), Theorem Proving in Higher Order Logics. IX, 409 pages. 2005.

Vol. 3602: R. Eigenmann, Z. Li, S.P. Midkiff (Eds.), Languages and Compilers for High Performance Computing. IX, 486 pages. 2005.

Vol. 3599: U. Aßmann, M. Aksit, A. Rensink (Eds.), Model Driven Architecture. X, 235 pages. 2005.

Vol. 3598: H. Murakami, H. Nakashima, H. Tokuda, M. Yasumura, Ubiquitous Computing Systems. XIII, 275 pages. 2005.

Vol. 3597: S. Shimojo, S. Ichii, T.W. Ling, K.-H. Song (Eds.), Web and Communication Technologies and Internet-Related Social Issues - HSI 2005. XIX, 368 pages. 2005.

Vol. 3596: F. Dau, M.-L. Mugnier, G. Stumme (Eds.), Conceptual Structures: Common Semantics for Sharing Knowledge. XI, 467 pages. 2005. (Subseries LNAI).

Vol. 3595: L. Wang (Ed.), Computing and Combinatorics. XVI, 995 pages. 2005.

Vol. 3594: J.C. Setubal, S. Verjovski-Almeida (Eds.), Advances in Bioinformatics and Computational Biology. XIV, 258 pages. 2005. (Subseries LNBI).

Vol. 3593: V. Mařík, R. W. Brennan, M. Pěchouček (Eds.), Holonic and Multi-Agent Systems for Manufacturing. XI, 269 pages. 2005. (Subseries LNAI).

Vol. 3592: S. Katsikas, J. Lopez, G. Pernul (Eds.), Trust, Privacy and Security in Digital Business. XII, 332 pages. 2005.

Vol. 3591: M.A. Wimmer, R. Traunmüller, Å. Grönlund, K.V. Andersen (Eds.), Electronic Government. XIII, 317 pages. 2005.

Vol. 3590: K. Bauknecht, B. Pröll, H. Werthner (Eds.), E-Commerce and Web Technologies. XIV, 380 pages. 2005.

Vol. 3589: A M. Tjoa, J. Trujillo (Eds.), Data Warehousing and Knowledge Discovery. XVI, 538 pages. 2005.

Vol. 3588: K.V. Andersen, J. Debenham, R. Wagner (Eds.), Database and Expert Systems Applications. XX, 955 pages. 2005.

Vol. 3587: P. Perner, A. Imiya (Eds.), Machine Learning and Data Mining in Pattern Recognition. XVII, 695 pages. 2005. (Subseries LNAI).

Vol. 3586: A.P. Black (Ed.), ECOOP 2005 - Object-Oriented Programming. XVII, 631 pages. 2005.

Vol. 3584: X. Li, S. Wang, Z.Y. Dong (Eds.), Advanced Data Mining and Applications. XIX, 835 pages. 2005. (Subseries LNAI).

Vol. 3583: R.W. H. Lau, Q. Li, R. Cheung, W. Liu (Eds.), Advances in Web-Based Learning – ICWL 2005. XIV, 420 pages. 2005.

Vol. 3582: J. Fitzgerald, I.J. Hayes, A. Tarlecki (Eds.), FM 2005: Formal Methods. XIV, 558 pages. 2005.

Vol. 3581: S. Miksch, J. Hunter, E. Keravnou (Eds.), Artificial Intelligence in Medicine. XVII, 547 pages. 2005. (Subseries LNAI).

Vol. 3580: L. Caires, G.F. Italiano, L. Monteiro, C. Palamidessi, M. Yung (Eds.), Automata, Languages and Programming. XXV, 1477 pages. 2005.

Vol. 3579: D. Lowe, M. Gaedke (Eds.), Web Engineering. XXII, 633 pages. 2005.

Vol. 3578: M. Gallagher, J. Hogan, F. Maire (Eds.), Intelligent Data Engineering and Automated Learning - IDEAL 2005. XVI, 599 pages. 2005.

Vol. 3577: R. Falcone, S. Barber, J. Sabater-Mir, M.P. Singh (Eds.), Trusting Agents for Trusting Electronic Societies. VIII, 235 pages. 2005. (Subseries LNAI).

Vol. 3576: K. Etessami, S.K. Rajamani (Eds.), Computer Aided Verification. XV, 564 pages. 2005.

Vol. 3575: S. Wermter, G. Palm, M. Elshaw (Eds.), Biomimetic Neural Learning for Intelligent Robots. IX, 383 pages. 2005. (Subseries LNAI).

Vol. 3574: C. Boyd, J.M. González Nieto (Eds.), Information Security and Privacy. XIII, 586 pages. 2005.

Vol. 3573: S. Etalle (Ed.), Logic Based Program Synthesis and Transformation. VIII, 279 pages. 2005.

Vol. 3572: C. De Felice, A. Restivo (Eds.), Developments in Language Theory. XI, 409 pages. 2005.

Vol. 3571: L. Godo (Ed.), Symbolic and Quantitative Approaches to Reasoning with Uncertainty. XVI, 1028 pages. 2005. (Subseries LNAI).

Vol. 3570: A. S. Patrick, M. Yung (Eds.), Financial Cryptography and Data Security. XII, 376 pages. 2005.